Fungal Biology

***Series Editors*:**
Vijai Kumar Gupta, PhD
Molecular Glycobiotechnology Group, Department of Biochemistry,
School of Natural Sciences, National University of Ireland Galway,
Galway, Ireland

Maria G. Tuohy, PhD
Molecular Glycobiotechnology Group, Department of Biochemistry,
School of Natural Sciences, National University of Ireland Galway,
Galway, Ireland

For further volumes:
http://www.springer.com/series/11224

Aakash Goyal • Chakravarthula Manoharachary
Editors

Future Challenges in Crop Protection Against Fungal Pathogens

Springer

Editors
Aakash Goyal, Ph.D., F.I.C.N., F.S.A.B.
Biodiversity and Integrated Gene Management Program
International Center for Agriculture Research in the Dry Area
Rabat, Morocco

Chakravarthula Manoharachary, M.Sc., Ph.D., D.Sc.
Department of Botany
Osmania University
Hyderabad, Telangana, India

ISSN 2198-7777 ISSN 2198-7785 (electronic)
ISBN 978-1-4939-1187-5 ISBN 978-1-4939-1188-2 (eBook)
DOI 10.1007/978-1-4939-1188-2
Springer New York Heidelberg Dordrecht London

Library of Congress Control Number: 2014946211

Printed on acid-free paper

Springer is part of Springer Science+Business Media (www.springer.com)

Preface

World agriculture has shown phenomenal growth in recent past due to biotechnological application, innovations in agricultural technologies, development of disease resistant varieties, adoption of plant protection measures including integrated disease management and adaptation/development of crop varieties to suit climatic changes. Globally, wheat, rice, maize, other cereals, pulses, oil seed crops, fiber crops, cash crops, and other such economically important plants have been offering food and nutritional security to world population. However the growing population world over and particularly in developing countries require around 50 % more crop production than the existing food production. Problems of hunger, poverty, malnutrition, and economic crisis arising out of unpredicted growth of world population are to be solved. It is estimated that more than 800 million people will be suffering due to inadequate food supply. Approximately 10 % global food production is lost due to plant diseases caused by fungi, bacteria, viruses, mycoplasma, nematodes, and others.

Fungal diseases cause huge losses in crop yields thus affecting world economy and causing shortage of food. The late blight of potato in 1840 caused by *Phytophthora infestans* in Ireland and Europe has been a bolt from the blue, causing a million deaths of starvation and more than a million tried to migrate. Coffee rust in Ceylon, Great Bengal famine in 1943 caused by *Cochliobolus miyabeanus* in rice and southern corn leaf blight during 1970–1971 in the USA caused by *Cochliobolus heterostrophus* have emphasized the role of fungi affecting crops. The control of plant pathogens has become a problem because their populations are variable in time and space. Accurate identification of the pathogen, proper estimate of the severity of disease, impact on crop productivity, recognition of virulence mechanisms, host–pathogen interaction, inoculum, potential, epidemiological aspects, role of environmental/meteorological conditions, and related issues and challenges are of utmost importance.

Incidence of crop diseases can be minimized by reducing pathogen inoculum, inhibition, or inactivation of virulence strategies, and also by prevention of genetic diversity in the crop and conventional breeding mechanism for resistance, as it is facilitated by marker-assisted selection. The transgenic approach and its modification with gene that offer resistance will also offer crop protection to disease.

Approximately 700 viruses are reported to cause devastating diseases in crop plants and often possess wide host range. Pathogenic bacteria belonging to several genera are reported to infect crop plants and cause huge losses in yield. Pathogenic Xanthomonads are reported to cause 350 and above plant diseases. Interesting problems of disease epidemics, distribution of fungal inoculum, failure to combat certain diseases and related issues or challenges are more in tropics and remain as unsolved.

It is predicted that world's population will be 8.3 billion by 2030 and it represents a global challenge to meet the requisite food demand. In this contest crop protection is a necessity in order to achieve greater crop productivity by the use of fungicides and pesticides. The future challenges to be taken in this area are as follows: (1) Concern for environment and public safety. (2) Tackling problem of fungicide resistance. (3) Novel modes of action. However the future success of disease control using fungicides will depend on maintaining a major commitment to research and field experiment database. Integrated disease management, through well-thought disease management practices and IPM to become sustainable approach in crop protection, it must integrate the rational and environmental friendly safe use of chemical control products. Modern disease diagnostic tools may also help in the proper use of fungicides.

A number of microbial and fungal bio-agents, non-fungicidal chemicals, and many other practices will help in the induction of resistance in the crop plants. Thermolabile genes have been identified in crop plants and it is essential to evaluate the effect of such genes in the tolerance of temperatures. Transgenic resistance can be achieved by modifying the plants with resistance imparting transgenes derived from either pathogens/host or other sources. Suppression of disease cum disease-causing pathogens forms the key factor in healthy and vigorous growth of plant. This can be achieved by enhancing beneficial and antagonistic microbes and fungi which probably balance the disease resistance and nutrient supply to the host plant. This has a positive effect on plant's vitality and resistance level. Exclusion of pathogens through plant quarantines is the first step towards reducing the pathogen inoculum followed by healthy and good cultural practices (crop rotation, phytosanitation, maintaining soil health, etc.) besides judicious use of fungicides, exploiting gene pools of plants in relation to breeders programme of disease resistance, combating virulence, improving plant performance, and maintaining genetic diversity in crop plant which has overriding importance.

Plant pathogens create even more problems, challenges, and issues for achieving global food security and some are listed below:

1. There is no database of many damaging pathogens that currently exist.
2. Measurement of the severity of disease symptoms is often subjective and qualitative rather than objective and quantitative.
3. Prediction of obtainable yields needs to be on scientific basis as the globe has witnessed many failures.

4. Scientifically accurate data has to be generated in the field of inoculum quality, its multiplication, virulence, effectiveness, and spread along with innovation of disease forecasting methodologies and creation of models.
5. It is necessary to know the potentiality of most dangerous plant pathogens that are genetically variable.
6. Problems in predicting the origin of next generation plant pathogens.
7. To face the unplanned shocks from pathogens which have evolved new virulence.
8. Minor diseases have become major and one has to know more about them.
9. Strengthening of integrated disease management along with varieties performing well in all geographic regions of a country or globe along with their tolerance to different agroclimatic conditions/soil conditions and levels of disease tolerance.
10. Expectation on the role of climate change with reference to plant pathogens and crop protection.
11. Success rate of transgenics in crop protection and global food security.
12. Farmers need to be educated about crop protection practices including IDM and use of biocontrols agents along with their demonstration in field.
13. More information is essential about abiotic stress upon plant pathogens and also during process of crop protection.
14. Implications involved in the application of nanotechnology for crop protection.
15. Positive and negative impacts of organic farming in relation to crop production and crop protection.
16. Creation of global network of stake holders in crop protection.
17. Creation of proper and modern disease diagnostic tools which are less time consuming and scientifically accurate.
18. Regular review of plant quarantines and other regulations/legislations of crop protection.
19. Culturing of non-culturable plant pathogens.
20. Strengthening of studies on biodiversity, conservation, taxonomy, and control of plant pathogenic fungi.
21. Data strengthening through application of bioinformatic tools including transcriptomics, proteomics, metabolomics, and other aspects of phytopathogens besides the application of DNA-based assay.
22. Elaborated studies are to be made on immunological and molecular detection of plant pathogens.
23. Reliable identification of the causal organisms of disease to the level of species, formae specialis, pathovar, biovar, and races.

Basic and applied/advanced research in the above aspects will pave the way in understanding intricate problems associated with host–pathogen interaction and pathogen biology in offering solutions pertaining to crop protection.

This book has got 12 chapters emphasizing the issues and challenges of crop pathogens and plant protection. Each chapter has been written by experienced and internationally recognized scientists in the field. The topics have been assembled with basic and advanced knowledge in such a way that it will be useful to the beginner as well as to experienced scientists. The need for this kind of books/volume has become imminent as no such book has been published on these aspects. The chapters in this volume include new approaches, new knowledge, and worthy information.

We are grateful to Series Editors-in-Chief Dr. V. K. Gupta and Dr. Maria Tuhoy, Editor (Botany) Eric Stannard, Developmental Editor Elizabeth Orthmann, and others concerned with Springer for their help in various ways. Many minds have helped in the preparation of this volume to which we are indebted. We are grateful to all the contributors for their concern and concerted effects in making knowledge volume. Since the chapters have been independent, written by the author(s), there may be minor overlap or repetition; it is difficult to avoid at this stage.

It is our earnest hope that information presented in this book/volume will make a valuable contribution to the science of Plant Pathology. We believe and trust that it will stimulate further discussions in the pursuit of new knowledge. We also hope that it will be useful to all concerned.

Rabat, Morocco — Aakash Goyal
Hyderabad, Telangana, India — Chakravarthula Manoharachary

Contents

Contributors

R.K. Arora, Ph.D. Central Potato Research Station, Jalandhar, Punjab, India

Subhash Chander Bhardwaj, Ph.D. Regional Station, Directorate of Wheat Research, Shimla, Himachal Pradesh, India

Adrian Lester Cabral, Ph.D. Agriculture and Agri-Food Canada, Morden, MB, Canada

Syama Chatterton, Ph.D. Lethbridge Research Centre, Agriculture and Agri-Food Canada, Lethbridge, AB, Canada

Rinku Dey, Ph.D. Microbiology Section, Directorate of Groundnut Research (ICAR), Junagadh, Gujarat, India

Santhosh J. Eapen, Ph.D. Division of Crop Protection, Indian Institute of Spices Research, Kozhikode, Kerala, India

Thomas Fetch, Ph.D. Agriculture and Agri-Food Canada, Morden, MB, Canada

Nora A. Foroud, Ph.D. Lethbridge Research Centre, Agriculture and Agri-Food Canada, Lethbridge, AB, Canada

Belaghihalli N. Gnanesh, Ph.D. Agriculture and Agri-Food Canada, Morden, MB, Canada

Aakash Goyal, Ph.D. Biodiversity and Integrated Gene Management Program, International Center for Agriculture Research in the Dry Area, Rabat, Morocco

Tom Gräfenhan, Ph.D. Grain Research Laboratory, Canadian Grain Commission, Winnipeg, MB, Canada

Puttaswamy Hariprasad, M.Sc., M.Phil., Ph.D. Food Microbiology Department, Central Food Technological Research Institute, Mysore, Karnataka, India

B. Kangjam, M.Sc. School of Crop Protection, College of Post Graduate Studies, Central Agricultural University, Shillong, Meghalaya, India

K. Ramachandra Kini, M.Sc., Ph.D. Department of Studies in Biotechnology, University of Mysore, Mysore, Karnataka, India

J.D. Kongbrailatpam, M.Sc. School of Crop Protection, College of Post Graduate Studies, Central Agricultural University, Shillong, Meghalaya, India

Indra Kala Kunwar, M.Sc., D.Phil. Department of Botany, Osmania University, Hyderabad, Telangana, India

D. Lyngdoh, M.Sc. School of Crop Protection, College of Post Graduate Studies, Central Agricultural University, Shillong, Meghalaya, India

D. Majumder, Ph.D. School of Crop Protection, College of Post Graduate Studies, Central Agricultural University, Shillong, Meghalaya, India

Chakravarthula Manoharachary, M.Sc., Ph.D., D.Sc. Department of Botany, Osmania University, Hyderabad, Telangana, India

Brent McCallum, Ph.D. Agriculture and Agri-Food Canada, Morden, MB, Canada

Curt McCartney, Ph.D. Agriculture and Agri-Food Canada, Morden, MB, Canada

James G. Menzies, B.Sc., M.Sc., Ph.D. Agriculture and Agri-Food Canada, Morden, MB, Canada

Jennifer Mitchell Fetch, Ph.D. Agriculture and Agri-Food Canada, Brandon, MB, Canada

Anandaraj Muthuswamy, Ph.D. Indian Institute of Spices Research, Kozhikode, Kerala, India

Ganapathy Kuyyamudi Nanaiah, Ph.D. Plant Breeding, Directorate of Sorghum Research, Hyderabad, Andhra Pradesh, India

Chandra S. Nayaka, Ph.D. Department of Studies in Biotechnology, University of Mysore, Mysore, Karnataka, India

Siddapura Ramachandrappa Niranjana, M.Sc., M.Phil., Ph.D. Department of Studies in Biotechnology, University of Mysore, Mysore, Karnataka, India

Kamal Krishna Pal, Ph.D. Microbiology Section, Directorate of Groundnut Research (ICAR), Junagadh, Gujarat, India

Robert F. Park, B.Sc. (Hons.), Ph.D. Plant Breeding Institute, The University of Sydney, Narellan, NSW, Australia

H.S. Prakash, Ph.D. Department of Studies in Biotechnology, University of Mysore, Mysore, Karnataka, India

Pramod Prasad, Ph.D. Regional Station, Directorate of Wheat Research, Shimla, Himachal Pradesh, India

Mohinder Prashar, Ph.D. Maharashtra Hybrid Seeds Company Limited, Jalna, Maharashtra, India

Lana M. Reid, Ph.D. Eastern Cereal and Oilseed Research Centre, Agriculture and Agri-Food Canada, Central Experimental Farm, Ottawa, ON, Canada

Sanjeev Sharma, Ph.D. Division of Plant Protection, Central Potato Research Institute, Shimla, Himachal Pradesh, India

E.G. Suting, M.Sc. School of Crop Protection, College of Post Graduate Studies, Central Agricultural University, Shillong, Meghalaya, India

K.V.B.R. Tilak, Ph.D. Department of Botany, Osmania University, Hyderabad, Telangana, India

Sheryl A. Tittlemier, Ph.D. Grain Research Laboratory, Canadian Grain Commission, Winnipeg, MB, Canada

T. Kelly Turkington, Ph.D. Lacombe Research Centre, Agriculture and Agri-Food Canada, Lacombe, AB, Canada

Chapter 1
Fungal Diseases of Groundnut: Control and Future Challenges

Kamal Krishna Pal, Rinku Dey, and K.V.B.R. Tilak

1.1 Introduction

Groundnut (*Arachis hypogaea* L.) is an important food and oilseed crop. The plants are approximately 15–60 cm tall and produce pinnate leaves with two opposing pairs of leaflets, 2–5 cm long (Porter 1997a). The plant produces yellow flowers that form on non-vegetative branches and withers within 5–6 h after opening (Smith 1950). After pollination, pegs are produced from flowers. At the apex of the peg, pod production occurs. The mature pod may contain 1–5 seeds. The seed contains up to 55 % oil and approximately 25 % protein. Groundnut oil is used mainly for cooking and for producing edible fats and soaps. The cake remaining after extraction of oil is used mainly as animal and poultry feed. The seeds are consumed whole as raw, boiled, or roasted kernels or are processed into various confectionery preparations. The haulm constitutes a nutritious animal feed.

It is a native of South America, originating in central Brazil but is grown widely in the world between 40°N and 40°S latitudes. Groundnut is grown on over 26.4 million hectares worldwide with annual production of 35.6 million tons (FAO 2007). The yields of groundnut are generally lower in developing countries due to various constraints like diseases, pests, and erratic rainfall patterns. The application of crop-protection chemicals is low in these countries because of the high cost and lack of technical knowledge about the application methods. The crop is susceptible to various diseases caused by fungi, bacteria, viruses, nematodes, etc. More than 55 pathogens have been reported to affect groundnut crop (Ghewande et al. 2002).

K.K. Pal, Ph.D. (✉) • R. Dey, Ph.D.
Microbiology Section, Directorate of Groundnut Research (ICAR), Ivnagar Road, PB No. 5, Junagadh, Gujarat 362001, India
e-mail: kkpal9426476749@gmail.com

K.V.B.R. Tilak, Ph.D.
Department of Botany, Osmania University, Hyderabad, Telangana 500007, India

A. Goyal and C. Manoharachary (eds.), *Future Challenges in Crop Protection Against Fungal Pathogens*, Fungal Biology, DOI 10.1007/978-1-4939-1188-2_1,

Some diseases are widely distributed and cause economic losses while others are restricted in distribution and are not of much economic importance at the present time, but they may become major diseases in course of time if situation and climatic conditions favor. All parts of the groundnut plant are susceptible to diseases. Diseases in groundnut crops can occur throughout the plant's life, and therefore, disease management practices are necessary from emergence to the post-harvest period. Many diseases can reduce the quantity or quality of pods and seed.

A large number of diseases attack groundnut in India (Mayee and Datar 1988). The majority are caused by fungi and several of them are yield reducers in certain regions and seasons (Mayee 1995). Among the foliar fungal diseases, leafspots (early and late) and rust are economically important and can cause substantial yield losses in susceptible genotypes. Other foliar fungal diseases like *Alternaria* leaf blight, anthracnose, pepper spot and leaf scorch, *Phomopsis* leaf spot, *Phyllosticta* leaf spot, *Pestalotiopsis* leaf spot, leaf blight, *Phoma* leaf disease, *Myrothecium* leaf blight, *Drechslera* leaf blight, *Zonate* leaf spot, *Cylindrocladium* leaf spot, powdery mildew and *Sclerotium* leaf spot are not economically important diseases at present and hence control schedules are not yet developed for these diseases (Ghewande et al. 2002).

Among seed- and soil-borne diseases, collar or crown rot, stem rot, and dry root rot cause severe seedling mortality and reduce pod yields. Other seed- and seedling diseases like *Pythium* diseases, aflaroot/yellow mold, *Diplodia* collar rot, *Rhizoctonia* damping-off, *Cylindrocladium* black rot, *Fusarium* wilt, pod rot, and *Rhizopus* seed and seedling rot and *Rhizoctonia* limb rot, etc. are reported from groundnut growing regions (Ghewande et al. 2002). Diseases caused by soil-borne pathogens especially pose a threat to groundnut production due to similarity of symptoms, which leads to problems in diagnosis (Thiessen and Woodward 2012). The problem is compounded by the close association of the pods with the soil. Soil-borne diseases are especially complicated to manage due to the difficulty of dispersing fungicides through the groundnut canopy to the soil profile (Thiessen and Woodward 2012). Several soil-borne pathogens that affect groundnut are important to the Southwest United States, including *Botrytis cinerea*, *Pythium* spp., *Rhizoctonia solani*, *Sclerotinia minor* and *S. sclerotiorum*, *Sclerotium rolfsii*, and *Verticillium dahlia* (Thiessen and Woodward 2012).

1.2 Fungal Diseases of Groundnut

The diseases caused by fungi are more in number and severity as compared to other disease causing agents. The fungal diseases can be classified into foliar diseases; seed and seedling diseases; stem, root, and pod diseases.

Fig. 1.1 Early leaf spot of groundnut caused by *Cercospora arachidicola*. (**a**) Symptom at upper surface. (**b**) Symptom at lower surface. (**c**, **d**) Symptoms at advanced stage

1.2.1 Foliar Diseases

1.2.1.1 Early Leaf Spot

It is caused by *Cercospora arachidicola* Hori. This is a very widely seen foliar disease of groundnut. Early leaf spot is characterized by circular spots that are brown to reddish brown on the upper surface of leaflet and a lighter shade of brown on the lower surface of leaflet (Fig. 1.1). The lesions on the upper surface of leaflet are often surrounded by a yellow halo. The lesions often coalesce under severe attack of the disease resulting in premature senescence and shedding of leaflets.

Disease development is favored by temperatures between 25 and 30 °C, prolonged wetness of leaves and high relative humidity. The conidia are disseminated by wind and insects leading to secondary infection.

In India, losses in yield due to leaf spots have been estimated to be in the range of 15–59 % (Ghewande et al. 2002). Besides the loss in pod and kernel yield, the value of fodder is also adversely affected.

Management of Early Leaf Spot

- Deep burying of crop residues and removal of volunteer plants can reduce the primary source of inoculum.

- Intercropping cereals like pearlmillet or sorghum with groundnut is helpful in reducing intensity of leaf spot.
- Varieties tolerant to early leaf spot like ICGS 44, ICGS 76, Somnath, CSMG 84-1 can be grown in areas where early leaf spot is severe.
- Foliar application of Carbendazim (0.05 %)+Mancozeb (0.2 %) at 2–3 weeks interval, two or three times starting from the initiation of the disease can be handy in minimizing the incidence of disease.

1.2.1.2 Late Leaf Spot

The causal agent of Late Leaf Spot is *Phaeoisariopsis personata* (Berk. & M.A. Curtis) van Arx or *Cercosporidium personatum* (Berk. & M.A. Curtis) Deighton. The symptoms for incidence of late leaf spot are appearance of dark brown to almost black circular spots on the lower leaflet surface which are darker than early leaf spots and have a feathery margin. Under severe situation, the affected leaflets become chlorotic, then necrotic, followed by the coalescence of the lesions resulting in premature senescence and shedding of the leaflets (Subrahmanyam et al. 1992).

The disease is favored during prolonged wet weather, with high humidity and temperature in the range of 25–30 °C. Like early leaf spot, the conidia are disseminated by wind and insects leading to secondary infection.

Early leaf spot and late leaf spot are the most important foliar diseases affecting groundnut throughout the world (Shokes and Culbreath 1997). The diseases often occur together or one disease may be more predominant in a given location or year (Smith 1984).

Management of Late Leaf Spot

- A cereal–cereal–groundnut crop rotation should be followed.
- Deep plowing should be done during land preparation.
- Infected crop residues should be removed and destroyed.
- Volunteer groundnut plants should be eliminated.
- The sowing date should be adjusted so that the most conducive environment for disease development can be avoided.
- Foliar application of Carbendazim (0.05 %)+Mancozeb (0.2 %) at 2–3 weeks interval, two or three times starting from the initiation of the disease.
- Cultivars tolerant to late leaf spot like ICGV 87160, ICGV 86590, ICGV 9202, ICGV 92093, ICGS1, TAG24, DRG12, CSMG84-1 may be cultivated.

For more than a decade now, researchers have focused their studies on the action of natural substances from yellow oleander and other plants on phytopathogenous fungi (Ambang et al. 2010). Integrating host resistance and METPS (methanolic extracts of *Thevetia peruviana* seeds) was found to efficiently protect groundnut against *Cercospora* leaf spots (CLS) (Ambang et al. 2011).

Vasavirama and Kirti (2012) demonstrated the potential of SniOLP and Rs-AFP2 genes in developing late leaf spot disease resistance in transgenic peanut. Peanut plants were transformed using a double gene construct with SniOLP (*Solanum nigrum* osmotin-like protein) and Rs-AFP2 (*Raphanus sativus* antifungal protein-2) genes under separate constitutive 35S promoters. Transgenic peanut plants expressing these genes showed enhanced disease resistance to late leaf spot based on a reduction in number and size of lesions on leaves.

1.2.1.3 Rust

The causal organism for rust disease of groundnut is *Puccinia arachidis* Spegazzini. Rust is characterized by the appearance of numerous tiny, reddish orange pustules (uredinia) on the lower surface of leaflets initially and on the upper surface later. Symptoms mainly appear on the leaflets but pustules can be seen on all the aerial parts of a plant. The lesions on the stem are elongated in shape. The infected leaves become necrotic and dry up but remain attached to the plant. When left unchecked, diseased groundnut plants take on a scorched appearance, quickly die, and shed most mature pods. Groundnut rust is known to perpetuate, spread, and produce severe disease outbreaks by means of urediniospores. The rust pathogen may also survive from season to season on volunteer groundnut plants.

Extended periods of cloudy, wet weather tend to favor the appearance of rust on groundnut. The disease spreads by wind movement, rains, and by insects. The disease appears at the same time as that of late leaf spot. These two foliar diseases prematurely defoliate plants, leading to losses in pod and haulm yield. At ICRISAT, in Patancheru, India, rust caused losses of over 50 % (Subrahmanyam and McDonald 1983). In addition to direct yield losses, rust can lower seed quality by reducing seed size and oil content (Ghewande et al. 2002).

Late leaf spot and rust are the two major foliar diseases that together could reduce pod and haulm yield by 70 % and in vitro digestibility of haulms by 22 % (Pande et al. 2003). Two genotypes (ICGV 9202 and 92093) were highly resistant to these foliar diseases maintaining high pod and haulm yield as well as high in vitro digestibility of haulms (>62.3 %) even under highest disease pressure.

Management of Rust

- The buildup of rust inoculum should be avoided by a cereal–cereal–groundnut crop rotation and eradication of volunteer groundnut plants.
- The sowing time can be adjusted so that the most conducive environment for rust (cloudy weather, high humidity) can be avoided.
- Suitable spray schedules should be developed for controlling the disease. Sprays of Bordeaux mixture and Dithiocarbamate have been found effective to control rust and late leaf spots. Spray of Chlorothalonil 0.2 % at regular intervals of 10–15 days, starting from 30 days after germination till 15 days before harvest, can be effective against rust and late leaf spot.
- Resistant/tolerant cultivars like ICGV 87160, ICGV 86590, ICGV 86325, TAG 24, ALR 3, VRI 5, CSMG 84-1, and ICGS 5 can be grown.

Fig. 1.2 (**a**, **b**) Alternaria leaf blight of groundnut caused by *Alternaria alternata*

1.2.1.4 Alternaria Leaf Spot

This disease is caused by *Alternaria alternata* (Fries) Keissler. The symptoms include appearance of small, chlorotic, water-soaked lesions on both surfaces of the leaflets (Subrahmanyam et al. 1992). The lesions are irregular in shape and brown in color. These lesions dry rapidly. Veins and veinlets adjacent to the lesions become necrotic. The affected leaflets show chlorosis and premature senescence under extreme disease severity.

1.2.1.5 Alternaria Leaf Blight

The causal organisms for this disease are *Alternaria alternata* (Fr.) Keissler, *Alternaria arachidis* Kulkarni or *Alternaria tenuissima* (Kunze ex Pers.) Wiltshire. The characteristic symptoms include the appearance of brown irregular shaped spots towards the leaf margins (Fig. 1.2).

1.2.1.6 Cercospora Leaf Blight

The causal organism for this disease is *Cercospora canescens* Ellis & Martin. The disease is characterized by the appearance of small lesions which enlarge into irregular shaped light brown spots (Fig. 1.3). Under long spells of wet weather, the spots merge resulting in blighting and defoliation.

The prevention and control measures include use of crop rotation and proper field sanitation practices, removal and destruction of infected plant tissues and avoiding working when the plants are wet.

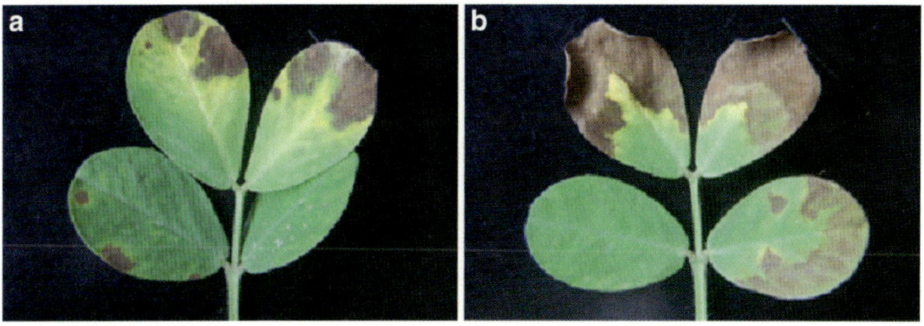

Fig. 1.3 Symptom of Cercospora leaf blight. (**a**) early stage; (**b**) late stage

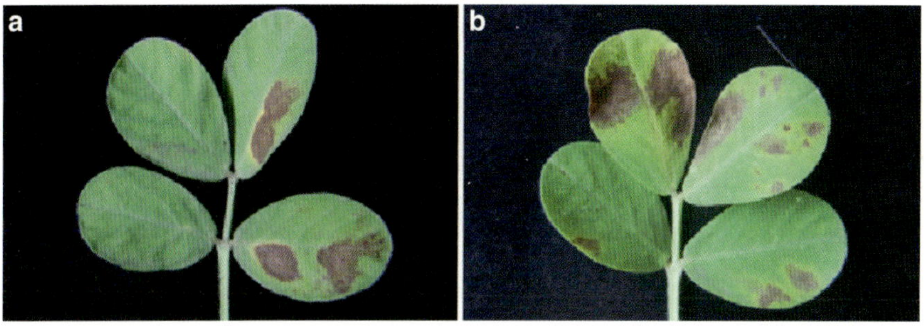

Fig. 1.4 (**a**, **b**) Symptoms of web blotch

1.2.1.7 Web Blotch

The causal organism for this disease is *Didymella arachidicola* (Chock.) Taber, Pettit & Philley. Irregular shaped lesions appear first on the upper surface of the leaves. The brown or dark brown lesions appear in a web or net-like pattern (Fig. 1.4). The lesions coalesce to form blotches. Defoliation occurs under severe disease condition. The disease progresses rapidly under wet weather.

1.2.1.8 Phyllosticta Leaf Spot

The causal organism for this disease is *Phyllosticta arachidis-hypogaea* Vasant Rao. The disease starts in the necrotic tissues and subsequently spreads to the living tissues of the leaves. Circular to irregular light brown lesions with dark brown margins appear on the leaflets (Fig. 1.5). These lesions may merge into irregular necrotic patches.

Fig. 1.5 (**a**, **b**) Phyllosticta leaf spot of groundnut

1.2.1.9 Scab

The scab disease of groundnut is caused by *Sphaceloma arachidis* Bit. & Jenk. Circular or irregular shaped lesions appear on both surfaces of the leaflets. The lesions on upper surfaces of the leaflets show tan color with raised margins and sunken centers while those on the lower surfaces have darker color without raised margins (Subrahmanyam et al. 1992). Lesions may develop on all parts of the plant. Lesions on the petioles and branches may develop into scabs giving the plant a burnt appearance. Under severe disease attack the plants become stunted and the stems become sinuous.

The disease was reported from Argentina, Brazil, Colombia, and Japan (Xu 2009). Wang et al. (2006) reported it to be a great threat to groundnut (*Arachis hypogaea*) production in south China. Wang et al. (2009) constructed a neighbor-joining tree and a minimum evolution tree based on the 18S rDNA sequences of *S. arachidis* and related fungal species. They concluded from their study that the pathogen was in close relationship with Basidiomycetes species. Kearney et al. (2002) established that infected residues from the previous peanut crop are a source of inoculum for onset and development of scab epidemics in the field.

1.2.1.10 Zonate Leaf Spot

Zonate leaf spot of groundnut is caused by *Cristulariella moricola* (Hino) Redhead. The disease is characterized by the appearance of small to large necrotic lesions on the leaflets. The smaller lesions have light brown center with dark brown margins whereas the larger lesions show zonate pattern on both the surfaces of the leaflets.

1.2.1.11 Powdery Mildew

The Powdery mildew of groundnut is caused by *Oidium arachidis* Chorin. The disease is characterized by the appearance of large powdery white patches on the upper

Fig. 1.6 (**a**, **b**) Symptom of Myrothecium leaf blight

surface of the leaflets. These patches are covered with sporulating fungal growth, which gives them a powdery white appearance. At later stages the centers of the spots become necrotic.

1.2.1.12 Myrothecium Leaf Blight

The causal organism for Myrothecium Leaf Blight is *Myrothecium roridum* Tode *ex* Fries. Appearance of round to irregular shaped lesions with gray or brown centers and brown margins takes place on both surfaces of the leaflets (Fig. 1.6). The lesions become large and coalesce to give blighted appearance to the leaves. Abundant black fruiting bodies, often arranged in concentric rings are formed on both leaf surfaces.

1.2.1.13 Pepper Spot and Leaf Scorch

This disease is caused by *Leptosphaerulina crassiasca* (Sechet) Jackson & Bell. This disease is widespread in many groundnut growing countries. The same fungus causes two different types of symptoms. In the Pepper Spot phase, very small necrotic spots appear on the leaflets (Fig. 1.7). The spots are circular to irregular and dark brown to black in color. Many spots may coalesce to give a net-like appearance to the leaflets (Subrahmanyam et al. 1992).

Scorching of the leaf is the common symptom of the disease. The symptom develops as V-shaped lesions starting from the tips of the leaflets mostly and sometimes from the margins. The necrotic lesions are dark brown in color with a yellow zone around the advancing disease margin (Fig. 1.8) and tend to break away from the leaflet margins. Leaf scorch is common early in the season. The disease is controlled by spraying fungicides used for other foliar fungal diseases of groundnut.

Fig. 1.7 Pepper leaf spot of groundnut

Fig. 1.8 (**a**, **b**) Leaf scorch of groundnut

Fig. 1.9 Sclerotium leaf spot of groundnut

1.2.1.14 Sclerotium Leaf Spot

The causal organism for this disease is *Sclerotium rolfsii* Saccardo. The disease is characterized by the appearance of necrotic gray ring spots which may develop holes (Fig. 1.9). Long periods of leaf wetness may result in the spots coalescing to

Fig. 1.10 (**a**, **b**) Anthracnose of groundnut caused by *Colletotrichum dematium*

give blighted appearance to the leaves. Sclerotia (about 0.5–0.8 mm in diameter) initially white, but later brownish in color can be seen on both leaflet surfaces. The other diseases caused by the pathogen are stem, root, and pod diseases which are more damaging as compared to the leaf disease.

1.2.1.15 Anthracnose

The causal organism of the disease is *Colletotrichum dematium* (Pers.) Grove. The disease is characterized by the appearance of wedge-shaped lesions on the tips of leaflets (Subrahmanyam et al. 1992). The lesions may also develop on the margins of the leaflets. The margins of the lesions show yellow zone (Fig. 1.10). The necrotic tissues turn brown and show fragmentation along the margins of the leaflets. The diseased tissues show abundance of fruiting bodies.

1.2.1.16 Choanephora Wet Blight

The disease is caused by *Choanephora cucurbitarum* (Berk. & Ravenel) Thaxt. The disease is characterized by the appearance of water-soaked lesions on the tips of the leaflets (Fig. 1.11). Later the lesions become necrotic and covered by brown mass of spores. In case of prolonged wet weather the disease may spread to the young shoots.

1.2.1.17 Pestalotiopsis Leaf Blight

The causal organism for this disease is *Pestalotiopsis arachidis* Satya. The disease appears in the form of dark brown circular lesions surrounded with light yellow haloes. Subsequently the lesions enlarge and coalesce to form necrotic areas along the margins of the leaflets (Fig. 1.12). Minute black fruiting bodies can be observed at the centers of the blighted areas when viewed through a hand lens.

Fig. 1.11 (**a**, **b**) Choanephora wet blight of groundnut

Fig. 1.12 (**a**, **b**) Pestalotiopsis leaf blight of groundnut

1.2.2 Seed and Seedling Diseases

1.2.2.1 Seed and Seedling Rots

The rotting of seeds and seedlings are caused by a number of organisms. The majority of these rots are caused by *Aspergillus niger*, *Phythium* spp., *Macrophomina phaseolina*, *Rhizoctonia solani*, *Sclerotium rolfsii*, *Rhizopus* spp., *Fusarium* spp., *Aspergillus flavus*, *Lasiodiplodia theobromae*, and *Penicillium* spp. (Subrahmanyam et al. 1992). These fungi are common in most agricultural soils and attack seeds and seedlings of many crops. Seedling disease in groundnut is favored by cool and wet soils which slow seed germination and seedling growth.

Failure of seedlings to emerge from the soil or sudden wilt and death of seedling shortly after emergence is symptom of seedling rot. The infected seeds and seedlings turn into black rotten masses and are covered with the fungal mycelia or spores. These diseases cause severe seedling mortality resulting in poor crop stand and reduce the yields from 25 to 50 % (Ghewande et al. 2002).

Rhizoctonia solani may cause seed decay prior to emergence. Under favorable conditions, the pathogen may attack emerged seedlings and dark, sunken lesions may be formed just below the soil surface.

Fig. 1.13 Collar rot of groundnut caused by *Aspergillus niger*. (**a**) Early symptom; (**b**) drying of the seedling at later stage

Management of Diseases

- Cultural practices like use of raised bed to promote drainage, planting high quality seed with good vigor, planting at recommended soil depths, etc. should be followed.
- Seed treatment with fungicide should be followed.

1.2.2.2 Crown/Collar Rot

The Crown Rot or Collar Rot of groundnut is caused by *Aspergillus niger* van Tieghem. The pathogen attacks the seeds and emerging seedlings but may also affect older plants from mid- to late-season. The diseased seeds fail to emerge and if they are lifted from the soil show black spore mass on the seed surface.

In postemergence infection, the emerged seedlings dry up rapidly (Fig. 1.13). The collar region shows black spore mass. Seedling infection occurs in the cotyledonary-hypocotyl region. The pathogen attacks mature plants also, but the symptoms are less apparent. In mature plants the branches dry out and break from the collar region. Black spore masses are also seen on the pod surfaces.

Aspergillus niger is present in most groundnut soils and is a common contaminant of groundnut seed. Extreme heat or fluctuations in soil moisture during the seedling stage, poor seed quality, seedling damage from pesticides or cultivation, and feeding by root and stem boring insects are stresses thought to aggravate the disease.

Management of Collar Rot

- Crop rotation should be followed.
- Deep sowing (more than 2 in.) should be avoided.
- Application of foliar fungicides should be practiced.
- Late sowing of crop should be avoided.

- High quality seed treated with Carbendazim 1–2 g/kg seed or Mancozeb 2–3 g/kg seed or Chlorothalonil or Captafol @ 2 g/kg seed should be planted.
- Adequate and uniform soil moisture should be maintained.
- Earthing-up of plants should be avoided.
- Resistant/tolerant varieties like J 11, JCG 88 and OG-52-1 can be grown.
- Seed treatment with *Trichoderma harzianum* or *T. viride* @ 4 g/kg seed and their soil application can be helpful.

1.2.2.3 Rhizoctonia Damping-Off

The disease is caused by *Rhizoctonia solani* Kuhn. The disease may be seed or soil-borne. In young seedlings the lesions are usually seen on the hypocotyl in the form of dark brown patches, below the soil surface. These lesions become dark in color, enlarge, and girdle the hypocotyl region, leading to typical symptom of damping-off. The affected areas show brown mycelial growth and dark brown sclerotia develop on the dead tissues.

1.2.2.4 Yellow Mold

The causal organism for this condition is *Aspergillus flavus* Link *ex* Fries. The pathogen attacks seeds and non-emerged seedlings and reduce them to shriveled dried masses covered with yellow or greenish yellow spores. The emerging seedlings decay because of rapid infection of the radicle and hypocotyl. Many strains of this fungus are capable of producing aflatoxins that render the seed unacceptable due to high toxicity for human or animal consumption. The production of aflatoxin by the pathogen strain results in aflaroot disease which is characterized by highly stunted plants with pale and greatly reduced leaves. Infected plants generally become stunted with symptoms of vein clearing chlorosis on the leaflets and lack a secondary root system, a condition known as "aflaroot." The fungus is capable of invading groundnut seeds before harvest, during postharvest drying, and during storage.

Management of Yellow Mold

- Drought stress to the crop should be avoided.
- Nematode infestation should be controlled.
- The crop should be harvested at proper maturity stage.
- The pods should be dried immediately after harvest to 6–8 % moisture content and damaged, infected pods and seeds should be discarded.
- Damage to the testa should be avoided during decortication.

1.2.2.5 Diplodia Collar Rot

The disease is caused by *Lasiodiplodia theobromae* (Pat.) Griffon & Maubi. The pathogen attacks young seedlings and mature plants. The infection takes place at or near the soil surface. The pathogen invades the stem and the infected plants wilt, dry and die within a few days. The base and root of the plants become black. Black pycnidia develop as pimple-like dots on the necrotic tissues.

Management of Diplodia Collar Rot

- Crop rotation with non-related crops lowers the fungal inoculum.
- Foliar diseases should be controlled by the application of fungicides.
- Deep plowing during land preparation is helpful.

1.2.3 Stem, Root, and Pod Diseases

1.2.3.1 Stem Rot

Stem rot of groundnut is caused by the fungus *Sclerotium rolfsii* Saccardo. This disease is also known as southern blight, white mold, southern stem rot, and Sclerotium rot. The fungus is ubiquitous and has a wide host range (Backman and Brenneman 1997; Shew et al. 1987; Punja 1985). It is a very serious disease of groundnut worldwide and causes substantial loss to plant stand and thereby yield. In extreme cases, the disease may cause up to 80 % yield loss; however, losses less than 25 % are more typical (Backman and Brenneman 1997). *Sclerotium rolfsii* is a Deuteromycete, in the group "Mycelia Sterilia." The fungus is characterized by white mycelia, and round, brown sclerotia. The mycelia of *S. rolfsii* survives best in sandy soils, whereas the sclerotia survive best in moist, aerobic conditions found at the soil surface (Punja 1985). The fungus spreads through infected soil, wind splashed rains and sclerotia.

The stem rot pathogen attacks all parts of the plant but the infection of the stem is the most destructive. The symptoms of the disease are wilting and yellowing of the main stem, the lateral branches, or entire plant. Disease may also occur in the early stage at the time of germination. The pathogen can colonize the germinating seeds and seedlings (Fig. 1.14) and white mycelial growth is seen at later stages at the base of the plants and in the soil surface surrounding the infected plants. The mycelia rapidly spread to other branches and plants during favorable weather with warm temperatures and high humidity. Abundant growth of sclerotia, initially white and then turning brown is seen in the infected area. The pathogen may infect pegs, pods, and roots also. In severe cases, the pods become covered with white mycelial growth. The fungus may infect the pods also. The infected pods show brown, water-soaked lesions.

Fig. 1.14 (**a–d**) Colonization of *Sclerotium rolfsii* at early seedling stage of groundnut

Management of Stem Rot

- Crop rotation should be followed to avoid groundnut following groundnut.
- Deep plowing with a moldboard plow should be followed during land preparation. Deep burial of surface organic matter and crop debris is beneficial.
- Growing groundnut on raised beds is helpful.
- High seed rates should be avoided because dense canopy favors the stem rot fungus.
- Soil should not be thrown on groundnut plants during cultivation.
- Leaf fall due to foliar diseases can be avoided by using fungicides. Fallen leaves serve as nutritional source for stem rot pathogen.
- Use of tolerant varieties like Dh 8, and ICGV 86590.
- Seed treatment with *Trichoderma harzianum* or *T. viride* @ 4 g/kg seed.
- Seed treatment with Carbendazim or Captan @ 2–3 g/kg seed.

1.2.3.2 Sclerotinia Blight

Sclerotinia blight is caused by *Sclerotinia minor* Jagger, and on rare occasions may be caused by *Sclerotinia sclerotiorum* (Lib.) de Bary (Porter and Melouk 1997; Woodward et al. 2006). The disease may cause yield losses of 10 % but in severe cases may cause losses exceeding 50 % (Porter and Melouk 1997; Butzler et al. 1998).

Sclerotinia minor and *S. sclerotiorum* are ascomycetes that produce white aerial mycelia and black, irregularly shaped sclerotia. The sclerotia produced by *S. minor* are smaller in size and more abundant in number as compared to the sclerotia of *S. sclerotiorum* (Thiessen and Woodward 2012).

The plants become infected by coming in contact with soil infested with *S. minor*. The symptoms of the disease are wilting of lateral branches followed by the appearance of green water-soaked lesions on the stem. These lesions later turn darker and sunken. The infection can spread to the main stem later. The foliage on the infected branches becomes dark brown and withers giving the plant a blighted appearance. During wet weather, white mycelial growth develops on the infected tissue. Developing pegs are infected. As a result of this disease severe pod loss takes place. Black sclerotia develop on the leaflets, branches, pegs, and the pods. Sclerotia also develop on the seeds inside the pods.

1.2.3.3 Botrytis Blight

Botrytis blight, also known as gray mold of groundnut is caused by *Botrytis cinerea* Pers. *ex* Fries. *Botrytis cinerea* Pers.: Fr. (anamorph) is a Deuteromycete that colonizes the plant quickly (Thiessen and Woodward 2012). This disease is seen first in the foliage of plants in contact with the soil surface. The infected tissues are rapidly decayed. The infected stems and leaves show a thin covering of gray sporulating fungal growth. The infection spreads to the pegs and pods. Black irregularly shaped sclerotia develop on the infected stems and pods. The pathogen may cause the wilt and death of the infected tissue or the entire plant.

1.2.3.4 Cylindrocladium Black Rot

The causal organism for this disease is *Cylindrocladium crotalariae* (Loos) Bell & Sobers. This disease is seen to appear in the fields in the form of patches. The plant symptom includes the chlorosis and wilting of leaves on the main stem. Overall, the plants appear chlorotic and have stunted growth. The infected branches show dense clusters of reddish fruiting bodies on the surface of the soil. The disease may infect pegs, pods, and roots also.

1.2.3.5 Verticillium Wilt

Verticillium wilt of groundnut is caused by the fungus *Verticillium dahliae* Kleb. or *Verticillium albo-atrum* Reinke & Berthier which also cause wilt diseases of many crop plants. The fungus produces microsclerotia, as overwintering structures, which can survive in soil for long periods of time, thereby making management practices difficult.

The fungus produces white fluffy mycelia and conidia that are hyaline and single-cellular (Melouk and Damicone 1997). The microsclerotia develop on plant debris and may remain dormant for long periods of time. The dormant microsclerotia germinate under favorable conditions and stimulation by root exudates. The fungus enters through the plant roots and spreads through the vascular system (Thiessen and Woodward 2012). The infection of the vascular system leads to marginal chlorosis, loss of turgidity, and curling in the leaves. This is followed by yellowing of leaves, necrosis, and defoliation at later stages. The infected plants initially show temporary wilting during the midday but recover during the night, but later the wilting becomes permanent (Subrahmanyam et al. 1992). The roots of the infected plants show brown discoloration of the vascular tissues and show rotting in severe cases. Drought stress accelerates symptom development. Well-watered plants may survive for a long period.

Management of *Verticillium* Wilt

- Fields should be kept clean from soil or crop debris from infested fields.
- Adequate irrigation should be provided to prevent moisture stress.
- Crop rotations with susceptible crops such as cotton and potato should be avoided.
- Long rotations with nonhosts such as corn, sorghum, etc. may be beneficial.

1.2.3.6 Charcoal Rot

This disease is caused by *Macrophomina phaseolina* (Tassi) Goidanich. The disease is seen in the form of water-soaked lesions in the hypocotyl region of the young plants near the soil surface. Later the lesions enlarge and girdle the hypocotyl and kill the plant. The disease infects older plants also. In older plants the infection starts near the soil surface and spreads to the stem and branches and downwards to the roots. The leaves of infected plants turn yellow and then brown at the margins, and appear scorched. The plants thus infected die. The disease may be sometimes restricted to the roots only. The lesions on the roots are initially water-soaked and later become brown. The lateral roots become black and rotted. The pods may be invaded. The inner surfaces of the pods become gray because of the production of microsclerotia.

The fungus causes charcoal rot in many agricultural crops. The fungus survives in soil and crop debris as microsclerotia for many years. Outbreaks of charcoal rot are sporadic and factors such as high soil temperature and low soil moisture are conducive to disease development.

Management of Charcoal Rot

- The promotion of healthy growth of plant by adequate irrigation, fertilization, and pest control may help reduce damage from charcoal rot.
- Seed treatment with *Trichoderma polysporum* or *T. viride* or *T. harzianum* @ 4 g/kg seed or seed treatment with Carbendazim @ 2 g/kg seed or Captafol or Thiram @ 3 g/kg seed is helpful.

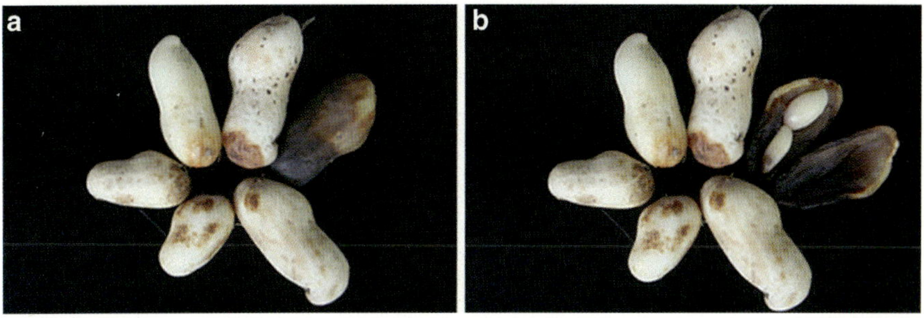

Fig. 1.15 Pod rot of groundnut. (**a**) Symptom on pod surface; (**b**) symptom inside the pod

1.2.3.7 Fusarium Wilt

The disease is caused by *Fusarium oxysporum* Schiechtend. *emend* Snyder & Hans. This disease usually occurs in drought affected plants. The characteristic symptom of the disease is wilting of the plant which may be sudden or gradual. During sudden wilting of the plant the leaves turn grayish-green in color and the plant dries. During gradual wilting, the leaves become chlorotic followed by defoliation and drying of the plants. No external symptoms of the disease are seen on the stem or root. But vascular discoloration is seen if the roots are cut longitudinally. The pathogen may infect the pegs and pods resulting in the pink discoloration of the inner surfaces of the pods.

1.2.3.8 Pod Rot

Pod Rot may be caused by various soil-borne pathogens namely *Pythium myriotylum* Dreschler, *Rhizoctonia solani* Kuhn, *Fusarium solani* (Mart.) Saccardo f. sp. *phaseoli* (Burkholder) Snyder & Hans., *Fusarium oxysporum* Schlechtend. *emend* Snyder & Hans., *Macrophomina phaseolina* (Tassi) Goidanich. The disease is characterized by the development of brown or black lesions on the surface of the pods (Fig. 1.15). The lesions enlarge and cause discoloration of the pod surface and in later stages the shells disintegrate and the kernels decay. The color and texture of the rotting tissues depend on the organism causing the disease. In case of *Fusarium* rot, the shells are pink or purple stained. Pod rots are also caused by *Sclerotinia* spp., *Verticillium* spp., *Sclerotium rolfsii*, and *Botrytis cinerea*, in addition to the other infections caused by them.

A complex of factors, in addition to the fungal pathogens, are probably responsible for severe outbreaks. These factors include excessive soil moisture, wide fluctuations in soil moisture, calcium deficiency, insect and nematode feeding, and irrigation with poor quality (salty) water. In a survey by Wheeler et al. (2005), approximately 40 % of fields in West Texas were found to contain *Pythium* spp., primarily *P. myriotylum*, *P. irregular*, and *P. ultimum*, which are capable of causing damage to the pod and the kernels and may cause significant yield loss. *Pythium* spp. may also cause diseases like damping-off, vascular wilt, and root rot of groundnut. Groundnut plants exhibiting root rot are generally stunted and may overcome the disease under favorable growing conditions (Beute 1997). Pythium pod rot is characterized by the browning and water soaking of pods followed by a brown to black appearance in the final stages of rot (Wells and Phipps 1997). Losses to the tune of 80 % have been reported due to Pythium pod rot (Beute 1997). Pod rot caused by *Pythium* spp. may also cause the junction of the peg and pod to become weakened, which may result in substantial loss at harvest (Lewis and Filonow 1990). The disease symptoms are more severe during frequent rains during pod development.

Rhizoctonia solani also causes a number of diseases besides pod rot, i.e., seed decay, damping-off, and root rot. *Rhizoctonia solani* is a ubiquitous fungus with a wide host range that may be difficult to differentiate from other seed decaying pathogens, making the management of *R. solani* diseases difficult (Thiessen and Woodward 2012).

Rhizoctonia solani Kuhn (anamorph) is a Deuteromycete that does not produce asexual spores; the teleomorph, *Thanatephorus cucumeris*, is a Basidiomycete (Thiessen and Woodward 2012). *Rhizoctonia solani* is capable of surviving saprophytically on a wide host range, including rotated crops and various weed species (Brenneman 1997). Infection of host tissue takes place by germinating sclerotia or hyphae in the soil. Various anastomosis groups (AG) of *Rhizoctonia* spp. occur; however, AG-4 is the most common cause of limb rot and pod rot in groundnut (Brenneman et al. 1994). Rhizoctonia pod rot is differentiated by a dry, brown rotted pod as opposed to the dark, greasy-appearing lesions caused by *Pythium* spp. The seeds may be infected and will harbor the fungus after drying and storing (Wells and Phipps 1997).

Management of Pod Rot

- Excessive irrigation and fertilizer application should be avoided.
- Crop rotation should be followed with a summer fallow and planting should be done on raised beds. Drainage should be improved in low-lying areas.
- Application of gypsum during pegging stage is helpful.
- Nematodes and pod-feeding insects should be controlled where they are a problem.
- Applications of fungicides containing metalaxyl at pegging stage may reduce levels of pod rot.

1.2.3.9 Blackhull

The causal organism of this disease is *Thielaviopsis basicola* (Berk. & Broome) Ferraris (syn. *Chalara elegans* Nag Raj & Kendrick). The disease is characterized by the appearance of black lesions on the surface of the pods which enlarge and coalesce and produce black spore mass, giving black appearance to the pods. The infection of the pegs leads to pod losses during harvesting. Usually the discoloration is superficial, but the decay may extend into the pod causing discoloration of the kernels.

The fungus survives for long periods in soil by producing resistant spores. The disease is favored by high soil pH (above 7.0), excessive soil moisture, and crop rotations with susceptible crops.

Management of Blackhull

- Excessive irrigation should be avoided.
- Crop rotation with alfalfa and cotton should be avoided.

1.3 Integrated Disease Management Practices

Integrated disease management (IDM) is an optimum blend of feasible and economically viable options of disease management for different agroclimatic regions depending on the occurrence and importance of the diseases (Ghewande et al. 2002). It includes use of resistant/tolerant varieties, cultural practices, use of biopesticides and biocontrol agents, chemical methods for need-based application, etc.

1.3.1 Cultural Practices

Soil-borne and foliar diseases of groundnut are best managed with the collective use of both cultural and chemical control measures. The cultural practices help in reducing the level of inoculums in soil which come in contact with the host plant. Deep tillage often reduces soil inoculums by burying the pathogen within the soil to impose anaerobic conditions (Punja 1985). Deep burying of crop residues, destruction of crop debris by burning, removal of volunteer groundnut plants, early planting with wider inter row spacing should be followed for managing leaf spots and rust. Similarly, following cultivation methods that do not pitch soil onto the crowns of groundnut plants reduces crown rot disease by limiting the contact of soil inoculum with the plant (Punja 1985). Intercropping with pearlmillet, sorghum, pigeonpea, and maize are beneficial for the management of early and late leaf spots and rust (Ghewande et al. 2002). Collar rot can be managed by mixed cropping with moth bean (*Phaseolus aconitifolius*) in alternate rows. Treating seeds with seed treatment chemicals reduces the inoculum load. There are many other cultivation practices which can reduce the incidences of disease, if followed systematically.

1.3.1.1 Crop Rotation

Crop rotation is a practice in which different crops are grown in a particular order in the same field over different seasons. Several benefits accrue from crop rotations, including limiting the buildup of fungal inoculum, weed control, and promoting good soil fertility (Peters et al. 2003). Groundnut is grown in rotation with different crops according to the cropping systems prevailing in different parts of the world. Rotating groundnut with cotton is practiced widely but is not desirable because of some of the pathogens which are common to both like *V. dahlia*, *R. solani*, and *Pythium* spp. Rotating groundnut with cereals or grass species such as corn, grain sorghum, or other pasture grasses may reduce both *R. solani* (Brenneman 1997) and *S. rolfsii* (Backman and Brenneman 1997). Crop rotation has been shown to reduce the inoculum density of *Pythium* spp. but has little effect on disease incidence (Beute 1997). Crop rotation with cotton, wheat, maize, onion, and garlic may reduce intensity of stem rot (Ghewande et al. 2002).

1.3.1.2 Irrigation Management

It is important to maintain adequate moisture in groundnut soil to limit the incidence of soil-borne diseases. Overwatering or flooding may increase the incidence of diseases caused by pathogens such as *Pythium* spp. which produces motile zoospores that travel in water (Thiessen and Woodward 2012). Drought should be avoided at pod formation and maturity stages. Heavy irrigation before harvesting may lead to pod rot if harvesting is delayed.

1.3.1.3 Host Resistance

Host resistance is an important aspect of managing foliar and soil-borne diseases. Exploiting the innate resistance in plants for breeding programmes and cultivating disease resistant cultivars would be the safest and most cost-effective strategy in disease management.

Until recently, host resistance to *Sclerotium rolfsii* was unavailable; however, tolerance to *S. rolfsii* may aid in management in conjunction with other control methods (Punja 1985). Groundnut plants with upright growth habits or with compact or open canopies show less incidence of disease than those with a more humid microclimate or more leaves in contact with the soil (Shew et al. 1987). In the USA, several partially resistant cultivars are available such as UF-MDR-98, C-99R, Georgia-07W, Georgia-03L, Georgia-02C, DP-1, and AP-3 (Branch and Brenneman 2009; Brenneman et al. 2005; Gorbet et al. 2004). The varieties tolerant to stem and pod rot are ICGV 87157 and ICGV 86590.

Control of pod rot through host resistance may be effective. Spanish cultivars, especially Toalson, may provide resistance to both *Pythium* spp. and *R. solani* (Beute 1997; Brenneman 1997). Resistance to Sclerotinia blight has been shown in the varieties Virginia 81B, Virginia 93B, Tamspan-90, and Southwest Runner (Porter and Melouk 1997), and in Tamrun OL07 (Baring et al. 2006).

For resistance of groundnut to other soil-borne diseases not much information is available. Growing cultivars with an upright growth habit may limit contact of the canopy with the soil, thereby reducing disease incidence (Backman and Brenneman 1997). According to Shew et al. (1987) employing phenological suppression may be a viable option when selecting groundnut cultivars due to the lack of resistance against various pathogens.

1.3.1.4 Soil Fertility

It is the capacity of a soil to provide crops with essential plant nutrients. It is determined by the physicochemical attributes of the soil and the cultural practices followed by the cultivators. Improving soil fertility also improves plant and soil health and reduces susceptibility to diseases. The concentration of soil nutrients also influences the incidence and severity of infections caused by soil-borne fungi (Thiessen and Woodward 2012).

The availability, deficient or excess, of nitrogen to plants is very important from disease point of view. Application of ammoniacal nitrogen may directly inhibit the germination and limit the mycelial growth of *S. rolfsii* (Punja 1985). Soil amendments with nitrogenous compounds or plant residues may also lead to the increase in population of antagonistic microbes, such as *Trichoderma* spp., *Gliocladium* spp., and *Penicillium* spp. (Porter and Melouk 1997) or may cause death of sclerotia (Punja 1985).

The availability of calcium in the soil also influences disease development. Optimum levels of calcium improve cell wall composition and make them more resistant to pathogen penetration (Agrios 2005). When disease pressure is low, higher levels of calcium in groundnut tissue may limit the disease development by *S. rolfsii* (Punja 1985). There are reports of calcium being used to prevent pod rot disease caused by *Pythium* spp., and *R. solani* (Beute 1997; Brenneman 1997; Walker and Csinos 1980). Calcium amendments have been shown to lessen disease incidence and severity (Csinos 1984).

1.3.2 Chemical Control Practices

Groundnut diseases can be managed best by effective utilization of both cultural and chemical control measures. Protectant fungicide applications prior to infection and curative fungicide applications just after infections occur are effective in reducing losses (Thiessen and Woodward 2012). Fungicides with broad-spectrum activity can be used for controlling wide range of fungal pathogens causing foliar as well as soil-borne diseases. Seed rots and seedling diseases of groundnut can be controlled by treating with Thiram @ 3 g/kg seeds or with Carbendazim @ 2 g/kg seeds. The Strobilurins, which are beta-methoxy acrylic acid derivatives, have broad-spectrum activity and show activity against various foliar and soil-borne pathogens (Bartlett et al. 2002). These compounds inhibit electron transport by binding to the QoI site of Cytochrome b. Other broad-spectrum fungicides such as Propiconazole and Tebuconazole (Triazole fungicides) inhibit sterol demethylation (Brenneman et al. 1994; Baird et al. 1991). There are various other mechanisms by which the fungicides inhibit the pathogens. Phenylamides, such as metalaxyl, inhibit nucleic acid synthesis by affecting RNA synthesis via RNA polymerase I while the fungicide flutolanil prevents respiration by inhibiting succinate dehydroginase synthesis (Thiessen and Woodward 2012). Pentachloronitrobenzene (PCNB), fungicide with aromatic hydrocarbon, causes lipid peroxidation, which leads to the loss of integrity of the cell membrane (Shim et al. 1998). A pyridinamine fungicide, Fluazinam, is a broad-spectrum fungicide with multisite activity that inhibits the respiration of fungi (Syngenta 2012).

A number of fungicides are used by groundnut growers. Broad-spectrum fungicides can be used to control a wide range of pathogens. Tebuconazole, a broad-spectrum systemic fungicide, is used to manage soil-borne Basidiomycetes such as *R. solani* and *S. rolfsii* (Brenneman et al. 1994; Baird et al. 1991). Azoxystrobin is another broad-spectrum fungicide which is used to control soil-borne Basidiomycetes and has limited activity on *Pythium* spp. (Grichar et al. 2000). Flutolanil, a systemic, curative fungicide, is used to control basidiomycetes such as *S. rolfsii* and *R. solani*, and is especially effective at controlling mycelia growth and infection cushion formation (Csinos 1987; Grichar 1995). Chlorothalonil, a multiaction protectant fungicide, is effective against foliar diseases but ineffective against most soil-borne pathogens (Porter 1997b). Oomycetes, such as *Pythium* spp. may be controlled by Metalaxyl and mefenoxam (Filonow and Jackson 1989). Iprodione inhibits the germination of spores and limits the fungal growth of *B. cinerea*, *S. minor*, and *S. sclerotiorum* (Langston et al. 2002). The fungicide Fluazinam has also been used to manage the Sclerotinia blight in groundnut (Butzler et al. 1998).

The timing of application of fungicides is equally important. The timing of application of fungicides should result in minimum loss of yield and minimum usage of the chemicals. In the United States, the initial applications are typically made 60 days after planting (DAP). Subsequent applications for controlling soil-borne pathogens are made between 90 and 120 DAP (Rideout et al. 2008). The frequency of applications may affect the disease development caused by soil-borne pathogens. In a study by Bowen et al., the number of spray applications was evaluated, and four applications in the growing season provided the greatest control (Bowen et al. 1997). The residual activity of fungicides affected disease development. The residual activity of flutolanil at pegging or pod development provided greater disease control than applications at planting (Csinos 1987).

1.3.3 Biological Control

Awareness about the health hazards and environmental concerns due to the indiscriminate use of pesticides resulted in the development of biological control agents. Native microorganisms with biocontrol and plant growth promoting potential are being tested for controlling fungal pathogens of groundnut. Bacteria isolated from the rhizosphere and belonging to a wide variety of genera have the potential to suppress diseases caused by a diversity of soil-borne plant pathogens. A significant reduction in the incidence of root rot caused by *Rhizoctonia* in *Rhizobium* and *Trichoderma* treatment was reported by Jayaraj and Ramabadran (1999). Combined application of *Rhizobium* and *Trichoderma harzianum* (ITCC-4572) successfully reduced the stem rot incidence and also increased the growth of the groundnut plants (Ganesan et al. 2007). Spray of neem seed kernel extract (5 %), or crude neem oil (2 %) is found to be effective against foliar pathogens. Seed treatment with *Trichoderma viride* or *T. harzianum* is found to be effective against seed and soil-borne pathogens @ 4 g/kg seed. Soil application of *Trichoderma viride* or *T. harzianum* @ 25–62.5 kg/ha, preferably in conjunction with organic amendments such as castor cake or FYM can be used effectively against seed and soil-borne diseases (Ghewande et al. 2002).

1.4 The Future Challenges

The large number of fungal diseases affecting groundnut crop is a serious challenge to researchers concerned with improving productivity of groundnut. With the increase in demand for groundnut as food crop and also as oilseeds crop, the productivity of groundnut needs to be increased substantially. There is also the urgency to enhance domestic production of edible oils to lessen the burden of importing oils. Incidences of major fungal diseases can reduce the productivity to as high as 50 %. The development of effective control measures can minimize the yield loss and enhance productivity in a given situation. With the change in the climatic conditions and reports of incidences of minor diseases becoming virulent, many more diseases can be potential threats to enhancing the groundnut productivity.

Concerted and multidisciplinary research efforts are needed to develop disease resistant cultivars, as this is one of the best options. But the task of developing resistant varieties against polyphagous fungal pathogens in groundnut is daunting and appears a distant dream. Introgression of multiple disease resistant polygenic traits in agronomic backgrounds would also be difficult. The narrow genetic base of the present groundnut cultivars has complicated the breeding efforts further. More sources of resistance need to be identified from different genetic backgrounds. Groundnut germplasm should be systematically explored for obtaining sources of resistance. Similarly, wild species, known for resistance to biotic and abiotic stresses, should be widely utilized in crop improvement programmes to transfer the resistant traits. Research efforts are also needed to identify existing and future physiological races of the fungal pathogens. An understanding of the life cycles and taxonomic relationships of the pathogens in the future and emerging climatic change scenarios will also be essential as ultimately disease is the outcome of the interaction among the host, pathogen, and environment.

High cost of chemical fungicides; pollution to soil, air and water; development of resistance to these chemicals and detrimental effects of chemicals on plant, animal, and human health are drawing the attention of researchers towards ecologically safer and environment friendly disease management practices. There has been renewed interest in the use of cultural practices such as intercropping and crop rotation, tillage, manuring, etc. The effects of changing climatic conditions on new varieties of groundnut and new strains of pathogens and the cultural practices needed to manage the host-pathogen-environment interaction needs to be investigated in future.

Besides, new ecologically competent biocontrol agents need to be identified for managing foliar and soil-borne fungal pathogens of groundnut. Options of using endophytic microorganisms for management of biotic stresses can also be explored on evolutionary perspective and on the concept of use of resident antagonists. Of late, efforts are being made to develop soils naturally suppressive to soil-borne fungal pathogens particularly *Aspergillus niger*, *Sclerotium rolfsii*, and *Aspergillus flavus* by enhancing the population of DAPG-producing fluorescent pseudomonads.

As having cultivars resistant/tolerant to soil-borne and foliar fungal diseases is a distant dream using conventional breeding techniques, use of biotechnological tools to transfer alien genes for developing transgenic cultivars seems to be a promising option. But with more stringency and less acceptability of GM crops, transgenic approach will also face uncertain future.

With the limited options available, IDM system consisting of resistant cultivars, improved tillage practices, and reduced application of fungicides, is well-supported and likely to continue in future attempts at control of the groundnut pathogens (Gremillion 2007). Each component of IDM system will have to be effective for successful disease management. Groundnut germplasm will need to be diversified to support breeding programmes and the future of host resistance. Efforts like the diversification of groundnut germplasm will help prevent the potentially serious threat of host resistance breakdown in pathogen populations (Gremillion 2007). The genetic diversity of the pathogens will need to be studied in great detail to develop long-term disease management strategies.

References

Agrios GN (2005) Effect of host-plant nutrition. In: Plant pathology, 5th edn. Elsevier Academic, Burlington, MA, pp 257–262

Ambang Z, Ngoh Dooh JP, Essono G, Bekolo N, Chewachong G, Asseng CC (2010) Effect of *Thevetia peruviana* seeds extract on *in vitro* growth of four strains of *Phytophthora megakarya*. Plant Omics J 3:70–76

Ambang Z, Ndongo B, Essono G, Ngoh JP, Kosma P, Chewachong GM, Asanga A (2011) Control of leaf spot disease caused by *Cercospora* sp. on groundnut (*Arachis hypogaea*) using methanolic extracts of yellow oleander (*Thevetia peruviana*) seeds. Austr J Crop Sci 5:227–232

Backman PA, Brenneman TB (1997) Stem rot. In: Kokalis-Burelle N, Porter DM, Rodriguez-Kabana R, Smith DH, Subrahmanyam P (eds) Compendium of peanut diseases, 2nd edn. APS, St. Paul, MN, pp 36–37

Baird RE, Brenneman TB, Bell DK, Murphy AP (1991) The effects of the fungicide propiconazole (Tilt) on the groundnut shell mycobiota. Mycol Res 95:571–576

Baring MR, Simpson CE, Burow MD, Black MC, Cason JM, Ayers J, Lopez Y, Melouk HA (2006) Registration of 'Tamrun OL07' peanut. Crop Sci 46:2721–2722
Bartlett DW, Clough JM, Godwin JR, Hall AA, Hamer M, Parr-Dobrzanski B (2002) The strobilurin fungicides. Pest Manag Sci 58:649–662
Beute MK (1997) *Pythium* diseases. In: Kokalis-Burelle N, Porter DM, Rodriguez-Kabana R, Smith DH, Subrahmanyam P (eds) Compendium of peanut diseases, 2nd edn. APS, St. Paul, MN, pp 27–30
Bowen L, Hagan AK, Weeks JR (1997) Number of tebuconazole applications for maximizing disease control and yield of peanut in growers' fields in Alabama. Plant Dis 81:927–931
Branch WD, Brenneman TB (2009) Field evaluation for the combination of white mould and tomato spotted wilt disease resistance among peanut. Crop Prot 28:595–598
Brenneman TB (1997) Rhizoctonia diseases. In: Kokalis-Burelle N, Porter DM, Rodriguez-Kabana R, Smith DH, Subrahmanyam P (eds) Compendium of peanut diseases, 2nd edn. APS, St. Paul, MN, pp 30–31
Brenneman TB, Sumner HR, Chandler LR, Hammond JM, Culbreath AK (1994) Effect of application techniques on performance of propiconazole for peanut disease control. Peanut Sci 21:134–138
Brenneman TB, Culbreath AK, Holbrook CC (2005) Screening cultivars and advanced germplasm for multiple disease resistance. In: Proceedings of the American Peanut Research and Education Society 37: article 30
Butzler TM, Bailey J, Beute MK (1998) Integrated management of Sclerotinia blight in peanut: utilizing canopy morphology, mechanical pruning, and fungicide timing. Plant Dis 82: 1312–1318
Csinos AS (1984) Evaluation of the insecticide chlorpyrifos for activity against southern stem rot on peanuts. Peanut Sci 11:98–102
Csinos AS (1987) Control of southern stem rot and *Rhizoctonia* limb rot of peanut with flutolanil. Peanut Sci 14:55–58
Filonow AB, Jackson KE (1989) Effect of metalaxyl plus PCNB or metalaxyl plus tolclofos-methyl on peanut pod rot and soil populations of *Pythium* spp. and *Rhizoctonia solani*. Peanut Sci 16:25–32
Food & Agriculture Organization of the United Nations (2007) Statistical database of Food and Agriculture Organization of the United Nations (FAO). www.fao.org/waicent/portal/statistics_en.asp
Ganesan S, Kuppusamy RG, Sekar R (2007) Integrated management of stem rot disease (*Sclerotium rolfsii*) of groundnut (*Arachis hypogaea* L.) using *Rhizobium* and *Trichoderma harzianum* (ITCC-4572)
Ghewande MP, Desai S, Basu MS (2002) Diagnosis and management of major diseases of groundnut. Bulletin, National Research Centre for Groundnut, Junagadh, Gujarat, p 36
Gorbet DW, Kucharek TA, Shokes FM, Brenneman TB (2004) Field evaluations of peanut germplasm for resistance to stem rot caused by *Sclerotium rolfsii*. Peanut Sci 31:91–95
Gremillion (2007) Contributions to management of diseases of peanut (*Arachis hypogaea*) through Bolivian-derived host resistance, integrated disease management and knowledge of pathogen variability. Ph.D. thesis, University of Georgia, Athens, Georgia
Grichar WJ (1995) Management of stem rot of peanuts (*Arachis hypogaea*) caused by *Sclerotium rolfsii* with fungicides. Crop Prot 14:111–115
Grichar WJ, Besler BA, Jaks AJ (2000) Use of azoxystrobin for disease control in Texas peanut. Peanut Sci 27:83–87
Jayaraj J, Ramabadran R (1999) *Rhizobium-Trichoderma* interaction in vivo and in vitro. Indian Phytopathol 52:190–192
Kearney MIT, Marinelli A, Oddino C, March GJ (2002) Transmission and dispersal of *Sphaceloma arachidis* by crop debris and seed from infected peanut. Peanut Sci 29:13–17
Langston DB Jr, Phipps PM, Stipes RJ (2002) An algorithm for predicting outbreaks of Sclerotinia blight of peanut and improving the timing of fungicide sprays. Plant Dis 86:118–126

Lewis PI, Filonow AB (1990) Reaction of peanut cultivars to *Pythium* pod rot and their influence on populations of *Pythium* spp. in soil. Peanut Sci 17:90–95

Mayee CD (1995) Current status and future approaches for management of groundnut disease in India. Indian Phytopathol 48:389–401

Mayee CD, Datar VV (1988) Diseases of groundnut in the tropics. Rev Trop Plant Pathol 5:169–198

Melouk HA, Damicone JP (1997) Verticillium wilt. In: Kokalis-Burelle N, Porter DM, Rodriguez-Kabana R, Smith DH, Subrahmanyam P (eds) Compendium of peanut diseases, 2nd edn. APS, St. Paul, MN, pp 37–38

Pande S, Bandyopadhyay R, Blümmel M, Narayana Rao J, Thomas D, Navi SS (2003) Disease management factors influencing yield and quality of sorghum and groundnut crop residues. Field Crop Res 84:89–103

Peters RD, Sturz AV, Carter MR, Sanderson JB (2003) Developing disease-suppressive soils through crop rotation and tillage management practices. Soil Till Res 72:181–192

Porter DM (1997a) The peanut plant. In: Kokalis-Burelle N, Porter DM, Rodriguez-Kabana R, Smith DH, Subrahmanyam P (eds) Compendium of peanut diseases, 2nd edn. APS, St. Paul, MN, pp 1–2

Porter DM (1997b) Botrytis blight. In: Kokalis-Burelle N, Porter DM, Rodriguez-Kabana R, Smith DH, Subrahmanyam P (eds) Compendium of peanut diseases, 2nd edn. APS, St. Paul, MN, pp 10–11

Porter DM, Melouk HA (1997) Sclerotinia blight. In: Kokalis-Burelle N, Porter DM, Rodriguez-Kabana R, Smith DH, Subrahmanyam P (eds) Compendium of peanut diseases, 2nd edn. APS, St. Paul, MN, pp 34–36

Punja ZK (1985) The biology, ecology, and control of *Sclerotium rolfsii*. Annu Rev Phytopathol 23:97–127

Rideout SL, Brenneman TB, Culbreath AK, Langston DB Jr (2008) Evaluation of weather-based spray advisories for improved control of peanut stem rot. Plant Dis 92:392–400

Shew BB, Wynne JC, Beute MK (1987) Field, microplot, and greenhouse evaluations of resistance to *Sclerotium rolfsii* in peanut. Plant Dis 71:188–191

Shim MY, Starr JL, Keller NP, Woodard KE, Lee TA Jr (1998) Distribution of isolates of *Sclerotium rolfsii* tolerant to pentachloronitrobenzene in Texas peanut fields. Plant Dis 82:103–106

Shokes FM, Culbreath AK (1997) Early and late leaf spots. In: Kokalis-Burelle N, Porter DM, Rodriguez-Kabana R, Smith DH, Subrahmanyam P (eds) Compendium of peanut diseases, 2nd edn. APS, St. Paul, MN, pp 17–20

Smith BW (1950) *Arachis hypogaea*. Aerial flower and subterranean fruit. Am J Bot 37:802–815

Smith DH (1984) Early and late leaf spots. In: Porter DM, Smith DH, Rodriguez-Kabana R (eds) Compendium of peanut diseases. APS, St. Paul, MN, pp 5–7

Subrahmanyam P, McDonald D (1983) Rust disease of groundnut. ICRISAT Information Bulletin no. 13, Patancheru, AP

Subrahmanyam P, Wongkaew S, Reddy DVR, Demski JW, McDonald D, Sharma SB, Smith DH (1992) Field diagnosis of groundnut diseases. ICRISAT Information Bulletin no. 36, Patancheru, AP

Syngenta (2012) Omega fungicide. http://www.syngentacropprotection.com/prodrender/index.aspx?prodid=632

Thiessen LD, Woodward JE (2012) Diseases of peanut caused by soil-borne pathogens in the southwestern United States. ISRN Agronomy vol 2012, Article ID 517905, 9 page. doi:10.5402/2012/517905

Vasavirama K, Kirti PB (2012) Increased resistance to late leaf spot disease in transgenic peanut using a combination of PR genes. Funct Integr Genomics 12:625–634

Walker ME, Csinos AS (1980) Effect of gypsum on yield, grade and incidence of pod rot in five peanut cultivars. Peanut Sci 7:109–113

Wang ZR, Wang KC, Fang SM, Guo JM, Wang YS, Xiong WJ, Lin HL, Wang DH (2006) Screening of peanut germplasm for resistance to peanut scab. Fujian Agric Sci Technol 21(4):313–316

Wang CT, Wang XZ, Tang YY, Chen DX, Zhang JC, Cui FG, Yu SL (2009) Cloning of the 18S rDNA sequence from *Sphaceloma arachidis,* the causal pathogen of groundnut scab. J SAT Agric Res 7:1

Wells JC, Phipps PM (1997) Peanut disease guide North Carolina and Virginia. Center for Integrated Pest Management. http://ipm.ncsu.edu/peanuts/diseases/guide/contents.html

Wheeler TA, Howell CR, Cotton J, Porter D (2005) *Pythium* species associated with pod rot on West Texas peanuts and *in vitro* sensitivity of isolates to mefenoxam and azoxystrobin. Peanut Sci 32:9–13

Woodward JE, Brenneman TB, Kemerait RC, Culbreath AK, Clark JR (2006) First report of Sclerotinia blight caused by *Sclerotinia sclerotiorium* on peanut in Georgia. Plant Dis 90: article 111

Xu XJ (ed) (2009) Compendium of diseases, pests and weeds in peanut field in China. China Agricultural, Beijing, 443 pp

Chapter 2
Plant Growth Promoting Rhizobacteria in Crop Protection and Challenges

Rinku Dey, Kamal Krishna Pal, and K.V.B.R. Tilak

2.1 Introduction

Plant beneficial microorganisms are increasingly being used in sustainable agriculture. Beneficial microorganisms are used with the aim of improving crop yields by augmenting nutrient availability, enhancing plant growth and providing protection to plants from diseases and pests. The bacteria residing in the rhizosphere of plants and which bring about enhancement in growth and yield of crop plants are widely referred to as plant growth promoting rhizobacteria (PGPR).

PGPR can mediate plant growth by different direct and indirect mechanisms (Glick 1995). Some of the mechanisms commonly observed are (1) increased availability of nutrients due to solubilization/mobilization; (2) biological nitrogen fixation; (3) providing protection to plants from diseases and pests by producing antibiotics, siderophores, hydrogen cyanide, etc. (Medeiros et al. 2005; Keel and Maurhofer 2009); (4) production of plant hormones like IAA, cytokinins, gibberellic acid, etc.; (5) improving the tolerance to stresses like salinity, drought, etc.; (6) lowering of ethylene levels in plants by production of the enzyme 1-aminocyclopropane-1-caroxylate (ACC) deaminase (Glick et al. 1999).

Over the years workers have added newer definitions of PGPR. According to Vessey (2003), numerous species of soil bacteria which flourish in the rhizosphere of plants, but which may grow in, on, or around plant tissues, and stimulate plant growth by a plethora of mechanisms are collectively known as PGPR. Gray and

R. Dey, Ph.D. • K.K. Pal, Ph.D.
Microbiology Section, Directorate of Groundnut Research (ICAR),
Ivnagar Road, PB No. 5, Junagadh, Gujarat 362001, India

K.V.B.R. Tilak, Ph.D. (✉)
Department of Botany, Osmania University, Hyderabad,
Telangana 500007, India
e-mail: tilakkvbr@gmail.com

A. Goyal and C. Manoharachary (eds.), *Future Challenges in Crop Protection Against Fungal Pathogens*, Fungal Biology, DOI 10.1007/978-1-4939-1188-2_2,

Smith (2005) went a step further and separated PGPR into extracellular (ePGPR) organisms, existing in the rhizosphere, on the rhizoplane, or in the spaces between cells of the root cortex, and intracellular (iPGPR), which exist inside root cells.

Several PGPR inoculants have been commercialized. These inoculants result in improvement of crop growth and yield or provide protection to the crop from pests and diseases. Several microbial inoculants are used as biofertilizers, which improve the uptake of nutrients like nitrogen, phosphorus, potassium, sulphur, iron, etc. The genera commonly used as biofertilizers are *Rhizobium*, *Bacillus*, *Pseudomonas*, etc. The genera commonly used as biocontrol agents are *Pseudomonas*, *Bacillus*, *Burkholderia*, *Agrobacterium*, *Streptomyces*, etc. These organisms suppress plant disease by production of antibiotics, siderophores, or by induction of systemic resistance or any other mechanism (Tenuta 2003). Biofertilizers have been an alternative to mineral fertilizers to increase the yield and plant growth in sustainable agriculture (Canbolat et al. 2006). The current trend is the development of a consortium of beneficial microorganisms which will offer multiple beneficial effects including growth promotion, yield enhancement and protection from diseases and pests. Understanding the interaction between consortium of microbial inoculants and plant systems will pave way to harness more benefits from microbial inoculants for improving plant growth and yield (Raja et al. 2006).

2.2 PGPR as Biocontrol Agents of Plant Diseases

There are several mechanisms by which PGPR bring about control of plant diseases. The most commonly used methods are competition and production of metabolites. The metabolites include antibiotics, siderophores, HCN, cell wall-degrading enzymes, etc. (Enebak et al. 1998; Kloepper 1993). Many mechanisms may simultaneously act in a single strain towards providing biocontrol of diseases. Kloepper et al. (1992) mentioned about two types of resistances in plants. Induced systemic resistance (ISR) or systemic acquired resistance (SAR) is defined as the activation of chemical and physical defenses of the plant host by an inducer which could be a chemical or a microorganism, leading to the control of several pathogens.

There are several reports of antagonism of pathogenic fungi by PGPR (Table 2.1). *Pseudomonas* strains MRS23 and CRP55b inhibited the growth of pathogenic fungi, i.e. *Aspergillus* sp., *Fusarium oxysporum* f. sp. *ciceri* and *Rhizoctonia solani* under culture condition (Goel et al. 2002).

There are several reports of reduction of disease incidences by application of PGPR. *Bacillus* spp. isolated from healthy cabbage, kale, and radish reduced black rot incidence in kale and cabbage caused by *Xanthomonas campestris* pv. *campestris* (*Xcc*), in greenhouse and field experiments (Assis et al. 1996). Later, Monteiro et al. (2005) reported that four of these *Bacillus* strains produced lipopeptides active against *Xcc* during its late growth phase. Lipopeptides can also stimulate ISR in plants, probably by interacting with plant cell membranes and inducing temporary

Table 2.1 PGPR having potential biocontrol properties

PGPR	Target pathogen	Disease	Crop
Pseudomonas fluorescens F113, Pf-5, Q2-87, CHA0, etc.	*Pythium ultimum*, *Pythium aphanidermatum*, and *Pythium* sp.	Damping off	Cotton
		Damping off	Tomato
	Rhizoctonia solani	Damping off	Tomato
	Fusarium oxysporum	Root rot	Cotton
Pseudomonas fluorescens strain PfA506	*Erwinia amylovora* strain 153nal super(R)	Fire blight	Apple
Agrobacterium radiobacter	*Agrobacterium tumefaciens*	Crown gall	Dicot plants
Bacillus subtilis AU195	*Aspergillus flavus*	Aflatoxin contamination	Groundnut
Bacillus amyloliquefaciens FZB42	*Fusarium oxysporum*	Wilt	Tomato
Bacillus subtilis 168	*Aspergillus niger*	collar rot	Groundnut
Bacillus subtilis QST713	*Botrytis cinerea*, *Rhizoctonia solani*	Damping off	Grape, cotton
Bacillus subtilis BBG100	*Pythium aphanidermatum*	Damping off	Papaya
P. fluorescens HV37aR2	*Pythium ultimum*	Damping off	Cotton
P. fluorescens HV37aR2	*Pythium ultimum*	Damping off	Cotton
Pseudomonas fluorescens 2-79, 30-84	*Gaeumannomyces graminis* var. *tritici*	Take-all	Wheat
Pseudomonas fluorescens Pf-5	*Pythium ultimum*, *Rhizoctonia solani*	Damping off	Cotton
P. cepacia	*R. solani* and *Pyricularia oryzae*	Damping off and rice blast	Cotton, rice
Bacillus cereus UW85	*Phytophthora medicaginis*, *Pythium aphanidermatum*	Damping off	Alfalfa
Pseudomonas fluorescens strain 97	*Pseudomonas syringae* pv. *phaseolicola*	Halo blight	Beans
Pseudomonas cepacea	*Sclerotium rolfsii*	Stem rot	Beans
Bacillus subtilis	*Blumeria graminis* f. sp. *hordei*	Powdery mildew	Barley
Pseudomonas sp. (WCS 417r)	*Burkholderia caryophylli*	Fusarium wilt	Carnation
Pseudomonas fluorescens	*Pythium ultimum*	Damping off	Cotton
Bacillus subtilis	*Meloidogyne incogita*	Root knot	Cotton
Pseudomonas cepacea	*Rhizoctonia solani*	Damping off	Cotton

(continued)

Table 2.1 (continued)

PGPR	Target pathogen	Disease	Crop
Pseudomonas putida (89B-27)	*Colletotrichum lagenarium*	Anthracnose	Cucumber
Pseudomonas cepacea	*Pythium ultimum*	Damping off	Cucumber
Pseudomonas sp.	*Aspergillus* sp., *Curvularia* sp., *Fusarium oxysporum*, *Rhizoctonia solani*	Wilt	Green gram
Pseudomonas aeruginosa, *Bacillus subtilis*	*Meloidogyne javanica*	Root knot	Mung bean
Pseudomonas fluorescens, *Burkholderia* sp.	*Rhizoctonia solani*	Rice sheath blight	Rice
Pseudomonas fluorescens strain Pf1 and Fp7	*Rhizoctonia solani*	Rice sheath blight	Rice
S. marcescens 90-1, *Bacillus pumilus* SE34	*Peronospora tabacina*	Blue mold	Rice
Aeromonas caviae	*Rhizoctonia solani* and *Fusarium oxysporum* f. sp. *vasinfectum*		Cotton
	Rhizoctonia solani		Bean
Enterobacter agglomerans, *Bacillus cereus*	*Rhizoctonia solani*		Cotton
Paenibacillus illinoisensis	*Rhizoctonia solani*		Cucumber
Serratia marcescens	*Sclerotium minor*		Lettuce
Bacillus spp.	*Sclerotium sclerotiorum*		

Source: Data from: Pal and Gardener (2006) and Bouizgarne (2013)

alterations in the plasma membrane which could raise plant defenses (Ongena et al. 2009). Phenaminomethylacetic acid produced by *Bacillus methylotrophicus* BC79 was reported to be a new kind of substance never found in *Bacillus methylotrophicus* (Shan et al. 2013). Culture filtrate of BC79 showed biocontrol efficiency against rice blast.

Vegetatively propagated crops like plantation and horticultural crops are often susceptible to soil-borne diseases which are difficult to control. The Fusarium wilt of banana caused by *Fusarium oxysporum* f. sp. *cubense* is a very destructive disease worldwide (Figueiredo et al. 2010). Application of endophytic and epiphytic bacteria, single culture or in mixtures, as root or substrate treatments, significantly improved the growth of micropropagated banana plantlets and controlled fusarium wilt (Mariano et al. 2004). *Bacillus amyloliquefaciens* Ba33 was used as a soil disinfector and an antiviral agent against tobacco mosaic virus (TMV) (Shen et al. 2012). Application of mixture of PGPR, more than one genera or species, is more desirable and effective means for controlling plant diseases, as compared to single cultures. The different members in a mixture will have additive or synergistic effects and therefore will result in better control of diseases.

Some bacteria reside in arbuscular and ectomycorrhizal systems and either assist mycorrhiza formation or promote the functioning of their symbiosis (Figueiredo et al. 2010). These bacteria are known as mycorrhiza helper bacteria (MHB). MHB present three significant functions: nutrient mobilization from soil minerals, fixation of atmospheric nitrogen, and plant protection against root pathogens (Frey-Klett et al. 2007). The MHB mentioned by this group were *Pseudomonas fluorescens, P. monteilii, Bacillus coagulans, B. subtilis, Paenibacillus brasiliensis, Rhizobium leguminosarum*, and *Bradyrhizobium japonicum*.

Several workers have successfully tried using biocontrol agents along with synthetic pesticides for disease control and yield enhancement. These treatments may reduce the application of chemical pesticides to crop plants. Corn seeds when bacterized with *Paenibacillus macerans* along with the seed-treatment with fludioxonil and metalaxyl M reduced incidences of pathogens, promoted germination and grain yield (Luz 2003). Similarly, Bugg et al. (2009) used *Bacillus*-based treatments along with seed-treatment practices.

Biocontrol agents need to be formulated if they have to be commercialized. The formulation should be cheap and should not pose any threat to human, animal or plant life or to the environment. Screening for new agents should consider the biology and ecology of the pathosystem, as well as agricultural practices associated with the crop (Fravel 2007). Raj et al. (2003a, b) studied the comparative performance of formulations of PGPR in growth promotion and suppression of downy mildew in pearl millet. The formulations contained two different strains of bacilli with chitosan as a carrier. Formulations LS256 and LS257 besides being the best growth promoters were also the most efficient resistance inducers. Among the application methods tested, soil amendment was found to be the most suitable and desirable way of delivering the formulations. The study demonstrates a potential role for plant growth promoting rhizobacterial formulations in downy mildew management. A few examples of PGPR and biocontrol products are: *Agrobacterium radiobacter* K1026 (Nogall®), *Bacillus pumilus* QST 2808 (Sonata® TM), *B. pumilus* GB34 (YieldShield®), *B. subtilis* GBO3 (Kodiak®), *Pantoea agglomerans* C9-1 (BlightBan C9-1®), *P. agglomerans* E325 (Bloomtime®), *Pseudomonas aureofaciens* Tx-1(Spot-Less®T), *P. syringae* ESC-10 and ESC-11 (Bio-save®), *P. fluorescens* A506 (BlightBan®), *P. chlororaphis* MA 342 (Cedomon®), *Streptomyces griseoviridis* K61 (Mycostop®) and *S. lydicus* WYEC 108 (Actinovate®) (Figueiredo et al. 2010). *B. subtilis* has great potential for use in agriculture and has been used in the formulation of commercial products for agricultural use in several countries (Lazzareti and Bettiol 1997). Several substances have been used in experimental formulations such as lactose, peptone, gum Arabic, xanthan, cellulose and others (Schisler et al. 2004). Formulations based on *Bacillus* are widely available because of their longer shelf life and tolerance to heat and desiccation.

2.3 PGPR Induced Systemic Resistance in Crop Plants Against Pests and Diseases

Plants have developed various strategies to combat aggressors (Van Loon et al. 1998). One of these strategies is the initiation of a defense reaction at the site of infection, which spreads throughout the plant resulting in the development of resistance. Induced resistance is defined as an enhancement of the plant's defensive capacity against a broad spectrum of pathogens and pests that is acquired after appropriate stimulation. The resulting elevated resistance due to an inducing agent upon infection by a pathogen is called ISR or SAR (Hammerschmidt and Kuc 1995). The induction of systemic resistance by rhizobacteria, which are nonpathogenic, is referred as ISR, whereas that by other agents is called SAR (Van Loon et al. 1998). SAR is commonly triggered by the elicitors of avirulent pathogens, such as microbial-associated molecular patterns (MAMPs) (Abramovitch et al. 2006), but it can also be induced by biological (nonmicrobial) and chemical compounds. Typically the ISR by PGPR do not cause any necrotic symptoms on the host plants, whereas SAR is expressed to a maximum level when the inducing organism causes necrosis (Cameron et al. 1994). The expression of induced resistance can be local or systemic when it is expressed at sites not directly exposed to the inducers agent (Stadnik 2000). ISR is quite similar to SAR, making the plant resistant to subsequent attacks of pathogenic organisms, such as viruses, bacteria and fungi (Bakker et al. 2007). SAR or ISR do not provide complete resistance to any particular pathogen, but provide substantial protection to plants for a long time to a broad range of pathogens. Some chemicals, such as SA or analogues [benzothiadiazole (BTH) and its derivatives, e.g. 2,6-dichloronicotinic acid], are known to induce SAR (Table 2.2) and have been successfully used in the field to control diseases (Vallad and Goodman 2004).

Table 2.2 Effect of some SAR elicitors on disease suppression potential

Crop	Pathogen	Disease	SAR elicitors	% Disease reduction
Monocots				
Maize	*Peronosclerospora sorghi*	Downy mildew	BTH	–35
Wheat	*Blumeria graminis* f. sp. *tritici*	Powdery mildew	BTH	–64
Dicots				
Tobacco	*Pseudomonas syringae* pv. *tabaci (tox+)*	Bacterial wildfire	BTH	–99
Tomato	*Pseudomonas syringae* pv. *tomato*	Bacterial speck	BTH	–47
Pepper	*Xanthomonas campestris* pv. *vesicartoria*	Bacterial spot	BTH	–64
Soybean	*Sclerotinia sclerotiarum*	White mold	INA	–46
Cotton	*Xanthomonas campestris* pv. *malvacearum*	bacterial blight	BTH	–42
Leguminous bean	*Uromyces appendiculatus*	rust	INA	–42
Peanut	*Cercosporidium personatum*	late leaf spot	INA	+52
Apple	*Erwinia amylovora*	fire blight	BTH	–73

Source: Data from: Vallad and Goodman (2004)
BTH benzo (1,2,3) thiadiazole-7-carbothiolic acid S-methylester, *INA* 2,6-dichloro isonicotinic acid

2.3.1 Induction of Systemic Resistance by PGPR Against Diseases and Pests

The use of PGPR for inducing systemic resistance against diseases has been demonstrated in field conditions (Vidhyasekaran and Muthamilan 1999; Viswanathan 1999). PGPR have been reported to induce resistance in plants against bacterial, fungal and viral diseases (Liu et al. 1995a, b; Maurhofer et al. 1998; Raj et al. 2003a, b; Halfeld-Vieira et al. 2006), and insect (Zehnder et al. 1997) and nematode pests (Sikora 1988). This type of induced resistance shows advantages such as: effectiveness against various pathogens; stability due to the action of different mechanisms of resistance, systemicity, energy economy; and metabolic utilization of genetic potential for resistance in all susceptible plants (Bonaldo et al. 2005). *Bacillus* and *Pseudomonas* are among the most studied genera of PGPR. Induced resistance was first analyzed in 1961 by pre-inoculation of tobacco plants with TMV (Ross 1961). This procedure protected the plant against other viruses and resulted in the conception of "Systemic Acquired Resistance" (SAR). The activation of defense mechanisms induced by fungi, bacteria, viruses, and nematodes can be achieved by different routes, which may occur alone or concomitantly (Bonaldo et al. 2005). The induction of resistance to disease is an added advantage to the promotion of plant growth and yield by the application of PGPR. The presence of the PGPR in the rhizosphere makes the entire plant, including the shoot, more resistant to pathogens (Figueiredo et al. 2010).

2.3.1.1 Diseases

PGPR have been reported to provide protection to plants from diseases by employing different mechanisms. These mechanisms include production of antibiotics like pyocyanine, pyrrolnitrin, 2,4- diacetylphloroglucinol (Pierson and Thomashow 1992); production of siderophores (Kloepper et al. 1980); competition for nutrients and space (Elad and Chet 1987); production of lytic enzymes like chitinases and β-1,3-glucanases (Potgieter and Alexander 1996; Velazhahan et al. 1999); HCN production (Defago et al. 1990), fluorescent pigments, etc.

The role of ISR in controlling diseases in plants has been demonstrated by many studies (Ramamoorthy et al. 2001). Application of PGPR strains as a seed-treatment resulted in a significant reduction in anthracnose disease caused by *Colletotrichum orbiculare* in cucumber (Wei et al. 1991, 1996). They showed that this plant could be used as a model for ISR. Induction of systemic resistance by *Pseudomonas putida* strain 89B-27 and *Serratia marcescens* strain 90-166 reduced *Fusarium* wilt of cucumber incited by *Fusarium oxysporum* f. sp. *cucumerinum* (Liu et al. 1995a). In sugarcane, Viswanathan and Samiyappan (1999a) established PGPR-mediated ISR against *Colletotrichum falcatum* causing red rot disease. *Pseudomonas fluorescens* 1-94 (Pf1-94) and *Pseudomonas fluorescens* (Pf4-92) strains isolated from rhizosphere soil of chickpea showed ISR against fusarium wilt of chickpea and charcoal rot (Srivastava et al. 2001).

PGPR induce systemic resistance against bacterial diseases as well. Treatment of cucumber seed with *Pseudomonas putida* strain 89B-27 and *Serratia marcescens* strain 90-166 decreased the incidence of bacterial wilt disease (Kloepper et al. 1993). Seed-treatment of cucumber with *P. putida* strain 89B-27, *Flavomonas oryzihabitans* strain INR-5, *S. marcescens* strain 90-166 and *Bacillus pumilus* strain INR-7 provided systemic protection against angular leaf spot caused by *Pseudomonas syringae* pv. *lachrymans* by reducing total lesion diameter compared with non-treated plants (Liu et al. 1995b; Wei et al. 1996).

There are reports of induction of systemic resistance in plants against viral diseases by PGPR. Seed-treatment with *P. fluorescens* strain 89B-27 and *S. marcescens* strain 90-166 reduced the number of cucumber mosaic virus (CMV)-infected plants and delayed the development of symptoms in cucumber and tomato (Raupach et al. 1996). Soil application also proved beneficial. Soil application of *P. fluorescens* strain CHAO resulted in induced systemic protection against inoculation with tobacco necrosis virus (TNV) in tobacco (Maurhofer et al. 1998). Thus, there are ample reports of PGPR ISRs in plants against bacterial, fungal and viral diseases.

2.3.1.2 Insect Pests

There are few reports on the induction of systemic resistance in crop plants against insect pests. Fluorescent pseudomonads have been found to influence the growth and development of different stages of insects. *Pseudomonas maltophila* affected the growth of larval stage of *Helicoverpa zea*, leading to more than 60 % reduction in adult emergence while pupae and adults that emerged from bacteria-infected larvae were smaller (Bong and Sikorowski 1991). Certain PGPR strains activate octadecanoid, shikimate and terpenoid pathways. This in turn alters the production of volatiles in the host plant leading to the attraction of natural enemies (Bell and Muller 1993). Qingwen et al. (1998) reported an increase in polyphenol and terpenoid content in cotton plants treated with *Pseudomonas gladioli*, which affected the relative growth rate, consumption rate and digestibility of feed by *Helicoverpa armigera*. *Serratia marcescens* strain 90-166 was found quite effective in reducing the populations of striped cucumber beetle, *Acalyma vittatum* and the spotted cucumber beetle, *Diabrotica undecimpunctata howardi* on cucumber and its efficacy was better than application of the insecticide esfenvalerate (Zehnder et al. 1997). Attempts have been made to transfer the insecticidal crystal protein from *Bacillus thuringiensis* to fluorescent pseudomonads, keeping in view the efficient root colonization ability and endophytic nature of some fluorescent pseudomonads. Transgenic *P. cepacia* strain 526 with the crystal protein gene has consistently shown insecticidal activity against tobacco hornworm (Stock et al. 1990). PGPR formulations comprising of bacterial strain mixtures having the capability to induce chitinase in plant play an important role in hydrolyzing chitin, the structural component in gut linings of insects and would lead to better control of insect pest (Broadway et al. 1998). Identification of entomopathogenic PGPR strains that have the capability to colonize phylloplane in a stable manner will be a breakthrough in the management of foliar pests (Otsu et al. 2004). *Pseudomonas fluorescens* CHA0 is a root-associated PGPR that suppresses soil-borne fungal diseases of crops. Remarkably, the pseudomonad is also endowed with systemic and oral activity against pest insects which depends on the production of the insecticidal Fit toxin (Pechy-Tarr et al. 2013). The toxin gene (*fitD*) is part of a virulence cassette encoding three regulators (FitF, FitG, FitH) and a type I secretion system (FitABC-E). *P. fluorescens* CHA0 hence can actively induce insect toxin production in response to the host environment, and FitH and FitG are key regulators in this mechanism. Thus, application of PGPR may be useful for management of insect pests as well.

2.3.1.3 Nematodes

Though studies on induction of systemic resistance by PGPR against nematode pests in crop plants are few, PGPR strains have been used successfully as biological control agents for sugar beet and potato cyst nematode (Sikora 1992). *P. fluorescens* induced systemic resistance against *Heterodera schachtii* and inhibited early root penetration in sugar beet (Oostendorp and Sikora 1990). Application of the bacterium *P. chitinolytica* reduced the root-knot nematode infection in tomato crop (Spiegel et al. 1991), while the level of infestation of root-knot nematode *Meloidogyne incognita* on tomato was reduced with fewer galls and egg masses in the soil following root-dipping with *P. fluorescens* strain Pf1 (Santhi and Sivakumar 1995).

2.4 Application of PGPR Mixtures

Application of mixed cultures are often better suited as biological control agents as compared to single ones. The mixed cultures closely mimic the natural environment and might broaden the spectrum of biocontrol activity and enhance the efficacy and reliability of control (Duffy and Weller 1995). The enhancement in biological control abilities of mixed cultures may be due to different mechanisms of action and synergism between the PGPR cultures. Chitinase-producing *Streptomyces* spp. and *Bacillus cereus* isolates used in combination with antibiotic-producing *P. fluorescens* and *Burkholderia* (*Pseudomonas*) *cepacia* isolates have shown a synergistic effect on the suppression of rice sheath blight caused by *Rhizoctonia solani* (Sung and Chung 1997). Similarly, combination of *P. fluorescens* strains Pf1 and FP7 gave effective control of rice sheath blight disease when compared to each strain applied singly (Nandakumar 1998). A combination of two chitinolytic bacterial strains viz., *Paenibacillus* sp. 300 and *Streptomyces* sp. 385 in the ratio of 1:1 or 4:1 was more effective than when they were applied individually for the control of *Fusarium* wilt of cucumber caused by *F. oxysporum* f. sp. *cucumerinum* (Singh et al. 1999). Biocontrol mixtures should be formulated very carefully. The individual strains in the mixture should be compatible with each other and should not inhibit the other strains.

2.5 Broad Spectrum of PGPR Activity

Literature shows many instances of PGPR ISR against a broad range of diseases and pests. Same PGPR strain may induce resistances against many bacterial and fungal diseases and sometimes against insect pests as well in the same crop. Seed-treatment with *P. fluorescens* strain WCS 417 protected radish through induction of systemic resistance against the fungal root pathogen *F. oxysporum* f. sp. *raphani*, avirulent bacterial leaf pathogen *P. syringae* pv. *tomato* and fungal leaf pathogens *Alternaria brassicicola* and *F. oxysporum* (Hoffland et al. 1996). Seed-treatment of *S. marcescens* strain 90-166 showed ISR in cucumber against anthracnose, CMV, bacterial angular leaf spot and cucurbit wilt diseases (Kloepper et al. 1993; Liu et al. 1995a, b). The same strain was also reported to be effective in controlling the striped cucumber beetle, *Acalyma vittatum* and spotted cucumber beetle, *Diabrotica undecimpunctata howardi* (Zehnder et al. 1997). PGPR can also induce ISR against different pathogens in different crops. *P. fluorescens* strain Pf1 induces resistance against different pathogens in different crops, viz., *Rhizoctonia solani* (Nandakumar 1998), *Colletotrichum falcatum* in sugarcane (Viswanathan 1999) and *Pythium aphanidermatum* in tomato (Ramamoorthy et al. 1999). Thus, it would be prudent to select a PGPR having a broad spectrum of activity involving plant growth promotion and induction of resistance against multiple diseases and pests.

2.6 Induction of ISR by Endophytic PGPR

Apart from the colonization of rhizosphere and rhizoplane, some PGPR colonize the internal tissues of plants and are reported to be endophytes. Endophytic bacteria reside within the living plant tissues without doing substantive harm or gaining benefit other than residency (Kado 1992). Endophytic bacteria have the advantage of the protected environment inside the living plant tissues and are potential candidates for inducing ISR in plants. Endophytic bacteria brought about significant control against *F. solani* in cotton and *Sclerotium rolfsii* in beans (Pleban et al. 1995). Seed-treatment of tomato with endophytic bacterium *Bacillus pumilus* strain SE 34 prevented the entry of vascular wilt fungus *F. oxysporum* f. sp. *radicis-lycopersici* into the vascular stele and the mycelial growth was restricted to the epidermis and outer root cortex (Benhamou et al. 1998). Two endophytic tomato root colonizing strains, *Bacillus amyloliquefaciens* CM-2 and T-5 enhanced the growth of tomato seedlings along with the biocontrol of tomato bacterial wilt caused by *Ralstonia solanacearum* (Tan et al. 2013). Biological control of wheat stripe rust by an endophytic *Bacillus subtilis* strain E1R-j in greenhouse and field trials was reported by Li et al. (2013). The biocontrol agent inhibited the germination of urediniospore and reduced the rate of diseased leaves. The use of endophytic PGPR for induction of resistance will be more useful in vegetatively propagated crops like sugarcane, banana, etc. Viswanathan and Samiyappan (1999a) revealed the utility of endophytic *P. fluorescens* strain EP1 isolated from stalk tissues of sugarcane in inducing systemic resistance against red rot caused by *Colletotrichum falcatum*. The endophytic bacteria survives in the vegetatively propagated plant parts and move from one crop to the succeeding crop through vegetative propagation.

2.7 Mechanisms of ISR by PGPR

The PGPR employ several mechanisms for bringing about ISR in plants. These mechanisms may involve strengthening or fortification of the cell wall or elicitation of chemicals for defense against the invasion of disease causing agents.

2.7.1 Structural Modification of Cell Wall in Plants

Plant growth promoting rhizobacteria induce structural modification of the cell wall in response to pathogenic attack (Benhamou et al. 1996b; M'Piga et al. 1997). Treatment of pea seeds with *P. fluorescens* strain 63-28 resulted in formation of structural barriers, viz., cell wall apposition (papillae) and deposition of newly formed callose and accumulation of phenolic compounds at the site of penetration of invading hyphae of *Pythium ultimum* and *F. oxysporum* f. sp. *pisi* (Benhamou et al. 1996a). Seed-treatment of tomato using *Bacillus pumilus* strain SE 34 also induced strengthening of cell walls in tomato against *F. oxysporum* f. sp. *radicis-lycopersici* (Benhamou et al. 1998). This type of rapid defense reaction does not allow the pathogen to invade and also offers the host plant sufficient time to employ other defense mechanisms to fight the pathogens.

2.7.2 PGPR-Mediated Biochemical Changes in the Host Plants

Biochemical and physiological changes have been reported in plants upon application of PGPR. ISR may be due to accumulation of pathogenesis-related (PR) proteins (M'Piga et al. 1997), synthesis of phytoalexin and other secondary metabolites (Zdor and Anderson 1992). ISR by *P. fluorescens* strain CHAO against TNV in tobacco was associated with accumulation of PR proteins namely β-1,3 glucanases and endochitinases (Maurhofer et al. 1994). Involvement of these lytic enzymes was reported by Benhamou et al. (1996b) in the induction of resistance by *P. fluorescens* strain 63-28. These lytic enzymes accumulated at the site of penetration of the fungus, *F. oxysporum* f. sp. *pisi* resulting in the degradation of fungal cell wall. Pathogenesis-related peroxidase and chitinase proteins have been found to induce systemic resistance. In sugarcane, PGPR-mediated ISR against *C. falcatum*, enhanced levels of chitinase and peroxidase and specific induction of two new chitinase isoforms were found when inoculated with *C. falcatum* (Viswanathan and Samiyappan 1999a, b).

PGPR induce systemic resistance in plants through means other than the production of PR proteins also (Pieterse et al. 1996). The plants produce other enzymes of the defense including peroxidases, phenylalanine ammonia-lyase (PAL), and polyphenol-oxidase (PPO). While peroxidase and PPO are catalysts in the formation of lignin, PAL and other enzymes are involved in the formation of phytoalexins (Figueiredo et al. 2010). The phytoalexins are secondary metabolites, antibiotics of low molecular weight produced by plants in response to physical, chemical, or biological stress. They are able to prevent or reduce the activity of pathogens, the rate of production dependent on the genotypes of host and/or pathogen (Daniel and Purkayastha 1995). *P. fluorescens* strains WCS 417r and WCS 358r induced protection in both wild type *Arabidopsis* and transgened *Arabidopsis* with NahG-gene (coding for salicylate hydrolase) without activating PR gene expression (Van Wees et al. 1997). Accumulation of phytoalexin in response to *Pseudomonas* sp. strain WCS 417r treatment in carnation resulted in protection of carnation from wilt disease (Van Peer et al. 1991). Zdor and Anderson (1992) recorded increased peroxidase activity as well as an increase in the level of mRNAs encoding for phenylalanine ammonia-lyase (PAL) and chalcone synthase in the early stages of interaction between bean roots and various bacterial endophytes. The enzymes produced by antagonistic strains have a crucial role to play in disease resistance. The production of enzymes related to pathogenesis (PR proteins) by strains of rhizobacteria is considered as one of the most important property of the antagonistic strains (Saikia et al. 2004). These enzymes are chitinases, lipoxygenases, peroxidases, and glucanases. Plants express the activity of peroxidase during pathogen–host interaction (Saikia et al. 2006). Peroxidase enzyme has been implicated in the oxidation of phenols, lignification (Saparrat and Guillen 2005), plant protection (Hammerschmidt et al. 1982), and elongation of plant cells (Goldberg et al. 1986). Similarly, another enzyme lipoxygenase also contributes to the defense reactions involving the inhibition of growth of the pathogen and induction of phytoalexins (Li et al. 1991). The extent of activity and accumulation of these enzymes depends mainly on the inducing agent, besides the genotype of the plant, physiological conditions, and the pathogen (Tuzun 2001). Certain proteins involved in plant growth and development were up-regulated, such as xyloglucan endotransglycosylase (Wang et al. 2013). Proteins involved in defense were also up-regulated, including peroxidases, glutathione S-transferases and kinases. These proteins associated with disease resistance characteristics were induced in rice plants after exposure to *Bacillus cereus* NMSL88. There are reports of induction of disease resistance by rhizobia also. Hemissi et al. (2013) reported enhanced defense responses of chickpea plants against *Rhizoctonia solani* by pre-inoculation with Rhizobium strains Pch Azm and Pch S.Nsir2. The reduction in infection was accompanied by enhanced level of defense-related enzymes, PAL and peroxidase (POX). An increased level of phenol content was also recorded in the roots of bacterized plants grown in the presence of pathogen.

The defense mechanisms induced by PGPR against insect pests are different. Treatment with PGPR brings about some physiological changes in the host plant that prevent the insects from feeding. Due to PGPR treatment, there was a shift in the metabolic pathway in cucumber plants away from the cucurbitacin synthesis and towards that of other plant defense compounds, resulting in fewer beetles being attracted (Zehnder et al. 1997). In controlling nematodes, PGPR induce resistance by altering root exudates or inducing the host to produce repellents that affect nematode attraction or recognition of the host (Oostendorp and Sikora 1990) and altering the syncytial development or sex ratio in the root tissue (Wyss 1989). Seed-treatment with PGPR strains resulted in increased chitinase enzyme activity and phenolic content in rice, which correlated with the reduced nematode infestation (Swarnakumari 1996). The application of PGPR can thus form an important component of integrated pest management practices in agriculture.

2.7.3 Pathogen-Associated Molecular Patterns

To cause a disease, the invading pathogen must access to the plant interior. But in the process plant also can sense the presence of the pathogens by recognizing the several bio-molecules of pathogens called pathogens associated molecular patterns (PAMPs). Once pathogen penetrates the rigid cell wall of the plant, it comes in touch with the host plasma membrane wherein they encounter the plant extracellular surface receptors which in turns recognizes the PAMPs. On the onset of this reception, activation of plant defenses against the invading pathogens starts with a dramatic cellular reprogramming and initiate PAMP triggered immunity (PTI). This PTI helps the plant to gain a hold over the pathogen and restricts their further proliferation. Thus, to cause disease, the pathogenic microbes must suppress PTI, activated in the plant system. To do so, the pathogens start interfering with the recognition at the plasma membrane or by secreting the effectors proteins into the plant cell cytosol that alters the signaling processes leading to manifestation of disease symptoms. However, if microbes succeeded in subverting the PTI, plant develops more specialized mechanisms to detect and inactivate invading microbes called effector-triggered immunity (ETI) (Chisholm et al. 2006). In this mechanism, plant resistance (R) proteins recognize the bacterial proteins, directly or indirectly, involved in subverting the PTI system activated earlier. It has been discovered that there is remarkable similarities between the molecular mode of PAMP perception in animal and plants. Over the last decades, a number of PAMPs has been identified including lipopolysaccharides (LPS), harpin and flagellin in Gram-negative bacteria; cold shock protein in both Gram-negative and –positive bacteria; transglutaminase, elicitin, β-glucans in Oomycetes; invertase in yeast; chitin and ergosterol in all fungi; xylanase in *Trichoderma,* etc. (Numberger et al. 2004). The role of plasma membrane receptor proteins in recognizing the PAMs and subsequent immunity has been studied in details. It has been proposed that PTI is induced on recognition of the microbial PAMPs and subsequent induction in the transcription of the pathogen-responsive genes, transcription of MAP kinase, production of reactive oxygen species along with the deposition of callose at the site of infection (Numberger et al. 2004).

The recognition of flagellin (protein present in flagella) as PAMP by plant has been studied in details. Though the central region of the flagellin is variable, the highly conserved regions at N and C terminals across eubacterial species facilitated it to become an excellent PAMP. In *Arabidopsis*, a 22 amino acid peptide (flg22) of the highly conserved N terminus region triggered the PTI. The flagellin receptor protein in *Arabidopsis*, FLS2, is a receptor like kinase (RLK) and mutant plant lacking this receptor is insensitive to flagellin which demonstrates the importance of receptors. Besides, flagellin, protein elongation factor Tu (EF-Tu) is one of the most abundant proteins and acts as PAMP in many plants (Chisholm et al. 2006). The possible mechanisms of PAMP-mediated disease suppression is shown in Fig. 2.1.

Once pathogenic microbes could overcome the PTI of plant, it secretes the effector molecules into the cytosol and thereby suppresses the PAMP triggered immunity. In bacteria, type III secretion system (TTSS) is present and it can directly deliver the effector protein into the plant cell. A number of effector proteins in different microbes have been identified. In *Pseudomonas syringae*, 20-30 effector proteins, including AvrRpt2 (protease), AvrB, AvrRpm1, HopPtoD2 (protein phosphatase) and AvrPtoB (E3 ligase), have been found during development of disease symptoms. These effector molecules inhibit the host defense responses initiated by PAMP recognition process (Fig. 2.1).

There are three plant-signaling molecules; salicylate (SA), jasmonate (JA) and ethylene; which regulate the plant defense against the invading microbes. The SA and JA defense pathways are mutually antagonistic and the bacterial pathogen takes advantage of this and overcome the SA-mediated defense responses.

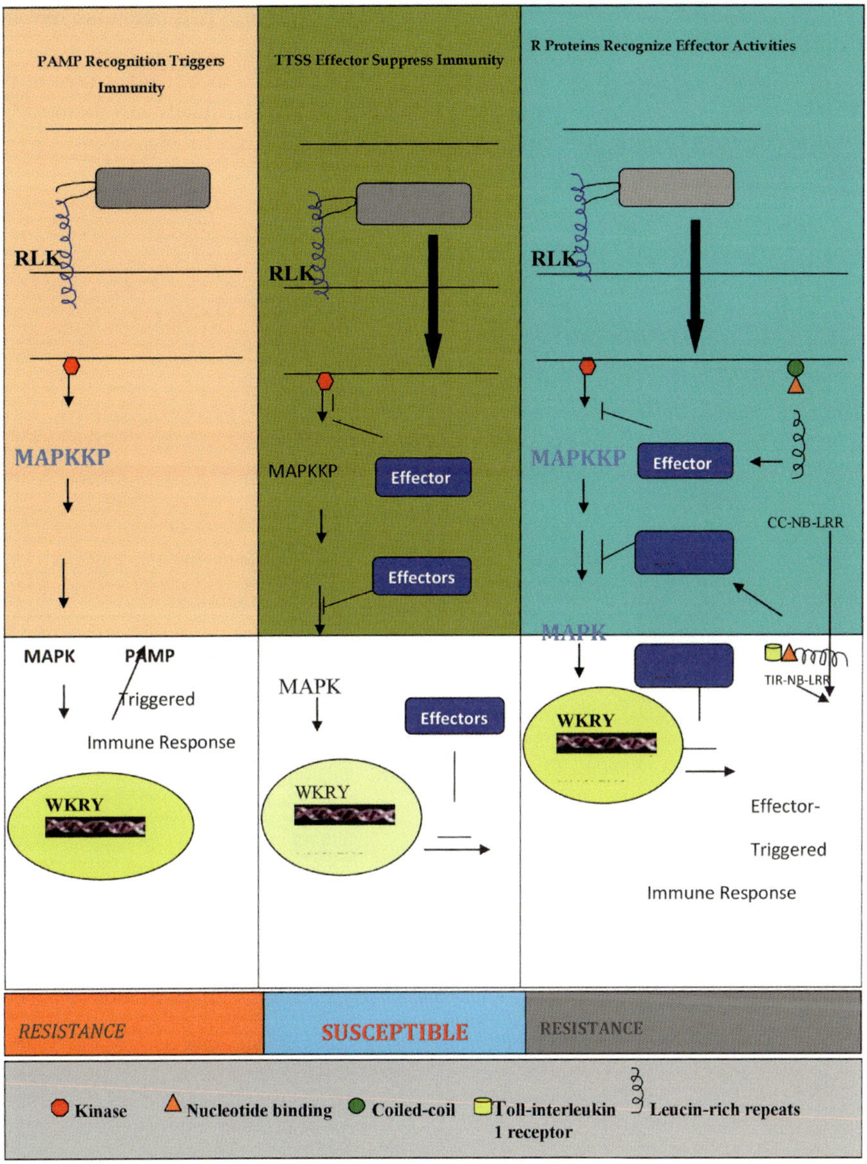

Fig. 2.1 Proposed model for the evolution of bacterial resistance in plants. (*Source*: Data from: Chisholm et al. 2006)

During infection, *Pseudomonas* pathogen produces coronatine which is similar to JA and thus overcome the SA pathway (He et al. 2004). Multiple effector proteins are found to be involved to manipulate the jasmonate pathway in *Pseudomonas syringae*. Majority of the effectors possess enzymatic activity and thus modify a number of host proteins to induce bacterial virulence. Besides, bacterial effectors,

effectors molecule have also been found in fungal and viral pathogenesis like in Oomycetes pathogen *Phytophthora infestans*.

The major focus in future would be on identification of novel plant receptors which would recognize the pathogen effector proteins and inactivate them as a disease control strategies.

2.8 Determinants of PGPR Imparting ISR

2.8.1 Lipopolysaccharides

The LPS present in the outer membrane of bacterial cells are important determinants of ISR in many PGPR strains (Table 2.3). The LPS of *P. fluorescens* strains WCS 374 and WCS 417 induced systemic resistance in radish against *F. oxysporum* f. sp.

Table 2.3 Bacterial determinants and types of host resistance induced by biocontrol agents

Bacterial strain	Plant species	Bacterial determinant	Type
Pseudomonas aeruginosa strain 7NSK2	Tobacco	Salicylic acid	SAR
	Bean	Salicylic acid	SAR
	Tomato	Phenazine and Salicylic acid	SAR
Bacillus amyloliquifaciens	Sugar beet	Lipopolysaccharide	ISR
Pseudomonas fluorescens	Tomato	Massetolide A	ISR
P. fluorescens strain P3	Tobacco	Salicylic acid	ISR
Pseudomonas fluorescens			
CHAO	Tobacco	Siderophore	SAR
	Arabidopsis	Antibiotics (DAPG)	ISR
WCS374	Radish	Lipopolysaccharide	ISR
		Siderophore	ISR
		Iron regulated factor	ISR
WCS417	Carnation	Lipopolysaccharide	ISR
	Arabidopsis	Lipopolysaccharide	ISR
	Radish	Lipopolysaccharide	ISR
		Iron regulated factor	ISR
	Tomato	Lipopolysaccharide	ISR
Pseudomonas putida WCS 358	*Arabidopsis*	Lipopolysaccharide	ISR
		Siderophore	ISR
Pseudomonas putida BTP1	Bean	Iron regulated metabolite Cx	ISR
Serratia marcescens 90-166	Cucumber	Siderophore	ISR
Bacillus mycoides strain Bac J	Sugar beet	Peroxidase, chitinase and β-1,3-glucanase	ISR
Bacillus pumilus 203-6 and 203-7	Sugar beet	Peroxidase, chitinase and β-1,3-glucanase	ISR
Bacillus subtilis GB03, IN937a	*Arabidopsis*	2,3-butanediol	ISR
Pseudomonas putida	Bean	Hexenal	ISR

Source: Data from: Pal and Gardener (2006)

raphani (Leeman et al. 1995). They further explained that the O-antigen side chain of the LPS might have triggered the induction of defense mechanism in plants. However, the LPS of *P. putida* strain WCS 358 having O-antigen side chain did not induce systemic resistance in radish. Van Wees et al. (1997) also obtained similar results where he reported that LPS of WCS 417r and mutant of WCS 417r lacking O-antigen side chain of LPS elicit defense mechanism in *Arabidopsis*. These studies indicated that LPS was not the only determining factor in ISR but other factors were also involved and also elicitation of ISR by LPS was different in different host plants.

2.8.2 Lipopeptides

Some lipopeptides that are produced by bacteria, especially by plant growth promoting rhizobacteria, have been found to induce systemic resistance in plants. Desoignies et al. (2013) investigated the putative action of *Bacillus amyloliquifaciens* lipopeptides in achieving rhizoctonia biocontrol through the control of the virus vector *Polymyxa betae*. Lipopeptides were shown to effectively induce systemic resistance in both the roots and leaves of sugar beet, resulting in a significant reduction in *P. betae* infection. Two classes of bacterial biosurfactant were found to be elicitors of ISR: rhamnolipids and cyclic lipopeptides (cLPs). Massetolide A from *Pseudomonas fluorescens* elicited ISR and enabled *Phytophthora infestans* on tomato to be controlled (Tran et al. 2007). The ISR activity of surfactin was associated, in treated plants, with the accumulation of antifungal compounds (phytoalexins) (Adam 2008) and with the stimulation of the lipoxygenase pathway, leading to the synthesis of fungitoxic oxylipins (Ongena et al. 2007). The induction of systemic resistance by cLPs is not yet clear, but a study by Henry et al. (2011) strongly suggests that the plant cell recognition of surfactin is mediated through interaction with lipids at the plasma membrane level, rather than through specific protein receptors

2.8.3 Siderophores

Siderophore production is an important feature in the suppression of plant pathogens. Siderophores are low molecular weight compounds produced by PGPR under iron-limited conditions. Siderophores act as determinants of ISR under iron starved conditions. The LPS of *P. fluorescens* strains WCS 374 and WCS 417 were the major determinants of ISR in radish against *Fusarium* wilt under iron-replete conditions but not under iron-limited conditions (Leeman et al. 1996). It was found that pyoverdin-type pseudobactin siderophore produced by these bacteria was responsible for ISR. Press et al. (2001) reported the gene for catechol siderophore biosynthesis in *Serratia marcescens* 90-166 and associated it with induced resistance in cucumber against anthracnose. Thus, iron availability may determine the type of PGPR determinant responsible for ISR.

2.8.4 Salicylic Acid

Certain PGPR strains are capable of producing salicylic acid and are responsible for the induction of ISR in plants (Maurhofer et al. 1994). Introduction of *pchA* and *pchB* gene which encode for the synthesis of salicylic acid in *P. fluorescens* strain P3, rendered this strain capable of salicylic acid production and significantly improved its ability to induce systemic resistance in tobacco against TNV. Under conditions of iron limitation, *P. fluorescens* strain CHAO, naturally produced salicylic acid and also induced ISR in tobacco against TNV (Maurhofer et al. 1998).

Apart from these studies, contradictory observations have been also reported by workers. Mutants of *S. marcescens* strain 90-166 lacking in salicylic acid production were found to induce the same level of resistance in cucumber as the wild strain in cucumber and tobacco. Press et al. (1997) working with the salicylic acid producing strain 90-166 of *S. marcescens*, reported induction of resistance both in wild type tobacco and NahG-tobacco (tobacco plant transgened with NahG-gene encoding salicylic acid hydroxylase which converts salicylic acid to catechol). Van Wees et al. (1997) suggested that ISR induced by *P. fluorescens* strains WCS 417r and WCS 358r was independent of salicylic acid production in *Arabidopsis*.

These studies further emphasize the fact that different determinants of PGPR are involved in the induction of systemic resistance and this resistance varies with iron-limiting conditions, PGPR strains, host plants and their cultivars.

2.9 Formulation of PGPR

PGPR need to be formulated for large-scale application in crop fields. PGPR formulation helps in enhancing the shelf life, effective application and delivery of the bacterial cultures to the targeted site. Formulation also aids the packaging, transport and storage of the microbial product. Suslow (1980) reported the survival of PGPR in a dried formulation and the effectiveness of methyl cellulose in a powder formulation for coating sugar beet seed. The organic carriers used for formulation development include peat, talc, lignite, kaolinite, pyrophyllite, zeolite, montmorillonite, alginate, press mud, sawdust and vermiculite. Talc and Peat have been used as traditional carrier materials for effective formulations of PGPR. Vidhyasekaran and Muthamilan (1995) reported that the population of bacteria had been stable up to 240 days in talc-based and peat-based formulations. PGPR can be effectively formulated for systemic protection of crop plants against diseases. The most commonly used formulations of PGPR involve strains of *Pseudomonas fluorescens, P. aeruginosa, P. putida, Bacillus subtilis, B. amyloliquifaciens,* etc. *P. putida* strain 30 and 180 survived up to 6 months in talc-based formulations. The population load at the end of 6th month was 10^8 cfu/g of the product (Bora et al. 2004). Shelf life of *P. chlororaphis* (PA23) and *B. subtilis* (CBE4) in peat carriers was retained for more than 6 months (Nakkeeran et al. 2004).

The formulated products can be delivered through different methods of application like seed-treatment, seed-priming, soil application, foliar application, root-dip, sett-treatment in sugarcane, sucker-treatment in banana. Drum priming of carrot and parsnip seeds with *P. fluorescens* Pf CHAO proliferated well on the seeds and could be explored for realistic scale up of PGPR (Wright et al. 2003). Root-dipping of seedlings has been found effective for the control of soil-borne pathogens in case of transplanted plants. Dipping of *Phyllanthus amarus* seedlings in talc-based formulation of *B. subtilis* (BSCBE4) or *P. chlororaphis* (PA23) for 30 min prior to transplanting reduced stem blight of *P. amarus* (Mathiyazhagan et al. 2004). Foliar application of PGPR formulations are used for controlling foliar diseases. However, the leaf surface microclimate is subjected to frequent changes and should be considered while designing spray schedules. Preharvest foliar application of talc-based fluorescent pseudomonads strain FP7 supplemented with chitin at fortnightly intervals (5 g/L; spray volume 20 L/tree) on to mango trees from pre-flowering to fruit maturity stage induced flowering to the maximum, reduced the latent infection by *Colletotrichum gloeosporioides* beside increasing the fruit yield and quality (Vivekananthan et al. 2004). Application of PGPR formulations with strain mixtures perform better than individual strains for the management of pest and diseases of crop plants, in addition to plant growth promotion (Nakkeeran et al. 2005). Combination of iron chelating pseudomonad strains and inducers of systemic resistance suppressed Fusarium wilt of radish better than the application of individual strains (de Boer et al. 2003).

Microencapsulation of rhizobacteria has been tried in recent years as a formulation. Microcapsules of rhizobacteria consist of a cross linked polymer deposited around a liquid phase, where bacteria are dispersed (Nakkeeran et al. 2005). The process of microencapsulation involves mixing of gelatin polyphosphate polymer pair (81:19 w/w) at acidic pH with rhizobacteria suspended in oil (Charpentier et al. 1999). The microencapsulation technique has not picked up in a big way. The cost factor could be a reason. This formulation needs to be tested in large-scale field trials in order to be adopted for commercial use.

2.9.1 Frequency of Application

The effectiveness of application of PGPR formulation remains for a certain time followed by a decline over time. This determines the number of applications of PGPR formulations needed to maintain the resistance levels in crop plants (Dalisay and Kuc 1995). Different methods of application have different durability. Foliar sprays of *P. fluorescens* formulations should be given at every 15 days intervals for managing rice foliar diseases (Vidhyasekaran et al. 1997). Experiments conducted by Nayar (1996) indicated that induction of defense mechanisms using *P. fluorescens* persisted up to 60 days by seed-treatment, 30 days by root-dipping and 15 days by foliar spray. The duration of the induced resistance varies from crop to crop and strain to strain of PGPR. The induction of resistance by PGPR persisted for 90 days of crop growth in sugarcane (Viswanathan 1999).

2.10 Challenges

Though PGPR have a potential scope in commercialization, the threat of certain PGPR (*P. aeruginosa*, *P. cepacia* and *B. cereus*) to infect human beings as opportunistic pathogens has to be clarified before large-scale acceptance (Nakkeeran et al. 2005). Potential biocontrol agents have to pass through several tests in order to be commercially viable. After thorough, large-scale field testing at multiple locations, differing in soil and climatic conditions, these agents can be recommended for registration with the government agencies. The technology must be transferred to some firms which can take up the mass production of the product and finally it must be adopted by the end users i.e. the farming community. The biocontrol agent should not pose any threat to human and animal health and should not be an environmental hazard.

The knowledge of ecology of the introduced PGPR strains is sometimes lacking which may be a serious impediment to the establishment and multiplication of the PGPR strains. The interaction of the introduced strains with the native flora and fauna will also be a deciding factor in the success of the biocontrol agent.

PGPR formulations are usually produced at small entrepreneurial levels or at the fermentation units of research stations, but seldom at very large industrial firms. Hence, technologies for production of biofertilizers and biopesticides at very large levels are not suitably developed. Moreover, IPR issues have not been dealt with suitably in case of these bioproducts. Ambiguities prevail with respect to registration/licensing/patenting of these products with the law differing in different countries.

PGPR have been discovered and researched for last two-three decades, but till date widespread use of these products is yet to be seen. Availability of good quality biofertilizers and biopesticides to the farmers is still an issue along with lack of awareness about the products and their benefits. The available products have less shelf life and should be used properly because of the biological nature of the products. The issue of quality control should be dealt with stringency to ensure quality products to the end users. Very often, locally formulated products are available in the market in plenty but quality of those products cannot be ascertained along with tangible benefit by the farmers.

2.10.1 Constraints to Commercialization

The success of any biological agents depends on availability of quality formulation with good shelf life, marketing and perceived acceptability and demand of the end users. The factors limiting the successful commercialization of biological agents are as follows:

- Reliability and authenticity of the selection of the biocontrol agent.
- Concerns about the possible ecological consequences of the intended commercialization of the biocontrol PGPR.

- Lack of awareness about the biological agents and their target pathogens.
- Risk associated with the mass multiplication of the biocontrol agents in industrial scale fermenters.
- Concerns of inconsistent performance of PGPR biocontrol agents in managing disease and pests.
- Chances of mutation and loss of desirable traits in the biocontrol agents.
- Lack of awareness among the farmers about the potential of the biocontrol agents in managing diseases and pests.
- Competition from the spurious locally developed biocontrol agents.
- Procedural delays in registration of the products.
- Lack of proper delivery system for biocontrol PGPR.
- Concerns about stability and quality of the products.
- Stiff challenges from environment protection agencies and inherent difficulties in addressing their concerns.
- Perceived potential threats from few opportunistic human pathogens as biocontrol agents.

2.11 Conclusions

PGPR are beneficial to crop plants in many ways. Inoculation with PGPR results in improvement of plant growth, control of diseases and induction of systemic resistance. Tikhonovich and Provorov (2011) argued that utilization of appropriate preparations of beneficial microorganisms is the most promising strategy for maintaining agricultural productivity whilst reducing the inputs of inorganic fertilizers, herbicides and pesticides and that 'microbiology is the basis of sustainable agriculture'. Several strains of PGPR have broad spectrum activity against multiple diseases and also provide protection against insect and nematode pests. Endophytic PGPR have been found beneficial in growth promotion and disease control in vegetatively propagated crops. With the progress of agriculture towards sustainability, microbes will find greater use as biocontrol agents.

However, we should be realistic with cautions. Though tall claims have been made by researchers over the past several decades about the potential applications of a plethora of PGPR biocontrol agents in managing a number of disease and pests in many crop species, not much success has been achieved yet for commercialization and their application at field level. Concerted efforts will be required to demonstrate the benefits of the PGPR biocontrol agents to the farmers so that the eco-friendly agents can be popularized. Unless end users are convinced by the benefits of the biocontrol PGPRs by conducting trials of their own, the success stories will remain in the research laboratories only.

Acknowledgement One of the author's (KVBRT) is grateful to the National Academy of Sciences, India for offering financial assistance as Senior Scientist, Platinum Jubilee Fellowship. Sincere thanks are due to Dr. (Mrs.) I K Kunwar for critically going through the manuscript.

References

Abramovitch RB, Anderson JC, Martin GM (2006) Bacterial elicitation and evasion of plant innate immunity. Nat Rev Mol Cell Biol 7:601–611

Adam A (2008) Elicitation of induced systemic resistance in tomato and cucumber and activation of the lipoxygenase pathway by non-pathogenic rhizobacteria. PhD Thesis, University of Liège, Belgium

Assis SMP, Mariano RLR, Michereff SJ, Coelho RSB (1996) Biocontrol of *Xanthomonas campestris* pv. *campestris* on kale with *Bacillus* spp. and endophytic bacteria. In: Tang W, Cook RJ, Rovira A (eds) Advances in biological control of plant diseases. China Agricultural University Press, Beijing, pp 347–353

Bakker PAH, Pieterse CMJ, Van Loon LC (2007) Induced systemic resistance by fluorescent *Pseudomonas* spp. Phytopathology 97:239–243

Bell E, Muller JE (1993) Characterization of an Arabidopsis lipoxigenase gene responsive to methyl jasmonate and wounding. Plant Physiol 103:1133–1137

Benhamou N, Belanger RR, Paulitz TC (1996a) Induction of differential host responses by *Pseudomonas fluorescens* in Ri T-DNA transformed pea roots after challenge with *Fusarium oxysporum*f. sp. *pisi* and *Pythium ultimum*. Phytopathology 86:114–178

Benhamou N, Kloepper JW, Quadt-Hallmann A, Tuzun S (1996b) Induction of defense-related ultrastructural modifications in pea root tissues inoculated with endophytic bacteria. Plant Physiol 112:919–929

Benhamou N, Kloepper JW, Tuzun S (1998) Induction of resistance against *Fusarium* wilt of tomato by combination of chitosan with an endophytic bacterial strain: ultra structure and cytochemistry of the host response. Planta 204:153–168

Bonaldo SM, Pascholati SF, Romeiro RS (2005) Induçao de resistência: naçoês básicas e perspectivas. In: Cavalcanti LS, di Piero RM, Cia P, Pascholati SF, Resende MLV, Romeiro RS (eds) Indução de resistência em plantas a patógenos e insetos. FEALQ, Piracicaba, pp 11–28

Bong CFJ, Sikorowski PP (1991) Effects of cytoplasmic polyhedrosis virus and bacterial contamination on growth and development of the corn earworm, *Helicoverpa zea*. J Invertebr Pathol 57:406–412

Bora T, Ozaktan H, Gore E, Aslan E (2004) Biological control of *Fusarium oxysporum* f. sp. *melonis* by wettable powder formulations of the two strains of *Pseudomonas putida*. J Phytopathol 152:471–475

Bouizgarne B (2013) Bacteria for plant growth promotion and disease management. In: Maheshwari DK (ed) Bacteria in agrobiology: disease management. Springer, Heidelberg. doi:10.1007/978-3-642-33639-3-2

Broadway RM, Gongora C, Kain WC, Sanderson JA, Monroy JA, Bennett KC, Warner JB, Hoffman MP (1998) Novel chitinolytic enzymes with biological activity against herbivorous insects. J Chem Ecol 24:985–998

Bugg K, Hairston W, Riggs J (2009) Succeeding in a traditional Ag-chemical company despite the "snake oil"/"foo-foo dust" concepts of biological-based products. In: Weller D, Thomashow L, Loper J, Paulitz T, Mazzola M, Mavrodi D, Landa BB, Thompson J (eds) 8th International PGPR workshop. Portland, p 17

Cameron RK, Dixon R, Lamb C (1994) Biologically induced systemic acquired resistance in *Arabidopsis thaliana*. Plant J 5:715–725

Canbolat MY, Barik KK, Cakmarci R, Sabin F (2006) Effects of mineral and biofertilizers on barley growth on compacted soil. Acta Agric Scand 56:324–332

Charpentier CA, Gadille P, Benoit JP (1999) Rhizobacteria microencapsulation: properties of microparticles obtained by spray drying. J Microencapsul 16:215–229

Chisholm ST, Coaker G, Day B, Staskawicz BJ (2006) Host-microbe interactions: shaping the evolution of the plant immune response. Cell 124:803–814

Dalisay RF, Kuc JA (1995) Persistence of induced resistance and enhanced peroxidase and chitinase activities in cucumber plants. Physiol Mol Plant Pathol 47:315–327

Daniel M, Purkayastha RP (1995) Handbook of phytoalexin metabolism and action. Marcel Dekker, New York, p 615

de Boer M, Bom P, Kindt F, Keurentjes JJB, van der Sluis I, vanLoon LC, Bakker PAHM (2003) Control of *Fusarium* wilt of radish by combining *Pseudomonas putida* strains that have different disease-suppressive mechanisms. Phytopathology 93:626–632

Defago G, Berling CH, Burger U, Hass D, Kahr G, Keel C, Voisard C, Wirthner P, Wuthrich B (1990) Suppression of black root rot of tobacco and other root diseases by strains of *Pseudomonas fluorescens*: potential applications and mechanisms. In: Hornby D (ed) Biological control of soilborne plant pathogens. CAB International, Wellingford, pp 93–108

Desoignies N, Schramme F, Ongena M, Legreve A (2013) Systemic resistance induced by *Bacillus* lipopeptides in *Beta vulgaris* reduces infection by the rhizomania disease vector *Polymyxa betae*. Mol Plant Pathol 14:416–421. doi:10.1111/mpp.12008

Duffy BK, Weller DM (1995) Use of *Gaeumannomyces graminis* var. *graminis* alone and in combination with fluorescent *Pseudomonas* spp. to suppress take-all of wheat. Plant Dis 79:907–911

Elad Y, Chet I (1987) Possible role of competition for nutrition in biocontrol of *Pythium* damping-off by bacteria. Phytopathology 77:190–195

Enebak SA, Wei G, Kloepper JW (1998) Effects of plant growth-promoting rhizobacteria on loblolly and slash pine seedlings. Forest Sci 44:139–144

Figueiredo MVB, Seldin L, Araujo FF, Mariano RLR (2010) Plant growth promoting rhizobacteria: fundamentals and applications. In: Maheshwari DK (ed) Plant growth and health promoting bacteria. Microbiology monographs 18. Springer, Berlin, pp 21–43. doi:10.1007/978-3-642-13612-2_2

Fravel D (2007) Commercialization of biocontrol agents for use against plant pathogens. In: IX Reunião Brasileira sobre Controle Biológico de Doenças de Plantas, Campinas, São Paulo, Brasil, CD-ROM, pp 1–2

Frey-Klett P, Garbaye J, Tarkka M (2007) The mycorrhiza helper bacteria revisited. New Phytol 176:22–36

Glick BR (1995) The enhancement of plant-growth by free-living bacteria. Can J Microbiol 41:109–117

Glick BR, Patten CL, Holgin G, Penrose DM (1999) Biochemical and genetic mechanisms used by plant growth promoting bacteria. Imperial College Press, London, p 267

Goel AK, Sindhu SS, Dadarwal KR (2002) Stimulation of nodulation and plant growth of chickpea (*Cicer arietinum* L.) by *Pseudomonas* sp. antagonistic to fungal pathogens. Biol Fertil Soils 36:391–396

Goldberg R, Imberty A, Liberman M, Prat R (1986) Relationships between peroxidatic activities and cell wall plasticity. In: Greppin H, Peneland C, Gaspar T (eds) Molecular and physiological aspects of plant peroxidases. University of Geneva, Geneva, pp 208–220

Gray EJ, Smith DL (2005) Intracellular and extracellular PGPR: commonalities and distinctions in the plant–bacterium signaling processes. Soil Biol Biochem 37:395–412

Halfeld-Vieira BA, Vieira JR Jr, Romeiro RS, Silva HSA, Baract-Pereira MC (2006) Induction of systemic resistance in tomato by autochthonus phylloplane residente *Bacillus cereus*. Pesq Agrop Bras 41:1247–1252

Hammerschmidt R, Kuc J (1995) Induced resistance to disease in plants. Kluwer Academic Publishers, Dordrecht, p 182

Hammerschmidt R, Nucckles F, Kuc I (1982) Association of enhance peroxidase activity with induced systemic resistance of cucumber to *Colletotrichum largenarium*. Physiol Plant Pathol 20:73–82

He P, Chintamanani S, Chen Z, Zhu L, Kunkel BN, Alfano JR, Tang X, Zhou JM (2004) Activation of a Col1-dependent pathway in Arabidopsis by Pseudomonas syringae type III effectors and coronatine. Plant J 37:589–602

Hemissi I, Mabrouk Y, Mejri S, Saidi M, Sifi B (2013) Enhanced defense responses of chickpea plants against *Rhizoctonia solani* by pre-inoculation with rhizobia. J Phytopathol 161:412–418

Henry G, Deleu M, Jourdan E, Thonart P, Ongena M (2011) The bacterial lipopeptide surfactin targets the lipid fraction of the plant plasma membrane to trigger immune-related defense responses. Cell Microbiol 13:1824–1837

Hoffland E, Hakulinem J, Van Pelt JA (1996) Comparison of systemic resistance induced by avirulent and nonpathogenic *Pseudomonas* species. Phytopathology 86:757–762

Kado CI (1992) Plant pathogenic bacteria. In: Balows A, Truper HG, Dworkin M, Harder W, Schleifer KH (eds) The prokaryotes. Springer, New York, pp 660–662

Keel C, Maurhofer M (2009) Insecticidal activity in biocontrol pseudomonads. In: Weller D, Thomashow L, Loper J, Paulitz T, Mazzola M, Mavrodi D, Landa BB, Thompson J (eds) 8th International PGPR workshop. Portland, p 51

Kloepper JW (1993) Plant growth-promoting rhizobacteria as biological control agents. In: Metting B (ed) Soil microbial technologies. Marcel Dekker, New York, pp 255–274

Kloepper JW, Schroth MN, Miller TD (1980) Effects of rhizosphere colonization by plant growth promoting rhizobacteria on potato plant development and yield. Phytopathology 70: 1078–1082

Kloepper JW, Tuzun S, Kuc J (1992) Proposed definitions related to induced disease resistance. Biocontrol Sci Technol 2:349–351

Kloepper JW, Tuzun S, Liu L, Wei G (1993) Plant growth-promoting rhizobacteria as inducers of systemic disease resistance. In: Lumsden RD, Waughn J (eds) Pest management: biologically based technologies. American Chemical Society Books, Washington, DC, pp 156–165

Lazzareti E, Bettiol W (1997) Tratamento de sementes de arroz, trigo, feijão e soja com um produto formulado a base de celulas e de metabolitos de Bacillus subtilis. Sci Agric 54:89–96

Leeman M, Van Pelt JA, Den Ouden FM, Heinsbroek M, Bakker PAHM, Schippers B (1995) Induction of systemic resistance against *Fusarium* wilt of radish by lipopolysaccharides of *Pseudomonas fluorescens*. Phytopathology 85:1021–1027

Leeman M, Van Pelt JA, Den Ouden FM, Heinsbroek M, Bakker PAHM, Schippers B (1996) Iron availability affects induction of systemic resistance to *Fusarium* wilt of radish by *Pseudomonas fluorescens*. Phytopathology 86:149–155

Li WX, Kodama O, Akatsuka T (1991) Role of oxygenated fatty acids in rice phytoalexin production. Agric Biol Chem 55:1041–1147

Li H, Zhao J, Feng H, Huang L, Kang Z (2013) Biological control of wheat stripe rust by an endophytic *Bacillus subtilis* strain E1R-j in greenhouse and field trials. Crop Protect 43:201–206

Liu L, Kloepper JW, Tuzun S (1995a) Induction of systemic resistance in cucumber against *Fusarium* wilt by plant growth promoting rhizobacteria. Phytopathology 85:695–698

Liu L, Kloepper JW, Tuzun S (1995b) Induction of systemic resistance in cucumber against bacterial leaf spot by plant growth promoting rhizobacteria. Phytopathology 85:843–847

Luz WC (2003) Combinac¸a˜o dos tratamentos biolo´gico e quı´mico de semente de milho. Fitopatol Bras 28:37–40

Mariano RLR, Medeiros FHV, Albuquerque VV, Assis SMP, Mello MRF (2004) Growth-promotion and biocontrol of diseases in fruits and ornamentals in the states of Pernambuco and Rio Grande do Norte, Northeastern Brazil. In: Kobayashi K, Gasoni L, Terashima H (eds) Biological control of soilborne plant diseases. JICA, Buenos Aires, pp 70–80

Mathiyazhagan S, Kavitha K, Nakkeeran S, Chandrasekar G, Manian K, Renukadevi P, Krishnamoorthy AS, Fernando WGD (2004) PGPR mediated management of stem blight of *Phyllanthus amarus* (Schum and Thonn) caused by *Corynespora cassiicola* (Berk and Curt) Wei. Arch Phytopathol Plant Protect 33:183–199

Maurhofer M, Hase C, Meuwly P, Metraux JP, Defago G (1994) Induction of systemic resistance of tobacco to tobacco necrosis virus by the root-colonizing *Pseudomonas fluorescens* strain CHAO: influence of the gacA gene and of pyoverdine production. Phytopathology 84: 139–146

Maurhofer M, Reimmann C, Sacherer SP, Heeb S, Haas D, Defago G (1998) Salicylic acid biosynthetic genes expressed in *Pseudomonas fluorescens* strain P3 improve the induction of systemic resistance in tobacco against tobacco necrosis virus. Phytopathology 88:678–684

Medeiros FHV, Silva G, Mariano RLR, Barros R (2005) Effect of bacteria on the biology of diamondback moth (*Plutella xylostella*) on cabbage (*Brassica oleraceae* var. *capitata*) cv. Midori. An Acad Pernamb Ciênc Agronôm 2:204–212

Monteiro L, Mariano RLR, Souto-Maior AM (2005) Antagonism of *Bacillus* spp. against *Xanthomonas campestris* pv. *campestris*. Braz Arch Biol Technol 48:23–29

M'Piga P, Belanger RR, Paulitz TC, Benhamou N (1997) Increased resistance to *Fusarium oxysporum* f. sp. *radicis-lycopersici* in tomato plants treated with the endophytic bacterium *Pseudomonas fluorescens* strain 63-28. Physiol Mol Plant Pathol 50:301–320

Nakkeeran S, Kavitha K, Mathiyazhagan S, Fernando WGD, Chandrasekar G, Renukadevi P (2004) Induced systemic resistance and plant growth promotion by *Pseudomonas chlororaphis* strain PA-23 and *Bacillus subtilis* strain CBE4 against rhizome rot of turmeric (*Curcuma longa* L.). Can J Plant Pathol 26:417–418

Nakkeeran S, Fernando WGD, Siddiqui ZA (2005) Plant growth promoting rhizobacteria formulations and its scope in commercialization for the management of pests and diseases. In: Siddiqui ZA (ed) PGPR: biocontrol and biofertilization. Springer, Dordrecht, pp 257–296

Nandakumar R (1998) Induction of systemic resistance in rice with fluorescent pseudomonads for the management of sheath blight disease. M.Sc. Dissertation. Tamil Nadu Agricultural University, Coimbatore (Tamil Nadu), India, p 105

Nayar K (1996) Development and evaluation of a biopesticide formulation for control of foliar disease of rice. Ph.D. Dissertation. Tamil Nadu Agricultural University, Coimbatore (Tamil Nadu), India, p 223

Numberger T, Brunner F, Kemmerling B, Piater L (2004) Innate immunity in plants and animals: striking similarities and obvious differences. Immunol Rev 198:249–266

Ongena M, Jourdan E, Adam A, Paquot M, Brans A, Joris B, Aripigny J-L, Thonart P (2007) Surfactin and fengycin lipopeptides of *Bacillus subtilis* as elicitors of induced systemic resistance in plants. Environ Microbiol 9:1084–1090

Ongena M, Henry G, Adam A, Jourdan E, Thonart P (2009) Plant defense reactions stimulated following perception of Bacillus lipopeptides. In: Weller D, Thomashow L, Loper J, Paulitz T, Mazzola M, Mavrodi D, Landa BB, Thompson J (ed) 8th International PGPR workshop. Portland, p 43

Oostendorp M, Sikora RA (1990) *In vitro* interrelationship between rhizosphere bacteria and *Heterodera schachtii*. Rev Nematol 13(3):269–274

Otsu Y, Matsuda Y, Mori H, Ueki H, Nakajima T, Fujiwara K, Matsumoto M, Azuma N, Kakutani K, Nonomura T, Sakuratani Y, Shinogi T, Tosa Y, Mayama S, Toyoda H (2004) Stable phylloplane colonization by entomopathogenic bacterium *Pseudomonas fluorescens* KPM-018P and biological control of phytophagous ladybird beetles *Epilachna vigintioctopunctata* (Coleoptera: Coccinellidae). Biocontrol Sci Technol 14:427–439

Pal KK, Gardener BM (2006) Biological control of plant pathogens. Plant Health Instructor 2:1117–1142

Pechy-Tarr M, Borel N, Kupferschmied P, Turner V, Binggeli O, Radovanovic D, Maurhofer M, Keel C (2013) Control and host-dependent activation of insect toxin expression in a root-associated biocontrol pseudomonad. Environ Microbiol 15:736–750

Pierson LS, Thomashow LS (1992) Cloning and heterologous expression of the phenazine biosynthetic locus from *Pseudomonas aureofaciens*. Mol Plant Microbe Interact 5:330–339

Pieterse CMJ, Van Wees SCM, Hoffland E, Van Pelt JA, Van Loon LC (1996) Systemic resistance in *Arabidopsis* induced by biocontrol bacteria is independent of salicylic acid accumulation and pathogenesis-related gene expression. Plant Cell 8:1225–1237

Pleban S, Ingel F, Chet I (1995) Control of *Rhizoctonia solani* and *Sclerotium rolfsii* in the greenhouse using endophytic *Bacillus* spp. Eur J Plant Pathol 101:665–672

Potgieter H, Alexander M (1996) Susceptibility and resistance of several fungi to microbial lysis. J Bacteriol 91:1526–1532

Press CM, Wilson M, Tuzun S, Kloepper JW (1997) Salicylic acid produced by *Serratia marcescens* 90-166 is not the primary determinant of induced systemic resistance in cucumber or tobacco. Mol Plant Microbe Interact 10:761–768

Press CM, Loper JE, Kloepper JW (2001) Role of iron in rhizobacteria mediated induced systemic resistance of cucumber. Phytopathology 91:593–598

Qingwen Z, Ping L, Gang W, Qingnian C (1998) On the biochemical mechanism of induced resistance of cotton to cotton bollworm by cutting of young seedling at plumular axis. Acta Phytophylacica Sin 25:209–212

Raj SN, Chaluvaraju G, Amruthesh KN, Shetty HS (2003a) Induction of growth promotion and resistance against downy mildew on pearl millet (*Penninsetum glaucum*) by rhizobacteria. Plant Dis 87:380–384

Raj SN, Deepak SA, Basavaraju P, Shetty SH, Reddy MS, Kloepper JW (2003b) Comparative performance of formulations of plant growth promoting rhizobacteria in growth promotion and suppression of downy mildew in pearl millet. Crop Protect 22:579–588

Raja P, Una S, Gopal H, Govindarajan K (2006) Impact of bio inoculants consortium on rice root exudates, biological nitrogen fixation and plant growth. J Biol Sci 6:815–823

Ramamoorthy V, Raguchander T, Samiyappan R (1999) Isolation, characterization and screening fluorescent pseudomonads for managing damping-off disease of major vegetable crops. In: Symposium on plant disease management for sustainable agriculture. Indian Phytopathological Society, Southern Zone. CPCRI, Kayankulam, Kerala, India, p 26

Ramamoorthy V, Viswanathan R, Raghuchander T, Prakasam V, Samiyappan R (2001) Induction of systemic resistance by plant growth promoting rhizobacteria in crop plants against pests and diseases. Crop Protect 20:1–11

Raupach GS, Liu L, Murphy JF, Tuzun S, Kloepper JW (1996) Induced systemic resistance in cucumber and tomato against cucumber mosaic cucumovirus using plant growth promoting rhizobacteria (PGPR). Plant Dis 80:891–894

Ross AF (1961) Localized acquired resistance to plant virus infection in hypersensitive hosts. Virology 14:340–358

Saikia R, Kumar R, Singh T, Srivastava AK, Arora DK, Gogoi DK, Lee MW (2004) Induction of defense related enzymes and pathogenesis related proteins in *Pseudomonas fluorescens*-treated chickpea in response to infection by *Fusarium oxysporum* f. sp. Ciceri. Mycobiology 32:47–52

Saikia R, Kumar R, Arora DK, Gogoi DK, Azad P (2006) *Pseudomonas aeruginosa* inducing rice resistance against *Rhizoctonia solani*: production of salicylic acid and peroxidases. Folia Microbiol 51:375–380

Santhi A, Sivakumar V (1995) Biocontrol potential of *Pseudomonas fluorescens* (Migula) against root-knot nematode, *Meloidogyne incognita* (Kofoid and White, 1919) Chitwood, 1949 on tomato. J Biol Control 9:113–115

Saparrat MCN, Guillen F (2005) Lignolitic ability and potential biotechnology applications of the South American fungus *Pleurotus lacioniatocrenatus*. Folia Microbiol 50:155–160

Schisler DA, Slininger PJ, Behle RW, Jackson MA (2004) Formulation of *Bacillus* spp. for biological control of plant diseases. Phytopathology 94:1267–1271

Shan H, Zhao M, Chen D, Cheng J, Li J, Feng Z, Ma Z, An D (2013) Biocontrol of rice blast by the phenaminomethylacetic acid producer of *Bacillus methylotrophicus* strain BC79. Crop Protect 44:29–37

Shen L, Wang F, Liu Y, Qian Y, Yang J, Sun H (2012) Suppression of tobacco mosaic virus by *Bacillus amyloliquefaciens* strain Ba33. J Phytopathol 161:293–294

Sikora RA (1988) Interrelationship between plant health promoting rhizobacteria, plant parasitic nematodes and soil microorganisms. Med Fac Landbouww Rijksuniv Gent 53(2b):867–878

Sikora RA (1992) Management of the antagonistic potential in agricultural ecosystems for the biological control of plant parasitic nematodes. Annu Rev Phytopathol 30:245–270

Singh PP, Shin YC, Park CS, Chung YR (1999) Biological control of *Fusarium* wilt of cucumber by chitinolytic bacteria. Phytopathology 89:92–99

Spiegel Y, Cohn E, Galper S, Sharon E, Chet I (1991) Evaluation of a newly isolated bacterium, *Pseudomonas chitinolytica* sp. nov., for controlling the root-knot nematode *Meloidogyne javanica*. Biocontrol Sci Technol 1:115–125

Srivastava AK, Singh T, Jana TK, Arora DK (2001) Induced resistance and control of charcoal rot in *Cicer arietinum* (chickpea) by *Pseudomonas fluorescens*. Can J Bot 79:787–795
Stadnik MJ (2000) Induc¸a˜o de resiste^ncia a Oı´dios. Summa Phytopathol 26:175–177
Stock CA, Mcloughlin TJ, Klein JA, Adang M (1990) Expression of a *Bacillus thuringiensis* crystal protein gene in *Pseudomonas cepacia* 526. Can J Microbiol 36:879–884
Sung KC, Chung YR (1997) Enhanced suppression of rice sheath blight using combination of bacteria which produce chitinases or antibiotics. In: Ogoshi A, Kobayashim K, Homma Y, Kodama F, Kondo N, Akino S (eds) Plant growth-promoting rhizobacteria-present status and future prospects: proceedings of the fourth international workshop on plant growth promoting rhizobacteria. Nakanishi Printing, Sapporo, Japan
Suslow TV (1980) Growth and yield enhancement of sugar beet by pelleting with specific *Pseudomonas* spp. Phytopathol News 12:40
Swarnakumari N (1996) Effect of chitin amendments and *Pseudomonas fluorescens* (Migula) on rice-root nematode, *Hirschmanniella oryzae* (Van Breda de hann, 1902) Luc and Goodey, 1963 in rice cv. ADT 38 (Oryza sativa L.). M.Sc. Dissertation. Tamil Nadu Agricultural University, Coimbatore, India, p 99
Tan S, Jiang Y, Song S, Huang J, Ling N, Xu Y, Shen Q (2013) Two *Bacillus amyloliquefaciens* strains isolated using the competitive tomato root enrichment method and their effects in suppressing *Ralstonia solanacearum* and promoting tomato plant growth. Crop Protect 43: 134–140
Tenuta M (2003) http://www.umanitoba.ca/afs/agronomists_conf/2003/pdf/tenuta_rhizobacteria.pdf
Tikhonovich IA, Provorov NA (2011) Microbiology is the basis of sustainable agriculture: an opinion. Ann Appl Biol 159:155–168
Tran H, Ficke A, Asiimwe T, Höfte M, Raaijmakers JM (2007) Role of the cyclic lipopeptide massetolide A in biological control of *Phytophthora infestans* and in colonization of tomato plants by *Pseudomonas fluorescens*. New Phytol 175:731–742
Tuzun S (2001) The relationship between pathogen-induced systemic resistance (ISR) and multigenic (horizontal) resistance in plants. Eur J Plant Pathol 107:85–93
Vallad GE, Goodman RM (2004) Systemic acquired resistance and induced systemic resistance in conventional agriculture. Crop Sci 44:1920–1934
Van Loon LC, Bakker PAHM, Pieterse CMJ (1998) Systemic resistance induced by rhizosphere bacteria. Annu Rev Phytopathol 36:453–483
Van Peer R, Niemann GJ, Schippers B (1991) Induced resistance and phytoalexin accumulation in biological control of *Fusarium* wilt of carnation by *Pseudomonas* sp. strain WCS 417r. Phytopathology 81:728–734
Van Wees SCM, Pieterse CMJ, Trijssenaar A, Van'T westende Y, Hartog F, Van Loon LC (1997) Differential induction of systemic resistance in *Arabidopsis* by biocontrol bacteria. Mol Plant Microbe Interact 10:716–724
Velazhahan R, Samiyappan R, Vidhyasekaran P (1999) Relationship between antagonistic activities of *Pseudomonas fluorescens* isolates against *Rhizoctonia solani* and their production of lytic enzyme. J Plant Dis Prot 106:244–250
Vessey JK (2003) Plant growth-promoting rhizobacteria as biofertilizers. Plant Soil 255:571–586
Vidhyasekaran P, Muthamilan M (1995) Development of formulations of *Pseudomonas fluorescens* for control of chickpea wilt. Plant Dis 79:782–786
Vidhyasekaran P, Muthamilan M (1999) Evaluation of powder formulation of *Pseudomonas fluorescens* Pf1 for control of rice sheath blight. Biocontrol Sci Technol 9:67–74
Vidhyasekaran P, Rabindran R, Muthamilan M, Nayar K, Rajappan K, Subramanian N, Vasumathi K (1997) Development of powder formulation of *Pseudomonas fluorescens* for control of rice blast. Plant Pathol 46:291–297
Viswanathan R (1999) Induction of systemic resistance against red rot disease in sugarcane by plant growth promoting rhizobacteria. Ph.D. Dissertation. Tamil Nadu Agricultural University, Coimbatore (Tamil Nadu), India p 175

Viswanathan R, Samiyappan R (1999a) Induction of systemic resistance by plant growth promoting rhizobacteria against red rot disease caused by *Colletotrichum falcatum* Went in sugarcane. Proc Sugar Tech Assoc India 61:24–39

Viswanathan R, Samiyappan R (1999b) Identification of antifungal chitinase from sugarcane. ICAR News 5:1–2

Vivekananthan R, Ravi M, Ramanathan A, Samiyappan R (2004) Lytic enzymes induced by *Pseudomonas fluorescens* and other biocontrol organisms mediate defense against the anthracnose pathogen in mango. World J Microbiol Biotechnol 20:235–244

Wang W, Chen L, Wu H, Zang H, Gao S, Yang Y, Xie S, Gao X (2013) Comparative proteomic analysis of rice seedlings in response to inoculation with *Bacillus cereus*. Lett Appl Microbiol 56(3):208–215. doi:10.1111/lam.12035

Wei G, Kloepper JW, Tuzun S (1991) Induction of systemic resistance of cucumber to *Colletotrichum orbiculare* by select strains of plant growth promoting rhizobacteria. Phytopathology 81:1508–1512

Wei L, Kloepper JW, Tuzun S (1996) Induced systemic resistance to cucumber diseases and increased plant growth by plant growth promoting rhizobacteria under field conditions. Phytopathology 86:221–224

Wright B, Rowse HR, Whipps JM (2003) Application of beneficial microorganisms to seeds during drum priming. Biocontrol Sci Technol 13:519–614

Wyss U (1989) Video assessment of root cell responses to Dorylaimid and Tylenchid nematodes. In: Veech JA, Dickson DW (eds) Vistas on nematology. Society of Nematologists, Inc., Hyattsville, pp 211–220

Zdor RE, Anderson AJ (1992) Influence of root colonizing bacteria on the defense responses in bean. Plant Soil 140:99–107

Zehnder G, Kloepper J, Yao C, Wei G (1997) Induction of systemic resistance in cucumber against cucumber beetles (Coleoptera: Chrysomelidae) by plant growth promoting rhizobacteria. J Econ Entomol 90:391–396

Chapter 3
Understanding the Mechanism Involved in PGPR-Mediated Growth Promotion and Suppression of Biotic and Abiotic Stress in Plants

Siddapura Ramachandrappa Niranjana and Puttaswamy Hariprasad

3.1 Introduction

Soil is a complex living food web, where a variety of microorganisms such as bacteria, actinomycetes, fungi, protozoa, and algae reside and are involved in key environmental processes such as degradation of organic matter and biogeochemical cycling of nutrients through which they participate in maintaining health and productivity of soil (Barea et al. 2004). As the plant depends on soil throughout their life for anchorage, water, minerals, and nutrition, soil property plays a key role in determining plant health. Under natural conditions, biotic and abiotic constituents of rhizosphere are different from that of bulk soil. Rhizosphere is defined by Hiltner (1904) as volume of soil surrounding the root, influenced chemically, physically, and biologically by the presence of living plant roots. According to Uren (2000), rhizosphere is longitudinal and radial gradients occurring with expanding root growth, nutrient and water uptake, exudation, and subsequent microbial growth. The extension of rhizospheric zone may vary with soil type, plant species and age, and other biotic and abiotic factors. Rhizosphere is highly favorable habitat for the proliferation of microorganisms and exerts a potential impact on plant health and soil fertility (Sorensen 1997). Root exudates are complex mixtures of carbon containing compounds (Carvalhais et al. 2011), which serve as primary source of nutrients, and support the dynamic growth and activities of various microorganisms within the vicinity of the roots. Hence, the structure and diversity of microbial population in the rhizosphere differs significantly from that of soil-borne microbial population (Mavingui et al. 1992; Maloney et al. 1997).

S.R. Niranjana, M.Sc., M.Phil., Ph.D. (✉)
Department of Biotechnology, University of Mysore, Mysore 570 006, Karnataka, India
e-mail: niranjanasr@rediffmail.com

P. Hariprasad, M.Sc., M.Phil., Ph.D.
Food Microbiology Department, Central Food Technological Research Institute, Mysore, Karnataka, India

A. Goyal and C. Manoharachary (eds.), *Future Challenges in Crop Protection Against Fungal Pathogens*, Fungal Biology, DOI 10.1007/978-1-4939-1188-2_3,

Root-colonizing microorganisms could be free-living, parasitic, or saprophytic and their diversity remains dynamic with a frequent shift in community structure and species abundance (Kunc and Macura 1988) and is mainly attributed to the quantity and composition of root exudates. An important group of bacterial communities that exert beneficial effects on plant growth upon root colonization were first defined by Joseph Kloepper and Milton Schroth and termed as plant growth-promoting rhizobacteria (PGPR) (Kloepper and Scroth 1978). These free-living and root-colonizing bacteria, when applied to seeds or roots, enhance the growth of the plant, reduce the damage from phytopathogens, and impart resistance against abiotic stress. Hence the possible way of utilizing these rhizobacteria to enhance agricultural productivity is exploited (Kloepper et al. 1991). Basically, PGPR are defined by three intrinsic characteristics (Barea et al. 2005): (1) they must be able to colonize the root; (2) they must survive and multiply in microhabitats associated with the root surface, in competition with other microbiota, at least for the time needed to express their plant promotion/protection activities; and (3) they must promote plant growth.

Strains of the genera such as *Aeromonas*, *Agrobacterium*, *Alcaligens*, *Azoarcus*, *Azospirillum*, *Azotobacter*, *Arthrobacter*, *Bacillus*, *Cellulomonas*, *Clostridium*, *Enterobacter*, *Erwinia*, *Flavobacterium*, *Gluconacetobacter*, *Klebsiella*, *Microbacterium*, *Micromonospora*, *Panibacillus*, *Pseudomonas*, *Rhizobia*, *Serratia*, *Streptomyces*, and *Xanthomonas* have been identified as PGPR, while the search for additional strains is still continued (Kloepper et al. 1989; Okon and Labandera-Gonzalez 1994; Glick et al. 1999; Murphy et al. 2003; Cezon et al. 2003; Lucy et al. 2004; Dey et al. 2004; Raj et al. 2004; Jaizme-Vega et al. 2004; Joo et al. 2004; Tripathi et al. 2005; Tahmatsidou et al. 2006; Aslantas et al. 2007; Hariprasad and Niranjana 2009; Hariprasad et al. 2009, 2013; Niranjana and Hariprasad 2012; Kumar et al. 2014). Each plant may harbor more than one type of PGPR at the same time and the same niche or at different time and different niche during its growth period (Tilak et al. 2005; Czaban et al. 2007; Kumar et al. 2014). Among the diverse range of PGPR identified, species of *Pseudomonas* and *Bacillus* have a wide distribution and are the most extensively studied group. Recent developments in metagenomics, i.e., the study of collective genome of an ecosystem provide insights of bacterial diversity in the rhizosphere including the cultivable and non-culturable rhizobacteria which revealed the involvement of non-culturable rhizobacteria in affecting plant health (Van Overbeek 2013).

Application of PGPR in agriculture improves plant performance under various kinds of stress environments and consequently, enhance yield by exerting various beneficial traits, also reduces the need for chemical fertilizers and pesticides, and contribute for sustainable agricultural production. Several mechanisms have been postulated to explain how PGPR stimulate plant growth which includes (Fig. 3.1) (1) the ability to produce or change the concentration of the plant hormones, indole acetic acid (IAA) (Mordukhova et al. 1991; Vessey 2003), gibberellic acid (Gutierrez-Mannero et al. 2001), cytokinins (Tien et al. 1979), ethylene (Arshad and Frankenberger 1991; Glick et al. 1995), and volatiles (Ryu et al. 2004); (2) asymbiotic nitrogen fixation (Boddey and Dobereiner 1995; Kennedy et al. 1997); (3) antagonism against phytopathogenic microorganisms by the production of

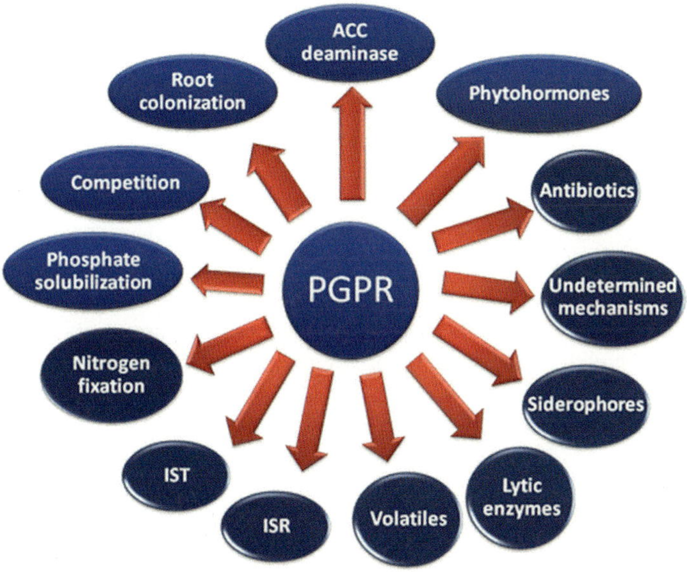

Fig. 3.1 Different mechanisms involved in PGPR-mediated growth promotion and suppression of biotic and abiotic stress in plants

siderophore (Wang et al. 1993; Raaijmakers et al. 1995), antibiotics (Ongena et al. 1999; Raaijmakers et al. 1997, 2002), Chitinase (Hallmann et al. 1999; Manjula and Podile 2001), β-1,3-glucanase (Tanaka and Watanabe 1995), and cyanide (Voisard et al. 1989); and (4) solubilization of mineral phosphate and other nutrients (de Freitas et al. 1997; Gyaneshwar et al. 2002). Induced systemic resistance (ISR) is one of the most studied mechanisms through which the rhizobacteria suppress the infection and disease development by elevating the host resistance mechanism against wide range of phytopathogens (Dobbelaere et al. 2003). Similarly, induced systemic tolerance (IST) has been proposed for PGPR-induced physiological and biochemical changes in plants that result in enhanced tolerance to abiotic stress (Yang et al. 2009). PGPR-mediated improvement of plant health is through expressing one or more traits individually or simultaneously which depends on various biotic and abiotic variables at rhizosphere (Glick et al. 1999; Rana et al. 2011; Hariprasad et al. 2011, 2013).

3.2 Root Colonization and Competition for Space and Nutrition

As the name indicates, only those bacteria residing in rhizosphere and rhizoplane are considered as rhizobacteria. Hence, it is reasonable to assume that at first PGPR must colonize the rhizosphere of the host plant to exhibit its beneficial effect efficiently. This basic stage in early life of PGPR is crucial which determines its fate in rhizosphere and also its effectiveness in improving plant health. Recent research has indicated that some of these PGPR entered the root and established endophytic association in stem, root, tubers, and other organs (Bell et al. 1995; Compant et al. 2005; Gray and Smith 2005). Researchers conducted various experiments under laboratory, greenhouse, and field conditions by using different techniques to reveal the mechanism of root colonization and factors affecting on it. According to their findings, colonization of rhizosphere by PGPR is not uniform and the size, composition, and distribution of PGPR community are determined by several abiotic and biotic factors such as soil pH, mineral, nutrient, and water content; species, genotype, and physiological state of the plant; difference in composition and amount of root exudates; and the presence of other microbial species (Yang and Crowley 2000; Goddard et al. 2001; Marschner and Timonen 2006; Albareda et al. 2006).

The amount and composition of exudates is largely affected by multiple factors such as plant species, plant age, root region, pH, temperature, surrounding microbes, and others (Rovira 1969; Baker 1987; Meharg and Killham 1995). About 49 % of photosynthetically fixed C is released into external environment in rhizosphere as root exudates (Kennedy 1999) which offers a carbon-rich diet to the rhizosphere microorganisms which includes organic acids (citrate, malate, succinate, pyruvate, fumarate, oxalate, and acetate) and sugars (glucose, xylose, fructose, maltose, sucrose, galactose, and ribose) constitute the main course, whereas variable amounts of α-aminoacids, nucleobases and vitamins (thiamin and biotin), fatty acids, enzymes, and sloughed off cells are also reported in root exudates (Rovira 1969; Baker 1987; Lugtenberg and Dekkers 1999). Some root exudates can also be effective as antimicrobial agents and some bacteria can utilize the compound of root exudates better than others, thus give ecological niche advantage to organisms that have adequate mechanism to survive in such environment (Walker et al. 2003; Bais et al. 2004, 2006; Barriuso et al. 2008).

Rhizobacteria may colonize the rhizosphere uniformly or its distribution may be restricted to certain regions of root. In naturally grown wheat plant, *Pseudomonas* spp. were relatively evenly distributed along the length of the root (Fig. 3.2), whereas filamentous bacteria occurred mostly in the area of the root cap and the mature zone of the root (Watt et al. 2006), and free-living diazotrophic bacteria colonize on root elongation zones and root hairs followed by the formation of biofilm (Assmus et al. 1995). *Kluyvera ascorbata* failed to colonize the growing tips of canola seedlings and was found only in upper two-third part of root surface (Ma et al. 2001). *Paenibacillus polymyxa* preferentially colonizes the young growing root tips of *Arabidopsis thaliana* seedlings (Timmusk et al. 2005). *Bacillus* sp. was reported to colonize spermosphere; on germination of seeds bacteria moved to the emerging radical and colonized the basal portion of the roots close to the seed-root junction, but they failed to colonize the growing root tips. Scanning electron microscope (SEM) observations revealed that the bacterial cells were arranged linearly and laterally on the growing root axis (Ugoji et al. 2005).

Chemotaxis of the bacteria towards root exudates plays a major role in the colonization of roots (Lugtenberg et al. 1996). According to Scher et al. (1985), only

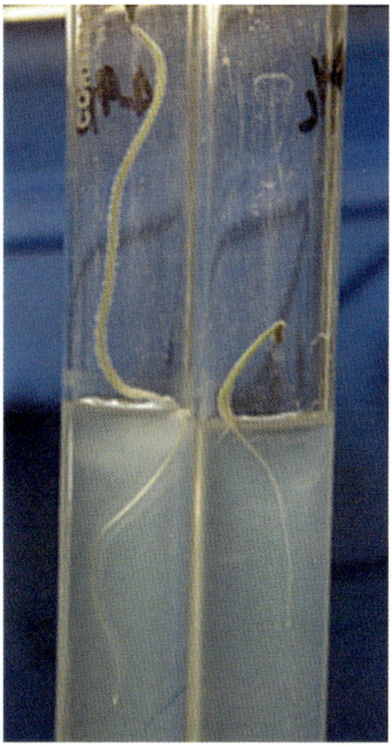

Fig. 3.2 Root colonization bioassay—aggressive root colonization by *Pseudomonas aeruginosa* strain 2apa (*left side*) and Control (*right side*)

motile strains of rhizobacteria in the presence of root exudates are able to colonize roots. Further, this phenomenon was evidenced by studies with *CheA* mutants of *Pseudomonas fluorescens*, defective in flagella-driven chemotaxis but still motile appeared to be impaired in competitive root tip colonization on tomato (de Weert et al. 2002). Irrespective of their motility, rhizobacteria can also move along the root by adhering to the expanding root surface (Mawdsley and Burns 1994), they are assisted by water percolation at the rhizosphere (Bowers and Parke 1993).

Specificity of the root exudates towards the rhizobacteria was studied by Mandimba et al. (1986). They demonstrated that bacterial strains isolated from maize rhizosphere showed chemotaxis towards maize root exudates but rhizobacterial isolates from rice did not respond to maize root exudates. Similarly, root exudates induce stronger chemotactic responses of PGPR as compared to other bacteria present in the rice rhizosphere (Bacilio-Jimenez et al. 2003). Our early experiments revealed that the rhizobacteria isolated from tomato rhizosphere are able to colonize similar group of plants such as chilli and brinjal (unpublished). Also, rhizobacteria isolated from the rhizospheric soil of *Raulwolfia* spp. from Western Ghat regions of Karnataka, India successfully colonized the tomato roots improved plant growth (Kumar et al. 2014).

It is now well established that rhizobacteria in its natural environment persist by forming biofilms. A number of microbial cell structures such as flagella or type IV pili, membrane polysaccharides, lipopolysaccharides (LPS) in particularly the O-antigen chain, and outer membrane proteins including adhesins are important in root colonization and biofilm formation (Vesper 1987; Skvortsov and Ignatov 1998; Dekkers et al. 1998; Vande Broek and Vanderleyden 1995; Lugtenberg et al. 2001;

Tans-Kersten et al. 2001; Hinsa et al. 2003). Bacterial cells adhere and proliferate on the surface of root as a colony and subsequently form biofilm as an extracellular polysaccharide matrix. Biofilm is a highly structured surface attached communities of cells encased in a self-produced extracellular matrix (Costerton et al. 1995). This process of colonization consists of the following steps: (1) transport of microbes to the surface, (2) initial attachment, (3) formation of microcolonies, and (4) biofilm maturation (de Weert and Bloemberg 2006). In the case of *Bacillus subtilis*, biofilm formation was induced by chemical signal released by plant root to which bacteria respond by stimulating biofilm synthesis. Strains with reduced ability or mutants for biofilm formation showed poor root colonization property and also reduced biocontrol activity (Chen et al. 2012). High levels of surfactin (antimicrobial agent) production were observed during colonization and biofilm formation (Nihorimbere et al. 2012). Surfactin produced by rhizobacteria act as biosurfactant and help in bacterial motility and root colonization (Bais et al. 2004). *Bacillus subtilis* mutants lacking surfactin were severely defective in swarming motility and showed poor root colonization in comparison with wild type (Kearns et al. 2004; Angelini et al. 2009).

In addition to the above said factors, density-dependent signalling in bacteria, i.e., "quorum sensing," also plays a major role in determining the density of root-colonizing bacteria in rhizosphere and regulate biofilm formation (Pierson et al. 1998). Acyl homoserine lactones (AHLs) are frequently reported as quorum sensing molecules produced by Gram-negative bacteria (Whitehead et al. 2001). In the case of Gram-positive bacteria it is protein/polypeptides, and c-butyrolactone in actinomycetes (Yamada and Nihira 1998). Plant roots are also reported to be capable of secreting compounds that may be structurally similar to Gram-negative AHLs (Toth et al. 2004; Teplitski et al. 2000) which evidenced that plants are also able to regulate the bacterial density at rhizosphere.

Apart from root colonization, PGPR should be able to compete with native microbial populations for space and available nutrients in the rhizosphere, through which it limits the growth and multiplication of other saprophytic and pathogenic microbes (Fig. 3.3) (Walsh et al. 2001; Whipps 2001; Bashan and de-Bashan 2005). Broadly, PGPR—other microbes competition at rhizosphere could be categorized as direct and indirect way. Indirect way involves occupying the space in the rhizosphere with some special aid such as biofilm formation, utilization of available nutrients efficiently (siderophore production for iron uptake), and breakdown of some complex form of nutrition into simple form and its uptake (phosphate solubilization). Direct way includes suppression of growth and multiplication of other microbes by producing antibiotics, lytic enzymes, etc. These direct and indirect mechanisms may express differentially which depends on various biotic and abiotic factors at rhizosphere, through which they successfully colonize the host roots and compete with other saprophytic or pathogenic microbes (Radjacommare et al. 2004; Yasmin et al. 2009; Sayyed and Chincholkar 2009; Chaiharn et al. 2009; Ramos-Solano et al. 2010; Hariprasad et al. 2011).

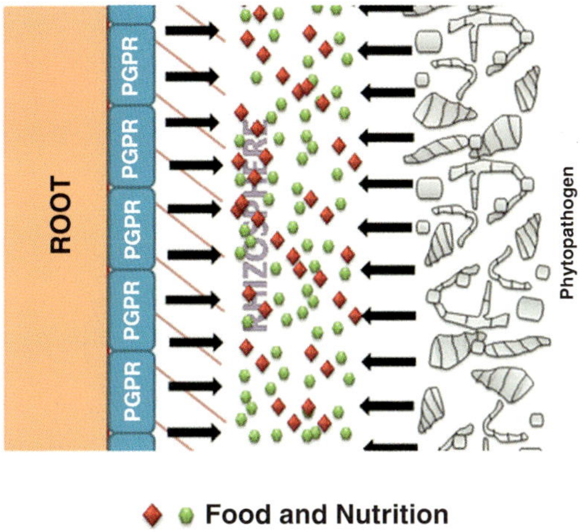

Fig. 3.3 Root colonization and competition for space and nutrition between PGPR and phytopathogens at rhizosphere

3.3 Phytohormones

Promotion of root growth is one of the major markers by which the beneficial effect of PGPR is measured. Phytohormones produced by PGPR within the vicinity of root stimulate profuse root growth, whether by elongation of primary roots or by proliferation of lateral and adventitious roots. The increased surface area of root improves the water, minerals, and nutrition-uptake capacity of plant from a large volume of soil, strengthens anchorage capacity thus establishing the chance of greater survival. Most of the PGPRs are known to produce IAA (Gaudin et al. 1994; Asghar et al. 2002), cytokinins (Vessey 2003), and gibberellins (Glick et al. 1998; Khan et al. 2006).

3.3.1 Indole Acetic Acid

The most common, best characterized, and physiologically most active auxin in plants is IAA. More specifically, IAA is a phytohormone controlling many important physiological process such as cell division, cell enlargement, tissue differentiation, apical dominance, root initiation, response to light and gravity in plants, initiation of adventitious and lateral roots, and both cell and vascular differentiation (Salisbury 1994; Taize and Zeiger 2006). Bacterial auxin production was normally associated with pathogenesis, especially with bacterial gall formation as observed in *Agrobacterium tumefaciens* (Kaper and Veldstra 1958). However it became apparent that many of the phytopathogens (not only gall inducing) as well as PGPR have the ability to synthesize IAA and through which it alters the architecture of root in beneficial way.

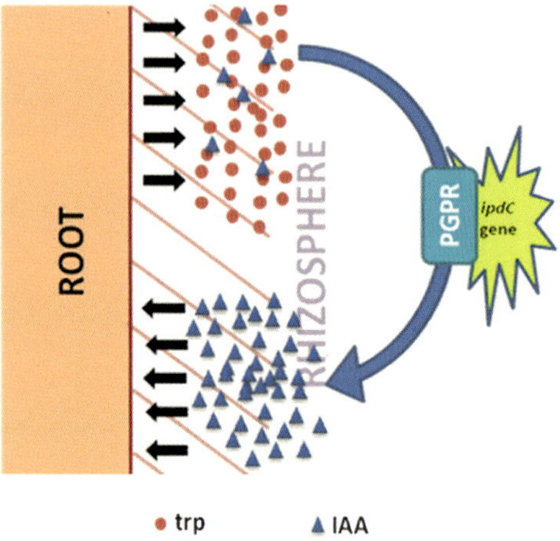

Fig. 3.4 Expression of rhizobacterial ipdC gene by lower concentration of IAA released from root and biosynthesis of IAA by rhizobacteria using tryptophan as precursor present in the root exudates

It has been estimated that 80 % of bacteria isolated from the rhizosphere can produce IAA (Patten and Glick 1996) and are more active in producing IAA than those from root-free soil because of rich supplies of substrates exuded from roots (especially tryptophan) (Kampert et al. 1975; Strzelczyk and Pokojska-Burdziej 1984; Martens and Frankenberger 1994; Kravchenko et al. 2004). Further, a positive correlation between L-tryptophan concentration and IAA production by different PGPR strains and their ability to increase the plant growth was reported (Asghar et al. 2002; Idris et al. 2007; Hariprasad et al. 2011).

The host plant can take an active part in the regulation of microbial IAA biosynthesis (Brandl and Lindow 1997). Low concentration of IAA released from plant in root exudates is sufficient to induce the expression of ipdC gene, which increases the rhizobacterial synthesis of IAA (Fig. 3.4) (Dobbelaere et al. 1999; Vande Broek et al. 1999). IAA biosynthesis in diverse bacteria is also related to environmental stress including acidic pH, somotic and matrix stress, and carbon limitation (Brandl and Lindow 1997; Patten and Glick 2002; Vande Broek et al. 2005).

There have been at least five different IAA biosynthetic pathways proposed using tryptophan as precursor (Patten and Glick 1996) which are as follows: (1) Indole-3-acetamide pathway (Morris 1995; Theunis et al. 2004), (2) Indole-3-pyruvate pathway (Patten and Glick 2002), (3) Tryptamine pathway (Hartmann et al. 1983; Perley and Stowe 1966), (4) Tryptophan side chain oxidase pathway (Oberhansli et al. 1991), and (5) Indole-3-acetonitrile pathway (Nagasawa et al. 1990; Kobayashi et al. 1993). Tryptophan-independent pathway was demonstrated in case of *Azospirillum brasilense* in the absence of tryptophan (Prinsen et al. 1993), but the enzymes involved in this pathway is yet to be identified. Indole-3-pyruvate pathway is described in a wide range of bacteria including pathogenic and beneficial bacterial group such as PGPR.

Several earlier research studies indicate that the IAA producing rhizobacteria had the potential to improve plant health which includes root growth, shoot growth, biomass, branches/tillers, flowering, pods/grains, and (Xie et al. 1996;

Asghar et al. 2002; Patten and Glick 2002; Vessey 2003; Araujo et al. 2005; Kaymak et al. 2008). Up to 50 % increase in root growth was reported in plants treated with IAA producing strains in comparison with mutants defective in IAA production (Patten and Glick 2002; Idris et al. 2007).

The impact of exogenous auxin on plant development ranges from positive to negative effects. The consequence for the plant is usually a function of the amount of IAA produced that is available to the plant and the sensitivity of the plant tissue to changes in IAA concentration. Exogenously supplied IAA above certain limit is known to suppress the root growth (Mulkey et al. 1982). Higher inoculum of IAA producing bacteria and IAA overproducing mutants were found to be suppressing growth of host root by activating ethylene biosynthesis pathway, and is achieved by inducing the expression of 1-aminocyclopropane-1-carboxylate acid (ACC) synthase enzyme in roots (Xie et al. 1996; Glick et al. 1998; Glick 2005; Spaepen et al. 2008). However few strains of rhizobacteria also have the ability to produce ACC deaminase which degrades ACC at normal IAA production level besides converting ACC to ammonia and α-ketobutyrate and utilize ACC as nitrogen source (discussed in Sect. 3.11.3).

3.3.2 *Gibberellins*

Gibberellins (GA) are tetracyclic diterpenoid acids that are involved in a number of developmental and physiological processes in plants. Gibberellins induced internode elongation in certain types of plants, such as dwarf and rosette species and grasses. Other physiological effects of gibberellins include changes in juvenility and flower sexuality, promotion of fruit set, fruit growth, and seed germination (Davies 1995; Crozier et al. 2000; Taize and Zeiger 2006). Gibberellins are also reported to be produced by several bacteria (Atzorn et al. 1988; Dobert et al. 1992; Janzen et al. 1992; Bastian et al. 1998; Gutierrez-Mannero et al. 2001).

The production of GA-like substances by rhizobacteria, as revealed by bioassay, was first described in *Azospirillum brasilense* (Tien et al. 1979). Further in 1988, it was confirmed by GC-MS analysis (Atzorn et al. 1988). Pectobutrazol-induced dwarf phenotype in alder (*Alnus glutinosa*) is reversed by spraying culture filtrate of PGPR containing GA (Gutierrez-Mannero et al. 2001). Involvement of rhizobacterial produced GA in enhancing plant growth was further evidenced by studies of Cassan et al. (2001). Where, *A. lipoferum* USA5b and *A. brasilense* Cd promoted sheath elongation of GA-deficient dwarf rice, when supplied with GA precursors. Application of GA-positive rhizobacterial strain also was found to increase the endogenous level GA in plants. Measurement of GA content using deuterated internal standards, and GCMS analysis, showed increased levels of GA_1, GA_{19}, GA_{20}, and GA_{44} in nodules formed by the rhizobacterial strains that enhanced growth (Dobert et al. 1992).

3.3.3 Cytokinins

Cytokinins are N6-substituted aminopurines which, when applied to plants, influence their physiological and developmental processes (Salisbury and Ross 1992). Cytokinins control the cell division, cell cycle, and stimulate developmental processes in plants (Srivastava 2002), also affect leaf senescence, nutrient mobilization, apical dominance, the formation and activity of shoot apical meristems, floral development, breaking of bud dormancy, and seed germination. Cytokinins also appear to mediate many aspects of light-regulated development, including chloroplast differentiation, the development of autotrophic metabolism, and leaf and cotyledon expansion (Werner et al. 2001; Taize and Zeiger 2006; Oldroyd 2007).

Cytokinin production by rhizobacteria was reported by several researchers (Ali et al. 2009; Horemans et al. 1986; Nieto and Frankenberger 1989; Taller and Wong 1989; Senthil Kumar et al. 2009; de Salamone et al. 2001). It has been reported that cytokinins produced by these microorganisms that live in close proximity to the roots also influence plant growth and development (Arshad and Frankenberger 1993). These microorganisms produce and secrete substantial amounts of cytokinins and/or cause the plant cells to synthesize plant hormones, including cytokinins (Akiyoshi et al. 1987).

Azospirillum brasilense, a cytokinin producing strain has been reported to enhance cell division in root tips of inoculated wheat (Molina-Favero et al. 2007). Application of bacterial extracts containing cytokinins (zeatin and zeatin riboside) enhanced cell division, fresh weight, and cotyledon size in dark as well as light grown cucumber cotyledons (Hussain and Hasnain 2009). Studies with cytokinin mutant strains CNT1 and CNT2 to promote the growth of radish plants was impaired compared to wild strain *Pseudomonas fluorescens* G20-18, which evidenced the involvement of bacterial produced cytokinin in plant growth promotion (Gracia de Salamone 2001).

3.3.4 Abscisic Acid

Abscisic acid (ABA) is a 15-carbon compound that resembles the terminal portion of some carotenoid molecules which plays primary regulatory roles in the initiation and maintenance of seed and bud dormancy and in the plant's response to stress, particularly water stress. In addition, ABA influences many other aspects of plant development by interacting, usually as an antagonist, with auxin, cytokinin, gibberellin, ethylene, and brassinosteroids (Ferguson and Lessenger 2006; Taize and Zeiger 2006).

Rhizobacteria such as *Corynebacterium* sp. (Hasegawa et al. 1984) and *Azospirillum brasilense* (Cohen et al. 2008) are reported to be capable of synthesizing ABA in defined culture media. Although these bacteria are known to synthesize ABA, but the biochemical mechanism involved is yet to be investigated completely, as only the presence of carotenoid cleavage oxygenase homologues was reported

(Marasco and Schmidt-Dannert 2008). *Azospirillum* spp. isolated from water-stressed conditions produced more ABA than strains isolated from well-watered plants. Further, rhizobacterial ability to produce ABA was increased, especially when exposed to osmotic stress (Cohen et al. 2008; Ilyas and Bano 2010).

Application of ABA producing PGPR can potentially alter the plant ABA concentration (Cohen et al. 2009). This change in ABA concentration may be due to bacterial ABA or plants are induced to produce ABA. Alleviated level of ABA in plant can also enhance the tolerance of plants to adverse environmental conditions like drought, salinity, and temperature and stimulate stomatal closure to limit transpirational water loss (Boiero et al. 2007; Cohen et al. 2009).

3.3.5 Ethylene

Ethylene, a gaseous plant hormone, acts as a messenger of biotic and abiotic stresses, acting as a negative regulator of plant growth. Ethylene also affects ripening and senescence in plants (Ferguson and Lessenger 2006; Taize and Zeiger 2006). Some bacteria are also able to produce ethylene when grown in medium supplemented with methionine (Boiero et al. 2007). But studies regarding rhizobacterial produced ethylene in suppressing plant growth are yet to be done. Although rhizobacteria are known to produce ethylene, recent attention has focused on rhizobacterial mediated decrease in plant ethylene level via the enzyme ACCd that degrades the ethylene precursor ACC. But none of the rhizobacteria known to produce ACCd activity was found to produce ethylene (Glick et al. 1998).

3.4 Volatiles in Growth Promotion

Plant growth promotion by volatiles produced by PGPR is the most recently identified mechanism. Ryu et al. (2003) demonstrated that PGPR strains release different volatile blends which stimulate plant growth. Volatiles produced by *Bacillus subtilis* and *B. amyloliquefaciens* were identified as 3-hydroxy 2-butanone and 2, 3-butanediol, which stimulated the growth of *Arabidopsis thaliana* in in vitro experiments as observed by an increase in the total leaf surface area. As evidenced by the studies of Lee et al. (2012), growth promotion elicited by bacterial volatiles is mediated by the ET and or cytokinin signalling pathways.

3.5 Biological Nitrogen Fixation

Nitrogen is an essential component of all life forms and is a paradox of nature because it cannot be assimilated by plants unless it is reduced to ammonia by diazotrophic microorganisms. PGPR strains are reported to fix atmospheric N_2 in soil

(Antoun et al. 1998; Riggs et al. 2001) and further it can be supplied to the associated host plant (Barbara and Thomas 1998). The process involves nitrogenase enzymes that reduce gaseous nitrogen into ammonia (NH_3) and ammonium (NH_4^+) (Chatterjee et al. 1997). Biological N_2 fixation by rhizobia and associative diazotrophic bacteria is a spontaneous process and one of the widely studied mechanisms by which plants benefit from the interacting partners. The bacteria benefit the plants by fixing N_2 in exchange for fixed carbon either provided directly to the bacteria or indirectly by releasing carbon as root exudates.

Symbiotic N_2 fixation to legume crops with the inoculation of effective PGPR is well known (Dobereiner 1997; Barea et al. 2005; Esitken et al. 2006). Various rhizobacterial species like *Azotobacter* spp. *Bacillus* spp. and *Beijerinckia* spp. have the capacity to fix atmospheric N_2 symbiotically. It is well established that N_2 is the major nutrient which determines the life in a particular region and N_2 fixed by the microbes is the only known mode of biological nitrogen fixation. Hence, the phenomenon of biological nitrogen fixation is well studied and many review articles and monographs are published which cover specific area (Gualtieri and Bisseling 2000; Schultze and Kondorosi 1998; Sessitsch et al. 2002; Franche et al. 2009).

On the other hand, nonsymbiotic biological N_2 fixation is basically carried out by free-living diazotrophs, belonging to the genera like *Azoarcus* (Reinhold-Hurek et al. 1993), *Azospirillum* (Bashan and de-Bashan 2010), *Burkholderia* (Estrada-de los Santos et al. 2001), *Gluconacetobacter* (Fuentes-Ramírez et al. 2001), and *Pseudomonas* (Mirza et al. 2006). Also even many PGPR showed their ability to fix N_2, but there is little evidence that these PGPR stimulate the growth of a specific host plant using nitrogenase activity (Vessey 2003). As our review is focused towards the mechanisms, this section is not included here as its mechanism is poorly understood.

3.6 Phosphate Solubilization

Phosphorus (P) is the second most important plant nutrient available in soil after nitrogen. Phosphate in the soil solution is readily absorbed by plant roots via an H^+–HPO_4^{2-} symporter and incorporated into a variety of organic compounds, including sugar phosphates, phospholipids, and nucleotides (Taize and Zeiger 2006). Though soils usually contain high amount of total phosphorous, most of the phosphorous occurs in an insoluble form as iron and aluminum phosphates in acidic soils and as calcium phosphates in alkaline soils (Goldstein 1986). However, organic matter on the other hand, is an important reservoir of immobilized phosphate that accounts for 20–80 % of soil phosphorous (Richardson 2001) and only a small portion is available to plants. Other organic P compounds in soil are in the form of phosphomonoesters, phosphodiesters including phospholipids and nucleic acids, and phosphotriesters.

Many of these P compounds are high molecular-weight material which must first be bioconverted to soluble ionic phosphate (HPO_4^{2-} and $H_2PO_4^-$) which can be assimilated by plants (Beever and Burns 1980; Glass 1989; Goldstein 1994). The ability of soil bacteria to solubilize complex form of P is frequently reported and common in the rhizosphere than in the bulk non-rhizospheric soil (Reyes et al. 2006).

Inorganic P is solubilized by the action of rhizobacterial produced organic and inorganic acids in which hydroxyl and carboxyl groups of acids chelate cations (Ca, Al, and Fe) and decrease the pH in basic soils (Kpomblekou and Tabatabai 1994). Carboxylic acids solubilize Al-P and Fe-P through direct dissolution of mineral phosphate as a result of anion exchange of PO_4^{3+} by carboxylic anions, or by chelation of both Fe and Al ions associated with phosphate (Khan et al. 2007; Henri et al. 2008). Phosphate from Ca-P complex results from the combined effect of pH decrease and carboxylic acids synthesis. Carboxylic anions produced have high affinity to calcium, solubilize more phosphorus than acidification alone (Staunton and Leprince 1996). Proton release can also decrease P sorption upon acidification which increases $H_2PO_4^-$ in relation to HPO_4^{2-} having higher affinity to reactive soil surfaces (Goldstein 1994; Illmer and Schinner 1995; Omar 1998; Whitelaw 2000; Villegas and Fortin 2002).

Also, different organic acids such as gluconic, 2-keto gluconic, lactic, isovaleric, isobutyric, acetic, oxalic, and citric acid were reported to be produced by rhizobacteria (Omar 1998; Rodriguez and Fraga 1999; Alikhani et al. 2006; Islam et al. 2007) and known to solubilize insoluble phosphates by lowering the pH, chelation of cations, and competing with phosphate for adsorption sites in the soil (Nahas 1996). Phosphorus desorption potential of different carboxylic anions lowers with decrease in stability constants of Fe- or Al-organic acid complexes (log *K*Al or log *K*Fe) in the order: citrate > oxalate > malonate/malatefaces > tartarate > lactate > gluconate > acetate > formiate (Ryan et al. 2001).

Soil with Ca-P as a major phosphorous source also has high buffering capacity (Ae et al. 1991). But several early reports suggest that the low buffering capacity of the screening media would lead to the isolation and designation of any bacteria that can lower the pH of the medium as phosphate solubilizing microorganisms do. But the bacteria probably will not be able to do this in soil because the soil is buffered. To resolve this problem, we added one more step in screening methodologies which includes estimation of available P in rhizosphere after treatment with selected rhizobacteria (Hariprasad and Niranjana 2009). However, acidification does not seem to be the only mechanism of P solubilization, as the ability to reduce the pH in some cases does not correlate with the ability to solubilize mineral phosphates (Subba Rao 1982). For instance, a genomic DNA fragment from *Enterobacter agglomerans* showed mineral phosphate solubilization activity in *Escherichia coli* JM109, although the pH of the medium was not altered (Kim et al. 1997).

Mineralization of organic to inorganic phosphate involves processes catalyzed by three groups of enzymes. (1) Nonspecific phosphatase, which performs dephosphorylation of phosphor-ester or phosphoanhydride bonds in organic matter; (2) Phytase, which specifically causes P release from phytic acid; and (3) Phosphonatase and C-P lyase enzymes that perform C-P cleavage in organophosphonates (Hilda and Fraga 1999; Yadav and Verma 2012). Our studies (Hariprasad and Niranjana 2009) revealed that phosphate solubilizing PGPR can be endowed with more than one mechanism and with it can degrade different forms of phosphate complex in nature. Hence, usage of PGPR with multiple mechanisms for P solubilization is advantageous over the PGPR with single mechanism.

3.7 Siderophore

Iron has an important role as a component of enzymes involved in the transfer of electron (redox reaction), such as cytochromes and also required for the synthesis of some of the chlorophyll-protein complexes in the chloroplast (Taize and Zeiger 2006). Iron is most commonly present in soils in the Fe^{3+} state contained in clays, oxides, and hydroxides, a form in which it is extremely insoluble (Tisdale et al. 1993), which is a limiting factor for the growth of plants and microorganisms even in iron-rich soils. The availability of iron in soil solutions is 10^{-18} M (Neiland 1981), a concentration which even cannot sustain the microbial growth. But several microbes overcome this problem by producing ferric ion-specific chelating agents widely known as siderophores (Lankford 1973).

Siderophores can be defined as small peptidic molecules containing side chains and functional groups that can provide a high-affinity set of ligands to coordinate ferric ions (Crosa and Walsh 2002). Most of the siderophores are water soluble and can be secreted extracellularly or produced inside bacterial cell which scavenge iron from the environment and to make the mineral available to the microbial cell (Winkelmann et al. 1987; Winkelmann 1991). Many bacteria are capable of producing more than one type of siderophore or have more than one iron-uptake system to take up multiple siderophores (Neiland 1981; Bultreys et al. 2003). According to iron coordinating functional group, structural features, and types of ligands, bacterial siderophore can be grouped into four types as follows: (1) Carboxylate, (2) Hydroxamates, (3) Phenol catecholates, and (4) Pyoveridines (Crowley 2006; Essen et al. 2007).

Pseudomonads are among the most widely studied of the rhizobacteria that produce abundant siderophores (Hofte et al. 1994). Fluorescent pseudomonads are known to produce a yellow-green, fluorescent, water-soluble pigment, which has a very high affinity for ferric iron known as pyoverdin (Leong et al. 1991). Each pyoverdin has three Fe-binding ligands, one of which is always ano-dihydroxy aromatic group derived from quinoline located in the chromophore. The other two are located in the peptide chain and are hydroxamic acids derived from ornithine (Budzikiewicz 1993). The structure of pyoverdines is variable among *Pseudomonas* spp. and even between strains of same species (Meyer et al. 2002; Visca et al. 2007). Other important siderophore produced by *Pseudomonas* are pyochelin, pseudomonine, and its precursor salicylic acid (Cox et al. 1981; Buysens et al. 1996; Mercado-Blanco et al. 2001; Sattely and Walsh 2008). Quinolobactin are known to be produced by certain *Pseudomonas* spp. when pyoverdin production was repressed and is utilized for iron acquisition (Mossialos et al. 2000). Wide arrays of other beneficial plant-associated bacterial genera, e.g., *Azotobacter, Bacillus, Enterobacter*, *Serratia*, *Azospirillum*, and *Rhizobium* secrete various types of siderophores (Glick et al. 1999).

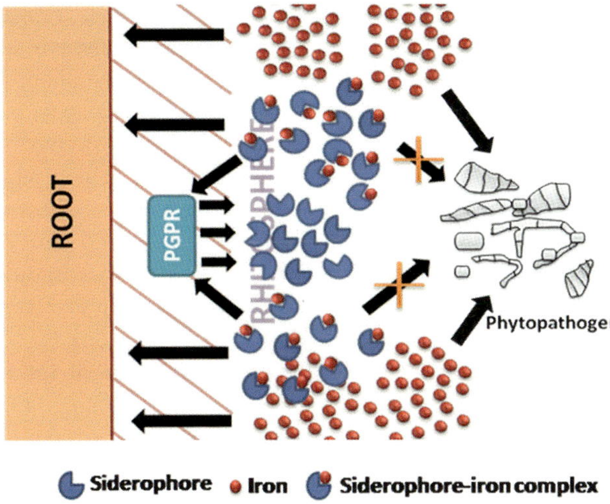

Fig. 3.5 Competition for iron between plant, PGPR, and phytopathogens. PGPR secreted siderophore complexes with available iron in rhizosphere. Siderophore-iron complex can be taken up by PGPR and plant, but phytopathogens are unable to use this complex

The synthesis of siderophores is repressed under conditions where iron is not limiting. But, under iron-limited conditions, the enzymes for synthesis of siderophores and the receptor protein in the outer membrane of PGPR are synthesized, and the siderophore is released. After the iron-siderophore complexes have formed, these soluble complexes are internalized via active transport into the cells by specific membrane receptors. The sensitivity of membrane transporter receptor varies with some only recognizing a self-synthesized molecule while others allow transport of siderophore complexes of different origins and structure (Dowling et al. 1994). Further, assimilation of iron from ferri-siderophore complex in cytoplasm involves the reduction of Fe^{3+} to Fe^{2+} for which the siderophore has no affinity (Crichton and Charloteaux-Wauters 1987; Glick et al. 1999).

A great deal of evidence exists that a number of plant species can absorb bacterial Fe^{3+}-siderophore complexes (Bar-Ness et al. 1991; Wang et al. 1993; Masalha et al. 2000). The rate and amount of iron taken up in this manner tends to be low, the concentrations utilized can be significant in sustaining plant growth (Glick et al. 1999). But according to the studies of Bar-Ness et al. (1992), two bacterial siderophores (pseudobactin and ferrioxamine B) were inefficient as iron sources for plants and rhizobacteria may compete for limited iron with the host plant.

Kloepper et al. (1988) for the first time reported that PGPR exert their plant growth-promoting activity by depriving native microflora. It has been suggested that siderophores act antagonistically by sequestering iron from the environment, restricting growth of the pathogen (Fig. 3.5) (Bashan and de-Bashan 2005). As evidenced by the work of Jagadeesh et al. (2001) increased concentration of siderophore is inversely proportional to the population of phytopathogen in rhizosphere, thus reducing the probability of infection and subsequent disease development. Phytopathogens in the rhizosphere may be unable to use siderophore produced by PGPR for iron acquisition, also PGPR produced siderophore may be having higher sensitivity and specificity for iron than that of phytopathogens derived.

Also several siderophores are known to induce ISR which suppresses the disease development in host plants which is independent of direct antagonistic activity (Bashan and de-Bashan 2005). In the case of *Serratia marcescens* strain 90-166, ISR inducing ability is reduced with increasing concentration of iron in the rhizosphere (Press et al. 1997) which is evidenced by the involvement of siderophores in inducing ISR. Possible involvement of rhizobacterial produced SA in inducing ISR was ruled out when *Serratia marcescens* strain 90-166 *ent*A mutant defective in siderophore production but able to produce SA failed to induce ISR (Press et al. 2001). But in radish, iron-regulated metabolites are involved in inducing ISR. But the same strain failed to induce ISR in *Arabidopsis* (Leeman et al. 1995; Van Wees et al. 1997; Ran et al. 2005). Whereas, *Pseudomonas* mutants which were defective in the biosynthesis of iron-regulated metabolites were found to be as effective as wild type in inducing ISR, suggesting these metabolites are not required in this interaction (Djavaheri 2007). Thus, the extent of disease suppression as a consequence of bacterial siderophore production is affected by several factors (Bashan and de-Bashan 2005), including the specific pathogen, PGPR, soil type, the crop, availability of iron, and the affinity of the siderophore for iron. For instance, siderophore-mediated suppression should be greater in neutral and alkaline soils than in acid soils (Baker et al. 1986). Thus, disease suppression under controlled laboratory conditions is only an indication of the efficacy of the biocontrol agent in the field.

3.8 Lytic Enzymes

Chitinase and β-1,3-glucanase are two major lytic enzymes produced by PGPR which are involved in degrading chitin and β-1,3-glucan, major constituent of cell wall in the case of fungi and oomycetes, respectively (Agrios 2005). Chitinase and β-1,3-glucanase enzymes produced by rhizobacteria have been postulated to play an important role in the biocontrol of fungal diseases (Kishore and Pande 2007; Kamil et al. 2007; Leelasuphakul et al. 2006) and was evidenced by in vitro studies where fungal/oomycetes growth was inhibited by rhizobacteria producing hydrolytic enzymes (Chet et al. 1990; Vaidya et al. 2003; Huang et al. 2005; Siwayaprahm et al. 2006). Also, antifungal and lysozyme activity of purified chitinase and β-1,3-glucanase from rhizobacteria revealed its potential to inhibit the fungal/oomycetes growth (Wang and Chang 1997; Vaidya et al. 2003; Huang and Chen 2004; Siwayaprahm et al. 2006; Khiyami and Masmali 2008; Chang et al. 2010; Hariprasad et al. 2011).

Antifungal properties of chitinolytic soil bacteria may enable them to effectively suppress the growth of fungi/oomycetes and colonize the specific niche and utilize the available nutrition. Additionally, the production of chitinase may be part of a lytic system that enables the bacteria to use living hyphae as the actual growth substrate. The antagonists were capable of growing on the expense of the hyphae of fungi/oomycetes indicating their potential for pathogen suppression. Where the antagonism takes place outside the limit of the rhizosphere. Once cell wall damage has occurred, the pathogen is more likely to be susceptible to attack by other biological, physical, and chemical agents. Scope of the chitinolytic microorganisms in management of phytopathogenic fungi has grown as chitin and crude fungal cell wall-supplemented applications were observed to increase the attainable levels of disease protection (Bell et al. 1998). The rationale of using chitin with chitinolytic isolates is a logic approach because bacterial isolates multiply and increase in their numbers by using chitin as a carbon source; meanwhile the chitinase produced inhibits the fungal pathogen growth at the rhizosphere (Kokalis-Burelle et al. 1992; Kishore et al. 2005; Hariprasad et al. 2011). Also, isolates of *Bacillus circulans*, *B. subtilis*, and *Serratia marcescens* that have been pre-induced for chitinase production by multiplication in chitin-supplemented peat formulations showed enhanced disease control activities in the rhizosphere and phylloplane of groundnut (Manjula and Podile 2001; Kishore et al. 2005). On the other hand, β-1,3-glucanase produced by PGPR is reported to degrade β-1,3-glucan a major cell wall constituent of oomycetes which is evidenced by the studies of Fridlender et al. (1993), where β-1,3-glucanase producing strain of *Pseudomonas cepacia* inhibited the rhizosphere proliferation of various soil-borne phytopathogens. Further, monomer or oligomer of chitin or β-1,3-glucan released from pathogen cell wall by the action of lytic enzymes may act as an elicitor which initiates transcription of the plant genes that encode the various components of the defense response (Fig. 3.6) (Takeuchi et al. 1990; Wu et al. 1997).

3.9 Antibiotics

From the past three decades, various plant root-colonizing bacterial species have been shown to be potent biological control agents in various plant pathogen systems and this is achieved by producing a set of chemically heterogeneous group of organic, low-molecular weight compounds (Fravel 1988; Thomashow et al. 1997; Duffy et al. 2003). Antibiotics produced by microbes are the key component of PGPR-mediated disease protection in plants especially soil-borne and seed-borne pathogens. Production of antibiotics has been described as a powerful mode of action which is exhibited by PGPR at rhizosphere through suppression and/or developmental activity of the phytopathogen (Handelsman and Stabb 1996; Glick et al. 2007).

PGPR are well known to produce diverse antimicrobial secondary metabolites responsible for their biocontrol activity which includes ammonia, butyrolactones,

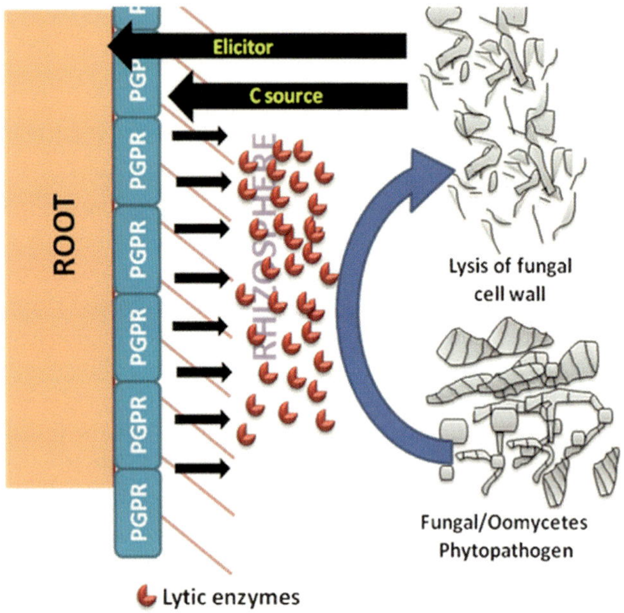

Fig. 3.6 Lytic enzymes such as chitinase and β-1,3-glucanase produced by PGPR lyses fungal or oomycetes cell wall by releasing monomers or oligomers which can be used by PGPR as C source, on the other hand degraded cell wall component acts as elicitor molecules which induce host defense response

2,4-diacetyl phloroglucinol, kanosamine, oligomycin A, oomycin A, phenazine, pyoluteorin, phloroglucinols, pyrrolnitrin, viscosinamide, xanthobaccin, zwittermycin A, HCN, polymixin, circulin, colistin, herbicolin-like compounds, thuricin, and volatile organic compounds (VOCs) (Lambert et al. 1987; Silo-Suh et al. 1994; He et al. 1994; Whipps 2001; Kumar et al. 2005; Rane et al. 2007; Maksimov et al. 2011). Cyclic lipopeptides (CLP) include surfactin, iturin A, and fengycins (Banat et al. 2000; Singh and Cameotra 2004). The above said antibiotics are known to possess antiviral, antimicrobial, insect and mammalian antifeedant, antihelminthic, phytotoxic, antioxidant, cytotoxic, antitumor, and plant growth-promoting activities. Under certain conditions, pyoverdin (siderophore) functions as diffusible biostatic or fungistatic antibiotics, whereas ferripyoverdin does not (Scher and Baker 1982).

Among the antibiotics named above, the six classes of antibiotic compounds for which the experimental evidence most clearly supports a function in their biocontrol activity are phenazines, phloroglucinols, pyoluteorin, pyrrolnitrin, CLPs (all of which are diffusible), and hydrogen cyanide (HCN, which is volatile). Mode of action of these antibiotics against phytopathogens is well established (Table 3.1).

Fluorescent pseudomonads have been shown to produce broad spectrum antibiotics. Pyoluteorin is an aromatic polyketide antibiotic consisting of a resorcinol ring, which is derived through polyketide biosynthesis (Nowak-Thompson et al. 1999).

Table 3.1 Antibiotics produced by PGPR and their mode of action against phytopathogens

PGPR	Antibiotics	Mode of action	References
Pseudomonas Actinobacteria *Streptomyces*	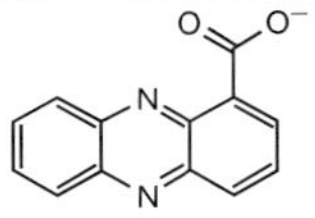 Phenazine	The antimicrobial activity of phenazine depends on the rate of oxidative reductive transformation of the compound coupled with the accumulation of toxic superoxide radicals in the target cells, and damages the lipid and other macromolecules	Hassett et al. (1993), Chin-A-Woeng et al. (2003), Mavrodi et al. (2006)
Pseudomonas	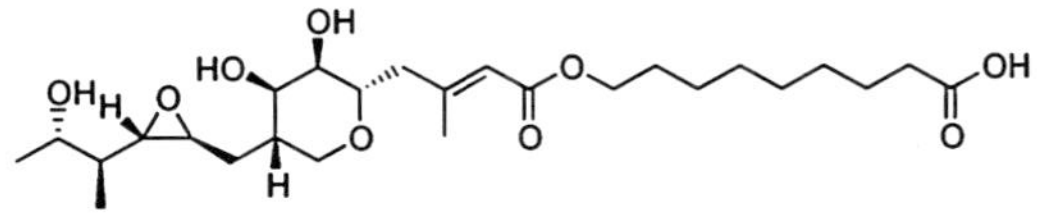 Mupirocin	Mupirocin is bacteriostatic at low concentrations and bactericidal at high concentrations. It inhibits isoleucyl-tRNA synthetase and prevents incorporation of isoleucine into newly synthesized proteins	Hughes and Mellows (1980), Kate and Bryant (2007)
Pseudomonas *Burkholderia*	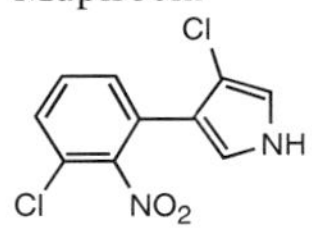 Pyrrolnitrin	Pyrrolnitrin is a chlorinated phenylpyrrole antibiotic most active against dermatophytic fungi. The primary site of action of pyrrolnitrin on *Saccharomyces cerevisiae* was the terminal electron transport system between succinate or reduced nicotinamide adenine dinucleotide (NADH) and coenzyme Q	Arima et al. (1964), Tripathi and Gottlieb (1969)
Pseudomonas	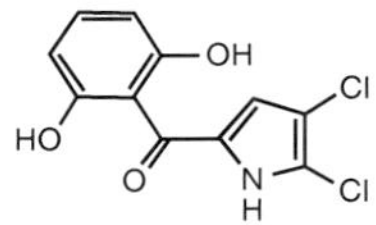 Pyoluteorin	Pyoluteorin (4,5-dichloropyrrol-2-yl 2,6-dihydroxyphenyl), a polyketide antibiotic, which effectively inhibits phytopathogenic fungi, and suppresses plant diseases. No mode of action has been published for pyoluteorin.	Howell and Stipanovic 1980; Maurhofer et al. 1992; Maurhofer et al. 1994; Nowak-Thompson et al. 1999
Pseudomonas	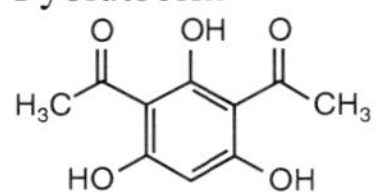 Phloroglucinol	Phloroglucinol causes disorganization in hyphal tips, including alteration (proliferation, retraction, and disruption) of the plasma membrane, vacuolization, and cell content disintegration (studied in *Pythium ultimum* var. *sporangiiferum*). At higher concentration it acts as phytotoxic (exact mode of action yet to be determined). In addition it also acts as inducing ISR	de Souza et al. (2003), Keel et al. (1992), Iavicoli et al. (2003), Bottiglieri and Keel (2006)

(continued)

Table 3.1 (continued)

PGPR	Antibiotics	Mode of action	References
Pseudomonas	HC≡N Hydrogen cyanide	HCN is a powerful inhibitor of many metal enzymes, especially copper containing cytochrome C oxidase of cytochrome oxidase pathway and is highly toxic to all aerobic microorganisms at picomolar concentrations	Bakker and Schippers (1987), Bashan and de-Bashan (2005)
Pseudomonas	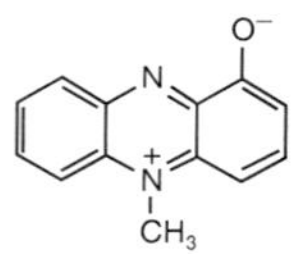 Pyocyanine	There are three different states in which pyocyanin can exist: oxidized, monovalently reduced, or divalently reduced. Mitochondria play a huge role in the cycling of pyocyanin between its redox states. Due to its redox-active properties, pyocyanin generates reactive oxygen species which target a wide range of cellular components and pathways. Pathways which are affected by pyocyanin include the electron transport chain, vesicular transport, and cell growth	Caltrider (1967), Hassan and Fridovich (1980)
Pseudomonas *Bacillus* *Azospirillum*	Volatile organic compounds	VOCs' action could be related to the ability of some volatiles to interfere with the levels and ratios of phytohormones known to be important factors in crown gall formation in agrobacterium-mediated gall formation	Zhang et al. (2007)
Bacillus	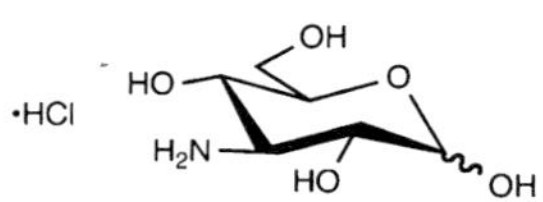 Kanosamine	Transported into fungal cells by glucose transport system and subsequently phosphorylated. The product is of intracellular metabolism; kanosamine-6-phosphate is inhibitor of enzyme glucosamine-6-phosphate synthase competitively with D-fructose 6-phosphate which leads to the morphological changes, inhibition of septa formation and cell agglutination	Janiak and Milewski (2001)

Stenotrophomonas *Xanthomonas*	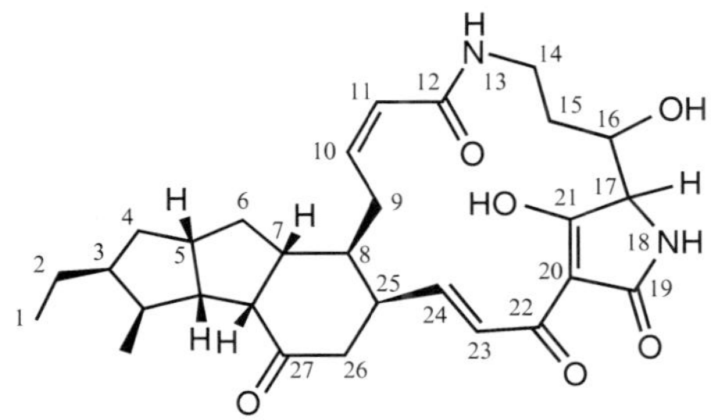 Xanthobaccin	A new tetramic acid-containing macrocyclic lactam exhibits a novel mode of action by disrupting sphingolipids important to the polarized growth of filamentous fungi	Nakayama et al. (1999), Lou et al. (2011)
Bacillus	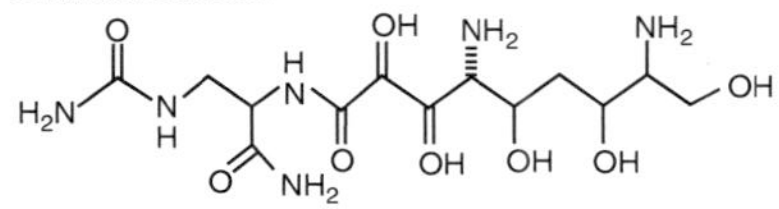 Zwittermicin A	Zwittermicin A is a linear aminopolyol that inhibits the growth of a variety of Gram-positive and Gram-negative eubacteria as well as certain ascomycete and basidiomycete fungi. The unique structure of zwittermicin A may reflect a novel mechanism of inhibition in target organisms (yet to be identified)	Haiyin et al. (1994), He et al. (1994), Stabb and Handelsman (1998), Silo-Suh et al. (1998)
Pseudomonas *Bacillus*	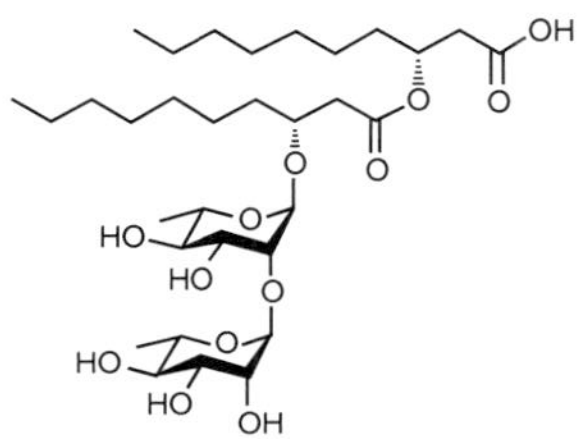 Rhamnolipids	Rhamnolipids are a class of glycolipid frequently cited as the best characterized of the bacterial surfactants. Rhamnolipids have long been reported to have antimicrobial properties (antifungal, antibacterial, and antiviral). Rhamnolipids have been suggested as antimicrobials able to remove *Bordetella bronchiseptica* biofilms. The mode of killing has been shown to result from intercalation of rhamnolipids into the cell membrane causing pores to form which result in cell lysis	Sotirova et al. (2008), Haba et al. (2003), Itoh et al. (1971)
Pseudomonas	Oomycin A	Heat stable amphipathic molecule of 700–800 Da effective against *Pythium ultimum*	Kim et al. (2000)

(continued)

Table 3.1 (continued)

PGPR	Antibiotics	Mode of action	References
Pseudomonas	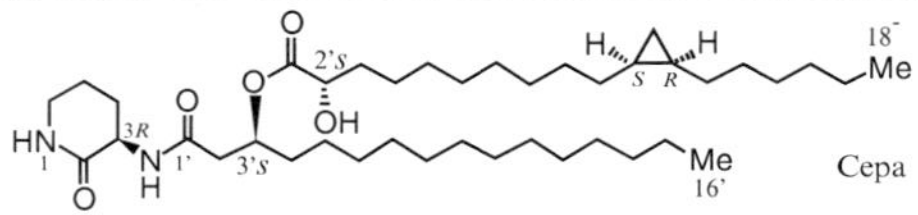 Cepa Cepaciamide A	Cepaciamide A is 3-amino-2-piperidinone-containing lipid, fungitoxic compound from *Pseudomonas cepacia* D-202, has been recognized as a biological control agent	Jiao et al. (1996), Toshima et al. (1999)
Pseudomonas	Ecomycins	A novel family of peptide antimycotics. The ecomycins have significant bioactivities against a wide range of human and plant pathogenic fungi	Miller et al. (1998)
Streptomyces	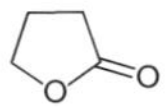 Butyrolactones	The butyrolactones bind to cytoplasmic receptor proteins and inhibit their binding to specific DNA targets. Most of these receptor proteins act as repressors, so that binding to g-butyrolactones induces expression of the target genes. Each receptor protein is highly specific for its cognate g-butyrolactone	Horinouchi (2002), Yamada (1999)
Paenibacillus	Cyclic lipopeptide	Cyclic lipopeptide (CLP) antibiotics consist of acyl side chains and peptides of various kinds, some including unusual amino acids. These characteristics confer a wide variety of biological activities and structural diversity on CLP antibiotics. Insert into plasma membrane and perturb their function which results in broad antibacterial and antifungal activities	Koch et al. (2002), Nielsen et al. (2000), Hashizume et al. (2008)
Pseudomonas	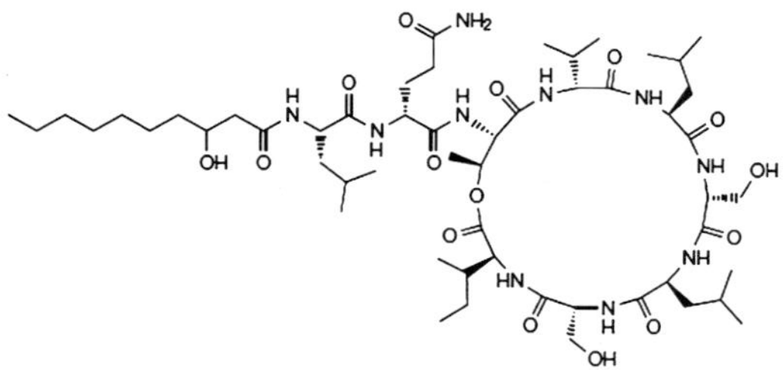 Viscosinamide	Antibiotic resulting in increased branching, swelling, and septation of hyphae, and reduced intracellular activity. The changes were suggested to be caused by altered intracellular ion (H^+ and Ca_2^+) contents, possible due to channel formation in the cell membrane	Thrane et al. (2000)

Phenazines are N-containing heterocyclic pigments synthesized by *Bravibacterium*, *Burkholderia*, *Pseudomonas*, and *Streptomyces* (Budzikiewicz 1993; Stevens et al. 1994). 2,4-diacetyl pholoroglucinol, a phenolic molecule is produced by many fluorescent pseudomonads and exhibits antifungal, antibacterial, antihelmenthic, and phytotoxic activities (Bangera and Thomashow 1996; Abbas et al. 2002). On the other hand, some metabolites, such as bacteriocins, are effective only against a specific group of closely related microorganisms (Riley and Wertz 2002).

Antagonists that produce antibiotics have a competitive advantage in occupying a particular niche and securing substrates as food sources because their antibiotics suppress the growth, multiplication, and other activities of other microorganisms. In some cases, antibiotics produced by rhizobacteria act as a determinant of ISR (Audenaert et al. 2002). Indeed, the first clear-cut experimental demonstration that a bacteria produced antibiotic could suppress plant disease in an ecosystem was made by Thomashow and Weller (1988).

As reviewed by Haas and Défago (2005), the contribution of antibiotic compounds to the biological control of root diseases has been documented through five experimental steps.

1. Diffusible or volatile secondary metabolites that are produced by biocontrol strains in vitro are purified and chemically identified. The inhibition of sensitive microorganisms by the pure compounds is then confirmed and quantified in vitro.
2. The antibiotic compound of interest is detected and quantified in the rhizosphere, which has been inoculated with the producer strain, through extraction and HPLC purification.
3. The structural and principal regulatory genes controlling the expression of the antibiotic compounds are identified and characterized. Non-producing and overproducing strains are constructed using molecular genetics techniques, and tested in microcosms that contain a chosen plant–pathogen system, with appropriate controls (i.e., the wild-type biocontrol strain or no added biocontrol agent).
4. Intrinsically poor biocontrol strains can acquire biocontrol activity by the introduction of antibiotic biosynthetic genes that are not present in the original strains.
5. The expression of antibiotic biosynthetic genes can be observed in the rhizosphere through the use of easily detectable reporter genes that are fused to structural genes for antibiotic biosynthesis.

3.10 Induced Systemic Resistance

Inducing the plants own defense mechanisms by prior application of biological inducer is a novel technique for plant protection. Induced disease resistance is the phenomenon by which a plant exhibits an increased level of resistance to infection by a pathogen after appropriate stimulation. This resistance response, first characterized by Ross (1961a,b), is expressed systemically throughout the plant and is effective against a broad spectrum of phytopathogens (Hammerschmidt and Kuc 1995). A well-characterized system of rhizobacteria-induced resistance between

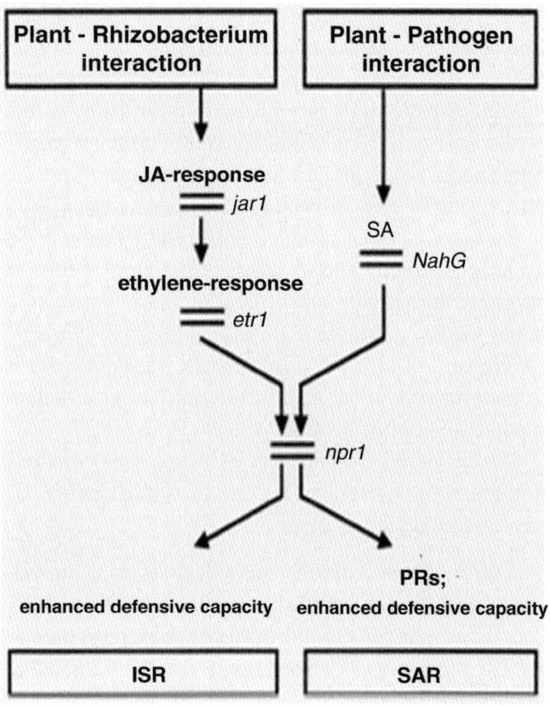

Fig. 3.7 Signalling in *Arabidopsis thaliana* leading to rhizobacteria-mediated induced systemic resistance (ISR) or to pathogen-induced systemic acquired resistance (SAR). *JA* jasmonate, *PRs* pathogenesis-related proteins, *SA* salicylate (courtesy, Van Loon et al. 1998)

Arabidopsis thaliana and *Pseudomonas fluorescens* strain WCS417 has been elaborated by Van Loon's group (Fig. 3.7) (Pieterse et al. 1996; Van Loon et al. 1998). The resulting elevated resistance due to an inducing agent against infection by a pathogen is called induced systemic resistance (ISR) or systemic acquired resistance (SAR) (Hammerschmidt and Kuc 1995). However, induction of systemic resistance by rhizobacteria is referred as ISR (Van Loon et al. 1998).

The ultimate goal of ISR and SAR is the same but the way they follow to achieve it is different. SAR requires accumulation of salicylic acid (SA) in the plant (Stitcher et al. 1997); ISR does not and, instead, is dependent on intact responses to ethylene and jasmonic acid (JA) (Pieterse et al. 1998). Advantage of ISR is its non-specificity in suppressing plant diseases, whereas classical biological control, in which antagonist selected is active against only one or few pathogens (Wei et al. 1991; Lyon and Newton 1997). Recently, Hariprasad et al. (2013) confirmed the non-specificity of ISR by inducing the resistance in tomato plants against root and foliar, fungal and bacterial pathogens. Once these natural plant resistance mechanisms are activated, increased defensive capacity is maintained for prolonged period against multiple pathogens (Tuzun 2001). Further, to conclude that the mechanism responsible for the protection through ISR requires the disease suppression shown to be plant mediated and that it is extended to plant parts that are not in contact with the inducing agent. A split root assay was followed by Liu et al. (1995) and Hariprasad et al. (2009) to demonstrate ISR to root pathogen where the inducing rhizobacteria applied to one part of a split root system did not move to the part inoculated with the pathogen. Further, suppression of foliar disease by PGPR clearly indicates the ISR is mediated through eliciting host defense response and not by direct antagonism.

Bacterial determinants of ISR includes LPS (Leeman et al. 1995; van Wees et al. 1997), 2,4-DAPG (Siddiqui and Shaukat 2003; Weller et al. 2007), siderophores (Maurhofer et al. 1994; Meziane et al. 2005), iron-regulated compounds (Press et al. 1997), SA (De Meyer et al. 1999), volatiles (Pieterse et al. 2002; Ryu et al. 2004), HCN (Defago et al. 1990), and other PGPR-derived macromolecules (Ongena et al. 2002). PGPR-mediated ISR has been demonstrated in many plant species against various phytopathogens. Various biochemical pathways of plants that are activated by PGPR were reviewed by Van Loon et al. (1998). Plant growth-promoting and bio-protecting bacteria triggered ISR fortifies structural barrier, such as thickened cell wall, suberization, and papillae formation due to the deposition of lignin and callose (Benhamou et al. 1996, 1998; M'Piga et al. 1997; Raj et al. 2012) and alters host physiology and metabolic responses, leading to an enhanced synthesis of plant defense chemicals upon challenge by pathogens and/or abiotic stress factors. Biochemical or physiological changes in plants include induced increased expression of defense-related enzymes such as peroxidase, phenylalanine ammonia lyase, polyphenol oxidase, lioxygenase (Van Peer et al. 1991; Zdor and Anderson 1992). These enzymes also bring about liberation of molecules that elicit the first steps of induction of resistance, also synthesize phytoalexins and phenolic compounds (Mauch et al. 1988; van Loon et al. 1998). Pathogenesis-related proteins such as PR-1, PR-2, chitinase, β-1, 3-glucanase, thaumatin-like protein (TLP), and some cell wall peroxidases are also known to play a major role in imparting ISR-mediated host resistance (van Peer et al. 1991; Zdor and Anderson 1992). Increase in host defense mechanism depends mainly on the inducing agent, plant genotype, physiological condition, and the pathogen (Tuzun 2001).

The molecular basis and the signalling pathways mediating the protective effect of ISR have been extensively studied and well described for the interaction by rhizobacteria. Plants are capable of differentially activating distinct defense-related pathways, depending on the inducing agent. SA, JA, and ET play an important role in this signalling network. Cross talk between SA-, JA-, and ET-dependent signalling pathways is thought to play an important role in fine tuning complex defense response (Bostock 1999; Galzebrook 1999; Pieterse and van Loon 1999). Previously it was reported that JA-dependent defence response is effectively inhibited by SA (Penninckx et al. 1996; Bowling et al. 1997) and vice versa (Niki et al. 1998). Later, studies of van Wees et al. (2000) evidenced that simultaneous activation of SAR and ISR resulted in an additive effect on the level of induced protection against phytopathogens.

The involvement of jasmonic acid (JA)/ethylene (ET) in inducing plant resistance by PGPR against a number of bacterial and fungal pathogens has been shown using the arsenal of *Arabidopsis* signal transduction mutants which demonstrated that ISR requires functional jasmonate and ethylene signalling (Van Loon et al. 1998; Pieterse et al. 2001) and is independent of SA and *PR* gene activation (Pieterse et al. 1996; Van Wees et al. 1997). The above findings were further supported by the studies where JA response mutant *jar1* and the ethylene response mutant *etr1*, that express normal levels of pathogen-induced SAR (Lawton et al. 1996; Pieterse et al. 1998), did not express ISR upon treatment with *Pseudomonas fluorescens* WCS417r, indicating that the ISR-signalling pathway requires components of the JA and ethylene response (Knoester et al. 1998). Blocking the response to either of these signal molecules renders plants more susceptible to pathogens (Knoester et al. 1998; Thomma et al. 1998; Vijayan et al. 1998; Hoffman et al. 1999). In *Arabidopsis*, both JA and ethylene have been shown to activate specific sets of defence-related genes and resistance against *P. syringae* pv. *tomato* DC3000 (Pieterse et al. 1998; Van Wees et al. 1999) which is similar to that of exogenously applied JA and ET (Boller 1991; Cohen et al. 1993).

NPR1 (Non-expressor of pathogenesis-related gene) is a key regulator of both SAR and ISR. In the SAR pathway, NPR1 regulates the SA-dependent expression of PR genes (Cao et al. 1994; Shah et al. 1997), whereas in the ISR pathway it is required for the expression of the JA- and ethylene-dependent enhanced defensive capacity (Pieterse et al. 1998). When *Arabidopsis* plants expressing *Pseudomonas fluorescens* WCS417r-mediated ISR were analyzed for the expression of well-characterized JA- and/or ethylene-responsive genes, none of them was unregulated locally or systemically in induced plants (van Wees et al. 1999). Similarly, the concentration of JA or ethylene was not found to be increased during PGPR-mediated ISR in tomato (Hariprasad et al. 2013). This suggested that ISR is not accompanied by major changes in the production of either JA or ethylene, but rather seems to be the result of sensitization of the tissue to these regulators. But, volatiles from *Bacillus amyloliquefaciens* strain IN937 is even independent of JA, SA, NPR1, and ethylene signalling pathways (Pieterse et al. 2002) which revealed that involvement of other mechanisms has to be studied in detail.

3.11 Induced Systemic Tolerance

Stress is an altered physiological condition caused by factors that tend to disrupt the equilibrium which involves physical and chemical change (Gaspar et al. 2002). Abiotic stress is the primary cause of crop loss worldwide, reducing average yields for most major crop plants by more than 50 %. They are caused by complex environmental conditions, e.g., salinity, drought, bright light, UV, too high and low temperatures, freezing, heavy metals, hypoxia, flooding, pesticides, and soil pH (Boyer 1982; Mahajan and Tuteja 2005; Mittler 2006).

One of the successful eco-friendly approaches to overcome the adverse effects of abiotic stresses is by using the biological agents, especially PGPR (Yang et al. 2009; Glick 2005; Arshad et al. 2008; Zahir et al. 2009). The term "induced systemic tolerance" (IST) has been proposed by Yang et al. (2009) for PGPR induced physical and chemical changes in plants that result in enhanced tolerance to abiotic stress. PGPR-mediated induction of IST depends on two crucial processes which are as follows, survival and root colonization of rhizobacteria by adapting to stressful environment and imparting tolerance to host plant against various stress.

3.11.1 Adaptations of Rhizobacteria to Abiotic Stress

Root colonization is one of the key characters exhibited by PGPR at rhizosphere. But, in order to exhibit its beneficial effect on host plant an efficient PGPR has to survive for a particular period in rhizosphere by adapting to various biotic and abiotic stress. Rhizobacterial adaptation to stress is a complex multilevel regulatory process in which many enzymes, proteins, and metabolites are involved.

Certain bacterial species living under extreme conditions (thermophiles and halophiles) adapt by elevating their optimum metabolic activity and membrane stability to higher temperature or salinity, respectively (Madigan and Oren 1999). Certain bacteria like *Pseudomonas* and *Azospirillum* survive under stress conditions by producing exopolysaccharides (EPS), which protect microorganisms from hydric stress and fluctuations in water potential by enhancing water retention and regulating the diffusion of carbon sources in microbial environment (Skvortsov and Ignatov 1998; Bleakley et al. 1988; Burdman et al. 2000; Sandhya et al. 2009). Also, EPS produced from rhizobacteria is known to modify soil structure by increasing soil macropores volume (Alami et al. 2000). Further, bacterial EPS mitigate saline stress by reducing the content of Na^+ available for plant uptake (Upadhyay et al. 2012). Salinity stress also has been reported to alter the cell envelope composition of the rhizobacteria resulting in changes in proteins, periplasmic glucans, and capsular, EPS and LPS, fatty acid composition, and cross-linkage of peptidoglycan (Jofré et al. 1998; Piuri et al. 2005) and is suggested that composition of the cell envelops play important role in osmoadaptation (López et al. 2000).

When subjected to osmotic stress conditions, bacteria synthesize one or various endogenous osmolytes (osmoprotectants) (K^+, glutamate, trehalose, proline, glycine betaine, proline betaine, and ectoine) (Blanco and Bernard 1994; Roessler and Müller 2001) or these osmolites are accumulated by active transport from surrounding medium (Sleator and Hill 2002). Osmoprotectants are highly soluble compounds that carry no net charge at physiological pH and are nontoxic at high concentrations. They raise osmotic pressure in the cytoplasm and also stabilize proteins and membranes under unfavorable environmental conditions. Most of the osmolites are involved in turgor maintenance only, while others protect cells and biological macromolecules against denatured effect of not only hyperosmotic stress, but also other stresses such as heating, freezing, and desiccation (Yancey 2005; Crowe 2007; Paul and Nair 2008).

The heat-shock response involves the induction of many proteins—called heat-shock proteins, or Hsps—in response to elevation of temperature (Neidhardt and Van Bogelen 1987). The bacterial heat-shock response is not limited to changes in temperature and is a general stress response. The heat-shock proteins include chaperones, involved in the proper folding of denatured proteins required for the degradation of irreversibly damaged proteins (Munchbach et al. 1999). Induction of this response improves thermotolerance, salt tolerance, and tolerance to heavy metals (Kusukawa and Yura 1988; Inbar and Ron 1993; Qi et al. 2004). Similarly cold-tolerant bacteria respond to a decrease in temperature by induction of cryoprotective protein (Koda et al. 2001).

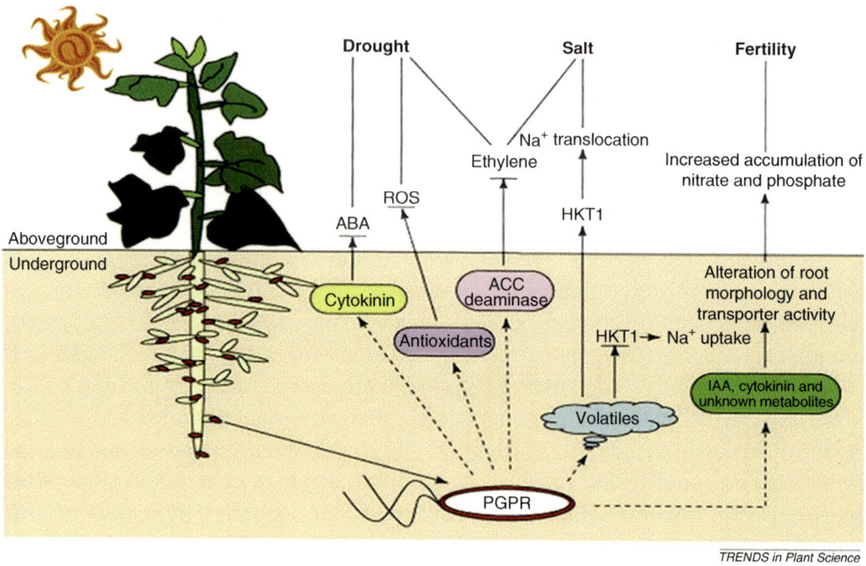

Fig. 3.8 Induced systemic tolerance (IST) elicited by PGPR against drought, salt, and fertility stresses underground (root) and aboveground (shoot). *Broken arrows* indicate bioactive compounds secreted by PGPR; *solid arrows* indicate plant compounds affected by bacterial components. Some PGPR strains, indicated in red on the plant roots, produce cytokinin and antioxidants such as catalase, which result in ABA accumulation and ROS degradation, respectively (Figueiredo et al. 2008; Kohler et al. 2008). Degradation of the ethylene precursor ACC by bacterial ACC deaminase releases plant stress and rescues normal plant growth under drought and salt stresses (Kohler et al. 2008; Mayak et al. 2004). The volatiles emitted by PGPR downregulate hkt1 expression in roots but upregulate it in shoot tissues, orchestrating lower Na^+ levels and recirculation of Na^+ in the whole plant under high salt conditions (Zhang et al. 2008). Production by PGPR of IAA or unknown determinants can increase root length, root surface area, and the number of root tips, leading to enhanced uptake of nitrate and phosphorous (Gyaneshwar et al. 2002; Mantelin and Touraine 2004; Adesemoye et al. 2008). *ABA* abscisic acid, *ACC* 1-aminocyclopropane-1-carboxylate, *HKT1* high-affinity K^+ transporter 1, *IAA* indole acetic acid, *IST* induced systemic tolerance, *PGPR* plant growth-promoting rhizobacteria, *ROS* reactive oxygen species. (Courtesy, Yang et al. 2009)

3.11.2 Alleviation of Abiotic Stress in Plants by Rhizobacteria

Recently Yang et al. (2009) reviewed the rhizobacteria-mediated IST in detail. Further, Grover et al. (2011) discussed different adaptive strategies followed by rhizobacteria during different stress conditions and alleviation tolerance to abiotic stress in plants. An overall mechanism involved in PGPR-mediated mitigation of abiotic stress in plant is shown in Fig. 3.8.

3.11.3 Modulating Ethylene Level Through ACC Deaminase Activity

Ethylene, a gaseous phytohormone, commonly appears to enhance root initiation and growth at lower level. If the ethylene concentration increases above a threshold level, it becomes detrimental for plant growth. Ethylene biosynthesis is also increased by stress conditions such as drought, flooding, chilling, exposure to ozone, or mechanical wounding (Ma et al. 1998; Abeles et al. 1992). Ethylene biosynthesis starts with the S-odenosylation of methionine to *S*-adenosylmethionine (SAM) followed by the closing of the cyclopropane ring to form 1-aminocyclopropane-1-carboxylate (ACC) which oxidatively cleaved to form ethylene. 1-aminocyclopropane-1-carboxylate deaminase (ACCd), an enzyme which cleaves ACC, the immediate precursor molecule of ethylene. Hence, many studies have been published on beneficial effects of PGPR containing ACCd activity on different plants. ACC deaminase is a cytoplasmically localized, multimeric enzyme containing 2–3 subunits with a monomeric subunit molecular mass of approximately 35–42 kDa (Jacobson et al. 1994; Glick 2005). It is a sulfhydryl enzyme which requires a cofactor pyridoxal 5-phosphate for enzymatic activity (Walsh et al. 1981; Glick et al. 1998). By analyzing the Km value of ACC, it was concluded that the enzyme does not have a particularly high affinity to ACC (Jacobson et al. 1994; Hontzeas et al. 2004). Usually, ACC levels in plants are typically in μM range, hence a small increase in the ACC concentration will result in the parallel increase in the rate of ACC cleavage. Plant root colonized with PGPR containing ACCd are dramatically more resistant to the injurious effect of stress ethylene that is synthesized. It has been reported that ACC deaminase containing bacteria promote plant growth under a variety of stressful conditions such as high salt (Saravanakumar and Samiyappan 2007; Nadeem et al. 2012), drought (Mayak et al. 2004; Ali et al. 2013), flooding (Grichko and Glick 2001), metals (Burd et al. 2000; Gerhardt et al. 2006; Rodriguez et al. 2008), organic contaminants (Gurska et al. 2009; Reed and Glick 2005; Gerhardt et al. 2006), and phytopathogens (Hao et al. 2007; Wang et al. 2000).

Downregulation of genes involved in ethylene induced plant stress response and upregulation genes involved in plant growth were reported in ACC-deaminase producing rhizobacteria treated canola roots (Hontzeas et al. 2004). Similarly, ACCd negative mutants of *Enterobacter cloacae* UW4 showed reduced activity of ACCd and subsequently its ability to promote the elongation of canola roots under gnotobiotic conditions was greatly diminished when compared to wild types (Li et al. 2000). In each of these cases ACC-deaminase containing bacteria markedly lowered the level of ACC in the stressed plants thereby limiting the amount of stress ethylene synthesis and hence the damage to the plant. These studies revealed the importance of ACCd activity of PGPR in mitigating abiotic stress by reducing ethylene level in root.

A model describing the role of ACCd in PGPR-mediated abiotic stress suppression was suggested by Glick et al. (1998) (Fig. 3.9). Briefly, under stress conditions in order to synthesize ethylene, ACC is produced in plant roots and a part is exuded

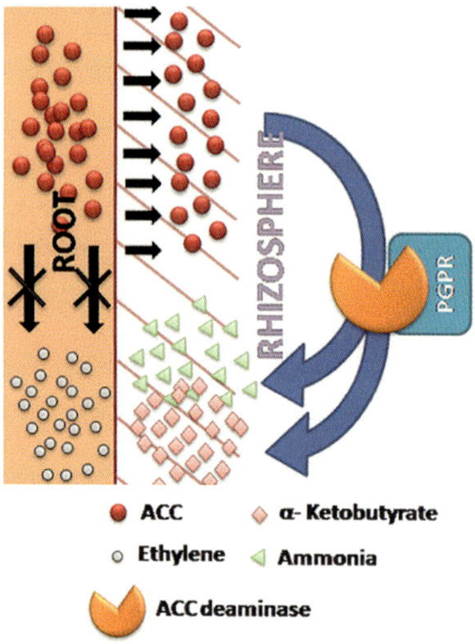

Fig. 3.9 Degradation of ACC which is an immediate precursor of ethylene by ACC deaminase producing PGPR reduced available ACC to synthesize ethylene in root. Degradation product ammonia is further utilized as N source by plant and PGPR

into rhizosphere. Rhizobacteria in close vicinity of the roots can take up some of this ACC and hydrolyze it by the activity of ACCd to ammonia and α-ketobutyrate. The uptake and subsequent hydrolysis of ACC by rhizobacteria decreases the amount of ACC outside of the plant. Further, to maintain the equilibrium between internal and external ACC levels, plant must exude large amount of ACC into the rhizosphere which is utilized by rhizobacteria after cleaving with ACC. This process leads to reduction of amount of ACC available to the decreased synthesis of ethylene and its inhibitory effect on root elongation is reduced.

3.12 Future Prospects and Challenges

Use of these PGPR in sustainable agriculture is directly related to understanding their mode of action of growth promotion and suppression of abiotic and biotic stress in plants. In spite of significant advancement made in understanding these mechanisms, their performance under field conditions are inconsistent and PGPR are yet to fulfill their promise as potential alternative agrochemicals. Hence more research should be carried out in order to develop strategies to maintain consistency of PGPR performance under field conditions. To achieve this, it is important to select PGPR based on their mode of action which is suitable to enhance the host plant health under particular conditions. Our review is helpful for beginners and also for those who are working in this area to conduct some advanced studies. As per our previous studies, in order to achieve the production in agriculturally important crops, it is important to use PGPRs which are endowed with multiple traits through which they can promote plant growth, mitigate abiotic stress, and also suppress the diseases nonspecifically. Even though several modes of actions are frequently reported survival of bacteria under stress conditions, rare mechanisms such as ABA production should be studied thoroughly.

References

Abbas A, Morrissey JP, Carnicero-Marquez P, Sheehan MM, Delany IR, O'Gara F (2002) Characterization of interactions between the transcriptional repressor PhlF and its binding site at the phlA promoter in *Pseudomonas fluorescens* F113. J Bacteriol 184:3008–3016

Abeles FB, Morgan PW, Saltveit ME (1992) Ethylene in plant biology, 2nd edn. Academic, San Diego, CA

Adesemoye AO, Torbert HA, Kloepper JW (2008) Enhanced plant nutrient use efficiency with PGPR and AMF in an integrated nutrient management system. Can J Microbiol 54:876–886

Ae N, Arihara J, Okada K (1991) Phosphorus response of chickpea and evaluation of phosphorus availability in Indian alfisols and vertisols. In: Johansen C, Lee KK, Sahrawat KL (eds) International Crops Research Institute for the semi arid tropics. Patancheru, Andhra Pradesh, India, pp 33–41

Agrios GN (2005) Plant pathology, 5th edn. Elsevier Academic, San Diego. ISBN 80120445653

Akiyoshi DE, Regier DA, Gordon MP (1987) Cytokinin production by *Agrobacterium* and *Pseudomonas* spp. J Bacteriol 169:4242–4248

Alami Y, Achouak W, Marol C, Heulin T (2000) Rhizosphere soil aggregation and plant growth promotion of sunflowers by an exopolysaccharide-producing *Rhizobium* sp. strain isolated from sunflower roots. Appl Environ Microbiol 66:3393–3398

Albareda M, Dardanelli MS, Sousa C, Megías M, Temprano F, Rodríguez Navarro DN (2006) Factors affecting the attachment of rhizospheric bacteria to bean and soybean roots. FEMS Microbiol Lett 259:67–73

Ali B, Sabri AN, Ljung K, Hasnain S (2009) Auxin production by plant associated bacteria: impact on endogenous IAA content and growth of *Triticum aestivum* L. Lett Appl Microbiol 48:542–547

Ali SZ, Sandhya V, Rao LV (2013) Isolation and characterization of drought-tolerant ACC deaminase and exopolysaccharide-producing fluorescent *Pseudomonas* sp. Arch Microbiol doi: 10.1007/s13213-013-0680-3

Alikhani HA, Saleh-Rastin N, Antoun H (2006) Phosphate solubilization activity of rhizobia native to Iranian soil. Plant Soil 287:35–41

Angelini TE, Roper M, Kolter R, Weitz DA, Brenner MP (2009) Bacillus subtilis spreads by surfing on waves of surfactant. Proc Natl Acad Sci U S A 106:18109–18113

Antoun H, Beauchamp CJ, Goussard N, Chabot R, Lalande R (1998) Potential of *Rhizobium* and *Bradyrhizobium* species as plant growth promoting rhizobacteria on non-legumes: effect on radishes (*Raphanus sativus* L.). Plant Soil 204:57–67

Araujo FF, Henning AA, Hungria M (2005) Phytohormones and antibiotics produced by *Bacillus subtilis* and their effects on seed pathogenic fungi and on soybean root development. World J Microbiol Biotechnol 21:1639–1645

Arima K, Imanaki H, Kousaka M, Fukuta A, Tamura G (1964) Pyrrolnitrin, a new antibiotic substance, produced by *Pseudomonas*. Agric Biol Chem 28:575–576

Arshad M, Frankenberger WT (1991) Microbial production of plant hormones. Plant Soil 133:1–8

Arshad M, Frankenberger WT (1993) Microbial production of plant growth regulators. In: Metting FB Jr (ed) Soil microbial ecology. Applications in agricultural and environmental management. Marcel Dekker, New York, pp 307–343

Arshad M, Shaharoona B, Mahmood T (2008) Inoculation with *Pseudomonas* spp. containing ACC-deaminase partially eliminates the effects of drought stress on growth, yield, and ripening of pea (*Pisum sativum* L.). Pedosphere 18:611–620

Asghar HN, Zahir ZA, Arshad M, Khaliq A (2002) Relationship between *in vitro* production of auxins by rhizobacteria and their growth-promoting activities in *Brassica juncea* L. Biol Fert Soils 35:231–237

Aslantas R, Cakmakci R, Sahin F (2007) Effect of plant growth promoting rhizobacteria on young apple tree growth and fruit yield under orchard conditions. Sci Hortic 111:371–377

Assmus B, Hutzler P, Kirchhof G, Amann R, Lawrence JR, Hartmann A (1995) In situ localization of Azospirillum brasilense in the rhizosphere of wheat with fluorescent labeled, rRNA-targeted oligonucleotide probes and scanning confocal laser microscopy. Appl Environ Microbiol 61:1013–1019

Atzorn R, Crozier A, Wheeler C, Sandberg G (1988) Production of gibberellins and Indole 3-acetic acid by *Rhizobium phaseoli* in relation to nodulation of *Phaseolus vulgaris* roots. Planta 175:532–538

Audenaert K, Pattery T, Cornelis P, Höfte M (2002) Induction of systemic resistance to *Botrytis cinerea* in tomato by *Pseudomonas aeruginosa* 7NSK2: role of salicylic acid, pyochelin, and pyocyanin. Mol Plant Microb Interact 15:1147–1156

Bacilio-Jimenez M, Aguilar-Flores S, Ventura-Zapata E, Perez-Campos E, Bouquelet S, Zenteno E (2003) Chemical characterization of root exudates from rice (*Oryza sativa*) and their effects on the chemotactic response of endophytic bacteria. Plant Soil 249:271–277

Bais HP, Park SW, Weir TL, Callaway RM, Vivanco JM (2004) How plants communicate using the underground information superhighway. Trends Plant Sci 9:26–32

Bais HP, Weir TL, Perry LG, Gilroy S, Vivanco JM (2006) The role of root exudates in rhizosphere interactions with plants and other organisms. Annu Rev Plant Biol 57:233–266

Baker KF (1987) Evolving concepts of biological control of plant pathogens. Annu Rev Phytopathol 25:67–85

Baker R, Elad Y, Sneh B (1986) Physical, biological and host factors in iron competition in soils. In: Swinburne TR (ed) Iron, siderophores, and plant diseases. Plenum Press, New York, pp 77–84

Bakker AW, Schipper B (1987) Microbial cyanide production in the rhizosphere in relation to potato yield reduction and *Pseudomonas* spp. mediated plant growth reduction. Soil Biol Biochem 19:452–458

Banat IM, Makkar RS, Cameotra SS (2000) Potential commercial applications of microbial surfactants. Appl Microbiol Biotechnol 53:495–508

Bangera MG, Thomashow LS (1996) Characterization of a genomic locus required for synthesis of the antibiotic 2,4-diacetylphloroglucinol by the biological control agent *Pseudomonas fluorescens* Q2-87. Mol Plant Microbe Interact 9:83–90

Barbara RH, Thomas H (1998) Life in grasses: diazotrophic endophytes. Trends Microbiol 6:139–144

Barea JM, Azcon R, Azcon-Aguilar C (2004) Mycorrhizal fungi and plant growth promoting rhizobacteria. In: Verma A, Abbott L, Werner D, Hampp R (eds) Plant surface microbiology. Springer, Heidelberg, pp 351–371

Barea JM, Pozo MJ, Azcon R, Azcon-Aguilar C (2005) Microbial co-operation in the rhizosphere. J Exp Bot 56:1761–1778

Bar-Ness E, Chen Y, Hadar Y, Marschner H, Römheld V (1991) Siderophores of *Pseudomonas putida* as an iron source for dicot and monocot plants. Plant Soil 130:231–241

Bar-Ness E, Hadar Y, Chen Y, Römheld V, Marschner H (1992) Short-term effects of rhizosphere microorganisms on Fe uptake from microbial siderophores by maize and oat. Plant Physiol 100:451–456

Barriuso J, Solano BR, Fray RG, Camara M, Hartmann A, Manero FJG (2008) Transgenic tomato plants alter quorum sensing in plant growth-promoting rhizobacteria. Plant Biotechnol J 6:442–452

Bashan Y, de-Bashan LE (2005) Bacteria. In: Hillel D (ed) Encyclopaedia of soils in the environment. Elsevier, Oxford, UK, pp 103–115

Bashan Y, de-Bashan LE (2010) How the plant growth promoting bacterium *Azospirillum* promotes plant growth—a critical assessment. Adv Agron 108:77–136

Bastian F, Cohen A, Piccoli P, Luna V, Baraldi R, Bottini R (1998) Production of indole-3-acetic acid and gibberellins A1 and A3 by *Acetobacter diazotrophicus* and *Herbaspirillum seropedicae* in chemically-defined culture media. Plant Growth Regul 24:7–11

Beever RE, Burns DJW (1980) Phosphorus uptake storage and utilization by fungi. Adv Bot Res 8:127–219

Bell CR, Dickie GA, Harvey WLG, Chan JWYF (1995) Endophytic bacteria in grapevine. Can J Microbiol 41:46–53
Bell AA, Liu L, Reidy B, Davis RM, Subbarao KV (1998) Mechanisms of subsurface drip irrigation-mediated suppression of lettuce drop caused by *Sclerotinia minor*. Phytopathology 88:252–259
Benhamou N, Kloepper JW, Quadt-Hallman A, Tuzun S (1996) Induction of defense related ultrastructural modifications in pea root tissues inoculated with endophytic bacteria. Plant Physiol 112:919–929
Benhamou N, Kloepper JW, Tuzun S (1998) Induction of resistance against *Fusarium* wilt of tomato by combination of chitosan with an endophytic bacterial strain: ultrastructure and cytochemistry of the host response. Planta 204:153–168
Blanco C, Bernard T (1994) Osmoadaptation in rhizobia: ectoine-induced salt tolerance. J Bacteriol 176:5210–5217
Bleakley BH, Gaskins MH, Hubbel DH, Zam SG (1988) Floc formation by *Azospirillum lipoferum* grown on polyb-hydroxybutyrate. Appl Environ Microbiol 54:2986–2995
Boddey RM, Dobereiner J (1995) Nitrogen fixation associated with grasses and cereals: recent progress and perspectives for the future. Fert Res 42:241–250
Boiero L, Perrig D, Masciarelli O, Pena C, Cassán F, Luna V (2007) Phytohormone production by strains of *Bradyrhizobium japonicum* and possible physiological and technological implications. Appl Microbiol Biotechnol 74:874–880
Boller T (1991) Ethylene in pathogenesis and disease resistance. In: Mattoo AK, Suttle JC (eds) The plant hormone ethylene. CRC Press, Boca Raton, FL, pp 293–314
Bostock RM (1999) Signal conflicts and synergies in induced resistance to multiple attackers. Physiol Mol Plant Pathol 55:99–109
Bottiglieri M, Keel C (2006) Characterization of PhlG, a hydrolase that specifically degrades the antifungal compound 2,4-diacetylphloroglucinol in the biocontrol agent Pseudomonas fluorescens CHA0. Appl Environ Microbiol 72:418–427
Bowers JH, Parke JL (1993) Colonization of pea (*Pisum sativum* L.) taproots by *Pseudomonas fluorescens*: effect of soil temperature and bacterial motility. Soil Biol Biochem 25:1693–1701
Bowling SA, Clarke JD, Liu Y, Klessig DF, Dong X (1997) The *cpr5* mutant of *Arabidopsis* expresses both NPR1-dependent and NPR1-independent resistance. Plant Cell 9:1573–1584
Boyer JS (1982) Plant productivity and environment. Science 218:443–448
Brandl MT, Lindow SE (1997) Environmental signals modulate the expression of an indole-3-acetic acid biosynthetic gene in *Erwinia herbicola*. Mol Plant Microb Interact 10:499–505
Budzikiewicz H (1993) Secondary metabolites from fluorescent pseudomonads. FEMS Microbiol Rev 104:209–228
Bultreys A, Gheysen I, Wathelet B, Maraite H, de Hoffmann E (2003) High-performance liquid chromatography analyses of pyoverdin siderophores differentiate among phytopathogenic fluorescent *Pseudomonas* species. Appl Environ Microbiol 69:1143–1153
Burd GI, Dixon DG, Glick BR (2000) Plant growth-promoting bacteria that decrease heavy metal toxicity in plants. Can J Microbiol 46:237–245
Burdman S, Jurkevitch E, Okon Y (2000) Recent advances in the use of plant growth promoting rhizobacteria (PGPR) in agriculture. In: Subba Rao NS, Dommergues YR (eds) Microbial interactions in agriculture and forestry, vol II. Science Publishers, Enfield, pp 229–250
Buysens S, Heungens K, Poppe J, Hofte M (1996) Involvement of pyochelin and pyoverdin in suppression of *Pythium*-Induced damping-off of tomato by *Pseudomonas aeruginosa* 7NSK2. Appl Environ Microbiol 62:865–871
Caltrider PG (1967) Protoanemonin. In: Corcoran JW, Hann F (eds) Antibiotics 1. Mechanism of action. Springer, Berlin, pp 671–673
Cao H, Bowling SA, Gordon AS, Dong X (1994) Characterization of an *Arabidopsis* mutant that is nonresponsive to inducers of systemic acquired resistance. Plant Cell 6:1583–1592
Carvalhais LC, Dennis PG, Fedoseyenko D, Hajirezaei MR, Boriss R, von Wirén N (2011) Root exudation of sugars, amino acids, and organic acids by maize as affected by nitrogen, phosphorus, potassium, and iron deficiency. J Plant Nutr Soil Sci 174:3–11

Cassan F, Bottini R, Schneider G, Piccoli P (2001) *Azospirillum brasilense* and *Azospirillum lipoferum* hydrolyze conjugates of GA20 and metabolize the resultant aglycones to GA1 in seedlings of rice dwarf mutants. Plant Physiol 125:2053–2058

Cezon R, Manero FJ, Gutierrez PA, Ramos B, Garica JAL (2003) Effect of two plant growth-promoting rhizobacteria on the germination and growth of pepper seedlings (*Capsicum annum*) cv. Roxy. Arch Agron Soil Sci 49:593–603

Chaiharn M, Chunhaleuchanon S, Lumyong S (2009) Screening siderophore producing bacteria as potential biological control agent for fungal rice pathogens in Thailand. World J Microbiol Biotechnol 25:1919–1928

Chang WT, Chen ML, Wang SL (2010) An antifungal chitinase produced by *Bacillus subtilis* using chitin waste as a carbon source. World J Microbiol Biotechnol 26:945–950

Chatterjee R, Allen RM, Ludden PW, Shah VK (1997) *In vitro* synthesis of the iron-molybdenum cofactor and maturation of the *nif*-encoded apodinitrogenase. J Biol Chem 272:21604–21608

Chen Y, Yan F, Chai Y, Liu H, Kolter R, Losick R, Guo J (2012) Biocontrol of tomato wilt disease by Bacillus subtilis isolates from natural environments depends on conserved genes mediating biofilm formation. Environ Microbiol 15:848–864

Chet I, Ordentlich A, Shapira R, Oppenheim A (1990) Mechanisms of biocontrol of soil-borne plant-pathogens by rhizobacteria. Plant Soil 129:85–92

Chin-A-Woeng TFC, Bloemberg GV, Lugtenberg BJJ (2003) Mechanisms of biological control of phytopathogenic fungi by *Pseudomonas* spp. In: Stacey G, Keen NT (eds) Plant–microbe interactions, vol 6. The American Phytopathology Society, St. Paul, USA, pp 173–225

Cohen Y, Gisi U, Niderman T (1993) Local and systemic protection against *Phytophthora infestans* induced in potato and tomato plants by jasmonic acid and jasmonic methyl ester. Phytopathology 83:1054–1062

Cohen AC, Bottini R, Piccoli PN (2008) *Azospirillum brasilense* Sp 245 produces ABA in chemically-defined culture medium and increases ABA content in *Arabidopsis* plants. Plant Growth Regul 54:97–103

Cohen AC, Travaglia CN, Bottini R, Piccoli PN (2009) Participation of abscisic acid and gibberellins produced by endophytic *Azospirillum* in the alleviation of drought effects in maize. Botany 87:455–462

Compant S, Duffy B, Nowak J, Clement C, Barka EA (2005) Use of plant growth-promoting bacteria for biocontrol of plant diseases: principles, mechanisms of action, and future prospects. Appl Environ Microbiol 71:4951–4959

Costerton JW, Lewandowski Z, Caldwell DE, Korber DR, Lappin-Scott HM (1995) Microbial biofilms. Annu Rev Microbiol 49:711–745

Cox CD, Rinehart KL Jr, Moore ML, Cook JC Jr (1981) Pyochelin: novel structure of an iron-chelating growth promoter for *Pseudomonas aeruginosa*. Proc Natl Acad Sci U S A 78:4256–4260

Crichton RR, Charloteaux-Wauters M (1987) Iron transport and storage. Eur J Biochem 164:485–506

Crosa JH, Walsh CT (2002) Genetics and assembly line enzymology of siderophore biosynthesis in bacteria. Microbiol Mol Biol Rev 66:223–249

Crowe JH (2007) Trehalose as a 'chemical chaperone': fact and fantasy. Adv Exp Med Biol 594:143–158

Crowley DA (2006) Microbial siderophores in the plant rhizosphere. In: Barton LL, Abadía J (eds) Iron nutrition in plants and rhizospheric microorganisms. Springer, Berlin

Crozier A, Kamiya Y, Bishop G, Yokota T (2000) Biosynthesis of hormones and elicitor molecules. In: Buchanan BB, Gruissem W, Jones RL (eds) Biochemistry and molecular biology of plants. American Society of Plant Physiology, Rockville, pp 850–929

Czaban J, Gajda A, Wroblewska B (2007) The motility of bacteria from rhizosphere and different zones of winter wheat roots. Pol J Environ Stud 16:301–308

Davies PJ (1995) Plant hormones: physiology, biochemistry, and molecular biology. Kluwer Academic, London, pp 6–7 de Freitas JR, Banerjee MR, Germida JJ (1997) Phosphate solubilizing rhizobacteria enhance the growth and yield but not phosphorus uptake of canola (*Brassica napus*). Biol Fertil Soils 24:358–364

De Meyer G, Capieau K, Audenaert K, Buchala A, Métraux JP, Höfte M (1999) Nanogram amounts of salicylic acid produced by the rhizobacterium *Pseudomonas aeruginosa* 7NSK2 activate the systemic acquired resistance pathway in bean. Mol Plant Microbe Interact 12:450–458

de Salamone G, Hynes RK, Nelson LM (2001) Cytokinin production by plant growth promoting rhizobacteria and selected mutants. Can J Microbiol 47:404–411

de Souza J, Arnould C, Deulvot C, Lamanceau P, Pearson VG, Raaijmakers JM (2003) Effect of 2,4 diacetyl phloro glucinol on *Pythium*: cellular responses and variation in sensitivity among propagules and species. Phytopathology 93:966–975

de Weert S, Bloemberg GV (2006) Rhizosphere competence and the role of root colonization in biocontrol. In: Gnanamanickam SS (ed) Plant-associated bacteria. Springer, Dordrecht, pp 317–333

de Weert S, Vermeiren H, Mulders IHM, Kuiper I, Hendrickx N (2002) Flagella-driven chemotaxis towards exudate components is an important trait for tomato root colonization by *Pseudomonas fluorescens*. Mol Plant Microbe Interact 15:1173–1180

Defago G, Berling CH, Burger U, Hass D, Kahr G, Keel C, Voisard C, Wirthner P, Wuthrich B (1990) Suppression of black root rot of tobacco and other root disease by strains of *Pseudomonas fluorescens*: potential applications and mechanisms. In: Schippers D, Cook RJ, Henis Y, Ko WH, Rovira AD, Schippers B, Scott PR (eds) Biological control of soil borne plant pathogens. CABI, Oxfordshire, UK, pp 93–108

Dekkers LC, van der Bij AJ, Mulders IHM, Phoelich CC, Wentwoord RAR, Glandorf DCM, Wijffelman CA, Lugtenberg BJJ (1998) Role of the O-antigen of lipopolysaccharide and possible roles of growth rate and of NADH: ubiquinone oxidoreductase (nuo) in competitive tomato root-tip colonization by *Pseudomonas fluorescens* WCS365. Mol Plant Microbe Interact 11:763–771

Dey R, Pal KK, Bhatt DM, Chauhan SM (2004) Growth promotion and yield enhancement of peanut (*Arachis hypogaea* L.) by application of plant growth-promoting rhizobacteria. Microbiol Res 159:371–394

Djavaheri M (2007) Iron-regulated metabolites of plant growth-promoting *Pseudomonas fluorescens* WCS374: their role in induced systemic resistance. PhD thesis, University Utrecht, The Netherlands

Dobbelaere S, Croonenborghs A, Thys A, Vande Broek A, Vanderleyden J (1999) Phyto stimulatory effect of *Azospirillum brasilense* wild type and mutant strains altered in IAA production on wheat. Plant Soil 212:155–164

Dobbelaere S, Vanderleyden J, Okon Y (2003) Plant growth-promoting effects of diazotrophs in the rhizosphere. Crit Rev Plant Sci 22:107–149

Dobereiner J (1997) Biological nitrogen fixation in tropics: social and economic contribution. Soil Biol Biochem 29:771–774

Dobert RC, Rood SB, Blevins DG (1992) Gibberellins and the legume-*Rhizobium* symbiosis. I. Endogenous gibberellins of Lima Bean (*Phaseolus lunatus* L.) stems and nodules. Plant Physiol 98:221–224

Dowling DN, Boesten B, Gill PR Jr, OíGara F (1994) Developing concepts in biological control: a molecular ecology approach. In: OíGara F et al (eds) Molecular ecology of rhizosphere microorganisms. Verlagsgesellschaft, Weinheim, Germany, pp 57–64

Duffy B, Schouten A, Raaijmakers JM (2003) Pathogen self defense: mechanisms to counteract microbial antagonism. Annu Rev Phytopathol 41:501–538

Esitken A, Pirlak L, Turan M, Sahin F (2006) Effects of floral and foliar application of plant growth promoting rhizobacteria (PGPR) on yield, growth and nutrition of sweet cherry. Sci Hortic 110:324–327

Essen SA, Johnsson A, Bylund D, Pederson K, Lundstrom US (2007) Siderophore production by *Pesudomonas stutzeri* under aerobic and anaerobic conditions. Appl Environ Microbiol 73:5857–5864

Estrada-de los Santos P, Bustillos-Cristales R, Caballero-Mellado J (2001) *Burkholderia*, a genus rich in plant-associated nitrogen fixers with wide environmental and geographic distribution. Appl Environ Microbiol 67:2790–2798

Ferguson L, Lessenger JE (2006) Plant growth regulators. In: Lessenger JE (ed) Agricultural medicine. Springer, New York, pp 156–166

Figueiredo MVB, Burity HA, Martínez CR, Chanway CP (2008) Alleviation of drought stress in the common bean (*Phaseolus vulgaris* L.) by co-inoculation with *Paenibacillus polymyxa* and *Rhizobium tropici*. Appl Soil Ecol 40:182–188

Franche C, Lindstrom K, Elmerich C (2009) Nitrogen-fixing bacteria associated with leguminous and non-leguminous plants. Plant Soil 321:35–59

Fravel DR (1988) Role of antibiosis in the biocontrol of plant diseases. Annu Rev Phytopathol 26:75–91

Fridlender M, Inbar J, Chet I (1993) Biological control of soilborne plant pathogens by a β-1,3-glucanase producing *Pseudomonas cepacia*. Soil Biol Biochem 25:1211–1221

Fuentes-Ramírez LE, Bustillos-Cristales R, Tapia-Herńandez A, Jiménez-Salgado T, Wang ET, Martínez-Romero E, Caballero-Mellado J (2001) Novel nitrogen-fixing acetic acid bacteria, *Gluconacetobacter johannae* sp. nov. And *Gluconacetobacter azotocaptans* sp. nov., associated with coffee plants. Int J Syst Evol Microbiol 51:1305–1314

Galzebrook J (1999) Genes controlling expression of defence responses in *Arabidopsis*. Curr Opin Plant Biol 2:280–286

Gaspar T, Franck T, Bisbis B, Kevers C, Jouve L, Hausman JF, Dommes J (2002) Concepts in plant stress physiology. Application to plant tissue cultures. Plant Growth Regul 37:263–285

Gaudin V, Vrain D, Jouanin L (1994) Bacterial genes modifying hormonal balance in plant. Plant Physiol Biol 32:11–29

Gerhardt KE, Green berg BM, Glick BR (2006) The role of ACC deaminase in facilitating the phytoremediation of organic metals and salt. Curr Trends Microbiol 2:61–73

Glass ADM (1989) Plant nutrition: an introduction to current concepts. Jones and Bartlett, Boston

Glick BR (2005) Modulation of plant ethylene levels by the enzyme ACC deaminase. FEMS Microbiol Lett 251:1–7

Glick BR, Karaturovic DM, Newell PC (1995) A novel procedure for rapid isolation of plant growth promoting pseudomonads. Can J Microbiol 41:533–536

Glick BR, Penrose DM, Li J (1998) A model for lowering of plant ethylene concentrations by plant growth-promoting rhizobacteria. J Theor Biol 190:63–68

Glick BR, Patten CL, Holquin G, Penrose DM (1999) Biochemical and genetic mechanisms used by plant growth promoting bacteria. Imperial College Press, London

Glick BR, Todorovic B, Czarny J, Cheng Z, Duan J, McConkey B (2007) Promotion of plant growth by bacterial ACC deaminase. Crit Rev Plant Sci 26:227–242

Goddard VJ, Bailey MJ, Darrah P, Lilley AK, Thompson IP (2001) Monitoring temporal and spatial variation in rhizosphere bacterial population diversity: a community approach for the improved selection of rhizosphere competent bacteria. Plant Soil 232:181–193

Goldstein AH (1986) Bacterial phosphate solubilization: historical perspective and future prospects. Am J Altern Agric 1:51–57

Goldstein AH (1994) Involvement of the quinoprotein glucose dehydrogenises in the solubilization of exogenous phosphates by Gram-negative bacteria. In: Gorini AT, Yagil E, Silver S (eds) Phosphate in microorganisms: cellular and molecular biology. American Society for Microbiology, Washington, DC, pp 197–203. ISBN 1-55581-265-1

Gracia De Salome IE, Hynes RK, Nelson LM (2001) Cytokinin production by plant growth promoting rhizobacteria and selected mutants. Can J Microbiol 47:404–411

Gray EJ, Smith DL (2005) Intracellular and extracellular PGPR: commonalities and distinctions in the plant-bacterium signaling processes. Soil Biol Biochem 37:395–412

Grichko VP, Glick BR (2001) Amelioration of flooding stress by ACC deaminase-containing plant growth-promoting bacteria. Plant Physiol Biochem 39:11–17

Grover M, Ali SZ, Sandhya V, Rasul A, Venkateswarlu B (2011) Role of microorganisms in adaptation of agriculture crops to abiotic stresses. World J Microbiol Biotechnol 27:1231–1240

Gualtieri G, Bisseling T (2000) The evolution of nodulation. Plant Mol Biol 42:181–194

Gurska J, Wang W, Gerhardt KE, Khalid AM, Isherwood DM, Huang XD, Glick BR, Greenberg BM (2009) Three year field test of a plant growth promoting rhizobacteria enhanced phytoremediation system at a land farm for treatment of hydrocarbon waste. Environ Sci Technol 43:4472–4479

Gutierrez-Mannero FJ, Ramos-Solano B, Probanza A, Mehouachi J, Tadeo FR, Talon M (2001) The plant growth promoting rhizobacteria *Bacillus pumilus* and *Bacillus licheniformis* produce high amounts of physiologically active gibberellins. Physiol Plant 111:206–211

Gyaneshwar P, Kumar GN, Parekh LJ, Poole PS (2002) Role of soil microorganisms in improving P nutrition of plants. Plant Soil 245:83–93

Haas D, Défago G (2005) Biological control of soilborne pathogens by fluorescent pseudomonads. Nat Rev Microbiol 3:307–319

Haba E, Pinazo A, Jauregui O, Espuny MJ, Infante MR, Manresa A (2003) Physicochemical characterization and antimicrobial properties of rhamnolipids produced by *Pseudomonas aeruginosa* 47T2 NCBIM 40044. Biotechnol Bioeng 81:316–322

Haiyin H, Silo-Suh LA, Handelsman J, Clardy J (1994) Zwittermicin A, an antifungal and plant protection agent from *Bacillus cereus*. Tetrahedron Lett 35:2499–2502

Hallmann J, Rodriguez-Kabana R, Kloepper JW (1999) Chitin-mediated changes in bacterial communities of the soil, rhizosphere and within roots of cotton in relation to nematode control. Soil Biol Biochem 31:551–560

Hammerschmidt R, Kuc JA (1995) Lignification as a mechanism for induced systemic resistance in cucumber. Physiol Plant Pathol 20:61–71

Handelsman J, Stabb EV (1996) Biocontrol of soil-borne plant pathogens. Plant Cell 8: 1855–1869

Hao Y, Charles TC, Glick BR (2007) ACC deaminase from plant growth promoting bacteria affects crown gall development. Can J Microbiol 53:1291–1299

Hariprasad P, Niranjana SR (2009) Isolation and characterization of phosphate solubilizing rhizobacteria to improve plant health of tomato. Plant Soil 316:13–24

Hariprasad P, Navya HM, Chandranayak S, Niranjana SR (2009) Advantage of using PSIRB over PSRB and IRB to improve plant health in tomato. Biol Control 50:307–316

Hariprasad P, Divakara ST, Niranjana SR (2011) Isolation and characterization of chitinolytic rhizobacteria for the management of *Fusarium* wilt in tomato. Crop Prot 30:1606–1612

Hariprasad P, Chandrashekar S, Brijesh Singh S, Niranjana SR (2013) Mechanism of plant growth promotion and disease suppression by *Pseudomonas aeruginosa* strain 2apa. J Basic Microbiol (DOI: 10.1002/jobm.201200491).

Hartmann A, Singh M, Klingmuller W (1983) Isolation and characterization of *Azospirillum* mutants excreting high amounts of indoleacetic acid. Can J Microbiol 29:916–923

Hasegawa S, Poling SM, Maier VP, Bennett RD (1984) Metabolism of abscisic-acid bacterial conversion to dehydrovomifoliol and vomifoliol dehydrogenase-activity. Phytochemistry 23:2769–2771

Hashizume H, Igarashi M, Sawa R, Adachi H, Nishimura Y, Akamatsu Y (2008) A new type of tripropeptin with anteiso-branched chain fatty acid from *Lysobacter* sp. BMK333-48F3. J Antibiot 61:577–582

Hassan HM, Fridovich I (1980) Mechanism of the antibiotic action pyocyanine. J Bacteriol 141:156–163

Hassett DJ, Woodruff WA, Wozniak DJ, Vasil ML, Cohen MS, Ohman DE (1993) Cloning of *sodA* and *sodB* genes encoding manganese and iron superoxide dismutase in *Pseudomonas aeruginosa*: demonstration of increased manganese superoxide dismutase activity in alginate-producing bacteria. J Bacteriol 175:7658–7665

Henri F, Laurette NN, Annertte D, John Q, Wolfgang M, Francois-Xavier E, Dieudonné N (2008) Solubilization of inorganic phosphates and plant growth promotion by strains of *Pseudomonas fluorescens* isolated from acidic soils of cameron. Afr J Microbiol Res 2:171–178

Hilda R, Fraga R (1999) Phosphate solubilizing bacteria and their role in plant growth promotion. Biotechnol Adv 17:319–339

Hiltner L (1904) Uber neuere erfahrungen und problem auf dem gebeit der bodenbakteriologie und unter besonderer berucksichtigung der grund ungung und brache. Arb Dtsc Landwirt Ges 98:59–78

Hinsa SM, Espinosa-Urgel M, Ramos JL, O'Toole GA (2003) Transition from reversible to irreversible attachment during biofilm formation by Pseudomonas fluorescens WCS365 requires an ABC transporter and a large secreted protein. Mol Microbiol 49:905–918

Hoffman T, Schmidt JS, Zheng X, Bent AF (1999) Isolation of ethylene-insensitive soybean mutants that are altered in pathogen susceptibility and gene-for-gene disease resistance. Plant Physiol 119:935–949

Hofte M, Woesyne MV, Verstraete W (1994) Role of siderophores in plant growth promotion and plant protection by fluorescent pseudomonads. In: Manthey JA et al (eds) Biochemistry of metal micronutrients in the rhizosphere. Lewis Publishers, Boca Raton, FL, pp 81–92

Hontzeas N, Zoidakis J, Glick BR, Abu-Omar MM (2004) Expression and characterization of 1-aminocyclopropane-1-carboxylate deaminase from the rhizobacterium *Pseudomonas putida* UW4: a key enzyme in bacterial plant growth promotion. Biochim Biophys Acta 1703:11–19

Horemans S, Koninck KD, Neuray J, Hermans R, Vlassak K (1986) Production of plant growth substances by *Azospirillum* sp. and other rhizosphere bacteria. Symbiosis 2:341–346

Horinouchi S (2002) A microbial hormone, A-factor, as a master switch for morphological differentiation and secondary metabolism in *Streptomyces griseus*. Front Biosci 7:2045–2057

Howell CR, Stipanovic RD (1980) Suppression of *Pythium ultimum*-induced damping-off of cotton seedlings by *Pseudomonas fluorescens* and its antibiotic, pyoluteorin. Phytopathology 70:712–715

Huang C, Chen CY (2004) Gene cloning and biochemical characterization of chitinase CH from *Bacillus cereus*-28-9. Ann Microbiol 54:289–297

Huang C, Wang T, Chung S, Chen C (2005) Identification of an antifungal chitinase from a potential biocontrol agent, *Bacillus cereus* 28-9. J Biochem Mol Biol 38:82–88

Hughes J, Mellows G (1980) Interaction of pseudomonic acid A with *Escherichia coli* B isoleucyl-tRNA synthetase. Biochemistry 191:209–219

Hussain A, Hasnain S (2009) Cytokinin production by some bacteria: Its impact on cell division in cucumber cotyledons. Afr J Microbiol Res 3:704–712

Iavicoli A, Boutet E, Buchala A, Métraux JP (2003) Induced systemic resistance in *Arabidopsis thaliana* in response to root inoculation with *Pseudomonas fluorescens* CHA0. Mol Plant Microbe Interact 16:851–858

Idris EES, Iglesias DJ, Talon M, Borriss R (2007) Tryptophan-dependent production of Indole-3-Acetic Acid (IAA) affects level of plant growth promotion by *Bacillus amyloliquefaciens* FZB42. Mol Plant Microbe Interact 20:619–626

Illmer P, Schinner F (1995) Solubilization of inorganic calcium phosphates-solubilization mechanisms. Soil Biol Biochem 27:257–263

Ilyas N, Bano A (2010) *Azospirillum* strains isolated from roots and rhizosphere soil of wheat (*Triticum aestivum* L.) grown under different soil moisture conditions. Biol Fert Soil 46:393–406

Inbar O, Ron EZ (1993) Induction of cadmium tolerance in *Escherichia coli* K-12. FEMS Microbiol Lett 113:197–200

Islam Md T, Deora A, Hashidoko Y, Rahman A, Ito T, Tahara S (2007) Isolation and identification of potential phosphate solubilizing bacteria from the rhizosphere of *Oryza sativa* L. cv. BR29 of Bangladesh. Z Natureforsch 62:103–110

Itoh S, Honda H, Tomita F, Suzuki T (1971) Rhamnolipids produced by *Pseudomonas aeruginosa* grown on n-paraffin (mixture of C12, C13 and C14 fractions). J Antibiot 24:855–859

Jacobson CB, Pasternak JJ, Glick BR (1994) Partial purification and characterization of 1-aminocyclopropane-1-carboxylate deaminase from the plant growth promoting rhizobacterium *Pseudomonas putida* GR12-2. Can J Microbiol 40:1019–1025

Jagadeesh KS, Kulkarni JH, Krishnaraj PU (2001) Evaluation of the role of fluorescent siderophore in the biological control of bacterial wilt in tomato using Tn5 mutants of fluorescent *Pseudomonas* sp. Curr Sci 81:882–883

Jaizme-Vega MDC, Rodriguez-Romero AS, Guerra MSP (2004) Potential use of rhizobacteria from the *Bacillus* genus to stimulate the plant growth of micropropagated bananas. Fruit 59:83–90

Janiak AM, Milewski S (2001) Mechanism of antifungal action of kanosamine. Med Mycol 39: 401–408

Janzen R, Rood S, Dormar J, McGill W (1992) *Azospirillum brasilense* produces gibberellins in pure culture and chemically-medium and in co-culture on straw. Soil Biol Biochem 24:1061–1064

Jiao Y, Yoshihara T, Ishikuri S, Uchino H, Ichihara A (1996) Structural identification of Cepaciamide A, a novel fungitoxic compound from *Pseudomonas cepacia* D-202. Tetrahedron Lett 37:1039–1042

Jofré E, Fischer S, Rivarola V, Balegno H, Mori G (1998) Saline stress affects the attachment of *Azospirillum brasilense* Cd to maize and wheat roots. Can J Microbiol 44:416–422

Joo GJ, Kim YM, Lee IJ, Song KS, Rhee IK (2004) Growth promotion of red pepper plug seedlings and the production of gibberellins by *Bacillus cereus*, *Bacillus macroides* and *Bacillus pumilus*. Biotechnol Lett 26:487–491

Kamil Z, Rizk M, Saleh M, Moustafa S (2007) Isolation and identification of rhizosphere bacteria and their potential in antifungal biocontrol. Global J Mol Sci 2:57–66

Kampert M, Strzelczyk E, Pokojska A (1975) Production of auxins by bacteria isolated from the roots of pine seedlings (*Pinus silvestris* L.) and from soil. Acta Microbiol Pol 7:135–143

Kaper JM, Veldstra H (1958) On the metabolism of tryptophan by *Agrobacterium tumefaciens*. Biochim Biophys Acta 30:401–420

Kate M, Bryant L (2007) Topical antibiotics—more harm than good. Focus. p.1. Archived from the original on 2007-07-16

Kaymak HC, Yarali F, Guvenc I, Donmez FM (2008) The effect of inoculation with plant growth rhizobacteria (PGPR) on root formation of mint (*Mentha piperita* L.) cuttings. Afr J Biotechnol 24:4479–4483

Kearns DB, Chu F, Rudner R, Losick R (2004) Genes governing swarming in *Bacillus subtilis* and evidence for a phase variation mechanism controlling surface motility. Mol Microbiol 52: 357–369

Keel C, Schiner U, Maurhofer M, Voisard C, Laville J, Burger U, Wirthner P, Haas D, Defago G (1992) Suppression of root diseases by *Pseudomonas fluorescens* CHA0: importance of the bacterial secondary metabolite 2,4-diacetylphloroglucinol. Mol Plant Microbe Interact 5:4–13

Kennedy AC (1999) Bacterial diversity in agroecosystems. Agric Ecosyst Environ 74:65–76

Kennedy IR, Pereg-Gerk LL, Wood C, Deaker R, Glichrist K, Katupitiya S (1997) Biological nitrogen fixation in non-leguminous field corps: facilitating the evolution of an effective association between *Azospirillum* and Wheat. Plant Soil 194:65–79

Khan MS, Zaidi A, Wani PA (2006) Role of phosphate-solubilizing microorganisms in sustainable agriculture—a review. Agron Sustain Dev 26:1–15

Khan MS, Zaidi A, Wani PA (2007) Role of phosphate-solubilizing microorganisms in sustainable agriculture-a review. Agron Sustain Dev 27:29–43

Khiyami M, Masmali I (2008) Characterization of thermostable chitinase enzyme of *Bacillus licheniformis* isolated from red palm weevil gut. Aust J Basic Appl Sci 2:943–948

Kim KY, Mc Donald GA, Jordan D (1997) Solubilization of hydroxypatite by *Enterobacter agglomerans* and cloned *Escherichia coli* in culture medium. Biol Fertil Soils 24:347–352

Kim BS, Lee JY, Hwang BK (2000) *In vivo* control and in vitro antifungal activity of rhamnolipid B, a glycolipid antibiotic, against *Phytophthora capsici* and *Colletotrichum orbiculare*. Pest Manag Sci 56:1029–1035

Kishore GK, Pande S (2007) Chitin-supplemented foliar application of chitinolytic *Bacillus cereus* reduces severity of *Botrytis* gray mold disease in chickpea under controlled conditions. Lett Appl Microbiol 44:98–105

Kishore GK, Pande S, Podile AR (2005) Chitin-supplemented foliar application of *Serratia marcescens* GPS 5 improves control of late leaf spot disease of groundnut by activating defense-related enzymes. J Phytopathol 153:169–173
Kloepper JW, Scroth MN (1978) Plant growth-promoting rhizobacteria on radishes. In: Proceedings of the fourth international Conference on Pl pathogenic bacteria, vol 2. Station de Pathologie Vegetale et Phytobacteriologie, INRA, Angers, France, pp 879–882
Kloepper JW, Hume DJ, Scher FM, Singleton C, Tipping B, Laliberte M, Frauley K, Kutchaw T, Simonson C, Lifshitz R, Zaleska I, Lee L (1988) Plant growth-promoting rhizobacteria on canola (rape seed). Plant Dis 72:42–45
Kloepper JW, Lifshitz R, Zablotowicz RM (1989) Free-living bacterial inocula for enhancing crop productivity. Trends Biotechnol 7:39–43
Kloepper JW, Zablokovisz RM, Tipping EM, Lifshitz R (1991) Plant growth promotion mediated by bacterial rhizosphere colonizers. In: Keister DL, Cregan PB (eds) The rhizosphere and plant growth. Kluwer Academic, The Netherlands, pp 315–326
Knoester M, van Loon LC, van den Heuvel J, Hennig J, Bol JF, Linthorst HJM (1998) Ethylene-insensitive tobacco lacks non host resistance against soil-borne fungi. Proc Natl Acad Sci USA 95:1933–1937
Kobayashi M, Izui H, Nagasawa T, Yamada H (1993) Nitrilase in biosynthesis of the plant hormone indole-3-acetic acid from indole-3-acetonitrile—cloning of the Alcaligenes gene and site-directed mutagenesis of cysteine residues. Proc Natl Acad Sci 90:247–251
Koch B, Nielsen TH, Sorensen D, Andersen JB, Christophersen C, Molin S, Givskov M, Sorensen J, Nybroe O (2002) Lipopeptide production in *Pseudomonas* sp. strain DSS73 is regulated by components of sugar beet exudates via the Gac two component regulatory system. Appl Environ Microbiol 68:4509–4516
Koda N, Asaeda T, Yamade K, Kawahara H, Obata H (2001) A novel cryoprotective protein (CRP) with high activity from the ice nucleating bacterium, *Pantoea agglomerans* IFO12686. Biosci Biotechnol Biochem 65:888–894
Kohler J, Hernandez JA, Caravaca F, Roldan A (2008) Plant growth promoting rhizobacteria and arbuscular mycorrhizal fungi modify alleviation biochemical mechanisms in water-stressed plants. Funct Plant Biol 35:141–151
Kokalis-Burelle N, Backman PA, Rodriguez-kabana R, Ploper LD (1992) Potential for biological control of early leaf spot of peanut using *Bacillus cereus* and chitin as foliar amendments. Biol Control 2:321–328
Kpomblekou K, Tabatabai MA (1994) Effect of organic acids on release of phosphorous from phosphate rocks. Soil Sci 158:442–453
Kravchenko LV, Azarova TS, Makarova NM, Tikhonovich IA (2004) The effect of tryptophan of plant root metabolites on the phytostimulating activity of rhizobacteria. Mikrobiologia 73:195–198
Kumar RS, Ayyadurai N, Pandiaraja P, Reddy AV, Venkateswarlu Y, Prakash O, Shakthivel N (2005) Characterization of antifungal metabolite produced by a new strain of *Pseudomonas aeruginosa* PUPa3 that exhibits broad-spectrum antifungal activity and biofertilizing traits. J Appl Microbiol 98:145–154
Kumar SPP, Hariprasad P, Brijesh Singh S, Gowtham HG, Niranjana SR (2014) Structural and functional diversity of rhizobacteria associated with *Rauwolfia* spp. across the Western Ghat regions of Karnataka, India. World J Microbiol Biotechnol 30:163–173
Kunc F, Macura J (1988) Mechanism of adaptation and selection of microorganisms in the soil. In: Vancura V, Kunc F (eds) Soil microbial associations. Elsevier, Amsterdam, pp 281–299
Kusukawa N, Yura T (1988) Heat-shock protein GroE of *Escherichia coli*: key protective roles against thermal stress. Genes Dev 2:874–882
Lambert B, Leyns F, Van Rooyen L, Gossele F, Papon Y, Swings J (1987) Rhizobacteria of maize and their antifungal activities. Appl Environ Microbiol 53:1866–1871
Lankford CE (1973) Bacterial assimilation of iron. Crit Rev Microbiol 2:273–331

Lawton KA, Friedrich L, Hunt M, Weymann K, Delaney T, Kessmann H, Staub T, Ryals J (1996) Benzothiadiazole induces disease resistance in *Arabidopsis* by activation of the systemic acquired resistance signal transduction pathway. Plant J 10:71–82

Lee B, Farag MA, Park HB, Kloepper JW, Lee SH, Ryu CM (2012) Induced resistance by a long-chain bacterial volatile: elicitation of plant systemic defense by a C13 volatile produced by *Paenibacillus polymyxa*. PLoS One 7:e48744

Leelasuphakul W, Sivanunsakul P, Phongpaichit S (2006) Purification, characterization and synergistic activity of -1,3-glucanase and antibiotic extract from an antagonistic Bacillus subtilis. Enzyme Microb Technol 38:990–997

Leeman M, Van Pelt JA, Den Ouden FM, Heinsbroek M, Bakker PAHM, Schippers BS (1995) Induction of systemic resistance by Pseudomonas fuorescens in radish cultivars differing in susceptibility to Fusarium wilt, using a novel bioassay. Eur J Plant Pathol 101:655–664

Leong J, Bitter W, Koster M, Marugg JD, Weisbeek PJ (1991) Genetics of iron transport in plant growth-promoting *Pseudomonas putida* WCS358. In: Keister DL, Cregan PB (eds) The rhizosphere and plant growth. Kluwer Academic, Dordrecht, The Netherlands, pp 271–278

Li J, Ovakim D, Charles TC, Glick BR (2000) An ACC deaminase minus mutant of *Enterobacter cloacae* UW4 no longer promotes root elongation. Curr Microbiol 41:101–105

Liu L, Kloepper JW, Tuzun S (1995) Induction of systemic resistance in cucumber against fusarium wilt by plant growth promoting rhizobacteria. Phytopathology 85:695–698

López CS, Heras H, Garda H, Ruzal S, Sánchez-Rivas C, Rivas E (2000) Biochemical and biophysical studies of *Bacillus subtilis* envelopes under hyperosmotic stress. Int J Food Microbiol 55:137–142

Lou H, Chen M, Black SS, Bushell SR, Ceccarelli M (2011) Altered antibiotic transport in ompc mutants isolated from a series of clinical strains of multi-drug resistant *E. coli*. PLoS One 6(10):e25825

Lucy M, Reed E, Glick BR (2004) Applications of free living plant growth-promoting rhizobacteria. Antonie Van Leeuwenhoek 86:1–25

Lugtenberg BJJ, Dekkers LC (1999) What makes *Pseudomonas* bacteria rhizosphere competent? Environ Microbiol 1:9–13

Lugtenberg BJJ, van Der Bij A, Blemberg G, Woeng TCA, Dekkers K, Kravchenko L, Mulders I, Phoelich C, Simons M, Spink H, Tikhonovich I, De Weger L, Wijffelman C (1996) Molecular basis of rhizosphere colonization by *Pseudomonas* bacteria. In: Stacey G, Mullin B, Gresshoff PM (eds) Biology of plant-microbe interactions. Proceedings of eighth international symposium on mol plant-microbe interaction, Knoxville, TN, pp 433–440

Lugtenberg BJJ, Dekkers L, Bloemberg GV (2001) Molecular determinants of rhizosphere colonization by bacteria. Annu Rev Phytopathol 39:461–490

Lyon GD, Newton AC (1997) Do resistance elicitors offer new opportunities in integrated disease control strategies? Plant Pathol 46:636–641

M'Piga P, Belanger RR, Paulitz TC, Benhamou N (1997) Increased resistance to *Fusarium oxysporum* f. sp. *radicis lycopersici* in tomato plants treated with endophytic bacterium *Pseudomonas fluorescens* strain 63–28. Physiol Mol Plant Pathol 50:301–320

Ma JH, Yao JL, Cohen D, Mom's B (1998) Ethylene inhibitors enhance *in vitro* formation from apple shoot cultures. Plant Cell Rep 17:211–214

Ma W, Zalec K, Glick BR (2001) Biological activity and colonization pattern of the bioluminescence-labeled plant growth-promoting bacterium *Kluyvera ascorbata* SUD165/26. FEMS Microbiol Ecol 35:137–144

Madigan MT, Oren A (1999) Thermophilic and halophilic extremophiles. Curr Opin Microbiol 2:265–269

Mahajan S, Tuteja N (2005) Cold, salinity and drought stresses: an overview. Arch Biochem Biophys 444:139–158

Maksimov IV, Abizgil'dina RR, Pusenkova LI (2011) Plant growth promoting rhizobacteria as alternative to chemical crop protectors from pathogens (Review). Appl Biochem Microbiol 47:333–345

Maloney PE, van Bruggen AHC, Hu S (1997) Bacterial community structure in relation to the carbon environments in lettuce and tomato rhizospheres and in bulk soil. Microbial Ecol 34:109–117

Mandimba G, Heulin T, Bally R, Guckert A, Balandreau J (1986) Chemotaxis of free-living nitrogen-fixing bacteria towards maize mucilage. Plant Soil 90:129–139

Manjula K, Podile AR (2001) Chitin-supplemented formulations improve biocontrol and plant growth promoting efficiency of *Bacillus subtilis* AF 1. Can J Microbiol 47:618–625

Mantelin S, Touraine B (2004) Plant growth-promoting bacteria and nitrate availability impacts on root development and nitrate uptake. J Exp Bot 55:27–34

Marasco EK, Schmidt-Dannert C (2008) Identification of bacterial carotenoid cleavage dioxygenase homologues that cleave the interphenyl α, β double bond of stilbene derivatives via a monooxygenase reaction. Chem Biol Chem 9:1450–1461

Marschner P, Timonen S (2006) Bacterial community composition and activity in rhizosphere of roots colonized by arbuscular mycorrhizal fungi. In: Mukerji KG, Manoharachari C, Singh J (eds) Microbial activity in the rhizosphere. Soil Biology 7:139-154. Springer, Berlin

Martens DA, Frankenberger WT (1994) Assimilation of exogenous 2'-^{14}C-indole-3-acetic acid and 3'-^{14}C-tryptophan exposed to the roots of three wheat varieties. Plant Soil 166:281–290

Masalha J, Kosegarten H, Elmaci O, Mengel K (2000) The central role of microbial activity for iron acquisition in maize and sunflower. Biol Fert Soils 30:433–439

Mauch F, Mauch-Mani B, Boller T (1988) Antifungal hydrolasses in pea tissue: II. Inhibition of fungal growth by combinations of chitinases and two β-1,3-glucanases. Plant Physiol 87: 936–942

Maurhofer M, Keel C, Schnider U, Voisard C, Haas D, Defago G (1992) Influence of enhanced antibiotic production in *Pseudomonas fluorescens* strain CHA0 on its disease suppressive capacity. Phytopathology 82:190–195

Maurhofer M, Hase C, Meuwly P, Metraux JP, Defago G (1994) Induction of systemic resistance of tobacco to tobacco necrosis virus by the root-colonizing Pseudomonas fluorescens strain CHAO: influence of the gacA gene and of pyoverdine production. Phytopathology 84:139–146

Mavingui P, Laguerre G, Berge O, Heulin T (1992) Genetic and phenotypic diversity of *Bacillus polymyxa* in soil and wheat rhizosphere. Appl Environ Microbiol 58:1894–1903

Mavrodi DV, Blankenfeldt W, Thomashow LS (2006) Phenazine compounds in fluorescent *Pseudomonas* spp. biosynthesis and regulation. Annu Rev Phytopathol 44:417–445

Mawdsley JL, Burns RG (1994) Root colonization by a *Flavobacterium* species and the influence of percolating water. Soil Biol Biochem 26:861–870

Mayak S, Tirosh T, Glick BR (2004) Plant growth-promoting bacteria confer resistance in tomato plants to salt stress. Plant Physiol Biochem 42:565–572

Meharg AA, Killham K (1995) Loss of exudates from the roots of perennial ryegrass inoculated with a range of microorganisms. Plant Soil 170:345–349

Mercado-Blanco J, van der Drift KMGM, Olsson PE, ThomasOates JE, van Loon LC, Bakker PAHM (2001) Analysis of the pmsCEAB gene cluster involved in biosynthesis of salicylic acid and the siderophore pseudomonine in the biocontrol strain Pseudomonas fluorescens WCS374. J Bacteriol 183:1909–1920

Meyer JM, Geoffroy VA, Baida N, Gardan L, Izard D, Lemanceau P, Achouak W, Palleroni NJ (2002) Siderophore typing, a powerful tool for the identification of fluorescent and nonfluorescent pseudomonads. Appl Environ Microbiol 68:2745–2753

Meziane H, Van der Sluis I, Van Loon LC, Hofte M, Bakker PAHM (2005) Determinants of *Pseudomonas putida* WCS358 involved in inducing systemic resistance in plants. Mol Plant Pathol 6:177–185

Miller CM, Miller RV, Kenny DG, Redgrave B, Sears J, Condron MM, Teplow DB, Strobel GA (1998) Ecomycins, unique antimycotics from *Pseudomonas viridiflava*. J Appl Microbiol 84: 937–944

Mirza MS, Mehnaz S, Normand P, Prigent-Combaret C, Moe¨nne-Loccoz Y, Bally R, Malik KA (2006) Molecular characterization and PCR detection of a nitrogen-fixing *Pseudomonas* strain promoting rice growth. Biol Fertil Soils 43:163–170

Mittler R (2006) Abiotic stress, the field environment and stress combination. Trends Plant Sci 11:15–19

Molina-Favero C, Creus CM, Lanteri ML, Correa-Aragunde N, Lombardo MC, Barassi CA, Lamattina L (2007) Nitric oxide and plant growth promoting rhizobacteria: common features influencing root growth and development. Adv Bot Res 46:1–33

Mordukhova EA, Skvortsova NP, Kochetkov VV, Dubei-kovskii AN, Boronin AM (1991) Synthesis of the phytohormone indole-3-acetic acid by rhizosphere bacteria of the genus *Pseudomonas*. Mikrobiologiya 60:494–500

Morris RO (1995) Genes specifying auxin and cytokinin biosynthesis in prokaryotes. In: Davies PJ (ed) Plant hormones. Kluwer, Dordrecht, The Netherlands, pp 318–339

Mossialos D, Meyer JM, Budzikiewicz H, Wolff U, Koedam N, Baysse C, Anjaiah V, Cornelis P (2000) Quinolobactin, a new siderophore of Pseudomonas fluorescens ATCC 17400, the production of which is repressed by the cognate pyoverdine. Appl Environ Microbiol 66: 487–492

Mulkey TJ, Kuzmanoff KM, Evans ML (1982) Promotion of growth and shift in the auxin dose/response relationship in maize roots treated with the ethylene biosynthesis inhibitors aminoethoxyvinylglycine and cobalt. Plant Sci Lett 25:43948

Munchbach M, Dainese P, Staudenmann W, Narberhaus F, James P (1999) Proteome analysis of heat-shock protein expression in *Bradyrhizobium japonicum*. Eur J Biochem 264:39–48

Murphy JF, Reddy MS, Ryu CM, Kloepper JW, Li R (2003) Rhizobacteria-mediated growth promotion of tomato leads to protection against Cucumber mosaic virus. Phytopathology 93: 1301–1307

Nadeem M, Younis A, Riaz A, Hameed M (2012) Growth response of some cultivars of bermuda grass (*Cyanodon dactylon* L.) to salt stress. Pak J Bot 44:1347–1350

Nagasawa T, Mauger J, Yamada H (1990) A novel nitrilase, arylacetonitrilase, of Alcaligenes faecalis JM3—purification and characterization. Eur J Biochem 194:765–772

Nahas E (1996) Factors determining rock phosphate solubilization by micro-organisms. World J Microbiol Biotechnol 12:567–572

Nakayama T, Homma Y, Hashidoko Y, Mizutani J, Tahara S (1999) Possible role of xanthobaccins produced by *Stenotrophomonas* sp. strain SB-K88 in suppression of sugar beet damping-off disease. Appl Environ Microbiol 55:4334–4339

Neidhardt F, Van Bogelen RA (1987) Heat shock response. In: Neidhardt FC, Ingraham JL, Low KB, Magasanik B, Schaechter M, Umbarger E (eds) *Escherichia coli* and *Salmonella typhimurium*, vol 2. American Society of Microbiology, Washington, DC, pp 1334–1345

Neiland JB (1981) Microbial iron compounds. Annu Rev Biochem 50:715–731

Nielsen TH, Thrane C, Christophersen C, Anthoni U, Sorensen J (2000) Structure, production characteristics and fungal antagonism of tensin—a new antifungal cyclic lipopeptide from *Pseudomonas fluorescens* strain 96.578. J Appl Microbiol 89:992–1001

Nieto KF, Frankenberger WT Jr (1989) Biosynthesis of cytokinin by *Azotobacter chroococcum*. Soil Biol Biochem 21:967–972

Nihorimbere V, Cawoy H, Seyer A, Brunelle A, Thonart P, Ongena M (2012) Impact of rhizosphere factors on cyclic lipopeptide signature from the plant beneficial strain Bacillus amyloliquefaciens S499. FEMS Microbiol Ecol 79:176–191

Niki T, Mitsuhara I, Seo S, Ohtsubo N, Ohashi Y (1998) Antagonistic effects of salicylic acid and jasmonic acid on the expression of pathogenesis-related (PR) protein genes in wounded mature tobacco leaves. Plant Cell Physiol 39:500–507

Niranjana SR, Hariprasad P (2012) *Serratia*—an under estimated potential plant growth promoting and bioprotecting rhizobacteria. Biopestic Int 8:49–62

Nowak-Thompson B, Chaney N, Wing JS, Gould SJ, Loper JE (1999) Characterization of the pyoluteorin biosynthetic gene cluster of *Pseudomonas fluorescens* Pf-5. J Bacteriol 181: 2166–2174

Oberhansli T, Défago G, Haas D (1991) Indole-3-acetic-acid (IAA) synthesis in the biocontrol strain CHA0 of Pseudomonas fluorescens—role of tryptophan side-chain oxidase. J Gen Microbiol 137:2273–2279

Okon Y, Labandera-Gonzalez CA (1994) Agronomic applications of *Azospirillum*. An evaluation of 20 years worldwide field inoculation. Soil Biol Biochem 26:1591–1601
Oldroyd GED (2007) Nodules and hormones. Science 315(5808):52–53
Omar SA (1998) The role of rock phosphate solubilizing fungi and vesicular arbuscular mycorrhizae (VAM) in growth of wheat plants facilitated with rockphosphate. World J Microbiol Biotechnol 14:211–218
Ongena M, Daayf F, Jacques P, Thonart P, Benhamou N, Paulitz TC, Cornelis P, Koedam N, Belanger RR (1999) Protection of cucumber against *Pythium* root rot by fluorescent pseudomonads: predominant role of induced resistance over siderophore and antibiosis. Plant Pathol 48:66–76
Ongena M, Giger A, Jacques P, Dommes J, Thonart P (2002) Study of bacterial determinants involved in the induction of systemic resistance in bean by *Pseudomonas putida* BTP1. Eur J Plant Pathol 108:187–196
Patten CL, Glick BR (1996) Bacterial biosynthesis of indole-3-acetic acid. Can J Microbiol 42:207–220
Patten CL, Glick BR (2002) Role of *Pseudomonas putida* indole acetic acid in development of the host plant root system. Appl Environ Microbiol 68:3795–3801
Paul D, Nair S (2008) Stress adaptations in a plant growth promoting rhizobacterium (PGPR) with increasing salinity in the coastal agricultural soils. J Basic Microbiol 48:378–384
Penninckx IAMA, Eggermont K, Terras FRG, Thomma BPHJ, De Samblanx GW, Buchala A, Métraux J-P, Manners JM, Broekaert WF (1996) Pathogen-induced systemic activation of a plant defensin gene in *Arabidopsis* follows a salicylic acid–independent pathway. Plant Cell 8:2309–2323
Perley JW, Stowe BB (1966) On the ability of *Taphrina deformans* to produce indoleacetic acid from tryptophan by way of tryptamine. Plant Physiol 41:234–237
Pierson EA, Wood DW, Cannon JA, Blachere FM, Pierson LS III (1998) Interpopulation signaling via *N*-acyl homoserine lactones among bacteria in the wheat rhizosphere. Mol Plant Microbe Interact 11:1078–1084
Pieterse CMJ, van Loon LC (1999) Salicylic acid-independent plant defence pathways. Trends Plant Sci 4:52–58
Pieterse CMJ, Van Wees SCM, Hoffland E, Van Pelt JA, Van Loon LC (1996) Systemic resistance in *Arabidopsis* induced by biocontrol bacteria is independent of salicylic acid accumulation and pathogenesis-related gene expression. Plant Cell 8:1225–1237
Pieterse CMJ, Van Wees SCM, Van Pelt JA, Knoester M, Laan R, Gerrits H, Weisbeek PJ, Van Loon LC (1998) A novel signalling pathway controlling induced systemic resistance in Arabidobsis. Plant Cell 10:1571–1580
Pieterse CMJ, Ton J, Van Loon LC (2001) Cross-talk between plant defence signalling pathways: boost or burden? AgBiotechNet 3:ABN 068
Pieterse CMJ, Van Wees SCM, Ton J, Van Pelt JA, Van Loon LC (2002) Signaling in rhizobacteria-induced systemic resistance in *Arabidopsis thaliana*. Plant Biol 4:535–544
Piuri M, Sanchez-Rivas C, Ruzal SM (2005) Cell wall modifications during osmotic stress in Lactobacillus casei. J Appl Microbiol 98:84–95
Press CM, Wilson M, Tuzun S, Kloepper JW (1997) Salicylic acid produced by *Serratia marcescens* 90-166 is not the primary determinant of induced systemic resistance in cucumber or tobacco. Mol Plant Microbe Interact 6:761–768
Press CM, Loper JE, Kloepper JW (2001) Role of iron in rhizobacteria mediated induced systemic resistance of cucumber. Phytopathology 91:593–598
Prinsen E, Costacurta A, Michiels K, Vanderleyden J, Van Onckelen H (1993) *Azospirillum brasilense* indole-3-acetic acid biosynthesis: evidence for a non-tryptophan dependent pathway. Mol Plant Microb Interact 6:609–615
Qi F, Merritt J, Lux R, Shi W (2004) Inactivation of the *ciaH*gene in *Streptococcus mutans* diminishes mutacin production and competence development, alters sucrose-dependent biofilm formation, and reduces stress tolerance. Infect Immun 72:4895–4899

Raaijmakers JM, van der Sluis I, Koster M, Bakker PAHM, Welsbeek PJ, Schippers B (1995) Utilization of heterologous siderophores and rhizosphere competence of fluorescent *Pseudomonas* spp. Can J Microbiol 41:126–135

Raaijmakers JM, Weller DM, Thomashow L (1997) Frequency of antibiotics producing spp. in natural environments. Appl Environ Microbiol 63:881–887

Raaijmakers JM, Vlami M, de Souza JT (2002) Antibiotic production by bacterial biocontrol agents. Antonie Van Leeuwenhoek 81:537–547

Radjacommare R, Kandan A, Nandakumar R, Samiyappan R (2004) Association of the hydrolytic enzyme chitinase against *Rhizoctonia solani* in rhizobacteria-treated rice plants. J Phytopathol 152:365–370

Raj NS, Shett NP, Shetty HS (2004) Seed bio-priming with *Pseudomonas fluorescens* isolates enhances growth of pearl millet plants and induces resistance against downy mildew. Int J Pest Manage 50:41–48

Raj NS, Lavanya SN, Amruthesh KN, Niranjana SR, Reddy MS, Shetty HS (2012) Histo-chemical changes induced by PGPR during induction of resistance in pearl millet against downy mildew disease. Biol Control 60:90–102

Ramos-Solano B, García JAL, Garcia-Villaraco A, Algar E, Garcia-Cristobal J, Mañero FJG (2010) Siderophore and chitinase producing isolates from the rhizosphere of *Nicotiana glauca* Graham enhance growth and induce systemic resistance in *Solanum lycopersicum* L. Plant Soil 334:189–197

Ran LX, Li ZN, Wu GJ, Van Loon LC, Bakker PAHM (2005) Induction of systemic resistance against bacterial wilt in *Eucalyptus urophylla* by fluorescent *Pseudomonas* spp. Eur J Plant Pathol 113:59–70

Rana A, Saharan B, Joshi M, Prasanna R, Kumar K, Nain L (2011) Identification of multi-trait PGPR isolates and evaluating their potential as inoculants for wheat. Ann Microbiol 61: 893–900

Rane MR, Sarode PD, Chaudhar BL, Chincholkar SB (2007) Detection isolation and identification of phenzine-1-carboxylic acid produced by biocontrol strains of *Pseudomonas aeruginosa.* J Sci Ind Res 66:627–631

Reed MLE, Glick BR (2005) Growth of canola (*Brassica napus*) in the presence of plant growth-promoting bacteria and either copper or polycyclic aromatic hydrocarbons. Can J Microbiol 51:1061–1069

Reinhold-Hurek B, Hurek T, Gillis M, Hoste B, Vancanneyt M, Kersters K, De Ley J (1993) *Azoarcus* gen. nov., nitrogen-fixing proteobacteria associated with roots of Kallar grass (*Leptochloa fusca* (L.) Kunth) and description of two species *Azoarcus indigens* sp. nov. and *Azoarcus communis* sp. nov. Int J Syst Bacteriol 43:574–584

Reyes I, Valery A, Valduz Z (2006) Phosphate-solubilizing microorganisms isolated from rhizospheric and bulk soils of colonizer plants at an abandoned rock phosphate mine. Plant Soil 287:69–75

Richardson AE (2001) Prospects for using soil microorganism to improve the acquisition of phosphate by plant. Aus J Plant Physiol 28:897–906

Riggs PJ, Chelius MK, Iniguez AL, Kaeppler SM, Triplett EW (2001) Enhanced maize productivity by inoculation with diazotrophic bacteria. Aus J Plant Physiol 28:829–836

Riley MA, Wertz JE (2002) Bacteriocins: evolution, ecology and application. Annu Rev Microbiol 56:117–137

Rodriguez JA, Nanos N, Grau JM, Gil L, Lopez-Arias M (2008) Multiscale analysis of heavy metal contents in Spanish agricultural topsoils. Chemosphere 70:1085–1096

Roessler M, Müller V (2001) Osmoadaptation in Bacteria and Archaea: common principles and differences. Environ Microbiol 3:743–754

Ross AF (1961a) Localized acquired resistance to plant vinis infection in hypersensitive hosts. Virology 14:329–339

Ross AF (1961b) Systemic acquired resistance induced by localized virus infections in plants. Virology 14:340–358

Rovira AD (1969) Plant root exudates. Bot Rev 35:35–57
Ryan PR, Delhaize E, Jones DL (2001) Function and mechanism of organic anion exudation from plant roots. Annu Rev Plant Physiol Plant Mol Biol 52:527–560
Ryu C-M, Farag MA, Hu C-H, Reddy MS, Wei HX, Paré PW, Kloepper JW (2003) Bacterial volatiles promote growth in *Arabidopsis*. Proc Natl Acad Sci U S A 100:4927–4932
Ryu CM, Farag MA, Hu CH, Reddy MS, Kloepper JW, Pare PW (2004) Bacterial volatiles induce systemic resistance in *Arabidopsis*. Plant Physiol 134:1017–1026
Salisbury FB (1994) The role of plant hormones. In: Wilkinson RE (ed) Plant-environment interactions. Marcel Dekker, New York, pp 39–81
Salisbury FB, Ross CW (1992) Plant physiology. Wadsworth, Belmont, CA, pp 127–138
Sandhya V, Ali SZ, Grover M, Reddy G, Venkateswarlu B (2009) Alleviation of drought stresseffects in sunflower seedlings by the exopolysaccharides producing Pseudomonas putida strain GAP-P45. Biol Fertil Soils 46:17–26
Saravanakumar D, Samiyappan R (2007) ACC deaminase from Pseudomonas fluorescens mediated saline resistance in groundnut (Arachis hypogaea) plants. J Appl Microbiol 102: 1283–1292
Sattely ES, Walsh CT (2008) A latent oxazoline electrophile for N-O-C bond formation in pseudomonine biosynthesis. J Am Chem Soc 130:12282–12284
Sayyed RZ, Chincholkar SB (2009) Siderophore-producing *Alcaligenes feacalis* exhibited more biocontrol potential Vis-à-Vis chemical fungicide. Curr Microbiol 58:47–51
Scher FM, Baker R (1982) Effect of *Pseudomonas putida* and a synthetic iron chelator on induction of soil suppressiveness to *Fusarium* wilt pathogens. Phytopathology 72:1567–1573
Scher FM, Kloepper JW, Singleton CA (1985) Chemotaxis of fluorescent *Pseudomonas* spp. to soybean seed exudates *in vitro* and in soil. Can J Microbiol 31:570–574
Schultze M, Kondorosi A (1998) Regulation of symbiotic root nodule development. Annu Rev Genet 32:33–57
Senthil Kumar R, Parthibhan KT, Govinda Rao M (2009) Molecular characterization of *Jatropha* genetic resources through inter-simple sequence repeat (ISSR) markers. Mol Biol Rep 36:1951–1956
Sessitsch A, Howieson JG, Perret X, Antoun H, Martinez-Romero E (2002) Adavances in *Rhizobium* research. Crit Rev Plant Sci 21:323–378
Shah J, Tsui F, Klessig DF (1997) Characterization of a salicylic acid–insensitive mutant (*sai1*) of *Arabidopsis thaliana*, identified in a selective screen utilizing the SA-inducible expression of the tms2 gene. Mol Plant Microbe Interact 10:69–78
Siddiqui IA, Shaukat SS (2003) Suppression of root-knot disease by *Pseudomonas fluorescens* CHA0 in tomato: importance of bacterial secondary metabolite, 2,4-diacetylphloroglucinol. Soil Biol Biochem 35:1615–1623
Silo-Suh LA, Lethbridge BJ, Raffel SJ, He H, Clardy J, Handelsman J (1994) Biological activities of two fungistatic antibiotics produced by *Bacillus cereus* UW85. Appl Environ Microbiol 60:2023–2030
Silo-suh LA, Stab VE, Raffel SR, Handelsman J (1998) Target range of Zwittermicin A, an Aminopolyol antibiotic from *Bacillus cereus*. Curr Microbiol 37:6–11
Singh P, Cameotra SS (2004) Potential applications of microbial surfactants in biomedical sciences. Trends Biotechnol 22:142–146
Siwayaprahm P, Audtho M, Ohmiya K, Wiwat C (2006) Purification and characterization of a *Bacillus circulans* No. 41 chitinase expressed in *Escherichia coil*. World J Microbial Biotechnol 22:331–335
Skvortsov IM, Ignatov VV (1998) Extracellular polysaccharides and polysaccharide-containing biopolymers from *Azospirillum* species: properties and the possible role in interaction with plant roots. FEMS Microbiol Lett 165:223–229
Sleator RD, Hill C (2002) Bacterial osmoadaptation: the role of osmolytes in bacterial stress and virulence. FEMS Microbiol Rev 26:49–71

Sorensen J (1997) The rhizosphere as a habitat for soil microorganisms. In: van Elsa JD, Trevors JT, Welington EMH (eds) Modern soil ecology. Marcel Dekker, New York, pp 21–46

Sotirova AV, Spasova DI, Galabova DN, Karpenko E, Shulga A (2008) Rhamnolipid–biosurfactant permeabilizing effects on gram-positive and gram-negative bacterial strains. Curr Microbiol 56:639–644

Spaepen S, Dobbelaere S, Croonenborghs A, Vanderleyden J (2008) Effects of *Azospirillum brasilense* indole-3-acetic acid production on inoculated wheat plants. Plant Soil 312:15–23

Srivastava LM (2002) Plant growth and development: hormones and environment. Academic, San Diego

Stabb EV, Handelsman J (1998) Genetic analysis of zwittermicin A resistance in *Escherichia coli*: effects on membrane potential and RNA polymerase. Mol Microbiol 27:311–322

Staunton S, Leprince F (1996) Effect of ph and some organic anions on the solubility of soil phosphate: implications for P bioavailability. Eur J Soil Sci 47:231–239

Stevens PJG, Walker JTS, Shaw PW, Suckling DM (1994) Organosilicone surfactants: tools for horticultural crop protection. BCPC Pests Dis 2:755–760

Stitcher L, Mauch-Mani B, Metraux JP (1997) Systemic acquired resistance. Annu Rev Phytopathol 35:235–270

Strzelczyk E, Pokojska-Burdziej A (1984) Production of auxins and gibberellin like substances by mycorrhizal fungi, bacteria and actinomycetes isolated from soil and mycorhizosphere of pine (*Pinus silvestris* L.). Plant Soil 81:185–194

Subba Rao NS (1982) Advances in agricultural microbiology. Oxford and IBH Publishing, Delhi

Tahmatsidou V, O'Sullivan J, Cassells AC, Voyiatzis D, Paroussi G (2006) Comparison of AMF and PGPR inoculants for the suppression of *Verticillium* wilt of strawberry (*Fragaria ananassa* cv. Selva). Appl Soil Ecol 32:316–324

Taize L, Zeiger E (2006) Plant physiology, 4th edn. Sinauer Association, Sunderland, MA, http://4e.plantphys.net

Takeuchi Y, Yoshikawa M, Takeba G, Tanaka K, Shibata D, Horino O (1990) Molecular cloning and ethylene induction of mRNA encoding a phytoalexin elicitor-releasing factor, β-1,3-glucanase, in soybean. Plant Physiol 93:673–682

Taller BJ, Wong TY (1989) Cytokinins in *Azobacter vinelandii* culture medium. Appl Environ Microbiol 55:266–267

Tanaka H, Watanabe T (1995) Glucanases and chitinases of *Bacillus circulans* WL-12. J Ind Microbiol 14:478–483

Tans-Kersten J, Huang H, Allen C (2001) *Ralstonia solanacearum* needs motility for invasive virulence on tomato. J Bacteriol 183:3597–3605

Teplitski M, Robinson JB, Bauer WD (2000) Plants secrete substances that mimic bacterial N-acyl homoserine lactone signal activities and affect population density dependent behaviors in associated bacteria. Mol Plant Microbe Int 13:637–648

Theunis M, Kobayashi H, Broughton WJ, Prinsen E (2004) Flavonoids, NodD1, NodD2, and nod-box NB15 modulate expression of the y4wEFG locus that is required for indole-3-acetic acid synthesis in *Rhizobium* sp. strain NGR234. Mol Plant Microbe Interact 17:1153–1161

Thomashow LS, Weller DM (1988) Role of a phenazine antibiotic from *Pseudomonas fluorescens* in biological control of *Gaeumannomyces graminis* var. *tritici*. J Bacteriol 170:3499–3508

Thomashow LS, Bonsall RF, Weller DM (1997) Antibiotic production by soil and rhizosphere microbes in situ. In: Hurst CJ, Knudsen GR, McInerney MJ, Stetzenbach LD, Walter MV (eds) Manual of environmental microbiology, 1st edn. American Society of Microbiology, Washington, DC, pp 493–499

Thomma BPHJ, Eggermont K, Penninckx IAMA, Mauch-Mani B, Vogelsang R, Cammue BPA, Broekaert WF (1998) Separate jasmonate-dependent and salicylate-dependent defense-response pathways in *Arabidopsis* are essential for resistance to distinct microbial pathogens. Proc Natl Acad Sci U S A 95:15107–15111

Thrane C, Nielsen TH, Nielsen MN, Olsson S, Sorensen J (2000) Viscosinamide producing *Pseudomonas fluorescens* DR54 exerts biocontrol effect on *Pythium ultimum* in sugar beet rhizosphere. FEMS Microbiol Ecol 33:139–146

Tien TM, Gaskins MH, Hubbell DH (1979) Plant growth substances produced by *Azospririllum brasilense* and their effect on the growth of pearl millet (*Pennisetum americanum* L.). Appl Environ Microbiol 34:1016–1024

Tilak KVBR, Ranganayaki N, Pal KK, De R, Saxena AK, Nautiyal CS, Mittal S, Tripathi AK, Johri BN (2005) Diversity of plant growth and soil health supporting bacteria. Curr Sci 89:136–150

Timmusk S, Grantcharova N, Wagner EGH (2005) Paenibacillus polymyxa invades plant roots and forms bio-films. Appl Environ Microbiol 11:7292–7300

Tisdale SL, Nelson WL, Beaton JD, Havlin JL (1993) Soil fertility and fertilizers, 5th edn. Macmillan, New York, p 634

Toshima BH, Maru K, Saito M, Ichihara A (1999) Study on fungitoxic 3-amino-2-piperidinone-containing lipids: total syntheses of cepaciamides A and B. Teterahedron Lett 40:939–942

Toth IK, Newton JA, Hyman LJ, Lees AK, Daykin M, Ortori C, Williams P, Fray RG (2004) Potato plants genetically modified to produce N-acylhomoserine lactones increase susceptibility to soft rot erwiniae. Mol Plant Microbe Interact 17:880–87161

Tripathi RK, Gottlieb D (1969) Mechanism of action of the antifungal antibiotic pyrrolnitrin. J Bacteriol 100:310–318

Tripathi M, Munot HP, Schouche Y, Meyer JM, Goel R (2005) Isolation and functional characterization of siderophore-producing lead and cadmium-resistant *Pseudomonas putida* KNP9. Curr Microbiol 50:233–237

Tuzun S (2001) The relationship between pathogen-induced systemic resistance (ISR) and multigenic (horizontal) resistance in plants. Eur J Plant Pathol 107:85–93

Ugoji EO, Laing MD, Hunter CH (2005) Colonization of *Bacillus* spp. on seeds and in plant rhizoplane. J Environ Biol 26:459–466

Upadhyay SK, Maurya SK, singh DP (2012) Salinity tolerance in free living plant growth promoting rhizobacteria. Indian J Sci Res 3:73–78

Uren NC (2000) Types, amounts and possible functions of compounds released into the rhizosphere by soil grown plants. In: Pinton R, Varani Z, Nanniperi P (eds) The rhizosphere: biochemistry, and organic substances at the soil interface. Marcel Dekker, New York, pp 19–40

Vaidya RJ, Roy S, Macmil S, Gandhi S, Vyas P, Chhatpar HS (2003) Purification and characterization of chitinase from *Alcaligenes xylosoxydans*. Biotehnol Lett 25:715–717

Van Loon LC, Bakker PAHM, Pieterse CMJ (1998) Systemic resistance induced by rhizosphere bacteria. Annu Rev Phytopathol 36:453–483

Van Overbeek LS (2013) Exploration of hitherto-uncultured bacteria from the rhizosphere. In: de Bruijn FJ (ed) Molecular microbial ecology of the rhizosphere, vol 1 & 2. Wiley, Hoboken, NJ. doi:10.1002/9781118297674.ch16

Van Peer R, Niemann GJ, Schippers B (1991) Induced resistance and phytoalexin accumulation in biological control of Fusarium wilt of carnation by *Pseudomonas* sp. strain WCS417r. Phytopathology 81:728–734

Van Wees S, Pieterse CMJ, Trijssenaar A, Van´t Westende YAM, Hartog F, Van Loon LC (1997) Differential induction of systemic resistance in *Arabidobsis* by biocontrol bacteria. Mol Plant Microbe Interact 10:716–724

Van Wees SCM, Luijendijk M, Smoorenburg I, Van Loon LC, Pieterse CMJ (1999) Rhizobacteria-mediated induced systemic resistance (ISR) in *Arabidopsis* is not associated with a direct effect on expression of known defense-related genes but stimulates the expression of the jasmonate-inducible gene *Atvsp* upon challenge. Plant Mol Biol 41:537–549

Van Wees SCM, de Swart EAM, Van Pelt JA, Van Loon LC, Pieterse CMJ (2000) Enhancement of induced disease resistance by simultaneous activation of salicylate and jasmonate dependent defence pathways in *Arabidopsis thaliana*. Proc Natl Acad Sci 97:8711–8716

Vande Broek A, Vanderleyden J (1995) The role of bacterial motility, chemotaxis, and attachment in bacteria-plant interactions. Mol Plant Microbe Interact 8:800–810

Vande Broek A, Lambrecht M, Eggermont K, Vanderleyden J (1999) Auxins upregulate expression of the indole-3-pyruvate decarboxylase gene in *Azospirillum brasilense*. J Bacteriol 181: 1338–1342

Vande Broek A, Gysegom P, Ona O, Hendrickx N, Prinsen E, Van Impe J, Vanderleyden J (2005) Transcriptional analysis of the *Azospirillum brasilense* indole-3-pyruvate decarboxylase gene and identification of a cis-acting sequence involved in auxin responsive expression. Mol Plant Microbe Interact 18:311–323

Vesper SJ (1987) Production of pili (fimbriae) by *Pseudomonas xuorescens* and correlation with attachment to corn roots. Appl Environ Microbiol 53:1397–1403

Vessey JK (2003) Plant growth promoting rhizobacteria as biofertilizers. Plant Soil 255:571–586

Vijayan P, Shockey J, Levesque CA, Cook RJ, Browse J (1998) A role for jasmonate in pathogen defense of *Arabidopsis*. Proc Natl Acad Sci U S A 95:7209–7214

Villegas J, Fortin JA (2002) Phosphorous solubilization and pH changes as a result of the interactions between soil bacteria and arbuscular mycorrhizae fungi on a medium containing no3 as nitrogen source. Can J Bot 80:571–576

Visca P, Imperi F, Lamont IL (2007) Pyoverdine siderophores: from biogenesis to biosignificance. Trends Microbiol 15:22–30

Voisard C, Keel C, Haas D, Defago G (1989) Cyanide production by *Pseudomonas fluorescens* helps suppress black root rot of tobacco under gnotobiotic conditions. Eur Mol Biol Org J 8: 351–358

Walker TS, Bais HP, Grotwewold E, Vivanco JM (2003) Root exudation and rhizosphere biology. Plant Physiol 132:44–51

Walsh C, Pascal RA Jr, Johnston M, Raines R, Dikshit D, Krantz A, Honma M (1981) Mechanistic studies on the pyridoxal phosphate enzyme 1-aminocyclopropane-1-carboxylate deaminase from *Pseudomonas* sp. Biochemistry 20:7509–7519

Walsh UF, Morrissey JP, O'Gara F (2001) *Pseudomonas* for biocontrol of phytopathogens: from functional genomics to commercial exploitation. Curr Opin Biotechnol 12:289–295

Wang S, Chang W (1997) Purification and characterization of two bifunctional chitinase/lysozymes extracellularly produced by *Pseudomonas aeruginosa* K-187 in a shrimp and crab shell powder medium. Appl Environ Microbiol 63:380–386

Wang Y, Brown HN, Crowley DE, Szaniszlo PJ (1993) Evidence for direct utilization of a siderophore, ferrioxamine B, in axenically grown cucumber. Plant Cell Environ 16:579–585

Wang CKE, Glick BR, Defago G (2000) Effect of transferring 1-aminocyclopropane-1-carboxylic acid (ACC) deaminase genes into *Pseudomonas fluorescens* strain CHA0 and its gacA derivative CHA96 on their growth-promoting and disease-suppressive capacities. Can J Microbiol 46:898–907

Watt M, Hugenholtz P, White R, Vinall K (2006) Numbers and locations of native bacteria on field grown wheat roots quantified by fluorescence *in situ* hybridization. Environ Microbiol 71: 1291–1299

Wei G, Kloepper JW, Tuzun S (1991) Induction of systemic resistance of cucumber to *Colletotrichum orbiculare* by select strains of plant growth-promoting rhizobacteria. Phytopathology 81:1508–1512

Weller DM, Landa BB, Mavrodi OV, Schroeder KL, De La Fuente L, Bankhead SB, Molar RA, Bonsall RF, Mavrodi DM, Thomashow LS (2007) Role of 2,4-diacetylphloroglucinol producing fluorescent *Pseudomonas* spp. in plant defence. Plant Biol 9:4–20

Werner T, Motyka V, Strnad M, Schmulling T (2001) Regulation of plant growth by cytokinin. Proc Natl Acad Sci U S A 98:10487–10492

Whipps JM (2001) Microbial interaction and biocontrol in the rhizosphere. J Exp Bot 52: 487–511

Whitehead NA, Barnard AML, Slater H, Simpson NJL, Salmond GPC (2001) Quorum-sensing in Gram-negative bacteria. FEMS Microbiol Rev 25:365–404

Whitelaw MA (2000) Growth promotion of plants inoculated with phosphate solubilizing fungi. Adv Agron 69:99–151

Winkelmann G (ed) (1991) Handbook of microbial iron chelates. CRC Press, Boca Raton, FL

Winkelmann G, van der Helm D, Neilands JB (1987) Iron transport in microbes, plants and animals. VCH Press, Weinheim

Wu S, Fallon RD, Payne MS (1997) Over-production of stereoselective nitrile hydratase from *Pseudomonas putida* 5B in *Escherichia coli*: requirement for a novel downstream protein. Appl Microbiol Biotechnol 48:704–708

Xie H, Pasternak JJ, Glick BR (1996) Isolation and characterization of mutants of the plant growth-promoting rhizobacterium *Pseudomonas putida* GR12–2 that overproduce indoleacetic acid. Curr Microbiol 32:67–71

Yadav BK and Verma A (2012) Phosphate solubilization and mobilization in soil through microorganisms under arid ecosystems. In Ali M (ed) The functioning of ecosystems. ISBN:978-953-51-0573-2, In Tech, Available from:http://www.intechopen.com/books/the-functioning-of-ecosystems/phosphate-solubilization-and-mobilization-in-soil-through-microorganisms-under-arid-ecosystems

Yamada Y (1999) Autoregulatory factors and regulation of antibiotic production in *Streptomyces*. In: England R, Hobbs G, Bainton N, Roberts ML (eds) Microbial signalling and communication. Society for General Microbiology, Cambridge, pp 177–196

Yamada Y, Nihira T (1998) Microbial hormones and microbial chemical ecology. In: Barton DHR, Nakanishi K, Meto-Cohn O (eds) Comprehensive natural products chemistry, vol 8. Elsevier, Oxford, pp 377–413

Yancey PH (2005) Organic osmolytes as compatible, metabolic and counteracting cytoprotectants in high osmolarity and other stresses. J Exp Biol 208:2819–2830

Yang CH, Crowley DE (2000) Rhizosphere microbial community structure in relation to root location and plant iron nutritional status. Appl Environ Microbiol 66:345–351

Yang J, Kloepper JW, Ryu CM (2009) Rhizosphere bacteria help plants tolerate abiotic stress. Trends Plant Sci 14:1–4

Yasmin F, Othman R, Sijam K, Saad MS (2009) Characterization of beneficial properties of plant growth-promoting rhizobacteria isolated from sweet potato rhizosphere. Afr J Microbiol Res 3:815–821

Zahir ZA, Ghani U, Naveed M, Nadeem SM, Arshad M (2009) Comparative effectiveness of *Pseudomonas* and *Serratia* sp. containing ACC-deaminase for improving growth and yield of wheat under salt-stressed conditions. Arch Microbiol 191:415–424

Zdor RE, Anderson AJ (1992) Influence of root colonizing bacteria on the defense responses of bean. Plant Soil 140:99–107

Zhang H, Kim MS, Krishnamachari V, Payton P, Sun Y, Grimson M, Farag MA, Ryu CM, Allen R, Melo IS, Pare PW (2007) Rhizobacterial volatile emissions regulate auxin homeostasis and cell expansion in *Arabidopsis*. Planta 226:839–851

Zhang H, Kim MS, Sun Y, Dowd SE, Shi H, Pare PW (2008) Soil bacteria confer plant salt tolerance by tissue specific regulation of the sodium transporter HKT1. Mol Plant Microbe Interact 21:737–744

Chapter 4
Downy Mildew Disease of Pearl Millet and Its Control

H.S. Prakash, Chandra S. Nayaka, and K. Ramachandra Kini

4.1 Pearl Millet Introduction

Pearl millet (*Pennisetum glaucum*) is an important food and fodder crop of arid and semi-arid tropics of India and Africa. This crop is grown in an area of 27 m ha in the world with yield of 36 m tons. India is the largest producer of this crop in terms of area (9 m ha) and production (9 mt), with a productivity of 780 kg/ha, Rajasthan being the major contributor (51 %) followed by Maharashtra and Gujarat. Single-cross F1 hybrids based on cyto-nuclear male sterility (CMS) contributed significantly to pearl millet production in India. Though pearl millet is considered to be an orphan crop, it promises to be a staple food crop in the years to come due to shortage of water. The crop is nutritionally rich with a good balance of starch, protein and fat. Pearl millet provides 11–12 % of worlds' supply of protein (Yadav et al. 2011). It is also rich in iron, phosphorus, B-complex vitamins and fibre content.

Currently 65 % pearl millet area is under high yielding varieties (Yadav 2012). During last 25 years 115 improved pearl millet cultivars have been released. The productivity of pearl millet has improved 45 %, from 5.83 m tons during 1986–1990 to 8.48 m tons during 2006–2010 in spite of 18 % decline in crop area from 10.7 m ha to 9.1 m ha (http://www.agricoop.nic.in).

Pearl millet is infected by five major pathogens, *Sclerospora graminicola* causing downy mildew, *Claviceps fusiformis* causing ergot, *Moesziomyces penicillariae* causing smut, *Puccinia substriata* causing rust and *Pyricularia grisea* causing blast.

H.S. Prakash, Ph.D. (✉) • C.S. Nayaka, Ph.D. • K.R. Kini, M.Sc., Ph.D.
Department of Studies in Biotechnology, University of Mysore,
Manasagangotri, Mysore 570 006, Karnataka, India
e-mail: hasriprakash@gmail.com

A. Goyal and C. Manoharachary (eds.), *Future Challenges in Crop Protection Against Fungal Pathogens*, Fungal Biology, DOI 10.1007/978-1-4939-1188-2_4,

4.2 Downy Mildew

The downy mildew is a major production constraint in pearl millet cultivation. The disease appeared in epiphytotic proportion with the introduction of hybrids in late 1960s. In early 1970s the disease alone was responsible for almost 70 % yield loss in the hybrid HB3.

4.2.1 Taxonomic Status of Sclerospora graminicola

The pathogen was first described as *Protomyces graminicola* from *Setaria verticillata* by Saccardo in 1876. Schroeter (1879) renamed the pathogen as *Sclerospora graminicola. S. graminicola* is placed in Eusclerospora as it produces spores that germinate indirectly by zoospores in contrast to *Peronosclerospora graminicola* that produces directly germinating spores. The oomycetous fungi including *S. graminicola* are now placed in a separate phylum oomycota under the kingdom Stramenopila, as they are closely related to the golden-brown algae and diatoms (Dick 2001). Though oomycota behave like the true fungi in exhibiting apical growth and similar infection strategies, they are unrelated to the true fungi having plant-like features including cell walls composed of primarily of glucans and cellulose-like polymers, cell membrane composed of plant sterols, coenocytic hyphae, and motile asexual spores.

4.2.2 Symptoms

The pathogen causes systemic infection. The symptoms are seen throughout the growth stages, starting from coleoptile stage. The leaves initially show chlorosis which later on shows whitish downy growth on the lower surface due to the production of sporangia (Figs. 4.1 and 4.2). Early infection leads to seedling death. Delayed infection cause dwarfing of plants and some tillers may escape disease. Severely infected plants do not produce panicles. The floral parts of infected plants may get transformed totally or partially to leafy structures, hence the name ‘green ear’. The host genotype, time of expression and ambient conditions may influence type of symptoms (Singh and King 1988).

4.2.3 Biology of the Pathogen

Sclerospora graminicola is an obligate parasite (Kenneth 1970). The sporangiophores are short, stout, determinate, dichotomously branched, hypophyllous or amphigenous, clavate, non-septate and measure 15–22 × 12–21 μm with one to six

Fig. 4.1 (**a**, **b**) Typical downy mildew symptoms of pearl millet showing chlorosis and stunted growth in field condition

Fig. 4.2 Abaxial side of pearl millet leaf infected with *Sclerospora graminicola*

short pedicels. Sporangiophores emerge through stomata. Sporangia are produced on pedicels located at the tip of sporangiophore branches. Sporangia are hyaline, thin-walled, ellipsoid or broadly elliptic, papillate and 15–22 × 12–21 μm (Jouan and Delassus 1971). The zoospores are produced in sporangium, each sporangium produces four to eight reniform zoospores that measure 9–12 μm on long axis. Each zoospore is biflagellate, the anterior flagellum is of tinsel-type, while the posterior flagellum is of whiplash-type. Zoospores are wall-less, but retain a consistent but flexible shape.

The pathogen produces oospores through sexual reproduction. The mature oospores are light brown to reddish-brown with thick walls and measure 22–35 μm in diam. Presence of retentive oogonial wall fused with the oospore wall is characteristic of *Sclerospora* genus. The oospore wall has three distinct layers: the exosporium, the mesosporium and the endosporium. The oospores germinate directly by hyphal germ tubes. Michelmore et al. (1982) have shown that *S. graminicola* is heterothallic. So far the germination of oospores in vitro has not been demonstrated conclusively. Hence the viability of oospores can be tested only by using a vital staining technique (Shetty et al. 1978).

The mycelium of *S. graminicola* is systemic, coenocytic and highly branched (Ramakrishnan 1963). The intracellular haustoria may be simple or branched, globose or digitate (Weston 1929). The asexual zoospores help in secondary spread of downy mildew (Singh and Williams 1980). The sporulation is supported by the photosynthate accumulated during exposure to sunshine between successive crops of sporulation and relatively low temperatures and high relative humidities occurring in the early hours of the morning (02.00–04.00 h). Optimum sporangial production occurs at 20–25 °C and 95–100 % RH. Zoospores germinate by germ tubes and retain their infectivity for about 4 h at 30 °C (Singh and Gopinath 1985).

4.2.4 Infection, Colonization and Disease Spread

The oospores present in soil are the main source of primary infection. The oospores remain viable for up to 10 years (Borchhardt 1927; Nene and Singh 1976). The oospores infect through coleorrhizas, radicles and lower portions of the coleoptiles of seedling, roots and underground portions of stem bases. The germ tube produces an appressorium at the junction of epidermal cells, directly over the epidermal cells or over stomata.

The secondary spread is through the sporangia/zoospores. The apical meristem is the most vulnerable site of infection for zoospores. The zoospores released per sporangium vary from 1 to 12 (ICRISAT 1987). Zoospores emerge through a pore produced by the release of an operculum. Zoospores swim for 30–60 min, encyst, and then germinate by forming a germ tube. Sometimes zoospores may germinate within the sporangium (Shaw 1981). Zoospore release from the sporangia occurs at a wide temperature range (10–45 °C); liberation is optimum at 30 °C in about 2 h 40 min. Zoospores retain their infectivity for about 4 h at 30 °C and for a longer period at lower temperatures (Singh and Gopinath 1985). Low temperature, high relative humidity and well distributed rainfall are critical factors for infection by *S. graminicola* and for the development and spread of disease in the crop (Jeger et al. 1998; Gupta and Singh 1999). *Sclerospora graminicola* is also seed-borne, externally in the form of oospore contamination and internally as mycelial infection in embryo (Shetty et al. 1980). The seed-borne inoculum could be detected by washing test for oospores and embryo extraction procedure for internal mycelia. The seed transmission has been established under in vitro conditions. This aspect is of relevance in seed health certification, especially in germplasm exchange.

4.2.5 Pathogen Variability

Sclerospora graminicola shows a high degree of variability in its pathogenicity. The pathotypes could be distinguished based on reaction on differential hosts. The first epidemic of downy mildew appeared in pearl millet hybrid HB3 in 1971.

The variability in *S. graminicola* was first reported in 1973 when NHB3 was found susceptible at Gulbarga but resistant at Mysore (Shetty and Ahmad 1981). Later on several pathotypes of *S. graminicola* have been reported from India and Africa (Singh et al. 1993) and even locational differences in virulence have also been established in the International Pearl Millet Downy Mildew Nursery, the West African Downy Mildew Variability Nursery and the West African Downy Mildew Observation Nursery (Singh et al. 1992). It has been shown that some genotypes such as MBH110, NHB3 and 81B (ICMB-1) showed differential downy mildew reactions between locations and pathogen populations (Werder and Ball 1992).

The variability in pearl millet cultivar response to downy mildew is determined by host and pathogen genotypes (Ball 1983). West African isolates of the pathogen were generally more pathogenic than Indian isolates and cultivar ICH 105 could differentiate these two isolates. Substantial differences were also established between two isolates collected from different host cultivars at the same location in Upper Volta. Cultivars 700516 and MBH110 also showed differential responses to isolates. Distinct types of symptom expression were also observed which was found to be characteristic of cultivar genotype, independent of pathogen isolate. Both race specific and race non-specific resistance may coexist in this pathosystem.

Thakur et al. (2004a) identified five major groups of *S. graminicola* based on host differentials. Oospore collections from Africa and India showed differences in virulence under greenhouse tests in the UK (Ball 1983; Ball et al. 1986) and in India (Thakur et al. 1992); Singh and Singh (1987) provided evidence that *S. graminicola* is highly cultivar-specific. Disease monitoring field surveys of pearl millet crops in Maharashtra, India, during 1993–1996 (Thakur et al. 1999) indicated high vulnerability of several popular hybrids whereas open-pollinated cultivars recorded trace or no downy mildew incidence. Virulence and DNA fingerprinting analyses showed that isolate Sg 021 from hybrid MLBH 104 was quite distinct from those collected from other hybrids (Sastry et al. 1995).

Five pathotypes of *S. graminicola* were identified in India (Thakur and Rao 1997; Thakur 1999). The pathotypes were less virulent on some pearl millet genotypes indicating non-pathotype-specific resistance. These genotypes can serve as sources of stable resistance. A new pathotype 'Path-7' was described from Jodhpur, India (Thakur et al. 1998). Variation in single-oospore and single-zoospore isolates for virulence on a set of differential lines has been recorded (Thakur and Shetty 1993; Gwary et al. 2007) reported the occurrence of five pathogen populations in Nigeria and no cultivar was resistant to all the pathotypes. JuZheng et al. (1996) reported the existence of physiological specialization among oospore isolates of *S. graminicola* infecting *Setaria italica* from China.

Currently, 14 host differentials are identified to differentiate the pathotypes of *S. graminicola*. There are currently many pathotypes of *S. graminicola* prevalent in different parts of pearl millet growing regions of India, and new ones with higher virulence levels continue to appear with deployment of new cultivars (Thakur et al. 2004b; Pushpavathi et al. 2006) More than 300 isolates of *S. graminicola* were collected from farmers fields of India, out of which 27 isolates with distinct features of variation were further characterized based on host differentials and molecular

Table 4.1 Host differentials for the identification of *Sclerospora graminicola* pathotype

Pathotypes assigned	HB 3	KaluKombu	MBH 110	7042S	MLBH104	HHB-67
Path-I	√	x	√	√	√	√
Path-II	x	√	x	√	√	x
Path-III	√	X	√	√	x	√
Path-IV	√	X	√	√	√	√
Path-V	x	X	√	√	√	√
Path-VI	√	X	x	√	√	√
ND	√	x	√	√	x	√

ND not determined

markers (Sudisha et al. 2008). The seven pathotypes of *S. graminicola* are differentiated based on the reaction of six host genotypes (Table 4.1).

Sharma et al. (2010) have determined the genotypic diversity among 46 isolates of *S. graminicocla* collected from seven states in India during 1992–2005 through pathotyping and AFLP analysis. These isolates were classified in 21 pathotypes based on reaction on a set of nine pearl millet lines. The average linkage cluster analysis of virulence index clustered the 46 isolates into eight groups. Region-specific and temporal variation was observed. Cluster analysis of AFLP data clustered the test isolates into seven groups. Sharma et al. (2011) have identified Sg 492 from Aligarh and Sg510 isolate from Badaun in Uttar Pradesh as the highly virulent isolates based on the reaction on pearl millet cultivar 7042S in greenhouse conditions.

A genomic library of 'path-6' (ex 7042S) of *S. graminicola* had shown 8 % repetitive DNA content on colony hybridization of total fungal DNA with a genomic library (Sastry et al. 1997). The library also revealed partial methylation of GATC and CCGG sequences in the genome and presence of retrotransposable elements. Sudisha et al. (2008) have screened 27 isolates on host differentials and identified 6 pathotypes and also developed RAPD (Table 4.2) and ISSR (Table 4.3) markers to distinguish the six pathotypes (Figs. 4.3 and 4.4).

The genetic basis of host specificity in *S. graminicola* was studied in a host-pathogen cross-inoculation experiment (Sastry et al. 2001). Two pathotypes, Path-I and Path-5 selected from genetically uniform hybrid NHB 3 and genetically heterogeneous landrace population 700651, respectively, were maintained for 10 asexual generations by serial passage on the seedlings of their respective hosts. Pathogenicity test with Path-I indicated an increase in virulence over its new host 700651, compared with the adapted host, NHB 3. However, it was not true for Path-5 with NHB3. RAPD primers were identified to detect variations in Path-I and Path-5. The DNA fingerprinting profile of the isolates obtained after 10 generations revealed differences within the microsatellite probe (GATA), compared with the initial generation. The change in virulence in Path-I and its adaptation to the new host, 700651, was demonstrated by the change in RAPDs and DNA fingerprinting profile in the two extreme generations.

Table 4.2 Fingerprint patterns generated using RAPD in six pathotypes of *S. graminicola*

Sl. No.	Primers	Total no. of bands amplified	No. of polymorphic bands
1	OPA-05	06	05
2	OPA-07	07	05
3	OPA-11	05	05
4	OPA-14	08	06
5	OPA-15	08	06
6	OPB-02	06	04
7	OPB-03	09	06
8	OPB-06	07	05
9	OPB-09	12	12
10	OPB-12	07	07
11	OPC-07	09	06
12	OPC-10	07	06
13	OPC-12	07	05
14	OPD-09	09	06
15	OPS-02	06	06
16	OPS-09	08	06
17	OPL-18	08	06
18	OPM-04	07	06
19	OPN-20	10	09
20	OPX-09	06	05
	Total	152	112

Table 4.3 Fingerprint patterns generated using ISSR primers in six pathotypes of *Sclerospora graminicola*

Sequence	ISSR primer	Annealing temp (°C)	No. of bands generated	No. of polymorphic bands
(CT)8TC	814	40	15	12
(CT)8 AC	844 A	40	14	11
(CT)8GC	844 B	40	15	15
(CA)6 AC	17898 A	45	15	13
(CA)6GT	17898 B	45	14	12
(CA)6AG	17899 A	45	18	15
(CA)6GC	17899 B	45	19	19
(GA)6GG	HB 08	48	16	12
(GT)6GG	HB 09	48	14	11
(GA)6CC	HB 10	48	17	14
(GT)6CC	HB 11	48	16	13
(CAC)3GC	HB 12	45	18	14
(GAG)3GC	HB 13	45	14	11
(CTC)3GC	HB 14	45	13	13
(GTG)3GC	HB 15	48	18	13
(GA)9 T	ISSR 16	48	14	11
(GA)9C	ISSR 17	48	16	12
(TAG)4	ISSR 18	40–55	00	00
(GACA)4	ISSR 19	48	14	10
(GGAT)4	ISSR 20	50	17	13
			297	244

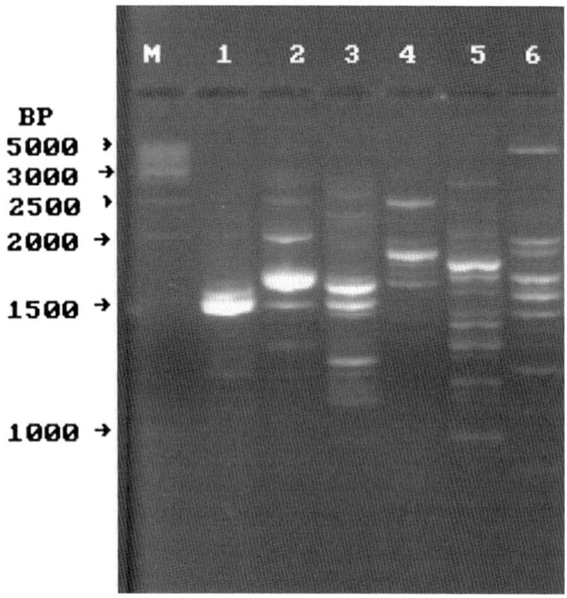

Fig. 4.3 Agarose gel electrophoresis shown DNA fingerprint of six pathotypes of *Sclerospora graminicola*

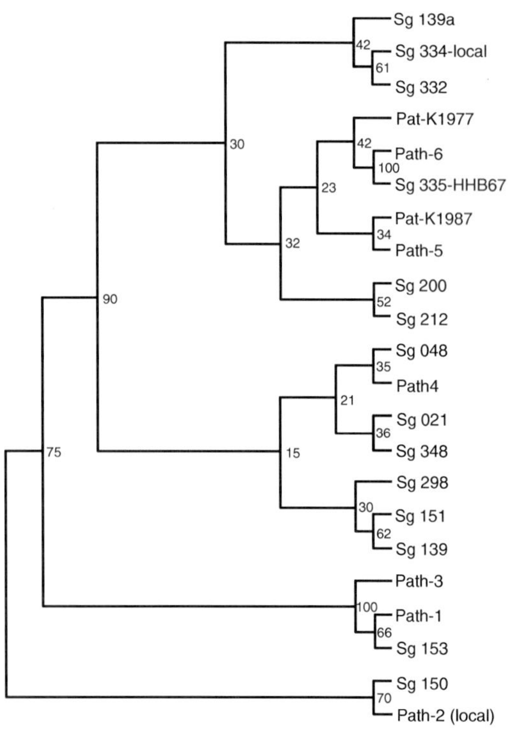

Fig. 4.4 Dendrogram based on ISSR polymorphisms of *Sclerospora graminicola* occurring on pearl millet by unweighted pair group method of averages (UPGMA) cluster analysis (*numbers* inside the branches are bootstrap values) (Data from Sudisha et al. 2008)

Different pathotypes of *S. graminicola* were also characterized based on cellular fatty acid composition (Geetha et al. 2002; Amruthesh et al. 2005), isoelectric focusing (IEF) and matrix-assisted laser desorption ionization (MALDI) techniques (Sharathchandra 2006) which also corroborates AFLP data. These isolates were grouped into four major clusters belonging to six pathotypes. Sequence Characterized Amplified Region (SCAR) primers were developed for pathotype-1 (Sudisha et al. 2009). PCR has amplified a single amplicon of 290 bp only in Pathotype-1. The novel sequences have been deposited in the GenBank of the NCBI (Accession number EF599095).

4.2.6 Control Strategies

Sclerospora graminicola is an obligate biotroph, multiplying only in living host tissue. The pathogen is seed, soil and air-borne. The crop is susceptible to infection throughout the vegetative phase of growth. Hence the control strategies should start from the seedling stage itself. An effective control strategy depends on a clear understanding of genetic basis of resistance to identify sources of resistance in breeding and an understanding of disease cycle to identify the vulnerable stage to break the disease cycle. Though several control strategies have been experimented, only two approaches i.e. deployment of resistance cultivars and chemical control strategies have been widely employed.

4.2.7 Chemical Control

The epidemics of downy mildew lead to the search of newer fungicides specifically targeting oomycetous fungi including downy mildew of pearl millet. Ciba-Geigy has developed Ridomil as a fungicide of choice to control downy mildew. This systemic fungicide was registered and released to market in 1980s. Metalaxyl 35 (Apron) seed treatment @ 6 g a.i. per kilogram of seeds protected the seedlings from downy mildew up to 30 days. Foliar spray with Metalaxyl 25 (Ridomil) at 30 days offered an effective control of downy mildew. Metalaxyl is recommended for the control downy mildew in farmers' fields and also in commercial seed production plots (Singh and Shetty 1990). Seed treatment with metalaxyl gave protection up to 30 days. After foliar treatment, metalaxyl residues persisted in all plant parts for 90 days depending on the concentration applied. More fungicide was taken up in soaking treatment than in slurry or dust treatments (Reddy et al. 1990). Field experiments by Gupta et al. (2012) with Apron (4.2 g a.i. per kg seed) and two sprays of Ridomil MZ 72 WP (2.88 kg a.i. per ha) at 20 and 40 days after sowing demonstrated effective control of downy mildew. Spray alone was more effective than seed treatment. The residue of metalaxyl or mancozeb was either nil or in traces in soil, grain and straw.

Three formulations of Strobilurin fungicides, azoxystrobin, kresoxim-methyl and trifloxystrobin inhibited sporulation, zoospores release and mobility at 0.1 to 2 ug mL^{-1} concentration. Azoxystrobin gave the best disease protection. Seed plus foliar spray gave 93 % protection, whereas foliar spray alone gave 91 % protection (Sudisha et al. 2005).

Sharathchandra et al. (2004) have tried a commercial aqueous chitosan formulation 'Elexa' at a concentration of 1:19 as seed soaking treatment for 6 h which gave 48 % protection against downy mildew. Foliar spay alone on 2-day-old seedlings gave 67 % protection. A combined seed and spray treatment gave 71 % protection in greenhouse and field conditions. Maximum resistance was observed after 24 h of Elexa application. Chaluvaraju et al. (2004) also obtained good control of downy mildew by seed and foliar application of phosphorous-based compounds like di-potassium hydrogen phosphate, 2,3,5 tri-iodo benzoic acid, phosphorous acid and its commercially available formulations alcomon-40 and potassium phosphate.

Deepak et al. (2005) have worked out the cost-return budget in the control of downy mildew of pearl millet and reported that less than 30 % disease control is uneconomical. Among the 15 commercially available and five experimental fungicides, anti-mildew activity of acylanilide series exhibited higher (over 95 %) protection effect.

4.2.8 Abiotic Elicitors

Several abiotic elicitors applied as seed treatment were effective in reducing the disease incidence in greenhouse and field conditions. Seed treatment with β-amino-butyric acid (BABA), a rare non-protein amino acid, at 50 mM concentration provided upto 75 % disease protection (Shailasree et al. 2001). The resistance induced in seedlings was long lasting through the vegetative and reproductive growth of pearl millet plants. This increased resistance was supposed to be due to the increased accumulation of defense-related proteins such as β-1,3-glucanase, phenylalanine ammonia lyase, peroxidase and hydroxyproline rich glycoproteins (Shailasree et al. 2007).

A total of 319 inducers were tested and these agents demonstrated considerable differences in their ability to induce downy mildew disease resistance at University of Mysore, India (Unpublished data; Table 4.4). The inducers tested were rated based on their performances after screening both under artificial green house and epiphytotic field conditions. The inducers included Amino acids (13), Antibiotics (1), Aromatic compounds (4), Biopolymer (2), Carbohydrates (21), Carbohydrate Specific proteins (2), Cell wall components (2), Commercial formulation (18), Endophytic actinomycetes (20), Fatty acids (24), Endophytic streptomycetes (37), Growth regulators (9), Metal salts (17), Micronutrients (3), Mineral salts (14), Nitric oxide donors (5), Organic acids (5), Phenolic acids (6), Plant extracts (58),

Table 4.4 Some important inducers tested for controlling downy mildew disease in pearl millet at University of Mysore

Chemicals	Name of the inducer	% DM protection
1. Amino acids	L-Phenyl alanine	68
	L-Leucine	68
	L-isoleucine	64
2. Antibiotics	Validamycin-A	63
3. Aromatic Compounds	Methyl jasmonate	44
	4-Hydroxybenaldehyde	27
4. Biopolymers	Chitin oligomers	61
	Curdlan	74
5. Carbohydrates	Laminarin	62
	Trehalose	63
	Mannitol	52
	Galactose	49
6. Carbohydrate specific proteins	*N*-Acetyl glucosoamine	56
	N-Acetyl neuramic acid	41
7. Cell wall components	*Colletotrichum dematium* Nml 237-04	52
	Pestalotiopsis sp. Gcr2-04	49
	Colletotrichum gloeosporioides Nml 230-04	45
8. Commercial formulation	Trichoshield	65
	Cerebroside B	58
	Bacillus sp. Mr 33	60
	Bacillus spp. Mr 35	61
	Nutri kelp	65
	Nutri care	62
9. Endophytic actinomycetes	Actinomycete Pgr 06-05	61
	Actinomycete Pgr 05-05	53
10. Endophytic fungi	*Penicillium* sp. Hcr 115-04	55
	Fusarium oxysporum Ais 231-04	58
	Fusarium oxysporum Pgr 32-04	50
11. Endophytic streptomycetes	Streptomycetes Pgr 05-05	71
	Streptomycetes Pgr 06-05	63
	Streptomycetes Pgr 30-05	56
12. Fatty acid	Eicosapentanoic acid	78
	Arachidonic acid	76
	Lineolic acid	66
13. Growth regulators	Indole 3 butryic acid	71
	Benzyl amino purine	63
	Abscisic acid	62
14. Metal salts	Calcium silicate	33
	Ammonium sulfamate	32
	Sodium silicate	30
15. Micronutrients	Meso inositol	61
	Myo-inositol	60
	Zinc sulphate	27

(continued)

Table 4.4 (continued)

Chemicals	Name of the inducer	% DM protection
16. Mineral salts	Copper sulphate	75
	Sodium carbonate	74
	Calcium nitrate	41
17. Nitric oxide donors	Sodium nitroprusside	60
	Iso sorbite dinitrite	59
	Glycerol trinitrate	48
18. Organic acids	Phosphorous acids	51
	Citric acid	37
	Gallic acid	37
19. Phenolic acids	Caffeic acid	63
	Coumaric acid	62
	T-Cinnamic acid	61
20. Plant extract	*Jatropha glandufolia*	73
	Lactuca sativa	73
	Tridax procumbens	64
21. Plant growth promoting rhizobacteria (PGPR)	*Azospirillum brasilense* Pgr 07	71
	Bacillus pumulis NI 1	67
	Bacillus pumulis INR 7	67
22. Plant growth promoting fungi (PGPF)	*Trichoderma viride Tstv 1*	52
	Trichoderma harzianum Pgr Th2	44
	Trichoderma lignorum Ts T11	42
23. Salicylic acid analogue	*para*-Hydroxy benzoic acid	41
	Methyl salicylate	73
	Benzoic acid	40
24. Secondary metabolite	Flavone	56
	Isoflavone	41
	Indoquinone	35
25. Vitamins	MSB	73
	Thiamine	72
	Roseoflavin	61

PGPR (13), PGPF (5), Salicylic acid analogues (11), Secondary metabolites (5) and Vitamins (7). Among all the inducers tested, 14 inducers offered more than 70 % protection against downy mildew, whereas other inducers offered less than 30–70 % disease protection in different methods of application.

Geetha and Shetty (2002) have demonstrated the induction of resistance by Benzothiadiazole (BTH), calcium chloride and hydrogen peroxide. BTH (0.75 %) gave the best protection of 78 %. Sarosh et al. (2005) have also reported the elicitation of defense-related enzymes like Pr-1a, B-1,3-glucanase, chitinase, POX, Lipoxygenase, chalcone synthase in terms of transcript accumulation in the susceptible genotype of pearl millet. Cow milk (10 %) and amino acid L-phenylalanine

Table 4.5 Downy mildew protection offered by different bioagents

Bioagent	Downy mildew protection (%)
Trichoderma harzianum UOM SAR1	67
Trichoderma viride UOM SAR 27	63
Chaetomium globosum UOM SAR 39	52
Pseudomonas fluorescens	80
Pseudomonas fluorescens UOM14	77
Pseudomonas fluorescens UOM80	71
Bacillus subtilis UOM SAR 4	70
Bacillus pumilus UOM SAR 16	68
Apron	87

Data from Raj (2005)

have elicitated the defense-related enzymes like phenyl alanine ammonia lyase (PAL), peroxidase (POX), in pearl millet against downy mildew disease (Sudisha et al. 2011) and offered 35 % and 68.6 % protection, respectively. Vitamins like pyridoxine, folic acid, riboflavin, niacin, D-biotin, menadione sodium bisulphate (MSB) also offered different levels of downy mildew protection (Pushpalatha et al. 2007). Soaking of seeds in 20 mM MSB for 6 h gave the best protection of 73 %. Maximum resistance was evident fourth day after sowing and persisted till the end of crop growth period.

Nitric oxide donors Nitroso-R-salt, 2-nitroso-1-naphthol and sodium nitro prusside (SNP) were tried as inducers of downy mildew resistance. Aqueous SNP seed treatment was very effective in induction of resistance both in greenhouse and field conditions, maximum resistance being evident in 3-day time gap. Primary defense responses like hypersensitive response, lignin deposition and enhanced defense-related enzyme activity were evident (Manjunatha et al. 2008; Deepak et al. 2007a) have demonstrated the induction of resistance against *S. graminicola* by a synthetic jasmonate analogue 1-oxo-indamoyl-L-isoleucine methyl ester based on greenhouse experiments and enhanced activity of PAL, POX and HRGP.

4.2.9 Biological Control

Umesha et al. (1998) tried biocontrol agents like pure culture and talc-based formulation of *Pseudomonas fluorescens* under greenhouse and field conditions. The seed treatment followed by foliar spray gave an effective control of the disease. Theradi Mani and Juliet Hepziba (2009) tried seed treatment with talc-based formulation of *Trichoderma* spp. (4 g/kg) and *P. fluorescens* (10 g/kg) and peat-based formulation of *Bacillus subtilis* (30 g/kg) and found that *P. fluorescens* treatment reduced the disease incidence to 9.50 % (Table 4.5).

4.3 Host-Plant Resistance

Several defense-responsive enzymes are associated with resistance mechanism of pearl millet against downy mildew. These include β-1,3-glucanase, chitinase, lipoxygenase, phenylalanine ammonia lyase, peroxidase, H^+-ATPase, superoxide dismutase, polyphenol oxidase etc. The involvement of plasma membrane H^+-ATPase in downy mildew disease resistance has been demonstrated by Madhu et al. (2001). Differential induction of superoxide dismutase in downy mildew resistant and susceptible genotypes of pearl millet upon inoculation with *S. graminicola* has been reported by Babitha et al. (2002a). Native PAGE analysis showed four isozymes of SOD, three of which (SOD-1, -2, and -4) were Cu/Zn SOD, and isozyme SOD-3 was Mn-SOD. Mn-SOD was further purified and partially characterized (Babitha et al. 2002b). Induction of Lipoxygenase (LOX) was also reported in resistant pearl millet seedlings due to infection with *S. graminicola*. Three of the six isozymes were purified, of which LOX-6 was attributed to the downy mildew resistance (Babitha et al. 2004). Higher levels of ribonuclease (RNase) enzyme activity and differential RNase isoenzymes profiles were evident in resistant cultivars (Shivakumar et al. 2000). Both constitutive and inducible lytic factors were observed in different resistant cultivars (Umesha et al. 2000). Pearl millet cells expressing hypersensitive reaction after inoculation with *S. graminicola* and arachidonic acid showed the differential accumulation of autofluorescent compounds in resistant and susceptible genotypes, with most accumulation occurring in resistant cells (Geetha et al. 1998).

In addition to the defense-related enzymes other cell wall associated defense proteins such as hydroxyproline rich glycoproteins (HRGPs) and polygalacturonase inhibitor proteins (PGIPs) have also been known to play a role in host resistance in pearl millet (Shailasree et al. 2004; Deepak et al. 2007b; Prabhu et al. 2012a). These proteins have been purified and characterized from pearl millet tissues (Deepak et al. 2007c; Prabhu et al. 2012b).

4.4 Resistance Breeding

Use of resistant cultivars is the most cost-effective method for the control of downy mildew. It is mandatory that the cultivars should have at least 8 % tolerance to downy mildew before notification.

The lack of diversity and inadequate downy mildew resistance in parental lines is a major bottleneck in breeding single-cross hybrids of pearl millet. Several downy mildew resistant male-sterile lines of pearl millet (Rai et al. 1998; Thakur et al. 2001) and male parents have been identified (Singh et al. 1997) and used in developing commercial hybrids. It is suggested that the male-sterile cytoplasm is not linked to downy mildew susceptibility and thus could be exploited commercially to broaden the cytoplasmic base of the male-sterile lines (Yadav 1996). Some of the resistant sources have been used in breeding of hybrids and open-pollinated varieties (Hash et al. 1999).

Wilson et al. (2008) identified several downy mildew resistant entries from sub-saharan African Countries based on multi-locational trials. Angarawai et al. (2008) reported that the inheritance of downy mildew in pearl millet is quantitative, highly heritable and would respond to selection. This could be further facilitated by modern biotechnological tools such as marker-assisted breeding techniques. Marker-assisted backcross breeding technique was employed to circumvent the breakdown of host resistance (Hash et al. 2003).

Reliable screening techniques using infector susceptible rows, test entries, check entries and scoring systems have been developed to identify sources of downy mildew resistance (Williams et al. 1981; Singh and Gopinath 1985; Singh et al. 1993). Singh (1995) identified accessions with 100 % resistance against major pathotypes occurring in India and West Africa and resistance sources were used to breed pollinators, male-sterile lines and cultivars. Several open-pollinated varieties and hybrids have been developed for cultivation in India (Singh 1995) and West Africa (Singh et al. 1993) by ICRISAT. A large number of resistant sources were identified from India/Western Africa based on laboratory, greenhouse and field screening (Singh 1990; Singh et al. 1997). Thakur et al. (2001) identified IP 18292 as the highly stable and resistant genotype for all the six pathotypes in India.

Several germplasm accessions of pearl millet having stable resistance to *S. graminicola* were selected from bulk germplasm from Nigeria. The severity of infection on the selected accessions ranged from <1 to 7 % compared with 42 to 69 % on the susceptible standard NHB3 (Singh and Shetty 1990). A number of pearl millet germplasm and breeding lines have been identified by field and greenhouse screening (Singh et al. 1997) and evaluated by multi-locational trials at 20 locations in Burkina Faso, India, Nigeria, Niger and Senegal under International Pearl Millet Downy Mildew Nursery (IPMDMN) to identify stable sources of downy mildew resistance.

The inheritance of downy mildew resistance in pearl millet is highly variable and inconsistent. This could be attributed to lack of homozygous resistant/susceptible genotypes, a lack of genetically pure pathogen isolate, and variable environmental conditions. The resistance mechanism is governed either by additive or dominant gene effects, or both (Rai et al. 2006; Shetty et al. 1998), while in others, both additive and non-additive gene effects with epistatic interactions were important (Deswal and Govila 1994). However, Singh and Talukdar (1998) have shown single dominant gene (*Rsg1*) controlled resistance.

4.4.1 Resistance Mechanism

4.4.1.1 Induced Resistance

Inoculation of pearl millet seedlings with a suboptimal concentration of *S. graminicola* induced resistance against subsequent infection by the pathogen (Kumar et al. 1998). Increased levels of β-1,3-glucanase and peroxidase activity was associated with the induction of resistance in pearl millet. Arachidonic acid also induced a hypersensitive response (HR) on coleoptile/root regions of 2-day-old pearl millet seedlings (Geetha et al. 1998). A time delay in the appearance of HR among genotypes was related to the degree of resistance to downy mildew. Pearl millet genotypes were categorized as highly resistant/resistant (HR in 13 h and above).

4.4.1.2 Recovery Resistance

Recovery resistance is a phenomenon in which the pathogen completes its life cycle without affecting the normal development of the plant (Singh and King 1988). A high level of recovery resistance (up to 95 %) was developed in ICMA1 (81A, a male-sterile line) and its maintainer line ICMB1 (81B), through pedigree selection for five generations (Singh and Talukdar 1996).

4.4.1.3 Resistance Gene Characterization

Plant disease resistance genes show significant similarity among their sequences with the presence of conserved motifs common to the nucleotide-binding site (NBS). Oligonucleotide degenerate primers designed from conserved NBS motifs encoded by different plant disease resistance genes can be used to amplify resistance gene analogues (RGAs) corresponding to the NBS sequences from the genomic DNA of various plant species. Using this approach 22 RGAs were cloned and sequenced from pearl millet (Sarosh 2002; Sarosh et al. 2011). Phylogenetic analysis of the predicted amino acid sequences grouped the RGAs into nine distinct classes. GenBank database searches with the consensus protein sequences of each of the nine classes revealed their conserved NBS domains and similarity to other known R genes of various crop species. One RGA 213 was mapped onto LG1 and LG7 in the pearl millet linkage map. Accumulation of the transcripts of this RGA during infection with *S. graminicola* in resistant pearl millet seedlings indicated its involvement in resistance mechanism against downy mildew (Ranjini et al. 2006). Further studies indicated that this RGA encodes a putative ser-thr protein kinase (Ranjini et al. 2011).

4.5 Future Challenges

Pearl millet is envisaged to gain the status of an important food and fodder crop in future because of its drought-tolerance ability and its nutritional quality. An effective strategy should be developed to manage the diseases in pearl millet crops in order to improve the productivity. The downy mildew disease is still the major biotic constraint in all pearl millet growing areas. Though extensive work has been done on this disease, still it is difficult to manage. Of late, blast disease is also becoming important in certain pearl millet growing regions. Hence this disease also needs immediate attention.

The future challenges are:

- Monitor the shift in virulence of *S. graminicola* with the introduction of new cultivars of pearl millet.
- Understand the basic mechanism of resistance using proteomic approaches.
- Identify suitable and reliable genetic markers for selection of promising resistant pearl millet cultivars and adapt them in the breeding programmes.
- Identify suitable biotic/abiotic elicitors; develop suitable formulation and delivery strategies for large-scale application.
- Exploit the newer biotechnological tools such as transgenic technology to incorporate genes offering resistance to downy mildew.

References

Amruthesh KN, Geetha NP, Hans LJ, de Neergaard E, Shetty HS (2005) Unsaturated fatty acids from zoospors of *Sclerospora graminicola* induce downy mildew resistance in pearl millet. Eur J Plant Pathol 111:125–127

Angarawai II, Kadams AM, Bello D, Mohammed SG (2008) Quantitative nature of downy mildew resistance in Nigerian elite pearl millet lines. J SAT Agric Res 6:1–3

Babitha MP, Bhat SG, Prakash HS, Shetty HS (2002a) Differential induction of superoxide dismutase in downy mildew-resistant and -susceptible genotypes of pearl millet. Plant Pathol 52:480–486

Babitha MP, Prakash HS, Shetty HS (2002b) Purification and partial characterization of manganese superoxide dismutase from downy mildew resistant pearl millet seedlings inoculated with *Sclerospora graminicola*. Plant Sci 163:917–924

Babitha MP, Prakash HS, Shetty HS (2004) Purification and properties of lipoxygenase induced in downy mildew resistant pearl millet seedlings due to infection with *Sclerospora graminicola*. Plant Sci 166:31–49

Ball SL (1983) Pathogenic variability of downy mildew (*Sclerospora graminicola*) on pearl millet I. Host cultivar reactions to infection by different pathogen isolates. Ann Appl Biol 104:41–51

Ball SL, Pike DJ, Burridge CY (1986) Characterization of *Sclerospora graminicola*. Annal Appl Biol 108:519–526

Borchhardt A (1927) Operations of the phytopathological section of the Agricultural Experiment Station in the eastern Steppe region in the year 1925. Dniepropetrovok (Ekaterinoslaff): 3–37, (Russian, Abs. in Botanic Centralbl, N.S. 11:440–441, 1928)

Chaluvaraju G, Puttalingaiah B, Shetty NP, Deepak SA, Amruthesh KN, Shetty HS (2004) Effect of some phosphorous-based compounds on control of pearl millet downy mildew disease. Crop Prot 23:595–600

Deepak SA, Chaluvaraju G, Basavaraju P, Amruthesh KN, Shetty HS, Oros G (2005) Response of pearl millet downy mildew (*Sclerospora graminicola*) to diverse fungicides. Int J Pest Manage 51:7–16

Deepak S, Raj SN, Shailasree S, Kini RK, Boland W, Shetty HS, Mithofer A (2007a) Induction ofresistance against downy mildew pathogen in pearl millet by a synthetic jasmonate analogon. Physiol Mol Plant Pathol 71:96–105

Deepak S, Shailasree S, Kini KR, Shetty HS, Hause B, Mithöfer A (2007b) Role of hydroxyproline rich glycoproteins in resistance of pearl millet against downy mildew pathogen *Sclerospora graminicola*. Planta 226:323–333

Deepak S, Shailasree S, Sujeeth N, Kini KR, Shetty HS, Mithofer A (2007c) Purification and characterization of proline/hydroxyproline rich glycoprotein from pearl millet coleoptiles infected with downy mildew pathogen *Sclerospora graminicola*. Phytochemistry 68:298–305

Deswal DP, Govila OP (1994) Genetics of resistance to downy mildew (*Sclerospora graminicola*) in pearl millet (*Pennisetum glaucum*). Indian J Agric Sci 64:661–663

Dick MW (2001) *Straminipolous* fungi. Kluwer Academic Publishers, New York

Geetha HM, Shetty HS (2002) Induction of resistance in pearl millet against downy mildew disease caused *Sclerospora graminicola* using benzothiadiazole, calcium chloride and hydrogen peroxide—a comparative evaluation. Crop Prot 21:601–608

Geetha S, Shetty SA, Shetty HS, Prakash HS (1998) *Sclerospora graminicola* and arachidonic acid-induced autofluorescence in downy mildew resistant and susceptible genotypes of pearl millet. Ann Appl Biol 33:136–139

Geetha NP, Amruthesh KN, Shetty HS (2002) Determination of cellular fatty acid components in five pathotypes of *Sclerospora graminicola*, the downy mildew pathogen of pearl millet. J Mycol Plant Pathol 32:345–353

Gupta GK, Singh D (1999) Role of edaphic factors in the development of downy mildew (*Sclerospora graminicola*) in pearl millet. J Agric Sci 133:61–68

Gupta SK, Ran Sharma Rai KN, Thakur RP (2012) Inheritance of foliar blast resistance in pearl millet (*Pennisetum glaucum*). Plant Breed 131(1):217–219

Gwary DM, Garba I, Gwary SD (2007) Evidence of pathogenic variation in *Sclerospora graminicola* populations from pearl millet regions of Nigeria. Int J Agric Biol 9(1):94–97

Hash CT, Singh SD, Thakur RP, Talukdar BS (1999) Breeding for disease resistance. In: Khairwal IS, Rai KN, Andrews DJ, Harinarayana G (eds) Pearl millet breeding. Oxford and IBH Publishing Co Pvt Ltd, New Delhi, pp 337–380

Hash CT, Sharma A, Kolesnikova-Allen MA, Serraj R,Thakur RP, Bidinger FR, Bhasker Raj AG, Rizvi SMH, Beniwal CR, Yadav HP, Yadav YP, Srikant, Bhatnagar SK, Yadav RS, Howarth CJ, Breese WA, Witcombe JR (2003) Marker-assisted breeding to improve pearl millet hybrid HHB 67: lab to field. In: the Wake of the Double Helix: From the Green Revolution to the Gene Revolution, abstracts and final program, 27–31 May 2003, Bologna (Abst.), p 82

ICRISAT (International Crops Research Institute for the Semi-Arid Tropics) (1987) Proc Internat Pearl Millet Workshop, 7–11 April 1986, ICRISAT Center, India, Patancheru, p 36

Jeger MJ, Gilijamse E, Bock CH, Frinking HD (1998) The epidemiology, variability and control of the downy mildews of pearl millet and sorghum, with particular reference to Africa. Plant Pathol 47:544–569

Jouan B, Delassus M (1971) Principales maladies des mis et sorghos observes au Niger (In Fr. Summaries In En. Es.). Agron Trop 26(8):830–860

JuZheng C, YaRu W, XiuJu C, ChunLan D, HuaiXiang L, YanChun S (1996) On the resistance of foxtail millet to downy mildew. II. Variability and physiological races of *Sclerospora graminicola* (Sacc.) Schröt. and its distribution in China. Acta Phytopathol Sin 26(2):145–151

Kenneth R (1970) Downy mildews of graminae in Israel. Indian Phytopathol 23:371–377

Kumar VU, Hindumathi CK, Ramachandra Kini K, Shetty HS (1998) Prior inoculation inducing resistance to downy mildew (*Sclerospora graminicola*) enhances growth, peroxidase and β-1,3-glucanase activity in pearl millet (*Pennisetum glaucum*). J Plant Pathol 80:203–209

Madhu D, Dharmesh MS, Chandrashekar A, Shetty HS, Prakash HS (2001) Role of H^{+}-ATPase in pearl millet downy mildew disease resistance. Plant Sci 161(4):799–806

Manjunatha G, Raj SN, Shetty NP, Shetty HS (2008) Nitric oxide donor seed priming enhances defense responses and induces resistance against pearl millet downy mildew disease. Pestic Biochem Physiol 91(1):1–11

Michelmore RW, Pawar MN, Williams RJ (1982) Heterothallism in *Sclerospora graminicola*. Phytopathology 72:1368–1372

Nene YL, Singh SD (1976) A comprehensive review of downy mildew and ergot of pearl millet. In Proc Consultants' Group Meetings on Downy Mildew and Ergot of Pearl Millet, 1–3 Oct 1975, ICRISAT Center, India, Patancheru, pp 15–53

Prabhu SA, Kini KR, Niranjan Raj S, Moerschbacher BM, Shetty HS (2012a) Polygalacturonase-inhibitor proteins in pearl millet: possible involvement in resistance against downy mildew. Acta Biochim Biophys Sin 45:415–423

Prabhu SA, Kini KR, Shetty HS (2012b) Partial purification and characterization of polygalacturonase-inhibitor proteins from pearl millet. Afr J Biotechnol 11:7000–7008
Pushpalatha HG, Mythrashree SR, Radhakrishna-Shetty, Geetha NP, Sharathchandra RG, Amruthesh KN, Shekar-Shetty H (2007) Ability of vitamins to induce downy mildew disease resistance and growth promotion in pearl millet. Crop Prot 26:1674–1681
Pushpavathi B, Thakur RP, Chandrashekara Rao K, Rao VP (2006) Characterization of *Sclerospora graminicola* isolates from pearl millet for virulence and genetic diversity. Plant Pathol J 22:28–35
Rai KN, Thakur RP, Rao AS (1998) Registration of pearl millet parental lines ICMA 92666 and ICMB 92666 with multiple disease resistance. Crop Sci 38(2):575
Rai KN, Kulkarni VN, Thakur RP, Haussmann BIG, Mgonja MA (2006) Pearl millet hybrid parents research: approaches and achievements. In: Gowda CLL, Rai KN, Reddy Belum VS, Saxena KB (eds), Hybrid parent research at ICRISAT, Internat Crops Research Institute for the Semi-Arid Tropics, Patancheru, pp 11–73
Raj SN (2005) Plant growth promoting rhizobacteria mediated induction of resistance in pearl millet against downy mildew-Histological, biochemical and molecular analysis. Ph.D. thesis, University of Mysore, Mysore
Ramakrishnan TS (1963) Diseases of millets. Indian Council of Agricultural Research, New Delhi
Ranjini P, Shailasree S, Kini KR, Shetty HS (2006) Transcripts of resistance gene candidates accumulate in pearl millet as a response to infection by *Sclerospora graminicola*. J Mycol Plant Pathol 36:411–414
Ranjini P, Shailasree S, Kini KR, Shetty HS (2011) Isolation and characterization of a NBS-LRR resistance gene analogue from pearl millet. Arch Phytopathol Plant Prot 44:1013–1023
Reddy MVB, Shetty HS, Reddy MS (1990) Mobility, distribution and persistence of metalaxyl residues in pearl millet (*Pennisetum americanum* (L) Leeke). Bull Environ Contam Toxicol 45:250–257
Sarosh BR (2002) Induced systemic resistance in pearl millet against downy mildew disease and molecular genetics of disease resistance. Ph.D. thesis submitted to University of Mysore, Mysore
Sarosh BR, Sivaramakrishnan S, Shetty HS (2005) Elicitation of defense related enzymes and resistance by L-methionine in pearl millet against downy mildew disease caused by *Sclerospora graminicola*. Plant Physiol Biochem 43:808–815
Sarosh BR, Niranjan Raj S, Shivaramakrishnan S, Shetty HS (2011) Isolation, cloning and characterization of resistance gene analogues in pearl millet based on conserved nucleotide-binding sites. J Phytopathol 159:382–389
Sastry JG, Ramakrishna W, Sivaramkrishna S, Thakur RP, Gupta VS, Ranjekar PK (1995) DNA fingerprinting detects genetic variability in the pearl millet downy mildew pathogen (*Sclerospora graminicola*). Theor Appl Genet 91:856–861
Sastry JG, Sivaramkrishnan S, Thakur RP, Gupta VS, Ranjekar PK (1997) Characterization of the genome *Sclerospora graminicola*, the causal fungus of downy mildew of pearl millet. World J Microbiol Biotechnol 13:683–687
Sastry JG, Sivaramakrishnan S, Rao VD, Thakur RP, Singru RS, Gupta VS, Ranjekar PK (2001) Genetic basis of host-specificity in *Sclerospora graminicola*, the pearl millet downy mildew pathogen. Indian Phytopathol 54(3):323–328
Schroeter J (1879) *Protomyces graminicola* Saccardo. Hedwigia 18:83–87
Shailasree S, Sarosh B, Vasanthi N, Shetty H (2001) Seed treatment with β-aminobutyric acid protects *Pennisetum glaucum* systemically from *Sclerospora graminicola*. Pest Manag Sci 57:721–728
Shailasree S, Ramachandra Kini K, Deepak S, Kumudini BS, Shetty HS (2004) Accumulation of hydroxyproline-rich glycoproteins in pearl millet seedlings in response to *Sclerospora graminicola* infection. Plant Sci 167:1227–1234
Shailasree S, Kini KR, Shetty HS (2007) Beta amino butyric acid induced resistance in pearl millet to downy mildew is associated with accumulation of defense related proteins. Australasian Plant Pathol 36:204–211

Sharathchandra RG (2006) Elicitation of systemic defense responses in pearl millet against downy mildew fungus and the associated molecular events. Ph.D. thesis, Department of Studies in Applied Botany and Biotechnology, University of Mysore, Mysore

Sharathchandra RG, Niranjan Raj S, Shetty NP, Amruthesh KN, Shekar Shetty H (2004) A chitosan formulation Elexa induces downy mildew disease resistance and growth promotion in pearl millet. Crop Prot 23:881–888

Sharma R, Rao VP, Varshney RK, Prasanth VP, Kannan S, Thakur RP (2010) Characterization of pathogenic and molecular diversity in *Sclerospora graminicola*, the causal agent of pearl millet downy mildew. Arch Phytopathol Plant Prot 43(6):538–551

Sharma R, Rao VP, Senthilvel S, Rajput SC, Thakur RP (2011) Virulence diversity in north Indian isolates of *Sclerospora graminicola*, the pearl millet downy mildew pathogen. J Plant Pathol 93(1):71–78

Shaw CG (1981) Taxonomy and evolution. In: Spencer DM (ed) The downy mildews. Academic, London, pp 17–29

Shetty HS, Ahmad R (1981) Physiologic specialization in *Sclerospora graminicola*. Indian Phytopathol 34:307–309

Shetty HS, Khanzada AK, Mathur SB, Neergard P (1978) Procedures for detecting seed-borne inoculums of *Sclerospora graminicola* in pearl millet (*Pennisetum typhoides*). Seed Sci Technol 6:935–941

Shetty HS, Mathur SB, Neergaard P (1980) *Sclerospora graminicola* in pearl millet seeds and its transmission. Trans Br Mycol Soc 74:127–134

Shetty SA, Shishupala S, Nagarathna KC, Kumar VU, Shetty HS (1998) Management of cereal downy mildews. In: Upadhyay RK, Mukerji KG, Rajak RL (eds) Book series in integrated pest management system in agriculture, I, Adithya Books Ltd, New Delhi, IARI, Delhi, PI 59

Shivakumar PD, Vasanthi NS, Shetty HS, Smedegaard-Petersen V (2000) Ribonucleases in the seedlings of pearl millet and their involvement in resistance against downy mildew disease. Eur J Plant Pathol 106:825–836

Singh SD (1990) Sources of resistance to downy mildew and rust in pearl millet. Plant Dis 74:871–874

Singh SD (1995) Downy mildew of pearl millet. Plant Dis 79:545–550

Singh SD, Gopinath R (1985) A seedling inoculation technique for detecting downy mildew resistance in pearl millet. Plant Dis 69:582–584

Singh SD, King SB (1988) Recovery resistance to downy mildew in pearl millet. Plant Dis 72:425–428

Singh SD, Shetty HS (1990) Efficacy of systemic fungicide metalaxyl for the control of downy mildew (*Sclerospora graminicola*) of pearl millet (*Pennisetum glaucum*). Indian J Agric Sci 60:575–581

Singh SD, Singh G (1987) Resistance to downy mildew in pearl millet hybrid NHB-3. Indian Phytopathol 40(2):178–180

Singh SD, Talukdar BS (1996) Recovery resistance to downy mildew in pearl millet parental lines ICMA 1 and ICMB 1. Crop Sci 36(1):201–203

Singh SD, Talukdar BS (1998) Inheritance of complete resistance to pearl millet downy mildew. Plant Dis 82:791–793

Singh SD, Williams RJ (1980) The role of sporangia in the epidemiology of pearl millet downy mildew. Phytopathology 70:1187–1190

Singh SD, Singh P, Andrew DJ, Talukdar BS, King SB (1992) Reselection of a pearl millet cultivar utilizing residual variability for downy mildew reaction. Plant Breed 109:54–59

Singh SD, King SB, Werder J (1993) Downy mildew disease of pearl millet. Information Bulletin no 37. ICRISAT, Patancheru

Singh SD, Wilson JP, Navi SS, Talukdar BS, Hess DE, Reddy KN (1997) Screening techniques and sources of resistance to downy mildew and rust in pearl millet. Information Bulletin no. 48. ICRISAT, Patancheru, p 110

Sudisha J, Amruthesh KH, Deepak SA, Shetty NP, Sarosh BR, Shetty HS (2005) Comparative efficacy of strobilurin fungicides against downy mildew disease of pearl millet. Pest Biochem Physiol 81(3):188–197

Sudisha J, Kumar SA, Shetty HS (2008) Characterization of downy mildew isolates of *Sclerospora graminicola* by using differential cultivars and molecular markers. J Cell Mol Biol 7:41–55

Sudisha J, Kumar SA, Niranjana SR, Shetty NP, Shetty HS (2009) Cloning and development of pathotype-specific SCAR marker associated with *Sclerospora graminicola* isolates from pearl millet. Aust Plant Pathol 38:216–221

Sudisha J, Kumar A, Amruthesh KN, Niranjana SR, Shetty HS (2011) Elicitation of resistance and defense related enzymes by raw cow milk and amino acids in pearl millet against downy mildew disease caused by *Sclerospora graminicola*. Crop Prot 30(7):794–801

Thakur RP (1999) Pathogen diversity and plant disease management. Indian Phytopathol 52(1):1–9

Thakur RP, Rao VP (1997) Variation in virulence and aggressiveness among pathotypes of *Scleropora graminicola* on pearl millet. Indian J Mycol Plant Pathol 50:41–47

Thakur RP, Shetty KG (1993) Variation in pathogenicity among single oospore isolates of *Sclerospora graminicola* the causal organism of downy mildew in pearl millet. Plant Pathol 42:715–721

Thakur RP, Shetty KG, King SB (1992) Selection for host-specific virulence in asexual populations of *Sclerospora graminicola*. Plant Pathol 41:626–632

Thakur RP, Rao, VP, Hash CT (1998) Emergence of a highly virulent pathotype of *Sclerospora graminicola* at Jodhpur, Rajasthan, India. International Sorghum Pearl Millet News Letter.

Thakur RP, Rao VP, Sastry JG, Sivaramakrishnan S, Amruthesh KN, Barbind LD (1999) Evidence for a new virulent pathotype of *Sclerospora graminicola* on pearl millet. Indian J Mycol Plant Pathol 29:61–69

Thakur RP, Rai KN, Rao VP, Rao AS (2001) Genetic resistance in pearl millet male-sterile lines to diverse pathotypes of *Sclerospora graminicola*. Plant Dis 85:621–626

Thakur RP, Rao VP, Wu BM, Subbarao KV, Shetty HS, Singh G, Lukose C, Panwar MS, Paco S, Hess DE, Gupta SC, Dattar VV, Panicker S, Pawar NB, Bhangale GT, Panchbhai SD (2004a) Host resistance stability to downy mildew in pearl millet and pathogenic variability in *Sclerospora graminicola*. Crop Prot 23:901–908

Thakur RP, Sivaramakrishnan S, Kannan S, Rao VP, Hess DE, Magil CW (2004b) Genetic and pathogenic variability among isolates of *Sclerospora graminicola* the downy mildew pathogen of pearl millet. In: Spencer-Phillips P, Jeger M (eds) Advances in Downy Mildew Research, Kluwer Academic Publishers, 2:179–192

Theradi Mani M, Juliet Hepziba S (2009) Biological management of pearl millet downy mildew caused by *Sclerospora graminicola*. Arch Phytopathol Plant Prot 42(2):129–135

Umesha S, Dharmesh SM, Shetty SA, Krishnappa M, Shetty HS (1998) Biocontrol of downy mildew disease of pear millet using *Pseudomonas fluorescens*. Crop Prot 17:387–392

Umesha S, Shylaja Dharmesh M, Shetty HS (2000) Lytic activity in pearl millet: its role in downy mildew disease resistance. Plant Sci 157(1):33–41

Werder J, Ball L (1992) Variability of pearl millet downy mildew (*Sclerospora graminicola*) and its implications for breeding for stable and durable resistance. Trop Pest Manag 38(1):43–47

Weston WH (1929) The occurrence of *Sclerospora graminicola* on maize in Wisconsin. Phytopathology 19:391–397

Williams RJ, Singh SD, Pawar MN (1981) An improved field screening technique for downy mildew resistance in pearl millet. Plant Dis 65:239–241

Wilson JP, Samoga MD, Nutsugah SK, Angarawai I, Fofana A, Traore H, Ahmadou I, Muuka FP (2008) Evaluation of Pearl millet for yield and downy resistance across seven countries in sub-saharan Africa. Afr J Agric Res 3(5):371–378

Yadav OP (1996) Downy mildew incidence of pearl millet hybrids with different male sterility inducing cytoplasms. Theor Appl Genet 92:278–280

Yadav OP (2012). Project Coordinator Review. Annual Workshop of All India Coordinated Pearl Millet Improvement programme, Indian Council of Agricultural Research, p15

Yadav RS, Sehgal D, Vadez V (2011) Using genetic mapping and genomics approaches in understanding and improving drought tolerance in pearl millet. J Exp Bot 62:397–408

Chapter 5
Research on Plant Pathogenic Fungi in the Genomics Era: From Sequence Analysis to Systems Biology

Anandaraj Muthuswamy and Santhosh J. Eapen

5.1 Introduction

Fungi are of immense importance to mankind, in a multitude of ways that reflect the diversity of this kingdom, which ranges from mushrooms to plant pathogens and biocontrol agents. With the recent advances in genome sequencing, we are on the verge of a flood of new genome data from a plethora of plant pathogens. The availability of whole-genome sequences has also fundamentally changed the methods used for the identification and mapping of genes. Current peer-reviewed literature is littered with many other exciting new tools and techniques that are being used in all areas of biology and medicine. Transcriptomics, proteomics and, more recently, metabolomics are three of these techniques that have impacted on fungal plant pathology. The relatively young discipline of bioinformatics became vital for managing the large amounts of data produced by these new techniques. Bioinformatics platforms have become essential tools for accessing "omics" data sets for the efficient mining and integration of biologically significant knowledge.

There are several recent reviews available on next-generation sequencing, genomics and bioinformatics (Mardis 2008; Morozova et al. 2009; Stahl and Lundeberg 2012; Li et al. 2013) and their application in studying plant pathogens and their interaction with host plants (Schneider and Collmer 2010; Studholme et al. 2011). In this review we summarize the current status of fungal genome projects including that of oomycetes and the usefulness of this vast genomic

A. Muthuswamy, Ph.D. (✉)
Indian Institute of Spices Research, P.B. 1701, Marikunnu Post, Kozhikode, Kerala 673012, India
e-mail: anandaraj@spices.res.in

S.J. Eapen, Ph.D.
Division of Crop Protection, Indian Institute of Spices Research, Kozhikode, Kerala, India

A. Goyal and C. Manoharachary (eds.), *Future Challenges in Crop Protection Against Fungal Pathogens*, Fungal Biology, DOI 10.1007/978-1-4939-1188-2_5,

information to determine gene function and to compare genomes from different species. Since this is such a fast-moving field, useful web sites and bioinformatics tools are listed that will enable the plant pathologists to employ them in their routine research.

5.2 Sequence Resources for Fungi

Recent high-throughput technological advances have provided opportunities to develop collections of sequence-based resources and related resource platforms for specific organisms. Comprehensively collected sequence data provide essential genomic resources for accelerating molecular understanding of biological properties and for promoting the application of such knowledge. Species-specific nucleotide sequence collections also provide opportunities to identify the genomic aspects of phenotypic characters based on genome-wide comparative analyses and knowledge of model organisms. Genomes of biocontrol strains and related species allow the identification of biocontrol factors in the respective strains. Next-generation sequencing technology coupled with reference genome sequence data allows us to discover variations among individuals, strains, and/or populations. Nucleotide polymorphisms are effectively identified by mapping sequence fragments onto a particular reference genome data set, a capability that is of immense importance in all genetic research. A necrotrophic fungal pathogen (*Pyrenophora teres* f. *teres*) of barley (*Hordeum vulgare*) has recently become the first eukaryotic phytopathogen to have a complete genome sequenced exclusively using short (75-nucleotide) Illumina reads (Ellwood et al. 2010). Short Illumina sequence reads from *Phytophthora infestans* and its closely related sister species were aligned against the high-quality reference sequence of strain T30–4 using the MAQ alignment software to study the genomic evolution in this group and to infer copy-number variations, SNPs, and signatures of positive selection (Raffaele et al. 2010).

There are a number of providers for fungal genome sequences and annotations. A list of web-accessible information resources providing genome sequences and annotations of various fungal species is provided in Table 5.1. Among the eukaryotes, the greatest numbers of sequenced genomes have been from the fungal kingdom (Haridas et al. 2011). The first genome sequence of a fungus was published in 1996 for *Saccharomyces cerevisiae*, which is now used as a model species. To date, several genome sequencing projects involving various fungal and oomycete plant pathogens have already been completed and many are in the pipeline (Table 5.2).

The proliferation of new sequencing technologies and assembly software and the decline in their cost have made it feasible for many research labs to now attempt sequencing and assembly of genomes of their interest. At Indian Institute of Spices Research (Kozhikode, India) two isolates of *Phytophthora capsici* from black pepper (05-06 and 98-93) were sequenced using Illumina/Roche 454 platforms and the cross-platform sequence data was de novo assembled and annotated structurally and functionally to curate all possible gene by gene information. Whole-genome

Table 5.1 Internet resources providing information on fungal genome sequencing projects

Organization/Resource	URL
Broad Institute	http://www.broad.mit.edu/annotation/fgi/
Genomes OnLine Database	http://www.genomesonline.org/
Genoscope, Sequencing National Centre	http://www.genoscope.cns.fr/spip/Fungi-sequenced-at-Genoscope.html
Joint Genome Institute	http://genome.jgi-psf.org/euk_home.html
National Center for Biotechnology Information	http://www.ncbi.nlm.nih.gov/genomes/FUNGI/funtab.html
Sanger Center	http://www.sanger.ac.uk/Projects/Fungi/
The Genome Center at Washington University (WU-GSC)	http://genome.wustl.edu/genomes/list/plant_fungi
The Institute for Genomic Research	http://www.tigr.org/tdb/fungal/index.shtml
The Sanger Institute fungal sequencing	http://www.sanger.ac.uk/Projects/Fungi/
Fusarium graminearum genome database (FGDB)	http://mips.helmholtz-muenchen.de/genre/proj/FGDB/
Neurospora crassa Genome Database (MNCDB)	http://mips.helmholtz-muenchen.de/genre/proj/ncrassa/
Ustilago maydis Database (MUMDB)	http://mips.gsf.de/genre/proj/ustilago
Saccharomyces Genome Database (SGD)	http://www.yeastgenome.org/
Aspergillus Database Repository (CADRE)	http://www.cadre-genomes.org.uk/

alignment of 05-06 and 98-93 sequence data with the reference genome revealed that their similarity was 95.35 % and 87.90 %, respectively, with the reference genome, indicating the wide variability existing in *P. capsici* isolates of Indian subcontinent (IISR 2013).

All these new techniques are massively parallel in nature and present new challenges in terms of bioinformatics support required. A broad outline of bioinformatics approaches for the analysis of NGS data and the tools available are given by Horner et al. (2009) and Haridas et al. (2011). Most of the current generation of bioinformatics tools for analysis of NGS data is command line driven and somewhat inaccessible to many biologists. More intuitive and simple graphic user interfaces are the current need of the hour to render the power of these new technologies available to a wider audience within the scientific community.

Table 5.2 Plant pathogenic fungi and oomycetes whose whole genomes have been completely sequenced

Pathogen	Disease	Genome size (Mb)	Status	Genome centre/lab
A. Fungi				
Alternaria brassicicola	Black spot on crucifers	30	Draft	• Washington University
Ashbya gossypii	Stigmatomycosis	8.74	Complete	• University of Basel • Syngenta AG
Aspergillus flavus	Grain mold	36	Draft	• J. Craig Venter Institute
Aspergillus niger ATCC 1015	Black mold	32	Draft	• DOE-JGI
Aspergillus niger CBS 513.88	Black mold	33.98	Complete	• DSM Food Specialties • Gene Alliance
Blumeria graminis f. sp. *hordei*	Barley powdery mildew	45	Complete	• Agencourt Bioscience • Blumeria Genome Sequencing Consortium
Botrytis cinerea	Gray mold rot of grapes	38.8	Complete	• Broad Institute
Cochliobolus heterostrophus	Southern corn leaf blight in maize	34.9	Draft	• DOE Joint Genome Institute
Fusarium graminearum	Fusarium head blight	36.33	Complete	• Broad Institute
Fusarium oxysporum f. sp. *lycopersici* (race 2)	Fusarium wilt of tomato	60	Draft	• Broad Institute
Fusarium verticillioides	Kernel and ear rot of maize	46 Mb	Draft	• Broad Institute
Gaeumannomyces graminis tritici	Take-all plant disease of cereal plants	41.98	Complete	• Broad Institute
Gibberella moniliformis	Kernel and ear rot of maize	41.98	Draft	• Broad Institute
Gibberella zeae	Fusarium head blight on wheat and barley	36.49	Complete	• Broad Institute
Grosmannia clavigera kw1407	Destruction of pine tree	32.5	Complete	• BC Genome Sciences Center
Magnaporthe grisea	Rice blast	41.5	Complete	• Broad Institute • North Carolina State University
Mycosphaerella fijiensis	Black leaf streak disease of banana	74.1	Complete	• DOE Joint Genome Institute
Mycosphaerella graminicola	*Septoria tritici* blotch	41.2	Complete	• DOE Joint Genome Institute
Nectria haematococca	Stem and root rot	51.27	Complete	• DOE Joint Genome Institute

(continued)

Table 5.2 (continued)

Pathogen	Disease	Genome size (Mb)	Status	Genome centre/lab
Phaeosphaeria nodorum	Disease of wheat	37.21	Complete	• Broad Institute
Phanerochaete chrysosporium	White rot	35.1	Draft	• DOE Joint Genome Institute
Puccinia graminis f. sp. *tritici*	Wheat stem rust	81.5	Draft	• Broad Institute
Pyrenophora tritici-repentis	Tan spot of wheat	40	Draft	• Broad Institute
Sclerotinia sclerotiorum	Large host range	38.53	Complete	• Broad Institute
Stagonospora nodorum	Glume blotch of wheat	37.1	Draft	• Broad Institute
Ustilago maydis	Corn smut	19.64	Complete	• Bayer • Broad Institute • LION Bioscience AG
Verticillium dahliae VdLs.17	Wilt in several plants	33.83	Draft	• Broad Institute
B. Oomycetes				
Hyaloperonospora parasitica	Downy mildew of *Arabidopsis thaliana*	75	Draft	• Washington University/VBI
Phytophthora capsici	Damping-off, stem and vine blight, wilting or fruit rot	–	Draft	• NCGR/DOE Joint Genome Institute
Phytophthora infestans	Potato late blight	240	Draft	• Broad Institute
Phytophthora ramorum	Sudden oak death	6.6	Draft	• DOE Joint Genome Institute
Phytophthora sojae	Soybean blight	95	Draft	• DOE Joint Genome Institute
Pythium ultimum	Pythium root rot	–	Draft	• JVCI/Michigan State University

The flood of new data from a large cohort of microbial genomes enables a rapid and systematic mining for gene identification, classification, and functional analysis. Whole-genome sequence information allows us to derive sets of important genomic features, including the identification of protein-coding or non-coding genes and constructs such as gene families, regulatory elements, repetitive sequences, simple sequence repeats (SSRs), and guanine–cytosine (GC) content. These data sets have become primary sequence material for the design of genome sequence-based platforms such as microarrays, tiling arrays or molecular markers, as well as for reference data sets for the integration of "omics" elements into a genome sequence.

Chromosome-scale comparisons identifying conserved similarities of gene coordinates facilitate documentation of segmental and tandem duplications in related species. Whole-genome comparisons identifying chromosomal duplication and conserved synteny among related species provide evidence for hypotheses on comparative evolutionary histories with regard to the diversification of species in a related lineage. New sequencing technologies have provided us with new opportunities to be addressed at the entire genome level in the fields of comparative genomics, metagenomics, and evolutionary genomics.

5.3 Comparative Genomics

The availability of genomes from several fungal species allows us to examine the demarcation of fungal species at the whole-genome level. Genomic data can aid fungal taxonomy by serving as a source of novel and unprecedented amounts of comparative data, as a resource for the development of additional diagnostic tools, and finally as a knowledge database about the biological differences between strains and species. Comparative genomics will greatly influence the way we understand these organisms, as comparative anatomy did in the eighteenth and nineteenth centuries. Such extensive genome comparisons have been made for *Aspergillus* (Rokas et al. 2007). Dedicated comparative genomics platforms like CFGP (http://cfgp.riceblast.snu.ac.kr/) are available for fungi (Park et al. 2008).

5.3.1 Molecular Markers

Studies on epidemiology, ecology, evolution, and taxonomy of plant pathogens rely on availability of genetic markers. Recent developments in genomics have opened up for newer opportunities to study the diversity and classification of fungi. With the progress of genome sequencing and large-scale EST analysis in various species, these sequence data sets have become quite efficient resources for designing molecular markers covering entire genomes. Hitherto we were relying on conventional markers such as RFLP (restriction fragment length polymorphism), RAPD (random amplified polymorphic DNA), SSRs, and AFLP (amplified fragment length polymorphism). More recently, it has become common to perform multilocus sequence analysis (MLSA), whereby partial or complete nucleotide sequences are determined for several housekeeping genes. A number of attempts to design polymorphic markers from accumulated sequence data sets have been made for various species. Novel high-throughput sequencing methods outperform earlier approaches in terms of resolution and magnitude. With the rapid development of NGS technologies, tremendous numbers of molecular markers like SSRs and SNPs have been identified. With the availability of genomic sequence data, SNP markers become more accessible for use in mapping, to help achieve a much better resolution. Genome-wide marker discovery by NGS has become more feasible using several new methods that facilitate studying association mapping, patterns of natural population structure and the decay of linkage disequilibrium.

5.3.2 *Evolutionary Genomics*

Understanding the evolutionary history of newly emerging or reemerging pathogens can help in management of emerging epidemics and suggest strategies to address future threats. Nucleotide sequence data offer the possibility of reconstructing patterns of descent among genotypes within a species, or among populations of one or more species (Goss et al. 2009; Stukenbrock et al. 2007). There is an increasing body of evidence to suggest that horizontal gene transfer (HGT), frequently observed in prokaryotes, is an important mechanism in eukaryotic genome evolution too. Of late, a number of fungal HGT events are being reported through bioinformatics-based methodologies. The implications HGT on the fungal tree of life have been recently reviewed by Fitzpatrick (2012).

Ancestral and derived states can be distinguished from sequence data using a combination of coalescent analysis to infer gene genealogies and Bayesian or maximum likelihood (ML) approaches to determine distributions for population parameters of interest (Grunwald and Goss 2011). Coalescent-based methods help in answering questions on the evolutionary origin of emerging pathogens as new species or subspecies, or are they evolved from local populations or are they new introductions to a host or environment, or are they variable populations as a result of diversification. Coalescent analysis based on genome-wide, high-density SNP data, or population genomic studies using high-throughput sequencing data will provide yet another level of analysis not currently available. If a barcoding system can be developed for use with NGS technology, many fungal species could now be sequenced simultaneously at a lower cost. Sequencing at lower levels of coverage or sequencing only targeted regions of DNA are practical strategies for studying population genetics, conservation genetics and molecular ecology. Some of the tools for estimation of the age of a population, species, or evolutionary lineages are listed in Table 5.3.

Table 5.3 Tools for population genetics studies

Tool	URL
BEAGLE	http://faculty.washington.edu/browning/beagle/beagle.html
Genetree	http://www.stats.ox.ac.uk/~griff/software.html
IM & IMa2	http://genfaculty.rutgers.edu/hey/software
InStruct	http://cbsuapps.tc.cornell.edu/InStruct.aspx
LAMARC	http://evolution.genetics.washington.edu/lamarc/lamarc_download.html
LDhat	http://ldhat.sourceforge.net/
MIGRATE	http://popgen.sc.fsu.edu/Migrate/Migrate-n.html
SequenceLD	http://www.maths.lancs.ac.uk/~fearnhea/software/Rec.html
STRUCTURE	http://cbsuapps.tc.cornell.edu/structure.aspx
TESS	http://membres-timc.imag.fr/Olivier.Francois/tess.html

5.4 Functional Genomics

5.4.1 Host–Pathogen Interaction

Many fungal genes involved in pathogenicity and genes involved in effector recognition and defense responses have been identified over the past decade. Complete genome sequences have generated new lists of virulence candidates based on homology, pathogen-specific paralog amplification, linkage with regions that are variable and/or enriched in known virulence genes, and other criteria (Schneider and Collmer 2010). Many hundreds of candidate cytoplasmic effector (CE) genes that would have previously escaped detection are now found out. Furthermore, with next-generation sequencing thousands of CEs can be invented for the pan-genome of various species. Genome mining and bioinformatics pipelines have streamlined the suite of effectors in important pathogen genomes, so researchers can make more targeted strikes on potentially important effectors. The current genomics-driven research though enables the discovery of many effector repertoires, urgently requires gene ontology (GO) terms to systematically accumulate information about them. The universal nature of GO terms will facilitate comparisons with other virulence factors in diverse systems. The Plant-Associated Microbe Gene Ontology (PAMGO) interest group has worked with the GO Consortium to generate more than 800 new GO terms which can address biological processes associated with host–microbe interactions (Torto-Alalibo et al. 2009). This combination of informatics and empirical studies will allow greater insight into effector function. Bioinformatics data mining tools and techniques are also being used for identifying and characterizing several genes helpful in the transcriptional and expression based study of pathogenesis like 1-aminocyclopropane-1-carboxylate deaminase in *P. sojae* (Singh and Kashyap 2012), glucanase inhibitor protein (Reena et al. 2010a) and elicitin (Vijeshkumar et al. 2013) in *P. capsici,* hydrophobins in *Aspergillus* (Littlejohn et al. 2012), transcription factors (Převorovsky et al. 2011), and pathogenesis related proteins (Prasath et al. 2014). Bioinformatics approaches are indispensable for predicting secreted proteins (Reena et al. unpublished; Brown et al. 2012; do Amaral et al. 2012) and miRNAs (Nan et al. 2012) from the available genomic information.

5.4.2 Expressed Sequence Tags and cDNA Clones

Expressed sequence tags (ESTs) are a sequence-based method for expression profiling. ESTs are created by partial "one-pass" sequencing of randomly picked gene transcripts that have been converted into cDNA. Since cDNA and EST collections can be acquired regardless of genomic complexity, this approach has been applied not only to model species but also to a number of other species.

As more and more EST data have become publicly available, the usage of ESTs has expanded to other areas, such as in silico genetic marker discovery, in silico gene discovery, construction of gene models, alternative splicing prediction, genome annotation, expression profiling, and comparative genomics. In comparison with whole-genome sequencing, EST technology is simpler and less costly, especially in the case of large genomes. Moreover, since ESTs represent "the expressed parts" of genomes, they are more immediately informative about the transcriptomes. On the other hand, ESTs are not suitable for the studies related to "the control parts" of genomes, such as promoters and transcription enhancing/inhibiting elements. In addition, information for rarely expressed genes is also difficult to mine from EST data.

The comprehensive and rapid accumulation of cDNA clones together with mass volume data sets of their sequence tags have become significant resources for functional genomics. ESTs derived from various kinds of tissues, including tissues from organisms in a range of developmental stages or under stress, could significantly facilitate gene discovery as well as gene structural annotation, large-scale expression analysis, genome-scale intraspecific and interspecific comparative analysis of expressed genes and the design of expressed gene-oriented molecular markers and probes for microarrays (Zhong et al. 2009). The sequence resources derived from full-length cDNAs can also help substantially in identifying transcribed regions in completed or draft genome sequences. Full-length cDNAs are also useful for determining the three-dimensional (3D) structures of proteins by X-ray crystallography and nuclear magnetic resonance (NMR) spectroscopy and for functional biochemical analyses of expressed proteins in the molecular interactions of protein–ligands, protein–proteins and protein–DNAs.

EST data mining requires bioinformatics resources such as databases, data retrieving tools and analysis algorithms. Because EST data collected from the cDNA libraries of a particular organism consist of redundant sequence data derived from the same gene locus or transcription unit, it is often necessary to perform EST grouping by transcription units and to assemble these groups in order to create a consolidated alignment and representative sequence of each transcript before further analysis that are performed computationally. Bioinformatics tools are also indispensible to deal with EST errors and contaminations. Many of these tools are freely available to the academic community. NCBI's dbEST, a public domain EST database (http://www.ncbi.nlm.nih.gov/dbEST/) includes a number of fungal species. ESTs from 18 species of plant pathogenic fungi and two species of phytopathogenic oomycete can be found in the COGEME database (Soanes and Talbot 2006). ESTs of *P. capsici* were processed and functionally annotated using in silico tools and a range of genes likely to be involved in pathogenesis, drug resistance, stress, host degradation, and genetic marker related proteins were identified (Reena et al. 2010b). In a similar approach, disease resistance genes were identified through downstream analysis of ginger ESTs (Karthika et al. 2013).

5.4.3 Transcriptomics

Transcriptomics, or quantitative gene expression profiling is the large-scale study of the transcriptome, giving a global view of all transcripts simultaneously. It helps in identification of the complete set of transcripts in a particular biological sample and estimation of their abundances under specific physiological conditions or at specific developmental stages. Northern blotting, real-time PCR, serial analysis of gene expression (SAGE), and microarrays were hitherto used for gene expression profiling. Application of next-generation sequencing technologies to cDNA sequencing is commonly called RNA-Seq. This method of transcriptome analysis is fast and simple because it does not require bacterial cloning of the cDNAs. RNA-seq technology, in conjunction with efficient bioinformatics tools, is now more widely used to support predicted gene models, extract differentially expressed genes, and find novel transcripts in de novo assemblies. In addition, being very sensitive, it allows the detection of low-abundance transcripts. Massively parallel short-read sequencing technologies have been highly efficacious for the discovery of novel small non-coding RNAs (ncRNAs) viz. micro-RNAs (miRNAs) and small interfering RNAs (siRNAs) which serve as posttranscriptional regulators of gene expression in a wide range of organisms (Morozova et al. 2009). miRNAs have been identified through high-throughput screening in *Sclerotinia sclerotiorum* (Zhou et al. 2012) and *Cryptococcus neoformans* (Nan et al. 2012).

Illumina and 454 sequencing platforms contributed large numbers of cDNA sequences from fungal (Spanu et al. 2010) and oomycete (Levesque et al. 2010; Miller et al. 2008) plant pathogens. Illumina genome analyzer based sequencing technology (Illumina, USA) yields huge amount of short reads with high coverage. Assembling such short reads is a challenging task, more so in the absence of reference sequences. The downstream analysis in transcriptomics involves transcriptome characterization and gene annotation using a set of differentially expressed genes. Bioinformatics tools dealing with transcriptome alignment, splicing event prediction, and de novo assembly are also available (Table 5.4).

Fungal transcriptomics has been reviewed by Vijai et al. (2007). For identifying putative key genes in plant–pathogen interactions, simultaneous transcriptome analysis of plant and pathogen by using de novo assembly was attempted in sorghum—*Bipolaris sorghicola* (Yazawa et al. 2013), rice—*Magnaporthe oryzae* (Kawahara et al. 2012), banana—*Fusarium oxysporum* f. sp. *cubense* (Li et al. 2012), banana—*Mycosphaerella musicola* (Passos et al. 2013). Recently, a large number of candidate pathogen response genes were identified by comparing the ginger and mango ginger transcriptomes and expression profiling based on read per exon kilobase per million (RPKM) (Prasath et al. 2014; IISR 2013). Many defense related genes differentially expressed in two species of *Piper* viz. *Piper nigrum* and *P. colubrinum* on challenging with *Phytophthora capsici* were identified through transcriptome data analysis (Johnson et al. unpublished; IISR 2013). SNP detection is also a common application of RNA-Seq.

Table 5.4 Bioinformatics tools/databases for functional genomics and transcriptome data analysis

Software	Description	URL
ABMapper	RNA-seq data alignment	http://hkbic.cuhk.edu.hk/software/abmapper
Bowtie	RNA-seq data alignment	http://bowtie-io.sourceforge.net/bowtie2/index.shtml
Cufflinks	Transcript assembly	http://cufflinks.cbcb.umd.edu/
DEGseq	Differential gene expression detection	http://www.bioconductor.org/packages/2.11/bioc/html/DEGseq.html
Infernal	RNA-seq data alignment	http://infernal.janelia.org/
Oases	De novo assembly	www.ebi.ac.uk/~zerbino/oases/
Tophat	RNA-seq data alignment and alternative splicing detection	http://tophat.cbcb.umd.edu/
Trans-AByss	De novo assembly	http://www.bcgsc.ca/platform/bioinfo/software/
Trinity	De novo assembly	http://trinityrnaseq.sourceforge.net/
FungiDB	Functional genomics of fungi	http://fungidb.org
PAMGO	Controlled vocabulary for the interaction of microbes with their hosts	http://pamgo.vbi.vt.edu/
PHI-base	Fungal pathogen host interactions	http://www.phi-base.org/
PLEXdb	Gene expression resources for plants and plant pathogens	http://www.plexdb.org
PathoPlant®	Signal transduction related to plant–pathogen interactions	http://www.pathoplant.de/
COGEME	Phytopathogenic fungi and oomycete EST database	http://cogeme.ex.ac.uk/
DFVF	Database of fungal virulence factors	http://sysbio.unl.edu/DFVF/

5.4.4 Proteomics

Detailed analysis of fungal biochemistry is now enabled by multiple technologies including protein mass spectrometry, genome and transcriptome sequencing and advances in bioinformatics. Unraveling intricacies of more complex proteomic interactions that involve fungi, oomycetes, PGPRs and symbiotic fungi, which strengthen plant defenses will generate valuable information on how pathosystems actually function in nature, and thereby provide clues to solving disease problems that engender major losses in crops every year. Yet, the assignment of function to fungal proteins, encoded either by in silico annotated, or unannotated genes, remains problematic. Recent reviews (González-Fernández et al. 2010; Gonzalez-Fernandez and Jorrin-Novo 2012) clearly illustrates that proteomics has also been exploited, but perhaps not to its potential, by the fungal phytopathogen community. A lack of genome sequence information has frustrated proteomics researchers and has largely contributed to this technique not fulfilling its potential. A combination of high-throughput, quantitative proteomics, allied to transcriptomic sequencing, are set to reveal much about protein function in fungi (Doyle 2011). They allow screening and analysis, at the sub-cellular level, of peptides and proteins resulting from plants, pathogens, and their interactions. They also highlight post-translational modifications to proteins, e.g. glycosylation, phosphorylation, or cleavage. Recent literature testifies the usefulness of novel, gel-free techniques in protein identification and quantification in pathogenic fungi like *Botrytis cinerea* (Gonzalez-Fernandez et al. 2013) and *P. infestans* (Lim et al. 2013).

5.4.5 Metabolomics

Metabolomics is the most recent of these techniques to emerge and is concerned with the non-targeted profiling of all metabolites in a given system. It consists of strategies to quantitatively identify cellular metabolites and to understand how trafficking of these biochemical messengers through the metabolic network influences phenotype. Metabolomics plays a key role in functional genomics and strain classification (Michael et al. 2006). The availability of data from fungal genome sequencing projects has facilitated the discovery and characterization of new compounds and their biosynthetic pathways. Despite their medical and agricultural importance, most putative gene clusters in fungal genomes have been predicted manually. Knowledge of the genetic background of endophytic natural product biosynthesis is discussed in detail by Staniek et al. (2008). Of late, the complex biosynthetic pathways associated with secondary metabolites are resolved by combining in silico pathway analysis with metabolite profiling (Forster et al. 2002; Nora et al. 2010). A few tools for predicting fungal secondary metabolites (SMURF, antiSMASH, etc.) biosynthesis are already available online (Khaldi et al. 2010; Blin et al. 2013). A list of proteomic tools currently in use is given in Table 5.5. Metabolomics studies on fungal plant pathogens are only just beginning to appear, although their potential to dissect many facets of the pathogen and disease will see its popularity increase quickly. There are a few metabolomic studies done on crops with fungal infection (Figueiredo et al. 2008; Parker et al. 2009; Aliferis and Jabaji 2012; Hong et al. 2012).

5.4.6 Systems Biology

The tools of systems biology can be successfully employed in understanding the complex ensemble of molecular interactions underlying pathogen–plant interactions. Systems biology enables studying virulence factors as pathosystem components, and pathosystems for their emergent properties (Schneider and Collmer 2010). Wherever genomic information is available, genome-scale metabolic reconstruction (GEMR), along with flux balance analysis, can be used to study host metabolic network phenotype during the interaction with pathogens. Alternatively, targeted metabolic reconstruction, where network reconstruction is guided by transcriptomic data instead of genomic information. A systems-level framework

Table 5.5 Bioinformatics resources commonly used in fungal proteomic studies

Program/Database	Description	URL
Mascot	Website for both peptide mass fingerprint and MS/MS database searches	http://www.matrixscience.com
SEQUEST	For correlating tandem mass spectra of peptides with amino acid sequences from protein and nucleotide databases	http://fields.scripps.edu/sequest/index.html
X!Tandem	Protein identification tool by matching tandem mass spectra with peptide sequences	http://www.thegpm.org/TANDEM/index.html
Expasy	A portal for proteomic resources	http://www.expasy.org/
UniprotKB/SwissProt/TrEMBL	A central hub for functional information on proteins and their annotation	http://www.uniprot.org/help/uniprotkb
Protein Information Resource (PIR)	A centralized resource for protein sequences and functional information	http://pir.georgetown.edu/
RCSB Protein Data Bank (RCSB PDB)	A portal for biological macromolecular structures	http://www.rcsb.org/pdb/download/download.do
EMBL-EBI's Protein Data Bank in Europe (PDBe)	European resource for the collection, organization and dissemination of 3D structural data on biological macromolecules	http://www.ebi.ac.uk/pdbe/
SWISS-2DPAGE	Data on proteins identified on various 2-D PAGE and SDS-PAGE reference maps	http://world-2dpage.expasy.org/swiss-2dpage/
FCPD	A database of fungal cytochrome P450	http://p450.riceblast.snu.ac.kr/
MPID	*Magnaporthe grisea* protein–protein interaction database	http://bioinformatics.cau.edu.cn/cgi-bin/zzd-cgi/ppi/mpid.pl
FPPI	Protein–protein interaction database of *F. graminearum*	http://csb.shu.edu.cn/fppi

will be highly useful for the interpretation and modelling of host–microbe interactions mediated by effectors. This approach is being applied successfully for the *Phytophthora infestans—Solanum tuberosum* pathosystem (Pinzón et al. 2011).

5.5 Metagenomics

Metagenomics (also referred to as environmental and community genomics) is the genomic analysis of microorganisms by direct extraction and cloning of DNA from an assemblage of microorganisms. The introduction of NGS platforms has enabled a great expansion of the field of metagenomics. As there is no need to clonally culture organisms before sequencing, any kind of organism can be captured in its natural environment, and its DNA can be subjected to sequencing. The tool is now applied to diverse habitats like ocean, soil, and human gut. NGS technologies enable identification and relative quantification of community members and offer new insights into fungal community ecology. These methods are currently taking over as the primary tool to assess fungal communities of plant-associated endophytes, pathogens, and mycorrhizal symbionts, as well as free-living saprotrophs. Species abundance in plant-associated fungal communities has been studied through metagenomics approaches (Unterseher et al. 2011). Lindahl et al. (2013) have reviewed the different stages involved in fungal community analysis, from field sampling via laboratory procedures to bioinformatics and data interpretation.

5.6 Conclusion

Next-generation sequencing is playing an increasingly high-profile role in transcriptomics, diagnostics, and epidemiology. Sequencing throughput is no longer the major limiting factor, but sequence assembly and annotation for complex genomes as well as data integration are the challenges. The data acquisition platforms for other "-omics", on the other hand, are under rapid development to catch up with the pace of genomic research. The standardization of data acquisition and storage formats using strictly controlled vocabulary is also important. There cannot be any doubt that NGS approaches are here to stay and will provide major stimuli for bioinformatics for many years to come. Nevertheless the field of phytopathology will be radically transformed within a decade by the inevitable flood of sequence data and the opportunities and challenges that it affords.

References

Aliferis KA, Jabaji S (2012) FT-ICR/MS and GC-EI/MS metabolomics networking unravels global potato sprout's responses to *Rhizoctonia solani* infection. PLoS One 7:e42576

Blin K, Medema MH, Kazempour D, Fischbach MA, Breitling R, Takano E, Weber T. antiSMASH 2.0—a versatile platform for genome mining of secondary metabolite producers. Nucleic Acids Res 2013;41(Web Server issue):W204–W212. doi: 10.1093/nar/gkt449.

Brown NA, Antoniw J, Hammond-Kosack KE (2012) The predicted secretome of the plant pathogenic fungus *Fusarium graminearum*: a refined comparative analysis. PLoS One 7(4):1–14

do Amaral A, Antoniw J, Rudd JJ, Hammond-Kosack KE (2012) Defining the predicted protein secretome of the fungal wheat leaf pathogen *Mycosphaerella graminicola*. PLoS One 7(12): 1–19

Doyle S (2011) Fungal proteomics: from identification to function. FEMS Microbiol Lett 321(1):1–9

Ellwood SR, Liu Z, Syme RA, Lai Z, Hane JK, Keiper F et al (2010) A first genome assembly of the barley fungal pathogen *Pyrenophora teres* f. *teres*. Genome Biol 11:R109

Figueiredo A, Fortes AM, Ferreira S, Sebastiana M, Choi YH, Sousa L, Acioli-Santos B, Pessoa F, Verpoorte R, Pais MS (2008) Transcriptional and metabolic profiling of grape (*Vitis vinifera* L.) leaves unravel possible innate resistance against pathogenic fungi. J Exp Bot 59: 3371–3381

Fitzpatrick D (2012) Horizontal gene transfer in fungi. FEMS Microbiol Lett 329(1):1–8

Forster J, Gombert AK, Nielsen J (2002) A functional genomics approach using metabolomics and *in silico* pathway analysis. Biotechnol Bioeng 79:703–712

Gonzalez-Fernandez R, Jorrin-Novo J (2012) Contribution of proteomics to the study of plant pathogenic fungi. J Proteome Res 11(1):3–16

González-Fernández R, Prats E, Jorrín-Novo J (2010) Proteomics of plant pathogenic fungi. J Biomed Biotechnol 2010:932527. doi:10.1155/2010/932527

Gonzalez-Fernandez R, Aloria K, Arizmendi J, Jorrin-Novo J (2013) Application of label-free shotgun nUPLC-MS(E) and 2-DE approaches in the study of *Botrytis cinerea* mycelium. J Proteome Res 12(6):3042–3056

Goss EM, Carbone I, Grunwald NJ (2009) Ancient isolation and independent evolution of the three clonal lineages of the exotic sudden oak death pathogen *Phytophthora ramorum*. Mol Ecol 18:1161–1174

Grunwald NJ, Goss EM (2011) Evolution and population genetics of exotic and re-emerging pathogens: novel tools and approaches. Annu Rev Phytopathol 49:249–267

Haridas S, Breuill C, Bohlmann J, Hsiang T (2011) A biologist's guide to de novo genome assembly using next-generation sequence data: a test with fungal genomes. J Microbiol Methods 86:368–375

Hong YS, Martinez A, Liger-Belair G, Jeandet P, Nuzillard JM, Cilindre C (2012) Metabolomics reveals simultaneous influences of plant defence system and fungal growth in *Botrytis cinerea*-infected *Vitis vinifera* cv. Chardonnay berries. J Exp Bot 63:5773–5785

Horner DS, Pavesi G, Castrignano T, De Meo PD, Liuni S, Sammeth M, Picardi E, Pesole G (2009) Bioinformatics approaches for genomics and post genomics applications of next-generation sequencing. Brief Bioinform 11(2):181–197

Indian Institute of Spices Research (2013) Annual report 2012–13. Kozhikode: IISR, 92 p

Karthika R, Rosana OB, Chandrashekar A, Prasath D. *In silico* identification of ginger ESTs potentially associated with disease resistance. In: International symposium on biotechnology and bioinformatics, 26 Jun 2013, SV University, Tirupathi

Kawahara Y, Oono Y, Kanamori H, Matsumoto T, Itoh T, Minami E (2012) Simultaneous RNA-seq analysis of a mixed transcriptome of rice and blast fungus interaction. PLoS One 7(11):e49423. doi:10.1371/journal.pone.0049423

Khaldi N, Seifuddin FT, Turner G, Haft D, Nierman WC, Wolfe KH, Fedorova ND (2010) SMURF: genomic mapping of fungal secondary metabolite clusters. Fungal Genet Biol 47:736–741

Levesque CA, Brouwer H, Cano L, Hamilton JP, Holt C, Huitema E et al (2010) Genome sequence of the necrotrophic plant pathogen *Pythium ultimum* reveals original pathogenicity mechanisms and effector repertoire. Genome Biol 11:R73

Li C, Deng G, Yang J, Viljoen A, Jin Y, Kuang R et al (2012) Transcriptome profiling of resistant and susceptible Cavendish banana roots following inoculation with *Fusarium oxysporum* f. sp. *cubense* tropical race 4. BMC Genomics 13:374

Li MW, Qi X, Ni M, Lam HM (2013) Silicon era of carbon-based life: application of genomics and bioinformatics in crop stress research. Int J Mol Sci 14:11444–11483

Lim S, Borza T, Peters R, Coffin R, Al-Mughrabi K, Pinto D, Wang-Pruski G. Proteomics analysis suggests broad functional changes in potato leaves triggered by phosphites and a complex indirect mode of action against *Phytophthora infestans*. J Proteomics 93:207

Lindahl B, Nilsson R, Tedersoo L, Abarenkov K, Carlsen T, Kjøller R, Kõljalg U, Pennanen T, Rosendahl S, Stenlid J, Kauserud H (2013) Fungal community analysis by high-throughput sequencing of amplified markers—a user's guide. New Phytol 199:288–299

Littlejohn KA, Hooley P, Cox PW (2012) Bioinformatics predicts diverse *Aspergillus* hydrophobins with novel properties. Food Hydrocoll 27(2):503–516

Mardis ER (2008) Next-generation DNA, sequencing methods. Annu Rev Genomics Hum Genet 9:387–402
Michael CJ, Gerald H, Jens N (2006) Fungal metabolite analysis in genomics and phenomics. Curr Opin Biotechnol 17:191–197
Miller JR, Delcher AL, Koren S, Venter E, Walenz BP, Brownley A et al (2008) Aggressive assembly of pyrosequencing reads with mates. Bioinformatics 24:2818–2824
Morozova O, Hirst M, Marra MA (2009) Applications of new sequencing technologies for transcriptome analysis. Annu Rev Genomics Hum Genet 10:135–151
Nan J, Yaping Y, Guilhem J, Jiao P, Xudong Z (2012) Identification and functional demonstration of miRNAs in the fungus *Cryptococcus neoformans*. PLoS One 7(12):1–10
Nora K, Fayaz TS, Geoff T, Daniel H, William CN, Kenneth HW, Natalie DF (2010) SMURF: genomic mapping of fungal secondary metabolite clusters. Fungal Gene Biol 47:736–741
Park J, Park B, Jung K, Jang S, Yu K, Choi J et al (2008) CFGP: a web-based, comparative fungal genomics platform. Nucleic Acids Res 36:D562–D571
Parker D, Beckmann M, Zubair H, Enot DP, Caracuel-Rios Z, Overy DP et al (2009) Metabolomic analysis reveals a common pattern of metabolic re-programming during invasion of three host plant species by *Magnaporthe grisea*. Plant J 59(5):723–737
Passos M, de Cruz V, Emediato F, de Teixeira CC, Azevedo VC, Brasileiro AC et al (2013) Analysis of the leaf transcriptome of *Musa acuminata* during interaction with *Mycosphaerella musicola*: gene assembly, annotation and marker development. BMC Genomics 14:78
Pinzón A, Rodriguez-R LM, González A, Bernal A, Restrepo S (2011) Targeted metabolic reconstruction: a novel approach for the characterization of plant–pathogen interactions. Brief Bioinform 12(2):151–162
Prasath D, Amruta B, Vijay M, Rosana OB, Jayasankar S, Anandaraj M (2014) Comparative study of pathogenesis related protein-5 of different *Zingiberaceae* species. Indian J Biotech 13(2): (in press)
Prasath D, Karthika R, Habeeba NT, Suraby EJ, Rosana OB, Shaji A, Eapen SJ, Deshpande U, Anandaraj M (2014). Comparison of the transcriptomes of ginger (*Zingiber officinale* Rosc.) and mango ginger (*Curcuma amada* Roxb.) in response to the bacterial wilt infection. *PLOS One* 9(6): e99731. doi:10.1371/journal.pone.0099731
Převorovsky M, Atkinson S, Ptáčkova M, McLean J, Gould K, Folk P, Půta F, Bähler J (2011) N-termini of fungal CSL transcription factors are disordered, enriched in regulatory motifs and inhibit DNA binding in fission yeast. PLoS One 6:1–11
Raffaele S, Farrer RA, Cano LM, Studholme DJ, MacLean D, Thines M et al (2010) Genome evolution following host jumps in the Irish potato famine pathogen lineage. Science 330: 1540–1543
Reena N, Anil P, Dhanya KP, Anandaraj M, Eapen SJ (2010a) Structural and docking studies of glucanase inhibitor protein from Phytopthora capsici with plant endo-β-1, 3 glucanses (Abstract). In: Prasath D et al (eds). Bioinformatics: an agricultural perspective. Indian Institute of Spices Research, Kozhikode, p 111
Reena N, Chandrasekar A, Riju A, Nima PL, Eapen SJ, Anandaraj M (2010b) Gene identification in *Phytophthora capsici* through expressed sequence tags. In: ISB 2010: Proceedings of the international symposium on biocomputing; 11–12 Oct 2010, ACM Digital Library. http://doi.acm.org/10.1145/1722024.1722043
Rokas A, Payne G, Fedorova ND, Baker SE, Machida M, Yu J et al (2007) What can comparative genomics tell us about species concepts in the genus *Aspergillus*? Stud Mycol 59:11–17
Schneider DJ, Collmer A (2010) Studying plant-pathogen interactions in the genomics era: beyond molecular Koch's postulates to systems biology. Annu Rev Phytopathol 48:457–479
Singh N, Kashyap S (2012) In silico identification and characterization of 1-aminocyclopropane-1-carboxylate deaminase from *Phytophthora sojae*. J Mol Model 18(9):4101–4111
Soanes DM, Talbot NJ (2006) Comparative genomic analysis of phytopathogenic fungi using expressed sequence tag (EST) collections. Mol Plant Pathol 7:61–70
Spanu PD, Abbott JC, Amselem J, Burgis TA, Soanes DM, Stüber K et al (2010) Genome expansion and gene loss in powdery mildew fungi reveal functional tradeoffs in parasitism. Science 330:1543–1546

Stahl PL, Lundeberg J (2012) Toward the single-hour high-quality genome. Annu Rev Biochem 81:359–378

Staniek A, Woerdenbag HJ, Kayser O (2008) Endophytes: exploiting biodiversity for the improvement of natural product-based drug discovery. J Plant Interact 3:75–93

Studholme DJ, Glover RH, Boonham N (2011) Application of high-throughput DNA sequencing in phytopathology. Annu Rev Phytopathol 49:87–105

Stukenbrock EH, Banke S, Javan-Nikkhah M, McDonald BA (2007) Origin and domestication of the fungal wheat pathogen *Mycosphaerella graminicola* via sympatric speciation. Mol Biol Evol 24:398–411

Torto-Alalibo T, Collmer CW, Gwinn-Giglio M (2009) The Plant-Associated Microbe Gene Ontology (PAMGO) Consortium: community development of new gene ontology terms describing biological processes involved in microbe-host interactions. BMC Microbiol 9(Suppl 1):S1

Unterseher M, Jumpponen A, Opik M, Tedersoo L, Moora M, Dormann C, Schnittler M (2011) Species abundance distributions and richness estimations in fungal metagenomics—lessons learned from community ecology. Mol Ecol 20(2):275–285

Vijai B, Lucia P, Wen-Sheng Z, You-Liang P (2007) Review: fungal transcriptomics. Microbiol Res 162:285–298

Vijeshkumar IP, Reena N, Anandaraj M, Eapen SJ, Johnson GK, Vinitha KB (2013) Amplification, cloning and in silico prediction of full length elicitin gene from *Phytophthora capsici*, the causal agent of foot rot disease of black pepper. J Plant Pathol 4:181

Yazawa T, Kawahigashi H, Matsumoto T, Mizuno H (2013) Simultaneous transcriptome analysis of sorghum and *Bipolaris sorghicola* by using RNA-seq in combination with *de novo* transcriptome assembly. PLoS One 8(4):e62460

Zhong SS, Leng YY, Friesen TL, Faris JD, Szabo LJ (2009) Development and characterization of expressed sequence tag-derived microsatellite markers for the wheat stem rust fungus *Puccinia graminis* f. sp. *tritici*. Phytopathology 99(3):282–289

Zhou J, Fu Y, Xie J, Li B, Jiang D, Li G, Cheng J (2012) Identification of microRNA-like RNAs in a plant pathogenic fungus *Sclerotinia sclerotiorum* by high-throughput sequencing. Mol Genet Genomics 287(4):275–282

Chapter 6
Pre and Post Harvest Diseases of Potato and Their Management

R.K. Arora and Sanjeev Sharma

6.1 Introduction

Potato, an important food crop plant has the potential to meet food demand of the fast growing human population. Management of diseases and pests is important to realize full potential of the crop. Potato can be affected by many diseases which affect at both pre and post harvest stage of the crop. Major fungal diseases which affect the crop are late blight, early blight, black scurf, fusarium dry rot, wart, powdery scab and charcoal rot. Such diseases are prevalent in many countries and cause a significant reduction in potato production. A brief description of these diseases and their management is given in this chapter.

6.2 Late Blight

Late blight is the most destructive disease which poses a great threat to potato cultivation worldwide. Worldwide loss due to *Phytophthora infestans* has been estimated to €12 billion per annum of which the losses in developing countries have been estimated around €10 billion per annum (Haverkort et al. 2009). A survey carried out to estimate the impact of late blight on potato yield and fungicide use in the USA revealed that use of the fungicides alone costs $77.1 million and an

R.K. Arora, Ph.D. (✉)
Central Potato Research Station, Post Bag No. 1, Model Town Post Office, Jalandhar, Punjab 144003, India
e-mail: rkacpri@yahoo.com

S. Sharma, Ph.D.
Division of Plant Protection, Central Potato Research Institute, Shimla, Himachal Pradesh, India

A. Goyal and C. Manoharachary (eds.), *Future Challenges in Crop Protection Against Fungal Pathogens*, Fungal Biology, DOI 10.1007/978-1-4939-1188-2_6,

Fig. 6.1 Potato leaves infected with late blight (*Phytophthora infestans*)

average cost of around $507 per ha which do not include non-fungicide control practices (Guenthner et al. 2001). Late blight is a potential threat and can raise the fear of famine in vast areas of Eastern Europe and Russia where millions of people are subsisting on potatoes (Mackin 1998). Information on various aspects of late blight has been reviewed by different workers (Erwin et al. 1983; Neiderhauser 1986; CIP 1989; Lucas et al. 1991; Ingram and Williams 1991; Arora and Khanna 1997; Singh and Shekhawat 1999; Fry 2008; Cooke et al. 2011).

6.2.1 Symptoms

The disease appears as water-soaked irregular pale green lesions mostly near tip and margins of leaves which rapidly grow into large brown to purplish black necrotic spots. A white mildew, which consists of sporangia and sporangiophores of the pathogen, can be seen on lower surface of the infected leaves especially around the edges of the necrotic lesions (Fig. 6.1). Light to dark brown lesions encircle the stems. The affected stems and petioles become weak at such locations and may collapse. Entire crop gives blackened blighted appearance especially under disease favourable conditions and may be destroyed within a week (Fig. 6.2). Tubers in soil become infected by rain borne sporangia coming from the diseased foliage. Late blight infected tubers show irregular reddish brown to purplish areas which extend into internal tissues of the tubers (Fig. 6.3). The infected tubers usually are hard, dry and firm but may get attacked by soft rot causing bacteria and rot in field and stores.

Fig. 6.2 A potato field infected with late blight

Fig. 6.3 Potato tubers infected with *Phytophthora infestans*

6.2.2 The Pathogen

Late blight is caused by *Phytophthora infestans* (Mont.) de Bary. It belongs to the oomycetes, a diverse group of eukaryotic microorganisms in a group called the Stramenopiles, clustering together with others in a super group, the Chromalveolata (Adl et al. 2005). The position of the oomycetes as a unique lineage of eukaryotes unrelated to true fungi but closely related to heterokont (brown) algae and diatoms, is well established through molecular phylogenies and biochemical studies (Baldauf et al. 2000). The pathogen is characterized by lemon shaped detachable, papilliate sporangia produced on sympodially branched sporangiophores of indeterminate growth. The sporangiophores exhibit a characterized swelling at junction where sporangia are attached with the sporangiophores. Ultrastructure of hyphae and haustoria has been studied in detail by Ehrlich and Ehrlich (1966).

Phytophthora infestans is heterothallic and requires two mating types A_1 and A_2 for sexual reproduction. Prior to 1984 the A_2 mating type was restricted to Mexico and Andean mountains the centre of origin of cultivated potatoes. First report of A_2 mating type outside Mexico was from Switzerland (Hohl and Iselin 1984) followed by reports from other countries (Malcolmson 1985; Mosa et al. 1989; Fry et al. 1989; Singh et al. 1994). This is considered as second migration of *P. infestans* outside Mexico (Fry et al. 1999), the first being from Europe and America during the historical potato famine around the year 1845. The new strains of the pathogen have been found to be more aggressive than the old population (Fry et al. 1999). Turkensteen and Mulder (1999) have reported that pathogen during the last 20 years have developed a shorter life cycle (by 30 %), ability to cause more leaf spots, shorter infestation period (6 instead of 8 h), tolerance to a greater temperature range (5–27 °C instead of 10–25 °C), form stem lesions more frequently, develops oospores and sporulation on tubers and is more inclined to develop resistance to fungicide metalaxyl.

Population of *P. infestans* in most countries has changed dramatically and original A_1 has almost been displaced by more virulent A_2 strain. Occurrence of both A_1 and A_2 strains at the same location has also opened up the possibility of development of thick walled oospores which could survive extreme winter (Medina and Platt 1999) or summer conditions. The oospores may act as an another source of primary inoculum, in addition to the already known sources such as infected seed tubers, waste heaps and volunteer plants etc. The impact of potato late blight increased greatly during the 1990s following the migration of more genetically diverse and more aggressive genotypes of *P. infestans* from Mexico (Goodwin et al. 1994). Recent work has indicated that the new *P. infestans* clones, especially the US-8 and US-14 genotypes, are more aggressive (Lambert and Currier 1997; Kirk et al. 2001, 2009). The new genotypes of late blight are 10 times more likely to produce infected sprouts than their predecessor, US-1 (Marshall and Stevenson 1996). Before 1980s, worldwide populations of *P. infestans* were dominated by a single clonal lineage, the US-1 genotype with Ib mtDNA haplotype (Goodwin et al. 1994). This lineage has since been displaced by the other haplotypes (Ristaino et al. 2001). Since 2002, in India most isolates studied were of Ia mtDNA haplotype (Chimote et al. 2010). Displacement of old Ib mtDNA haplotype population by new Ia haplotypes is consistent with the global trend of mt haplotype distribution. Migration and sexual recombination can play an important role in enhancing genetic diversity in *P. infestans*. Chowdappa et al. (2013) have recently reported that migration of 13_A2 genotype that was responsible for outbreak of destructive late blight epidemics in Karnataka state of India since 2009 and have suggested the importance of bio-security in agricultural trade.

6.2.3 Epidemiology

Persistence of *P. infestans* from year to year is a critical component of late blight epidemics. An understanding of pathogen survival is a major goal in developing management strategies for the disease. Infected seed tubers serve as overwintering and the primary source of inoculum (Kirk 2003). The pathogen over winters as mycelium in infected tubers, in refuse piles and volunteer plants or over summer in subtropical zones through tubers kept in cold stores (Pushkarnath and Paharia 1963; Boyd 1981). Potato tubers left in the field after harvest and cull potato tubers can produce volunteer plants which can carry over the pathogen to the next season (Zwankhuizen et al. 1998). Latent infection of potato tubers by *P. infestans* has been implied in development of the disease in Ecuadorian highlands (Kromann et al. 2008). Latent infection was demonstrated when the pathogen was detected with the aid of polymerase chain reaction (PCR) in asymptomatic tubers (Appel et al. 2001; Hussain et al. 2005; Sharma et al. 2010). Johnson and Cummings (2009) demonstrated presence of latent infection in seed tubers and production of viable sporangia of *P. infestans* after cold storage of infected potato tubers. Survival of pathogen as oospores in soil serves as another source of primary inoculum. However, its exact role and extent of contribution is not clear. Movement of pathogen from infected tubers to new plant could be indirect through soil. Tubers in soil get infected by contact with sporangia coming from infected haulms through rain water. The infection can also occur during washing of tubers.

Forecasting appearance of the disease in advance can greatly help potato farmers to take prophylactic sprays and prevent or delay appearance of the disease and thus reduce the losses. Weather conditions such as temperature, relative humidity, rainfall, dew, sunshine hours etc. have a direct effect on *P. infestans*. These parameters have been exploited to develop different models to forecast the disease both in India and elsewhere (Van Everdingen 1926; Beaumont 1947; Hyre 1954; Wallins 1962; Ulrich and Schrodter 1966; Krause et al. 1975; Braun and Fry 1981; Bhattacharyya et al. 1982). Similarly, potato late blight alert networks have been in operation in some countries with satisfactory results (Chow and Bernard 1999; Hensen et al. 2000). Different methods and weather criteria are required for forecasting potato blight for different regions. Based on local weather parameters a computerized forecast for late blight named as 'JHULSACAST' has been developed for western subtropical plains of India (Singh et al. 2000). Recently the JULSACAST model was modified to forecast late blight under Punjab conditions in India (Arora et al. 2012). JHULSACAST model was implemented in western Uttar Pradesh using Wireless Sensor Networks (WSN), which is web based, to forecast late blight and the model could forecast the same well in advance in comparison to other forecasting models tested. An internet based decision support system (PhytoPRE+2000) has been developed in Switzerland which is an improved version of PhytoPRE where weather conditions on major infection and sporulation period (MISP) have been incorporated (Cao et al. 1996). Comparing to original PhytoPRE programme, it can save about 30–50 % chemical usage.

International Potato Centre has been linked to disease forecasting models, Blitecast and Simcast to climate database in a geographical information system (GIS) to estimate global severity of potato late blight. Tropical highlands are the zone of high late blight severity. Major production zones with a low late blight severity include Western Plains of India, where irrigated potato is produced in the cool dry season, North Central China and North-Western USA. Average number of sprays calculated for different countries using GIS database of potato production compared with estimated current fungicide use revealed that the estimated number of sprays in developing countries whether from Blitecast or Simcast, predicted optimum number of sprays much higher as compared with the actual number observed. On the basis of GIS database it was suggested that an increased access to host resistance and fungicides in developing countries could have a strong economic impact on potato production (Hijmans et al. 2000). Decision support system for organic potato farming (Bio-PhytoPRE) has also been developed by Agroscope FAL Reckenholz to assist Swiss Organic potato producers to control late blight with reduced amounts of copper (Musa-Steenblock and Forrer 2005).

6.2.4 Management

Reduction of the primary sources of inoculum is the first step in management of late blight. Control of contaminated sources such as waste heaps, infected tubers, volunteer plants, disease in neighbouring fields and re-growth after haulms destruction can help in management of the disease (Turkensteen and Mulder 1999). In Switzerland, it has been estimated that onset of epidemic can be delayed by 3–6 weeks if all primary infection from early potato can be eliminated (Forrer et al. 2000). It has been shown that during most years late blight epidemics start from infected plants on dumps (Zwankhuizen et al. 2000), therefore, covering of dumps with black plastic sheet throughout the season and preventing seed tubers from becoming infected is an important step in reducing the primary inoculum (Cooke et al. 2011). The sheeting must be kept in place and remain intact until the tubers are no longer viable. This will prevent re-growth and the proliferation of spores on the piles, reducing the risk to nearby crops. Oospores are a threatening primary inoculums source, especially with short crop rotation. Sandy and clay soils contaminated with oospores remained infectious for 48 and 34 months, respectively (Turkensteen et al. 2000). Use of early-maturing cultivars, pre-sprouting the seed and early planting can help to manage late blight. Avoiding excess nitrogen and use of moderate nitrogen fertilization is often recommended as a cultural practice to delay the development of late blight. Use of systemic fungicides early in the season is an effective strategy to manage late blight if source of primary infection is infected seed (Hermansen and Naestad 2009). Increased application of nitrogen can lead to increase in disease severity and use of more and more fungicides. Higher dose of phosphorus and potassium has been found to give a higher yield in a late blight year (Roy et al. 2001).

Importance of oospores as soil-borne inoculum is determined both by their formation in plant tissue and their survival in soil. There is a correlation between crop rotation and late blight infections. Infection usually start early in fields which are not subjected to crop rotations. The decline in early infection was most pronounced in fields subjected to crop rotations for three or more year between the potato crops (Bodker et al. 2006; Hannukkala et al. 2007). This indicated that a sound crop rotation is important and is an effective way of reducing the risk of soil-borne infections of *P. infestans*. Choice of suitable cultivars, well aerated fields, pre-sprouting of tubers, early planting and use of resistant varieties are some of the measures against foliar blight while planting potatoes on large steep ridges, right time of mechanical weeding and harvesting, avoiding rapid shift of harvested tubers and long transports could minimize tuber blight (Meinck and Kolbe 1999).

Development of resistant cultivars has played an important role in the control of late blight. *Solanum demissum*, a hexaploid wild species, has extensively been used to confer resistance against *P. infestans*. Field resistance is polygenic and more durable. *Solanum bulbocastanum*, *S. microdontum*, *S. verrucosum* and *S. chacoense* have been used as a source of field resistance in breeding programmes. Since the pathogen is quite plastic and mutable matching races against major R genes develop readily and overcome the resistance of the new cultivars. However, major genes which have evolved naturally in *S. demissum* population for thousands of years where late blight occurs annually still hold their importance. A multilineal combination of 11 resistant genes (R genes) identified so far, into commercial varieties has significant potential in management of late blight (Niederhauser et al. 1996). Disease resistance in potato varieties together with use of fungicides can slow down the development of late blight. A variety with field resistance to late blight in tubers and a medium to high resistance in the foliage can help in reducing the use of fungicides.

Use of host density as a tool for management of late blight has also been used for control of late blight. Tuber yield from both resistant and susceptible cultivar increases when these were grown in mixture as compared to the single genotype stands (Garret and Mundt 2000). Strip cropping of potatoes significantly reduced late blight severity in organic production when planted perpendicular to the wind neighbored by grass clover (Bounes and Finckh 2008).

Spraying with an effective fungicide has been a standard practice for control of late blight. Bordeaux mixture, which consists of copper sulphate, hydrated lime and water was a standard fungicide for many years. Subsequently organic fungicides especially carbamates which controlled both early and late blight replaced Bordeaux mixture. Metalaxyl—a phenylamide group of fungicides specific to oomycetes however, revolutionized late blight control (Bruck et al. 1980). Since it was most effective its use increased rapidly and this became one of the major fungicides used world over but strains of *P. infestans* which do not respond to metalaxyl appeared worldwide (Dowley and O'Sullivan 1981; Gisi and Cohen 1996). In India, resistance to metalaxyl in *P. infestans* wild population was first observed in

Nilgiri hills of South India in 1989. Metalaxyl resistant strains appeared towards the end of summer crop season and their frequency increased to 13 % in autumn season (Arora et al. 1992a). Metalaxyl in mixture with unrelated contact fungicide however, could retard development of resistance in the pathogen (Gangawane et al. 1993). Cymoxanil mixtures have been found effective for managing metalaxyl resistant strains (Samoucha and Cohen 1988). A synergism between cymoxanil and mancozeb has also been reported by Evenhuis et al. (1996). Fluazinam (Shirlan), cyazofamid (Ranman) and mandipropamid (Revus) have been used for the disease management. Spraying with effective fungicides (cyazofamid and mandipropamid) before periods with high risk of infections can give very effective control of late blight. Studies conducted in Denmark in 2009 showed that use of cyazofamid and mandipropamid could be reduced by 30 % by adjusting the dose according to resistance level in a variety and used according to the infection pressure (Cooke et al. 2011). Application of sub-phytotoxic concentration of boron with reduced rate of fungicide propineb + iprovalicarb has been reported as more effective as compared to plants treated with fungicide alone (Frenkel et al. 2010).

Heavy dependence on fungicides could pose threat to environment and human population (Bradshaw et al. 2000). Biocontrol agents and biopesticides could be a safe option to the use of synthetic fungicides. Antagonism to *P. infestans* by some naturally occurring microorganisms such as *Trichoderma viride*, *Penicillium viridicatum*, *P. aurantiogriseum*, *Chaetomium brasiliense* (Arora et al. 1992b; Gupta et al. 2004), *Acremonium strictum* (CPRI 1998–1999), *Myrothecium verrucaria*, *Penicillium aurantiogriseum* (Roy et al. 1991), *Epiccocum purpurascens*, *Stachybotrys coccodes*, *Pseudomonas syringae*, *Fusarium graminearum* (Kim et al. 1996) and *Pythium ultimum* (Kuzuestova et al. 1995) have been observed in laboratory and field studies. The biocontrol agents in general have been found to be very effective under laboratory and glasshouse conditions but less effective under field conditions (Arora 2000b). However, an integrated use of biocontrol agents along with low dose of fungicides could help to reduce the quantity of fungicides used in the management of late blight (CPRI 2000–2001).

6.3 Early Blight

Early blight affects both potato and tomato. It is a ubiquitous disease of potato prevalent in many countries in Asia, Africa, Australia, Europe, North, Central and South America (Millar and Pollard 1976). The disease used to appear earlier than late blight in the USA hence the name early blight. However, the name is misleading because the disease rarely attacks young growing plants and more often affect mature old plants showing loss of vigour. The disease is particularly severe under alternate dry and wet climate where the annual loss from this disease could range between 10 and 25 %.

Fig. 6.4 Potato leaves infected with early blight (*Alternaria solani*)

6.3.1 Symptoms

Small, round, oval or angular, dark brown to black, dry and papery necrotic spots which have angular margins appear on leaves. These spots are generally limited by leaf veins. Concentric rings of raised and depressed tissue within the leaf spot give it a bull's eye or target board appearance. Early blight lesions are less prone to secondary infections. Leaf tissues around the spots often become chlorotic and yellow suggesting involvement of toxins. The leaf spots may coalesce and the affected field appears severely blighted (Fig. 6.4). The disease also affects the tubers. On tubers, the lesions are dark, sunken, circular to irregular in shape, shallow and separated by healthy tissue by purplish-brown dry cork layer.

6.3.2 The Pathogen

Early blight is caused by *Alternaria solani* Sorauer (Ellis and Martin). Other species of *Alternaria* which attack potato are *Alternaria alternata* (Fries) Keissler, and *A. consortialis* (Conners 1967). In Germany, the frequency of *Alternaria alternata* and *A. solani* is almost equal (Hauslanden and Bassler 2004) while in Poland the occurrence of *A. alternata* has been reported more than that of *A. solani* (Kapsa 2007). *A. solani* has septate mycelium and bears conidia on erect and septate conidiophores. Cultural characters vary widely on potato dextrose agar medium. Colonies of the fungus are spreading, grey brown to black occasionally with yellow red pigment in the media. The mycelium sporulates sparingly in media however sporulation can be induced by mutilation of mycelium or exposing the culture to different light sources (Singh 1967; Barksdale 1969). Conidia are obclavate, olive brown with tapering long filiform beak. The conidia are multicellular and possess 3–14 cross septa and 0–18 longitudinal septa (Western 1971). Spores germinate in water within an hour at optimum range of temperature between 24 and 30 °C. Genetic diversity among isolates of *A. solani* in South Africa indicated that fungus has high potential to adapt to resistant cultivars and fungicides (Waals et al. 2003).

6.3.3 Epidemiology

Alternaria solani survives in crop debris, soil, infected tubers or alternate hosts which act as primary source of inoculum. The disease is favoured by short rotation, continuous cropping of potatoes and tomato. Infection is favoured by warm temperature and alternating high relative humidity provided by heavy dew, light rains or irrigation. Temperature in the range of 25–30 °C is congenial for the disease (Barclay et al. 1973). A positive correlation exists between minimum temperature, afternoon relative humidity and rainfall with early blight (Behera et al. 2009). Actively growing, properly fertilized young plants do not exhibit the disease. A delay of 10–15 days between haulm destruction and harvest prevents infection in tubers. Late maturing cultivars are generally more resistant than the early varieties. Predisposition of plants to injury, poor nutrition or other stresses could favour disease development (Singh et al. 1987b).

6.3.4 Management

Cultivation of solanaceous crops, being collateral hosts, near potato fields must be avoided. Removal and destruction of diseased haulms from infected fields reduces sources of primary inoculum for the next crop. Applying recommended dose of fertilizers especially nitrogen ensures healthy and vigorous growth and less disease. Permitting tubers to mature in soil and avoiding bruises at harvest minimizes tuber infection. Sprinkler irrigation favours disease and should not be used more often than necessary. Crop sprayed with one percent urea at 45 days of growth improves plant vigour and prevents onslaught of early blight and other leaf spots. Fungicides such as maneb, zineb, mancozeb, captafol, chlorothalonil provide good control of the disease. First spray should be applied as soon as lower leaves develop the spots which coincide with the secondary spread of the disease. Use of thidiazuron (TBZ) a growth regulator having cytokinin like activity with mancozeb and chlorothalonil resulted in delay of disease progress (Pavlista 2003). Similarly, the use of Pyton Consento 450SC @2l/ha (Bernat 2004), Zoxamide + mancozeb, fenamidone + mancozeb (Osowski 2004); potassium or sodium bicarbonate alone or in combination with Nerol (Abd-El-Kareem 2007) and kresoxim-methyl (Chakraborty and Roy 2012) have been found effective against early blight. Franc et al. (1988) have developed a prediction model based on accumulated day-degree above 7.2 °C from planting. This model forecasts beginning of secondary spread of the pathogen. Fungicides used according to the forecast model helps in reducing the use of fungicides. An integrated management of both early and late blight by combined application of fungicides have been suggested by Shtienberg (2001).

Decision support systems are important tools in reducing the large amount of fungicides applied to suppress disease intensity. FAST, CUFAST, TOMCAST and PLANT-Plus are the potential useful tools for integrated management of early blight (Waals et al. 2003; Batista et al. 2006). Beside chemicals, biopreparations have shown potential against early blight. The biopreparation Gluticid (*Pseudomonas aeruginosa*) and alirin-B and gamair (*Bacillus subtilis*) were found as effective as mancozeb (Rodriguez-Maza and Stefanova-Narimova 2005; Bairambekov and Korneva 2009).

Resistance to *Alternaria solani* is available in *Solanum phureja* and *S. chacoense* which can be exploited in breeding varieties resistant to early blight. A few varieties such as Kufri Sindhuri developed in India shows good resistance to early blight. Four synthetic peptides viz. pep 6, pep 7, pep 11 and pep 20 have been found to inhibit both *A. solani* and *P. infestans* on potato leaves. Expression of these peptides in transgenic potato plants could lead to enhanced disease resistance against these pathogens (Ali and Reddy 2000). Wild species of *Solanum* are excellent sources of disease resistance genus that may be incorporated into *S. tuberosum* through breeding. Clone C545 exhibited improved resistance to early blight (Jansky and Rouse 2003). No cultivar has been reported immune or resistant to early blight but large number of cultivars across the world possesses moderately resistance to early blight. These are Quaggy Joe (Reeves et al. 1999), Alta Russet (Lynch et al. 2004), Sazava, Nikoleta, Marcela (Voral 2005), MegaChip (Groza et al. 2007), Picasso, Victoria, Provento, Ramus and Milva (Alexandrov 2008).

The shift in sensitivity of *A. solani* to strobilurin fungicides has been observed in the USA (Pasche et al. 2002). A highly significant and strong correlation among the isolates tested for fungicide cross-sensitivity was detected between azoxystrobin and pyraclostrobin (Pasche et al. 2004). Isolates possessing reduced sensitivity to azoxystrobin were also less sensitive to famoxadone and fenamidone (Pasche et al. 2005). The repeated exposure of *A. solani* populations to chlorothalonil resulted in considerable variability in sensitivity causing isolates to have decreased sensitivity to the fungicide at the end of growing season. Five out of seven field isolates of *A. solani* collected by the end of season were found significantly less sensitive to chlorothalonil than isolates collected at the beginning of the season in the USA (Holm et al. 2003). Similarly, isolates of *A. solani* have been shown to possess resistance to mancozeb fungicide in Jordan (Al-Mughrabi 2004). Recently the resistance to boscalid fungicide has been reported in the USA which is widespread in Idaho (Wharton et al. 2012).

Fig. 6.5 Black scurf of potato caused by *Rhizoctonia solani*

6.4 Black Scurf and Stem Canker

Black scurf of potato caused by *Rhizoctonia solani* is a well-known disease of potato with worldwide distribution. It causes damping off of seedlings, stem canker in growing plants and black scurf on potato tubers. The disease kills potato sprouts, delays crop emergence, reduces crop stand, affects tuber quality and marketability of the produce (Errampalli and Johnston 2001). Significant yield losses primarily due to reduced crop stand has been reported by some workers (Carling et al. 1989).

6.4.1 *Symptoms*

The most common symptoms are on tubers as black irregular lumpy encrustations of fungal sclerotia which stick to the surface of tubers (Fig. 6.5). Other symptoms on tubers could be cracking, malformation, pitting and stem end necrosis. The pathogen produces a growth regulating toxin that may be partially responsible for tuber malformation. The fungus may kill emerging sprouts in soil which results in reduced crop stand. Reddish brown lesions may develop on stems and often girdle them. Partial or complete girdling of the stems could result in resetting of plant tops, purple pigmentation, upward curling or rolling of leaves. Formation of aerial tubers in axis of leaves due to interference with starch translocation is often also observed in plants infected with *R. solani* (Fig. 6.6).

6.4.2 *The Pathogen*

Rhizoctonia solani Kuhn is the imperfect stage of the pathogen where as *Thanatephorus cucumeris* (Frank) Donk (syn. *Corticum vagum* Berk. & Curt) is the perfect basidial stage. The sexual or perfect stage appears on stem just above soil line as whitish grey mat or mycelial felt. These mats are often located above

Fig. 6.6 Formation of aerial tubers in axis of leaves due to *Rhizoctonia solani*

a lesion on the below ground portion of stem and are generally visible later in the growing season under favourable weather conditions. The isolates have also been characterized on basis of sclerotial patterns and cultural characteristics (Raj et al. 1974). Mycelium of the pathogen is generally dark brown in colour. The hyphae are large multinucleate and branch near distal septum of the cell. They show right angle branching and constriction at the point of origin and a prominent septal pore. Early infection are initiated by differentiation of hyphal tips to T-shaped branches followed by formation of cushion like structure and development of appressoria from where thin infection hyphae arise and penetrate the underlying stem or stolon tissues. These cushions serve as additional food basis for colonization of underground plant parts and are pre-requisite for development of lesions on stems or stolons (Keijer et al. 1996). *Rhizoctonia solani* populations are distinguished by anastomosis between hyphae of the isolates belonging to the same 'anastomosis group' (AG). Fourteen different anastomosis groups' viz., AG-1 to AG-10, AG-BI (Sneh et al. 1991), AG-11 (Carling et al. 1994), AG-12 (Carling et al. 1989), and AG-13 (Carling et al. 2002) have been recognized. Several AGs have been subdivided further into subgroups that differ for one or more biochemical, genetic or pathogenic characteristics (Carling and Leiner 1990; Johnk and Jones 1993; MacNish et al. 1993).

Rhizoctonia solani isolated from potato mostly belong to anastomosis group 3 (AG-3) (Bandy et al. 1988). Eight subgroups have been identified within group AG-3 based on variations in isozyme patterns (Laroche et al. 1992). However, other anastomosis groups (AG-1, AG-2-1, AG-2-2, AG-4, AG-5, AG-7 and AG-9) also have been isolated from potato stems, stolons, roots and tubers, as well as from soils in which potatoes were grown (Carling and Leiner 1986; Chand and Logan 1983; Abd-Elsalam et al. 2009). In Central Mexico, in addition to AG-3, isolates of AG -2-2, -5, and -7 have been collected from potato plants and / or tubers from fields.

6.4.3 Epidemiology

Seed-borne (i.e. tuber borne) inoculum is the main source of primary infection leading to stem canker symptoms on the underground plant parts. The pathogen is both soil and seed borne but the disease spreads to new growing areas through sclerotia-covered seed tubers (Tsror and Perezt-Alon 2005). The disease gets established in fields wherever the untreated infected tubers are used as seed. Secondary inoculum of the *Rhizoctonia* disease is soil borne and accomplished by *R. solani* mycelia and sclerotia already inhabiting soil where the potato crop is planted (Balali et al. 1995). Soil-borne infection emerges later in the season since the fungus needs some time to grow into proximity with its potato host (Carling and Leiner 1986). Sclerotia of the pathogen germinate between 8 and 30 °C and invade emerging sprouts or potato stems. Optimum temperature for germination of sclerotia is 23 °C and for development of stem lesions is 18 °C (Walker 1969). Sclerotial development on tubers is initiated depending on environmental conditions. Late harvested crop shows more black scurf incidence since maximum development of sclerotia takes place in the period between dehaulming and harvest of the crop.

6.4.4 Management

Use of healthy seed free from sclerotia of the pathogen helps in disease management (Frank and Leach 1980). The disease can best be managed in an integrated manner by following proper cultural practices together with seed disinfestations. Soil solarization with transparent polyethylene mulching during hot summer months in Indian subtropical plains has been found to be very effective for control of the soil-borne part of the disease (Arora et al. 1997). Crop rotation offers an effective protection against soil-borne inoculum of *R. solani* (Carling et al. 1989). Planting of potato should be carried out in relatively dry and warm soil to achieve rapid crop emergence and an appropriate crop rotation programme should be followed to manage the disease (Anderson 1982; Bandy et al. 1988; Carling and Leiner 1990). Shallow covering of seed tubers allows less opportunity for the fungus to attack the susceptible sprouts and thus less disease. Two to four year crop rotation with cereals and legumes leads to decline in the population levels of the *R. solani.* Cereals are good rotational crops since *R. solani* affecting cereals have different AGs and cannot affect potato (Anderson 1982).

Biocontrol agents such as *Trichoderma viride* (Arora 1999), *T. harzianum* (Mishra et al. 2000), *Bacillus subtilis* (Schmiedeknecht et al. 1998), non-pathogenic binucleate *Rhizoctonia* (Tsror et al. 2001), *Trichoderma atroviride* (Huang-Mc Breath 2001), *Gliocladium virens*, *G. catenulatum* and others have been identified to be effective against *R. solani*. Biocontrol products developed to manage the disease (Nieme and Lahdenpera 2000; Yakhin et al. 1998) have been found effective against black scurf disease. *Trichoderma* spp. are well documented for their ability to protect plants from *R. solani* infection by production of antibiotics, antifungal chemicals and hydrolytic enzymes. Some *Trichoderma* spp. have been reported to provide plant growth enhancement as well as crop protection (Verma et al. 2007). A bioformulation developed at Central Potato Research Institute from *T. viride* strain A-7 and evaluated in several field trials in India was found very effective when used as seed treatment before planting potatoes (Arora and Somani 2001). Efficacy of *Trichoderma viride, Bacillus subtilis* and *Bacillus cereus* in consortium for control of *Rhizoctonia solani* was evaluated in field trials in India and the biocontrol agents used in combination were found to control the disease better than the biocontrol agents used alone (Somani and Arora 2010). An integrated use of *Trichoderma viride* and boric acid significantly improved disease control (Arora 2008). Experimental results of Wilson et al. (2008) suggest that combining chemical control with antagonist treatment can protect potato during the whole growing season.

Various fungicides such as benomyl, thiabendazole, carboxin, pencycuron and azoxystrobin (Virgen-Callerus et al. 2000), fenpiclonil (Welsh and Callaghan 1996) are effective for control of the disease. Seed treatment with 3 % boric acid has been identified as a safe and effective chemical treatment for the control of black scurf (Arora et al. 2006a). Seed treatment with 3 % boric acid as atomized application on infected tubers was found more economical than the dip treatment for control of seed inoculum (Khanna and Sharma 1996). Spray treatment of boric acid was equally effective in washed or unwashed seed tubers (Arora 2005). A combination of soil solarization carried out in north Indian Plains during summer months together with seed treatment with 3 % boric acid or *Trichoderma viride* was found very effective for control of black scurf in *Rhizoctonia* infested soils (Arora 2000a; Arora et al. 2006b, 2008).

6.5 *Fusarium* Dry rot

Fusarium dry rot, an important post harvest disease of potato tubers, causes significant losses in storage and transit of both seed and table potatoes. The disease is distributed worldwide and occurs wherever potatoes are grown (Stevenson et al. 2001). Healthy tubers become infected through bruises and wounds occurring during harvest, handling and transport. The symptoms of dry rot generally become evident 1–2 months after storage of the tubers. The affected tubers if used as seed can cause fusarial wilt in field. *Fusarium* spp. are present in abundance in soil and can infect surfaces of cut tubers when used as seed. Planting of un-suberized cut pieces of potato tubers which gets infected with fusaria can result in seed piece decay. Under such conditions losses by *Fusarium* rots may go up to 50 % (Chelkowski 1989). *Fusarium* spp. are known to produce toxins that cause mycotoxicoses in humans and animals (Senter et al. 1991).

Fig. 6.7 *Fusarium* spp. developing in dry rot affected tubers

6.5.1 Symptoms

The disease symptoms on potato tubers are generally visible in about a month after storage. The symptoms appear as small brown lesions on surface of the affected tubers. The lesions subsequently enlarge, appear dark, sunken, and wrinkled producing white, pink, or blue pustules (Stevenson et al. 2001). In later stages a cavity often develops in the centre of the concentric ring and whitish, pinkish or dark brown growth of fungal mycelium may become visible (Fig. 6.7). Rotten tubers may shrivel and get mummified. Under high relative humidity the secondary organisms such as *Erwinia* spp. can invade the infected tubers and cause soft rot (El-Gholl et al. 1985). Exudates containing bacteria come out from such tubers and further rot the surrounding tubers.

6.5.2 The Pathogen

About thirteen species of *Fusarium* have been reported to cause dry rot of potatoes (Cullen et al. 2000; Gachango et al. 2011). *F. sulphureum* Schlechtend (syn. *F. sambucinum* Fuckel) is the most common pathogenic species in North America and some regions of Europe (Hanson et al. 1996; Mecteau et al. 2002; Stevenson et al. 2001), whereas *F. coeruleum* (Libert) Sacc. (syn. *F. solani* var. *coeruleum*) is considered to be the predominant causal agent in the United Kingdom (Hide et al. 1992; Peters et al. 2004) and *F. oxysporum* Schlechtend in plains of India (Singh et al. 1987a). Other pathogenic species associated with dry rot include *F. avenaceum* (Fr.) Sacc., *F. culmorum* (Wm. G. Sm.) Sacc., *F. acuminatum* Ellis & Everh.

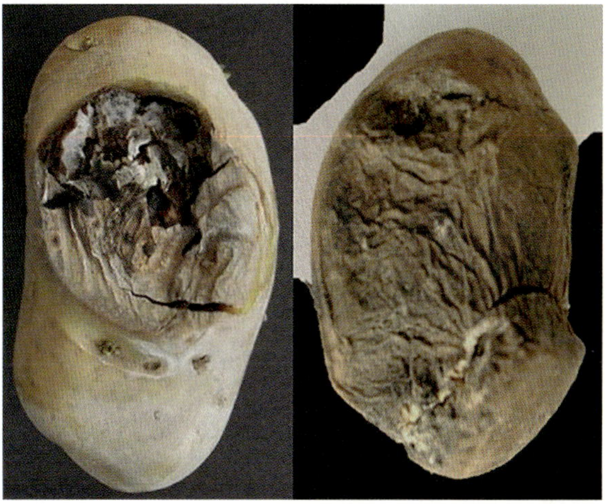

Fig. 6.8 Dry rot caused by *F. sambucinum*

F. crookwellense L. W. Burgess, P. E. Nelson & T. A. Toussoun, *F. equiseti* (Corda) Sacc., *F. graminearum* Schwabe, *F. scirpi* Lambotte & Fautrey, *F. semitectum* Berk. & Ravenel, *F. sporotrichioides* Sherb., and *F. tricintum* (Corda) Sacc. (Hide et al. 1992). Recently, *F. sambucinum* has been reported to cause dry rot of potatoes in cold stores in India (Fig. 6.8, Sagar et al. 2011).

White fluffy mycelial cushions may develop on surface of infected tubers. Mycelium of the pathogens consist of branched and septate hyphae which are present inter or intra cellular in the host tissue. Conidiophores arise from the mycelium and produce 1 to 4 celled sickle shaped conidia of variable size. Chlamydospores, the propagules of the pathogen which can survive in unfavourable conditions, develop in pustules. These may be intercalary or terminal. The fungus remains viable in soil for 9–12 months. *Fusarium* spp. has good saprophytic ability to survive in soil.

6.5.3 Epidemiology

Fusaria are always present in soil, in air, on implements, containers and it is practically not possible to eradicate them. They cannot infect intact periderm and lenticels of tubers. Cuts and wounds created during harvest, grading, transport and storage predispose them to infection. An increase in interval between haulm destruction and the harvest increases strength of tuber skin and is generally believed to reduce dry rots but the contrary view also exists Carnegie et al. (2001). Dry rot development is affected by tuber damage, degree of curing, tuber size and storage conditions. Use of herbicide paraquat, used for destruction of halums at maturity of the crop, has been observed to increase dry rots (Somani and Chohan 1994). The pathogen enters the tubers through wounds and proper wound healing could reduce the infection.

Tubers cured for wound healing at 21 °C with adequate aeration develop wound periderm in 3–4 days but it takes more time at lower temperature. Development of disease is also affected by moisture and temperature. The fungus grows well between 15 and 28 °C. *F. oxysporum* has been reported to become non-pathogenic below 10 °C. However, disease development continues at low temperature in cold stores. Storage period and relative humidity have been found to be positively correlated with dry rot whereas maximum temperature was negatively correlated (Singh 1986). Large sized tubers are more susceptible than small tubers. Susceptibility to tubers may also increase with tuber age during storage. No significant correlation exists between chemical composition of tubers and susceptibility to dry rots (Singh 1986; Percival et al. 1999). Some volatile compounds are produced by dry rot affected tubers and an early warning system based on sensors to detect these volatile compounds has also been developed by de Lacy-Costello et al. (2001).

Infected and rotting tubers are main source of spread of the inoculum of *Fusarium* spp. and results in soil infestation (Choiseul et al. 2001). *Fusarium* spores can survive in soil for several years and can infect the cut or damaged surfaces of seed potatoes whenever these come in contact with the spores infested soil. The pathogen may also get introduced to new locations through contaminated soil which adheres to the farm implements, through wind and irrigation water etc. Studies have demonstrated that transmission of *F. sulphureum* to progeny tubers was greater from highly contaminated seed than from rotting seed whereas in case of *F. coeruleum*, the disease transmitted readily from the rotted mother tuber (Carnegie et al. 2001; Choiseul et al. 2001). This was attributed to the different capacity of each species of the pathogen to sporulate underground on seed tubers and on stem bases. *F. coeruleum* sporulates profusely on the surface of rotting seed tubers, whereas *F. sulphureum* sporulates more readily on stem bases (Adams and Lapwood 1983; Choiseul et al. 2001).

6.5.4 Management

Avoiding bruises and damage to potato tubers by careful handling of the produce minimize the dry rots. This can be done by delaying harvesting for about 2 weeks after haulm destruction when skin of the tubers have matured. Harvesting on cold frosty morning predisposes potato to bruises. Bruises can also be avoided by taking suitable precaution with machinery, proper adjustment and padding etc. of the equipments. Washing of tubers to remove contaminated soil that adheres the tubers and drying these in shade can reduce the risk of infection. Harvested potatoes should be stored at around 13–18 °C and moderate humidity for 2–3 weeks for bruises to heal before putting the potato to cold stores. Planting of healthy seed, adopting sanitation measures to avoid soil contamination through farm implements, irrigation water and reducing soil inoculum through crop rotation and eliminating volunteer potatoes are some of the measures which can reduce the risk of dry rot.

Most potato cultivars are susceptible to *Fusarium* spp., though some are less susceptible than the others (Corsini and Pavek 1986; Lees et al. 1998). Breeding lines possessing higher degree of resistance to dry rot have been reported by Wharton and Kirk (2007). Transgenic potato plants constitutively over expressing beta 1-3-glucanase gene from *Nicotiana plumbaginifolia* have been developed for resistance against *F. oxysporum* (Libantova et al. 1998). Management of Fusaria through biocontrol agents such as *Trichoderma* spp. (Pinzon-Perea et al. 1999), *Pseudomonas fluorescens* (Schisler et al. 2000; Zin-Woo et al. 1998), *P. aeruginosa* (Gupta et al. 1999), *Bacillus subtilis* (Kim et al. 1995) have been found effective. Commercial biopreparations from *Pseudomonas fluorescens* have been developed (Ermakova and Shterushis 1994). Combination of biocontrol genera *Enterobacter* and *Pseudomonas* and two chitinolytic enzymes from *Trichoderma harzianum* had inhibitory effect on spore germination of *F. solani* (Lorito et al. 1993). Bacteria capable of binding to the fungal cell walls and expressing fungal genes coding cell wall degrading enzymes may act as powerful biocontrol agents.

Avoiding planting of cut tubers or treating the cut tubers with dithiocarbamates can reduce *Fusarium* seed piece decay (Rich et al. 1960). Tuber treatment with 1,200 ppm thiabendazole or benomyl can reduce the disease incidence (Leach 1976; Leach and Nelson 1975). However, resistance to thiabendazole in *Fusarium* has been reported by Hanson et al. (1996). Fungicides such as imazalil and mixtures containing TBZ have also found effective for control of dry rot (Carnegie et al. 1990, 1998).

6.6 Wart

Wart disease of potato caused by a fungus *Synchytrium endobioticum* (Schilberszky) Percival is an important and serious disease of potato worldwide (Franc 2001). It has been reported in Asia, Africa, Europe, Oceania, North America, and South America (EPPO 2006). It caused great damage to potato in Europe until immune varieties were introduced. The disease once established is difficult to eradicate since the resting sporangia can survive inter-host periods for up to 20 years (Hampson 1993). The disease can be managed by enforcing strict quarantine legislation and growing potato varieties immune to the wart. Potato is the principal host of the pathogen and plants other than potato apparently do not play an important role in the disease cycle of the pathogen. However, experimentally some of species of *Solanum* take infection upon artificial inoculation (Phadtare and Sharma 1971) and survival of the pathogen in such hosts could not be ruled out.

6.6.1 Symptoms

Warts are mostly spherical outgrowths or protuberances that appear on buds and eyes of tubers, stolons, or underground stems or at stem base. Tubers may get completely replaced by warts which desiccate or decay at harvest (Fig. 6.9). The disease

Fig. 6.9 Warty growths on underground potato stems and tubers

may appear occasionally on above ground stem, leaf or flowers. Underground galls are white to light pink when young and become brown or light black with age. Above ground galls are green to brown or black. The wart tissues are soft and spongy.

6.6.2 The Pathogen

The fungus *Synchytrium endobioticum* (Schilberszky) Percival is the cause of wart of potato. It is a member of *Chytridiales*. The fungus does not produce hyphae and is an obligate, holocarpic, endobiotic parasite (Hampson 1993a, b; Walker 1983). Numerous pathotypes of the fungus exist and are defined by their virulence on differential potato cultivars. More than 40 pathotypes have been described from Europe (Baayen et al. 2006). The fungus lack mycelium and has a thin walled summer sporangium stage and a thick walled 'winter' or resting sporangium stage. Both summer and winter sporangium produce an extended vesicle called sorus from where zoospores are produced. The zoospores are pear shaped and possess a posterior flagellum. These sporangia can remain viable in soil for 2–3 decades even in the absence of suitable hosts. The resting sporangia under wet soil conditions and temperature between 10 and 27 °C germinate to release haploid uni-nucleate zoospores. The zoospores swim in soil, encyst and infect epidermal cells of meristematic tissues of growing buds, stolons tips or leaf primordia by means of an infection peg within 1–2 h of their formation. After successful infection a uni-nucleate thallus develops within the infected cell which enlarges to form a prosorus. A vesicle develops from prosorus and contents of the prosorus pass on to the vesicle to form a sorus within an infected cell. The sorus divides repeatedly to form several sporangia in which zoospores develop. Finally wall of the sorus breaks releasing sporangia and zoospores in soil. New infection results from the zoospores. This process continues throughout the growing season. Growth of the fungus within host stimulate hypertrophy and hyperplasia of neighbouring host cells without actively infecting them which result in increase in meristematic activity and development of warts of variable size depending upon the degree of stimulation (Lapwood and Hide 1971).

6.6.3 *Epidemiology*

Warty growths disintegrate releasing abundant resting sporangia in soil which serve as primary inoculum. The pathogen spreads from one locality to another through infected seed tubers, infested soil adhering tubers, machinery and other carriers of contaminated soil. Resting sporangia survive passage through the digestive track of animals fed with the infected potatoes, and the contaminated manure, therefore, can disperse the inoculum. Earthworms have been found to serve as means of inoculum dispersal. The resting sporangia can also be dispersed by wind-blown soil or by flowing surface water. Wart is favoured by periodic flooding followed by drainage and aeration since free water is required for germination of sporangia and dispersal of zoospores. Temperature favourable for germination of resting sporangia to zoospores ranges between from 14 to 24 °C. Both summer sporangia and resting spores can germinate between 12 and 28 °C. Mean temperature below 18 °C and annual precipitation of about 70 cm favour disease development.

6.6.4 *Management*

Rotational crops such as bean and radish and intercropping of potato with maize have been found to reduce population of viable resting spores in soil (Singh and Shekhawat 2000). Amendment of infested soil with 4 and 8 % crabshell (w/w) reduces population of the pathogen (Hampson and Coombs 1995). Application of fungicides and chemicals to soil is costly and not practical (Hodgson et al. 1974; O'Brien and Rich 1976). Effective control of the disease has been achieved through cultivation of wart immune varieties. Many varieties resistant or immune to wart have been developed throughout the world. In resistant varieties the pathogen infects the plants but symptom development is suppressed while in immune varieties a hypersensitive reaction occurs upon infection with zoospores of the fungus get killed in the process. Development and introduction of wart immune varieties such as Kufri Jyoti, Kufri Bahar, Kufri Sherpa and Kufri Kanchan to wart infested region of Darjeeling Hills of India coupled with domestic quarantine had a great impact in containing wart in this region (Sharma et al. 1976; Singh 1998). The disease has been successfully managed by sanitation, long crop rotation, growing resistant and immune varieties and by enforcing strict quarantine legislation in countries of EPPO region (Mc Namera and Smith 1998), Canada (Hampson 1993a, b), Maryland USA (Putnam and Sindermann 1994) and India (Singh and Shekhawat 2000). However, periodic surveys need to be carried out to monitor viability of the pathogen in soil.

6.7 Powdery Scab

Powdery scab of potato causes scab like lesions on tubers and seriously reduces tuber quality and marketability, resulting in significant economic loss (Qu et al. 2011). It is prevalent worldwide (Harrison et al. 1997). It is a problem in northern Asia, Europe, North and South America, Australia and New Zealand (Walker 1969; Merz 2008). The disease is prominent in cool, wet climates and may cause extensive losses in seed and ware potato crops under such conditions (Wale 2000). Apart from potato and related species, powdery scab pathogen can affect other crop plants such as oilseed rape, sugar beet, spinach, and a large number of common weeds including chickweed, poppy, nettle and fat hen. However, the role of non-solanaceous hosts in contributing to the inoculum for the disease is not well established (Wale 2000).

6.7.1 Symptoms

Powdery scab is confined to underground parts of potato plants. Symptoms on potato tubers appear as purplish brown sunken lesions which later turn to scab like lesions. However, unlike common scab the lesions of powdery scab are round, raised, filled with powdery mass of spores and surrounded by ruptured remains of epidermis (Fig. 6.10). Under certain conditions wart like protuberances may develop (O'Brien and Rich 1976; Bhattacharyya and Raj 1978). The infected tuber may shrivel or develop dry rot type symptom in storage. The powdery mass consists of cytosori or balls of spores. Each spore ball contains several spores which adhere to one another along with their walls.

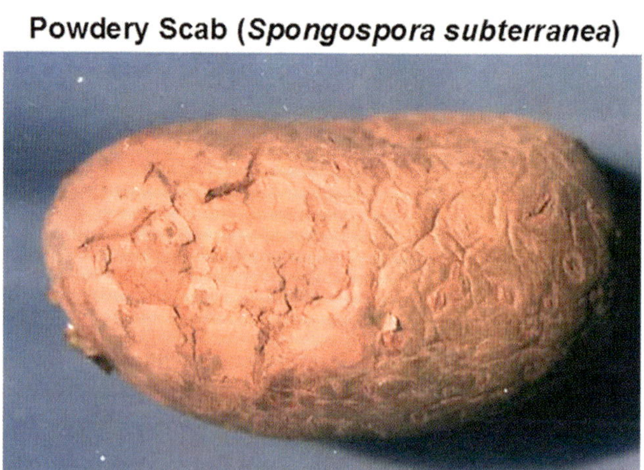

Fig. 6.10 Powdery scab lesions on potato tubers surrounded by ruptured remains of epidermis

6.7.2 The Pathogen

The disease is caused by *Spongospora subterranea* (Wallr.) Lagerh. f. sp. *subterranea* Tomlinson, a biotrophic pathogen belonging to plasmodiophorales, characterized of having multinucleate plasmodia and biflagellate zoospores and resting spores (Karling 1968; Harrison et al. 1997). The powdery mass consists of cytosori or balls of spores. Each spore ball contains several spores which adhere to one another along with their walls. The spores of the pathogen are yellow to brown, thin walled, polyhedral, uni-nucleate structures which germinate to produce a single primary zoospore. *S. subterranea* is an important vector of potato mop-top furovirus (Kirk 2008).

6.7.3 Epidemiology

The pathogen can survive for many years in a quiescent form as uni-nucleate or binucleate thick walled resting spores (Lahert and Kavanagh 1985). The spores can survive passage through the alimentary canal of farm animals (Morse 1914). The application of slurry or manure from stock fed on powdery scab infected tubers can provide an additional source of inoculum for the disease (Harrison et al. 1997). The pathogen survives winter as sporangia in infected potato tubers. It can also survive in soil up to 6 years. The zoospores of the pathogen penetrate roots, stolons, tubers and produce multinucleate sporangial plasmodium in the host. In roots, the plasmodium produces sporangia which further produce up to 8 secondary zoospores. The zoospores re-infect the host tissue and several such generations of zoospores may be produced in a single season under ideal environment. The plasmodium produces resting spore which can overcome winter and persist in tuber and soil for a long period. Powdery scab pustules also predispose the tubers to *P. infestans* pathogen (Bonde 1955). The cytosori can initiate the disease on the seed tubers in soil. The disease can spread through contaminated farm implements and irrigation water.

6.7.4 Management

The disease can be managed by avoiding conditions leading to flooding of the fields through proper drainage and by following crop rotation with non-solanaceous hosts. Growing trap crops such as *Datura stramonium* immediately before planting potato could reduce powdery scab incidence (Winter and Winiger 1983). Use of healthy seed from disease free area and avoiding planting of potatoes in fields having previous history of the disease can help in management of powdery scab. Light-skinned and red-skinned potato varieties in general are most susceptible (Christ 1993). Cultivars resistant to the disease have been developed in Germany, Russia and Chile (Manzer et al. 1964).

Control of powdery scab with fungicides is not successful under field conditions (Burnett 1991). However, low tuber infections were observed when infested soils were treated with soil fumigants such as methyl bromide, metam sodium, and chloropicrin. Seed treatments in general are not effective.

6.8 Charcoal Rot

Charcoal rot caused by *Macrophomina phaseolina* is an important disease of potato and many other vegetable crops in tropical and subtropical countries. The disease is favoured by temperature exceeding 28 °C (Chupp and Sherf 1960). The affected tubers rot both in field and in country stores and can cause severe losses under unusually warm wet weather. The disease is prevalent in the Mediterranean region, warmer areas of India and Peru (French 2001), Hawaii and Southern USA (O'Brien and Rich 1976). The fungus survives on a wide range of plants, at least 284 hosts have been recorded which are both cultivated and wild (French 2001).

6.8.1 Symptoms

Macrophomina phaseolina attacks growing potato plants and tubers both at harvest and storage. Affected plants in field exhibit stem blight or shallow rot similar to black leg and cause the affected foliage to wilt and turn yellow. Early symptoms on tubers develop around eyes, lenticels and stolon end where a dark light grey, soft, water soaked lesion develop on the surface of the tuber (Fig. 6.11). Subsequently, the lesions become filled with black mycelium and sclerotia of the pathogen. Secondary organisms may develop in such lesions especially under wet conditions causing significant losses (Pushkarnath 1976). Under low moisture the lesions may shrink and develop symptoms similar to dry rots.

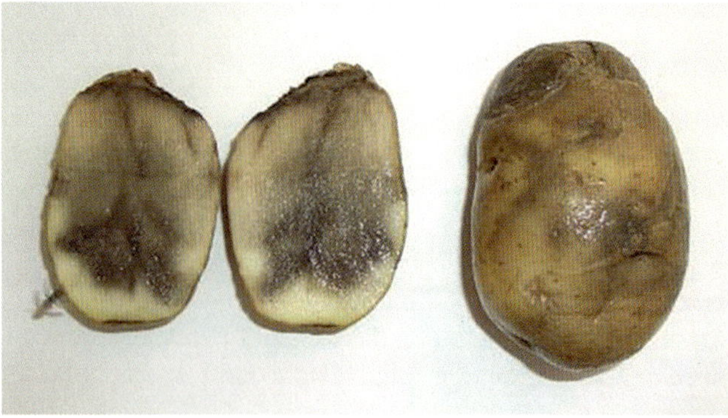

Fig. 6.11 Charcoal rot of potato tubers

6.8.2 The Pathogen

The pathogen *Macrophomina phaseolina* (Tassi) Goidanich Syn. *M. phaeoli* Maubl. (O'Brien and Rich 1976) develops smooth, hard 0.1–1.0 mm sized sclerotia within roots, stems, tubers and leaves. The perfect stage of the fungus is *Botryodiplodia solani-tuberosi* Thiram. (Thirumalachar and O'Brien 1977) which may develop in stems of potato, jute, sun hemp and maize. Pycnidia may develop on leaves and stems depending upon the strain of the fungus. Conidia are single, hyaline and ellipsoid to obovoid.

6.8.3 Epidemiology

The fungus *M. phaseolina* is a weak parasite. The pathogen survives in soil as sclerotia present in plant debris, on weeds and alternate host crops. Both soil and infected tubers serve as source of inoculum. *M. phaseolina* persists on dead or dying plant tissues and survives the unfavourable periods by forming microsclerotia (French 2001). The pathogen spreads through the infected seed tubers and through the infested soil carried along with the implement. Temperature around 30 °C is optimum for growth and infection of the fungus. Poor plant nutrition and wounds predispose the plants to charcoal rot. Temperature around 30 °C or above is very favourable for infection, the rot is slow at 20 to 25 °C and stops at 10 °C or below. Fungal growth stops in tubers placed in cold stores but it resumes the growth after cold storage.

6.8.4 Management

Soil temperature preceding harvest is crucial for development of charcoal rot. Disease can be managed through planting early-maturing cultivars, frequent irrigations to keep the soil temperature down and harvesting potato tubers before the soil temperature exceeds 28 °C (Thirumalachar 1955). Rotation with non-host crops, use of seed from disease free area, avoiding cuts and bruises at harvest are some of the measures which can be followed to further reduce the disease incidence. Use of biocontrol agent such as *Bacillus subtilis* through seed treatment has been reported to reduce the disease (Thirumalachar and O'Brien 1977). Resistance against charcoal rot has been identified in *Solanum chacoense* which could be utilized in developing varieties resistant to charcoal rot (French 2001).

References

Abd-El-Kareem F (2007) Potassium or sodium bicarbonate in combination with Nerol for controlling early blight disease of potato plants under laboratory, greenhouse and field conditions. Egypt J Phytopathol 35(1):73–86

Abd-Elsalam KA, Moslem MA, Bahkali AH (2009) First morpho-molecular identification of *Rhizoctonia solani* AG-7 from potato tuber-borne sclerotium in Saudi Arabia. Afr J Microbiol Res 3:952–956

Adams MJ, Lapwood DH (1983) Transmission of *Fusarium solani* var. *coeruleum* and *F. sulphureum* from seed potatoes to progeny tubers in the field. Ann Appl Biol 103:411–417

Adl SM, Simpson AGB, Farmer MA, Andersen RA, Anderson OR, Barta JR et al (2005) The new higher level classification of eukaryotes with emphasis on the taxonomy of protists. J Eurkaryot Microbiol 50:399–451

Alexandrov V (2008) Response of different potato varieties to causal agent of early blight disease (*Alternaria solani*). J Balkan Ecol 11(4):391–395

Ali GS, Reddy ASN (2000) Inhibition of fungal and bacterial plant pathogens by synthetic inhibition of disease progression. Mol Plant Microbe Interact 13:847–859

Al-Mughrabi KI (2004) Sensitivity of Jordanian isolates of *Alternaria solani* to mancothane. Phytopathol Mediterr 43(1):14–19

Anderson NA (1982) The genetics and pathology of *Rhizoctonia solani*. Annu Rev Phytopathol 20:329–347

Appel R, Adler N, Habermeyer J (2001) A method for the artificial inoculation of potato tubers with *Phytophthora infestans* and polymerase chain reaction assay of latently infected sprouts and stems. J Phytopathol 149:287–292

Arora RK (1999) Evaluation of bioagents for control of soil and tuber borne diseases of potato. Indian Phytopathol 52:310 (Abstr.)

Arora RK (2000a) Combined effects of soil solarization and seed treatments on control of potato black scurf. In: Paul Khurana SM, Shekhawat GS, Singh BP, Pandey SK (eds) Potato global research and development. Vol. I. Proceedings of global conference of potato, New Delhi, December 6–11, 1999. Indian Potato Association, Central Potato Research Institute, Shimla, pp 390–392

Arora RK (2000b) Bio-control of potato late blight. In: Paul Khurana SM, Shekhawat GS, Singh BP, Pandey SK (eds) Potato global research and development. Vol. I. Indian Potato Association, Central Potato Research Institute, Shimla, pp 620–623

Arora RK (2005) Efficacy of boric acid spray for control of black scurf in unwashed and washed potato tubers. Potato J 32:183–184

Arora RK (2008) Management of black scurf of potato with the integrated use of *Trichoderma viride* and boric acid. Potato J 35:130–133

Arora RK, Khanna RN (1997) *Phytophthora* blight of potato an age old problem. In: Agnihotri VP, Sarbhoy AK, Singh DV (eds) Management of threatening diseases of national importance. Malhotra Publishing House, New Delhi, pp 96–109

Arora RK, Somani AK (2001) A bioformation for control of black scurf disease of potato. J Indian Potato Assoc 28:88–89

Arora RK, Kamble SS, Gangane LV (1992a) A resistance to metalaxyl in *Phytophythora infestans* in Nilgiri hills of southern India. Phytophthora Newslett 18:8–9

Arora RK, Rose de Leema J, Paulraj F (1992b) Phylloplane microflora in biocontrol of potato late blight. In: National symposium on non-chemical methods for the management of plant diseases. Indian Phytopathological Society, Bhavnagar, p 49

Arora RK, Sharma J, Garg ID, Singh RK, Somani AK (2006a) Boric acid for control of tuber borne diseases. In: Technologies in aid of healthy seed production. Part II. Technical Bulletin No. 35, Central Potato Research Institute, Shimla, pp 1–4

Arora RK, Trehan SP, Sharma J, Khanna RN (1997) Soil solarization for improving potato health and production. Proceedings Indian Phytopathological Society–Golden Jubilee International Conference on Integrated Plant Disease Management for Sustainable Agriculture. Vol. II pp.1190–1191

Arora RK, Sharma J, Singh RK (2006b) Soil solarization. In: Technologies in aid of healthy seed production. Part I. Technical Bulletin No. 35, Central Potato Research Institute, Shimla, pp 1–7

Arora RK, Sunaina V, Somani AK (2008) Eco-friendly technologies to manage soil and tuber borne diseases of potato. In: Pandey SK, Chakrabarti SK (eds) Twenty steps towards hidden treasure—technologies that triggered potato revolution in India. Central Potato Research Institute, Shimla, pp 137–150

Arora RK, Ahmed I, Singh BP (2012) Forecasting late blight of potato in Punjab using JHULSACAST model. Potato J 39:173–176

Baayen RP, Cochius G, Hendriks H, Meffert JP, Bakker J, Bekker M et al (2006) History of potato wart disease in Europe—a proposal for harmonisation in defining pathotypes. Eur J Plant Pathol 116:21–31

Bairambekov SB, Korneva OG (2009) Using bio-preparations against early blight of potato. Zaschita-i-Karantin-Rastenii 8:30–31

Balali GR, Neate SM, Scott ES, Whisson DL, Wicks TJ (1995) Anastomosis group and pathogenicity of isolates of *Rhizoctonia solani* from potato crops in South Australia. Plant Pathol 44:1050–1057

Baldauf SL, Roger AJ, Wenk-Siefert I, Doolittle WF (2000) A kingdom level phologeny of eukaryotes based on combined protein data. Science 290:972–977

Bandy BP, Leach SS, Tavantzixs SM (1988) Anastomosis group 3 is the major cause of Rhizoctonia disease of potato in Maine. Plant Dis 72:596–598

Barclay GM, Murphy HJ, Manzer FE, Hutchinson FE (1973) Effects of differential rates of nitrogen and phosphorus on early blight of potatoes. Am Potato J 50:42–48

Barksdale TH (1969) Resistance of tomato seedlings to early blight. Phytopathology 59:443–446

Batista DC, Lima MA, Haddad F, Maffia LA, Mizubuti ESG (2006) Validation of decision support system for tomato early blight and potato late blight under Brazilain conditions. Crop Prot 25(7):664–670

Beaumont A (1947) The dependence of weather on the dates of outbreak of potato blight epidemics. Trans Br Mycol Soc 31:45–53

Behera B, Senapati S, Pattanaik KK, Biswal G (2009) Prediction of potato early blight in coastal tract of Orissa. J Plant Prot Environ 6(1):79–85

Bernat E (2004) New possibilities of chemical protection against early and late blights in potato crops using a fungicide Pyton Consento 450SC. Biuletyn-Instytutu Hodowli i Aklimatyazacji-Roslin 233:303–308

Bhattacharyya SK, Raj S (1978) Studies on powdery scab of potatoes: factors affecting disease development. J Indian Potato Assoc 5:1–6

Bhattacharyya SK, Raj S, Singh DS, Khanna RN, Ram S (1982) Forecasting late blight of potato in Indian hills. In: Nagaich BB, Shekhawat GS, Gaur PC, Verma SC (eds) Potato in developing countries. Indian Potato Association, Shimla, pp 414–424

Bodker L, Pedersen H, Kristensen K, Moller L, Lehtinen A, Hannukkala A (2006) Influence of crop history of potato on early occurrence and disease severity of potato late blight caused by *Phytophthora infestans*. In: Westerdijk CE, Schepers HTAM (eds) Proceedings of the 9th workshop of an European network for development of an integrated control strategy of potato late blight. PPO special report no. 11, pp 53–56

Bonde R (1955) The effect of powdery scab on the resistance of potato tubers to late blight rot. Maine Agriculture Experimental Station Bulletin, No. 538

Bounes H, Finckh MR (2008) Effects of strip intercropping of potatoes with non-hosts on late blight severity and tuber yield in organic production. Plant Pathol 57:916–927

Boyd AEW (1981) Development of potato blight (*Phytophthora infestans*) after planting infected seed tubers. Ann Appl Biol 77:259–270

Bradshaw NG, Elcock SJ, Turner JA, Hardwick NV (2000) Are potato blight fungicides being used rationally? In: Proceedings BCPC-international conference: pests and diseases, UK, 13–16 November 2000, vol 3, pp 847–852

Braun JA, Fry WE (1981) Analysis of potato blight epidemiology by simulation modeling. Phytopathology 71:612–616

Bruck RI, Fry WE, Apple AE (1980) Effect of metalaxyl an acylalanine fungicide on development of *Phytophthora infestans*. Phytopathology 70:567–601

Burnett F (1991) The Biology and Control of Powdery Scab (Spongospora subterranea) of potatoes. Aberdeen, UK: University of Aberdeen, PhD Thesis

Cao KQ, Fried PM, Ruckstuhl M, Forrer HR (1996) Ereignisorientierte Krautfaeuleprognose nit PhytoPRE+2000. Agrarforschung 3(7):325–328

Carling DE, Leiner RH (1986) Isolation and characterization of *Rhizoctonia solani* and binucleate *Rhizoctonia solani*-like fungi from aerial stems and subterranean organs of potato plants. Phytopathology 76:725–729

Carling DE, Leiner RH (1990) Effect of temperature on virulence of *Rhizoctonia solani* and other *Rhizoctonia* on potato. Phytopathology 80:930–934

Carling DE, Leiner RH, Westphale PC (1989) Symptoms, signs and yield reduction associated with Rhizoctonia disease of potato induced by tuberborne inoculum of *Rhizoctonia solani* AG-3. Am Potato J 66:693–701

Carling DE, Rothrock CS, MacNish GC, Sweetingham MW, Brainard KA, Winters SW (1994) Characterization of anastomosis group 11 (AG-11) of *Rhizoctonia solani*. Phytopathology 84: 1387–1393

Carling DE, Baird RE, Gitaitis RD, Brainard KA, Kuninaga S (2002) Characterization of AG-13, a newly reported anastomosis group of *Rhizoctonia solani*. Phytopathology 92:893–899

Carnegie SF, Ruthven AD, Lindsay DA, Hal TD (1990) Effects of fungicides applied to seed potato tubers at harvest or after grading on fungal storage diseases and plant development. Ann Appl Biol 116:61–72

Carnegie SF, Cameron AM, Lindsay DA, Sharp E, Nevison IM (1998) The effect of treating seed potato tuber with benzimidazole, imidazole and phenylpyrrole fungicides on the control of rot and blemish diseases. Ann Appl Biol 133:343–363

Carnegie SF, Cameron AM, Haddon P (2001) The effect of date of haulm destruction and harvest on the development of dry rot caused by *Fusarium solani*. Phytopathology 84:1387–1393

Chakraborty A, Roy R (2012) Efficacy of Kresoxim methyl 500 SC against late and early blight of potato. Ann Plant Prot Sci 20(1):248–249

Chand T, Logan C (1983) Cultural and pathogenic variation in potato isolates of *Rhizoctonia solani* in Northern Ireland. Trans Br Mycol Soc 81:585–589

Chelkowski J (1989) Toxinogenic of *Fusarium* species causing dry rot of potato tubers. In: Chelkowski J (ed) Fusarium mycotoxin, taxonomy and pathogenicity. Elsevier Publishing Co., New York, pp 435–440

Chimote VP, Kumar M, Sharma PK, Singh PH, Singh BP (2010) Characterization of changes in phenotype and genotype of *Phytophthora infestans* isolates from India. J Plant Pathol 92:669–677

Choiseul JW, Allen L, Carnegie SF (2001) The role of stem inoculum in the transmission of *Fusarium sulphureum* to potato tubers. Potato Res 44:165–172

Chow TL, Bernard G (1999) A versatile, fully automated, real time potato late blight unit. Comput Electron Agr 23:55–69

Chowdappa P, Kumar NBJ, Madhura S, Kumar MSP, Myers KL, Fry WE, Squires JN, Cooke DEL (2013) Emergence of 13_A2 blue lineage of *Phytophthora infestans* was responsible for severe outbreaks of late blight on tomato in south west India. J Phytopathol 161:49–58

Christ BJ (1993) Powdery scab of potatoes-what we know. In: Proceedings of 12th Annual National Potato Council Seed Seminar, pp 24–25

Chupp C, Sherf AT (1960) Vegetable diseases and their control. Ronald Press, New York

CIP (1989) Fungal diseases of potato. International Potato Centre, Lima, p 216

Conners IL (1967) An annotated index of plant diseases in Canada. Research Branch Publication, Canada Department of Agricultural, Toronto, p 1251

Cooke LR, Schepers HTAM, Hermansen A, Bain RA, Bradshaw NJ, Ritchie F et al (2011) Epidemiology and integrated control of potato late blight in Europe. Potato Res 54:183–222

Corsini DL, Pavek JJ (1986) *Fusarium* dry-rot resistant potato germplasm. Am Potato J 63:629–638

CPRI (1998–1999) Annual scientific report. Central Potato Research Institute, Shimla, pp 72–75

CPRI (2000–2001) Annual scientific report. Central Potato Research Institute, Shimla

Cullen DW, Lees AK, Toth IK, Bell KS, Duncan JM (2000) Detection and quantification of fungal and bacterial potato pathogens in plant and soil. Bulletin OEPP 30:485–488

de Lacy-Costello BPJ, Ewans P, Ewens RJ, Gunson HE, Jones PRH, Ratcliffe NM, Spencer-Phllips PTN (2001) Gas chromatography—mass spectrometry analysis of volatile compounds for potato tubers inoculated with *Phytophthora infestans* or *Fusarium coeruleum*. Plant Pathol 50:489–496

Dowley LJ, O'Sullivan E (1981) Metalaxyl resistant strains of *Phytophthora infestans* in Ireland. Potato Res 24:417–421

Ehrlich MA, Ehrlich HG (1966) Ultrastructure of hyphal and haustoria of *Phytophthora infestans* and hyphae of *Phytophthora erythroseptica*. Can J Bot 44:1495–1504

El-Gholl NE, Schubert TS, Alfieri SA, Graham AE (1985) Dry rot of potato tubers. Fla. Dept. Agric. & Consumer Serv., Div. of Plant Ind., Pathol. Circ. No. 267

EPPO (2006) A2 list of pests recommended for regulation as quarantine pests (version 2006-09). Online. *Quarantine info*. European and Mediterranean Plant Protection Organization (EPPO), Paris, France

Ermakova NE, Shterushis MV (1994) New biological preparation of RIT's against plant diseases. Zashchita Rastonii Moskva 12:18

Errampalli D, Johnston HW (2001) Control of tuber-borne black scurf [*Rhizoctonia solani*] and common scab [*Streptomyces scabies*] of potatoes with a combination of sodium hypochlorite and thiophanate methyl pre-planting seed tuber treatment. Can J Plant Pathol 23:68–77

Erwin DCS, Bartnicki-Garcia S, Tsao PH (1983) *Phytophthora*, its biology, taxonomy and pathology. The American Phytopathological Society, St. Paul

Evenhuis A, Scheepers HTAM, Bus CB, Stegeman W (1996) Synergy of cymoxanil and mancozeb when used to control potato late blight. Potato Res 39:551–559

Forrer HR, Hecker A, Steenblock T, Alfoidi T, Llockeretz W, Niggli U (2000) Hot water treatment of potato seed tubers—a practicable means to prevent primary foci and delay epidemics of potato late blight. IFOAM 2000: the world grows organic. In: Proceedings of 13th international IFOAM scientific conference, Basel, Switzerland, 28–31 August 2000

Franc GD (2001) Wart. In: Stevenson WR, Loria R, Franc GD, Weingartner DP (eds) Compendium of potato diseases. American Phytopathological Society, St. Paul, pp 46–47

Franc GD, Harrison MD, Lahman LK (1988) A simple day degree model for initiating chemical control of potato early blight in Colorado. Plant Dis 72:851–854

Frank JA, Leach SS (1980) Comparison of tuber-borne and soil-borne inoculum of *Rhizoctonia* disease on potato. Phytopathology 70:51–53

French ER (2001) Charcoal rot. In: Stevenson WR, Loria R, Franc GD, Weingartner DP (eds) Compendium of potato diseases. American Phytopathological Society, St. Paul, pp 46–47

Frenkel O, Yermiyahu U, Forbes GA, Fry WE, Shtienberg D (2010) Restriction of potato and tomato late blight development by sub-phytotoxic concentrations of boron. Plant Pathol 59:626–633

Fry W (2008) *Phytophthora infestans*: the plant (and R gene) destroyer. Mol Plant Pathol 9(3):385–402

Fry WE, Tooley PW, Spielman LJ (1989) The importance of perfect stage of *Phytophthora infestans* from the stand point of epidemiology and adaptation. In: Proceedings of fungal diseases of potatoes: report planning conference on fungal diseases of potato, CIP, Lima Peru, 21–25 September 1987, pp 117–130

Fry WE, Smart CD, Monti L, Leone A, Struik PC, Hide GA, Storey RMJ (1999) The return of *Phytophthora infestans*, a potato pathogen that just won't quit. In: Struik PC, Hide GA (eds) Proceedings of the 14th triennial conference of the European Association for Potato Research; Sorrento, Italy May 2–7, extra edition 42, pp 279–282

Gachango E, Kirk WW, Hanson LE, Rojas A, Tumbalam P, Shetty K (2011) First report of *Fusarium torulosum* causing dry rot of seed potato tubers in Michigan. Plant Dis 95:1194

Gangawane LV, Arora RK, Kamble SS (1993) Effect of passage on the development of metalaxyl resistance in *Phytophthora infestans*. In: Gangawane LV, Mukadam DS, Papdiwal PB, Shindi SR (eds) Proceedings of symposium chemical management of plant pathogen in Western India. Marathwada University, Aurangabad, pp 27–30
Garret KA, Mundt CC (2000) Host diversity can reduce potato late blight severity for focal and general patterns of primary inoculum. Phytopathology 90:1307–1312
Gisi U, Cohen Y (1996) Resistance to phenylamide fungicides: a case study with *Phytophthora infestans* involving mating types and race structure. Annu Rev Phytopathol 34:547–592
Goodwin SB, Cohen BA, Deahl KL, Fry WE (1994) Migration from northern Mexico as the probable cause of recent genetic changes in populations of *Phytophthora infestans* in the United States and Canada. Phytopathology 84:553–558
Groza HI, Bowen BD, Bussan AJ, Stevenson WR, Navarro F, Kichefski D et al (2007) MegaChip—a new potato variety of chipping. Am J Potato Res 84(4):343–350
Guenthner JF, Michale KC, Nolte P (2001) The economic impact of potato late blight on U.S. growers. Potato Res 44:121–125
Gupta CP, Sharma A, Dubey RC, Maheshwari DK (1999) *Pseudomonas aeruginosa*. (GRCI) as a strong antagonist of *Macrophomina phaseolina* and *Fusarium oxysporum*. Cytobios 99: 183–189
Gupta H, Singh BP, Mohan J (2004) Biocontrol of late blight of potato. Potato J 31:39–42
Hampson MC (1993) History, biology and control of potato wart disease in Canada. Can J Plant Pathol 15:223–244
Hampson MC, Coombs JW (1995) Reduction by potatoes wart disease with crushed crabshell: suppression or eradication. Can J Plant Pathol 17:69–74
Hannukkala AO, Kaukoranta T, Lehtinen A, Rahkonen A (2007) Late blight epidemics on potato in Finland 1933-2002; increased and earlier occurrence of epidemics associated with climate change and lack of rotation. Plant Pathol 56:167–176
Hanson LE, Schwager SJ, Loria R (1996) Sensitivity to thiabendazole in *Fusarium* species associated with dry rot of potato. Phytopathology 86:378–384
Harrison JG, Searle RJ, Williams NA (1997) Powdery scab disease of potato—a review. Plant Pathol 46:1–25
Hauslanden H, Bassler E (2004) Early blight disease in potatoes. What are the causes? Kartaffelbau 6:210–212
Haverkort AJ, Struik PC, Visser RGF, Jacobsen E (2009) Applied biotechnology to control late blight in potato caused by *Phytophthora infestans*. Potato Res 52:249–264
Hensen JG, Bodker L, Nielson BJ (2000) Decision support for the control of potato late blight. In: Proceeding of 17th Danish Plant Protection Conference. II. Site specific crop protection, decision support, pests and diseases, early blight, DJF Rapport, Markbrug, vol 24, pp 87–99
Hermansen A, Naestad R (2009) Bekjemping av potettorrate. Gartneryrket 107(7):20–23
Hide GA, Read PJ, Hall SM (1992) Resistance to thiabendazole in *Fusarium* species isolated from potato tubers affected by dry rot. Plant Pathol 41:745–748
Hijmans RJ, Forbes GA, Walker TS (2000) Estimating the global severity of potato late blight with GIS-linked disease forecast models. Plant Pathol 49:697–705
Hodgson WA, Pond DD, Munro J (1974) Diseases and pests of potatoes. Canadian Department of Agriculture, Ottawa
Hohl HR, Iselin K (1984) Strains of *Phytophthora infestans* from Switzerland with A2 mating type behavior. Trans Br Mycol Soc 83:529–530
Holm AL, Rivera VV, Secor GA, Gudmestad NC (2003) Temporal sensitivity of *Alternaria solani* to foliar fungicides. Am J Potato Res 80(1):33–40
Huang-Mc Breath J (2001) Biocontrol and growth promotion with cold tolerant *Trichoderma*. IPM Practitioner 23:1–6
Hussain S, Lees AK, Duncan JM, Cooke DEL (2005) Development of a species-specific and sensitive detection assay for *Phytophthora infestans* and its application for monitoring of inoculums in tubers and soil. Plant Pathol 54:373–382
Hyre BA (1954) Progress of forecasting late blight of potato and tomato. Plant Dis Report 38:245–253

Ingram DS, Williams PH (eds) (1991) Advances in plant pathology. *Phytophthora infestans*, the cause of late blight of potato, vol 7. Academic Press, New York
Jansky SH, Rouse DI (2003) Multiple disease resistance in interspecific hybrids of potato. Plant Dis 87(3):266–272
Johnk JS, Jones RK (1993) Differentiation of populations of AG-2-2 of *Rhizoctonia solani* by analysis of cellular fatty acids. Phytopathology 83:278–283
Johnson DA, Cummings TF (2009) Latent infection of potato seed tubers by *Phytophthora infestans* during long-term cold storage. Plant Dis 93:940–946
Kapsa J (2007) Application of the Burkard spore trap to determine a composition of the genus *Alternaria* in potato crops. Biuletyn Instytutu Hodowli i Aklimatyazacji-Roslin 244:223–229
Karling JS (1968) The plasmodiophorales, 2nd edn. Hafner Publishing Co, New York
Keijer J, Houterman PM, Dulhemans AM, Korsman MG (1996) Heterogeneity in electrophoretic karyotype within and between anastomosis groups of *Rhizoctonia solani*. Mycol Res 100: 787–797
Khanna RN, Sharma J (1996) Effect of boric acid treatment on seed and soil borne *Rhizoctonia solani* inocula and rhizosphere microflora. J Indian Potato Assoc 23:1–7
Kim BS, Cho KC, Cho KY (1995) Antifungal effects of plant pathogenic fungi and characteristics of antifungal substances produced by *Bacillus subtilis* SJ-2 isolated from sclerotia of *Rhizoctonia solani*. Kor J Plant Pathol 11:165–172
Kim BS, Kim K, Lee J, Lee Y, Cho KY, Kim BS (1996) Isolation and purification of several substances produced by *Fusarium graminearum* and their antimicrobial activities. Kor J Plant Pathol 11:158–164
Kirk WW (2003) Thermal properties of overwintered piles of cull potatoes. Am J Potato Res 80:145–149
Kirk HG (2008) Mop-top virus, relationship to its vector. Am J Potato Res 85:261–265
Kirk WW, Felcher KJ, Douches DS, Coombs J, Stein JM, Baker KM, Hammerschmidt R (2001) Effect of host plant resistance and reduced rates and frequencies of fungicide application to control potato late blight. Plant Dis 85:1113–1118
Kirk WW, Abu-El Samen F, Thumbalam P, Wharton P, Douehes D et al (2009) Impact of different US genotypes of *Phytophthora infestans* on potato seed tuber rot and plant emergence in a range of cultivars and advanced breeding lines. Potato Res 52:121–140
Krause RA, Massie IB, Hyre RA (1975) BLITECAST: a computerized forecast of potato blight. Plant Dis Rep 59:95–98
Kromann P, Taipe A, Andrade-Piedra JL, Munk L, Forbes GA (2008) Pre-emergence infection of potato sprouts by *Phytophthora infestans* in the highlands tropics of Ecuador. Plant Dis 92:569–574
Kuzuestova MA, Sschcherbakova LA, Ihnskaya LI, Filippov AV, Ozeretekovskay OL (1995) Mycelium extract of the fungus *Pythium ultimum* is an efficient preventive of *Phytophthora* infection. Microbiology 64:422–424
Lahert H, Kavanagh JA (1985) The fine structure of cytosorus of *Spongospora subterranea*, the cause of powdery scab of potato. Can J Bot 63:2278–2282
Lambert DH, Currier AJ (1997) Differences in tuber rot development for North American clones of *Phytophthora infestans*. Am Potato J 74:39–371
Lapwood DH, Hide GA (1971) Potato diseases. In: Western JH (ed) Diseases of crop plants. Willey, New York, pp 89–122
Laroche JP, Jabagi- Hare SH, Charent PH (1992) Differentiation of two anastomosis groups of *Rhizoctonia solani* by isozyme analysis. Phytopathalogy 82:1387–1393
Leach SS (1976) Increasing storability of potatoes by chemical treatments at the time of storage. Am Potato J 53:364
Leach SS, Nelson LW (1975) Elimination of *Fusarial* contamination on seed potatoes. Am Potato J 9:211–218
Lees AK, Bradshaw JE, Stewart HE (1998) Inheritance of resistance to *Fusarium* spp. and to *Phytophthora infestans* in crosses between Neotuberosum and Tuberosum potatoes estimated by seedling tests. Potato Res 41:267–275

Libantova J, Bauer M, Mlynarova L, Moracikova J (1998) Bekesiova 1. Tobacco and potato plants expressing basic vacuolar beta-1,3-glucanase from *Nicotiana Plumbaginifolia*. Biologia Bratislara 53:739–748

Lorito M, Di Pietro A, Hayes CK, Woo SL, Harman GE, Di Pletro A (1993) Antifungal, synergistic interaction between *Trichoderma harzianum* and *Enterobacter cloacae*. Phytopathalogy 83:721–728

Lucas JA, Shattock RC, Shaw DS, Coope R (eds) (1991) *Phytophthora*. British Mycological Society, Cambridge. Cambridge University Press, Cambridge

Lynch DR, Kawchuk LM, Chen Q, Konschuh M, Holley J, Fujimoto DK et al (2004) Alta Russet: an early—maturing high quality Russet cultivar for wedge cut French fry production. Am J Potato Res 81(3):195–201

Mackin J (1998) Potato late blight raise fears of a famine. Cornell Focus 7:12–17

MacNish GC, Carling DE, Brainard KA (1993) Characterization of *Rhizoctonia solani* AG-8 from bare patches by pectic isozyme (Zymogram) and anastomosis techniques. Phytopathology 83:922–927

Malcolmson JF (1985) *Phytophthora infestans* A2 compatibility type recorded in Great Britain. Trans Br Mycol Soc 85:531

Manzer FE, Akeley RV, Merrian D (1964) Resistance to powdery scab in *Solanum tuberosum* L. Am Potato J 41:374–376

Marshall KD, Stevenson WR (1996) Transmission of *Phytophthora infestans* from infected seed potato tubers to developing sprouts. Am J Potato Res 73:370–371

Mc Namera DG, Smith IM (1998) National control measures for *Synchytrium endobioticum*. Bulletin OEPP 20:507–511

Mecteau MR, Arul J, Tweddell RJ (2002) Effect of organic and inorganic salts on the growth and development of *Fusarium sambucinum*, a causal agent of potato dry rot. Mycol Res 106: 688–696

Medina MV, Platt HW (1999) Viability of oospores of *Phytophthora infestans* under field conditions in northeastern North America. Can J Plant Pathol 21:137–143

Meinck S, Kolbe H (1999) Control of leaf and tuber blight in ecological potato cultivation. Kartoffelbau 50:172–175

Merz U (2008) Powdery scab of potato—occurrence, life cycle and epidemiology. Am J Potato Res 85:241–246

Millar PR, Pollard HL (1976) Multilingual compendium of plant diseases. American Phytopathalogical Society, St. Paul

Mishra RC, Singh R, Singh HB, Dikshit A (2000) *In situ* efficacy of *Trichoderma harzianum* as mycoparasite on *Sclerotium rolfsii* and *Rhizoctonia solani*. Trop Agric 77:205–206

Morse WJ (1914) Powdery scab of potatoes. Maine Agr Exp Stat Tech Bull 227:89–104

Mosa AA, Kato M, Sato N, Kobayashi K, Ogoshi A (1989) Occurrence of A2 mating type of *Phytophthora infestans* on potato in Japan. Ann Phytopath Soc Jpn 55:615–620

Musa-Steenblock T, Forrer HR (2005) Bio-PhytoPRE- a decision support system for late blight control in organic potato production in Switzerland. In: Ende der Niche. Beitrage zur 8. Wissenschaftstagung zum Okologischer Landban, Kassel University Press, Kassel; pp 133–136

Neiderhauser JS (1986) Late blight of potato in Mexico, its place of origin and solution. Revista Mexicana de Fitopatologia 4:31–36

Niederhauser JS, Alvarez-Luna E, Mackenzie DR (1996) Retona. A new strategy in the control of potato late blight. Am Potato J 73:225–229

Nieme M, Lahdenpera ML (2000) *Gliocladium catenulatum* JI 446- a new biofungicide for horticultural crops. 17th Danish plant protection conference, Horticulture D. J. F.-Rapport, Hareburg, vol 12, pp 81–88

O'Brien MJ, Rich AE (1976) Potato diseases. U.S. Department of Agriculture Handbook, Washington, DC

Osowski J (2004) The effectiveness of different fungicides containing mancozeb in reducing development of early blight of potato in field and laboratory investigations. Biuletyn-Instytutu Hodowli i Aklimatyazacji-Roslin 233:295–302

Pasche JS, Wharam CM, Gudmestad NC (2002) Shift in sensitivity of *Alternaria solani* (potato early blight) to strobilurin fungicides. In: Proceedings of international conference on pests and diseases, Brighton, UK, 18–21 November 2002, pp 841–846
Pasche JS, Wharam CM, Gudmestad NC (2004) Shift in sensitivity of *Alternaria solani* in response to QoI fungicide. Plant Dis 88:181–187
Pasche JS, Piche LM, Gudmestad NC (2005) Effect of the F129L mutation in *Alternaria solani* on fungicides affecting mitochondrial respiration. Plant Dis 89(3):269–278
Pavlista AD (2003) Thidiazuron a cytokinin-like compound enhances fungicidal activity against early blight in potato. Acta Hortic 619:145–152
Percival GC, Karim MS, Dixon GR (1999) Pathogen resistance in aerial tubers of potato cultivars. Plant Pathol 48:768–776
Peters JC, Stroud GS, Lees A, Cullen DW, Sullivan L (2004) Results from a survey of *Fusarium* spp. on GB potatoes. In: Duvauchelle S (ed) Abstracts of the European Association of Potato Research, pathology section meeting, Lille, France
Phadtare SG, Sharma KP (1971) Additional hosts of *Synchytrium endobioticum*. Indian Phytopathol 24:389–392
Pinzon-Perea L, Salgado R, Martineg-Lopez G (1999) Antagonism between different isolates of *Trichoderma* spp. and *Fusarium oxysporum* f. sp. *dianthi* (Prill. and Del.) Snyd. and Hans. Fitopatol Colomb 23:7–11
Pushkarnath (1976) Potato in subtropics. Orient Longman, New Delhi
Pushkarnath, Paharia KD (1963) Survival of *Phytophthora infestans* on infected tubers in cold storage in the plains of India. Indian Potato J 5:48–51
Putnam ML, Sindermann B (1994) Eradication of potato wart from Maryland. Am Potato J 71:734–747
Qu XS, Wanner LA, Christ BJ (2011) Multiplex real-time PCR (TaqMan) assay for the simultaneous detection and discrimination of potato powdery and common scab diseases and pathogens. J Appl Microbiol 110:769–777
Raj S, Dull BL, Sahai D (1974) Cultural variations in potato isolates of *Rhizoctonia solani* Kuhn. Acta Phytopathol Acad Sci Hung 9:71–79
Reeves AF, Olanya OM, Hunter JH, Well JM (1999) Evaluation of potato varieties and selections for resistance to bacterial soft rot. Am J Potato Res 76:183–189
Rich AE, Campbell CS, Mullany RE, Manzer FE, Blood PT (1960) The influence of various fungicides and antibiotics on cut seed potatoes. Am Potato J 37:351
Ristaino JB, Grovers CT, Parra GR (2001) PCR amplification of the Irish potato famine pathogen from historic specimens. Nature 411:695–697
Rodriguez-Maza F, Stefanova-Narimova M (2005) Biological control of early blight (*Alternaria solani* Sorouer) in potato (*Solanum tuberosum*) crop in field conditions. Fitosanidad 9:35–37
Roy S, Singh BP, Bhattacharyya SK (1991) Biocontrol of late blight of potato. Phytophthora Newsl 17:18
Roy SK, Sharma RC, Trehan SP (2001) Integrated nutrient management by using farmyard manure and fertilizers in potato-sunflower-paddy rice rotation in the Punjab. J Agric Sci 137:271–278
Sagar V, Sharma S, Jeevalatha A, Chakrabarti SK, Singh BP (2011) First report of *Fusarium sambucinum* causing dry rot of potato in India. New Dis Rep 24:5
Samoucha Y, Cohen Y (1988) Synergistic interactions of cymoxanil mixtures in the control of metalaxyl resistant *Phytophthora infestans* of potato. Phytopathology 78:636–640
Schisler DA, Slininger PJ, Hanson LL, Loria R (2000) Potato cultivar, pathogen isolate and antagonist cultivation medium influence the efficiency and ranking of bacterial antagonist of *Fusarium* dry rot biocontrol. Sci Technol 10:267–269
Schmiedeknecht G, Bochow H, Jungo H (1998) Use of *Bacillus subtilis* as biocontrol agent-II. Biological control of plant diseases. Zeit Pflanzen Pflanzenschutz 105:376–386
Senter LH, Sanson DR, Corley DG, Tempesta MS, Rottinghaus AA, Rottinghaus GE (1991) Cytotoxicity of trichothecenemycotoxins isolated from *Fusarium sporotrichioides* (MC-

72083) and *Fusarium sambucinum* in baby hamster kidney (BHK-21) cells. Mycopathologia 113:127–131

Sharma KP, Phadtare SG, Kishore H (1976) Now you have wart immune varieties. Indian Farm 25:13–15

Sharma S, Jeevalatha A, Sagar V, Sharma NN, Chakrabarti SK, Singh BP (2010) Detection of latent infection of *Phytophthora infestans* in potato seed tubers. In: Annual meeting of Indian Phytopathological Society (NZ) and symposium on emerging plant diseases in North India: status and management strategies held at CCS HAU, Hisar w.e.f. 28-29 October 2010

Shtienberg G (2001) Integrated management of early and late blight of potatoes in Israel. Afr Crop Sci J 9:203–207

Singh BM (1967) Inducing sporulation in different strains of *Alternaria solani.* Effect of visible light. Mycopath Mycol Appl 31:144–150

Singh BP (1986) Studies on fusarial wilt and dry rot of potatoes (*S. tuberosum*). PhD Dissertation. Aligarh Muslim University, Aligarh, India

Singh PH (1998) Present status of wart diseases of potato in Darjeeling hills. J Indian Potato Assoc 25:135–138

Singh BP, Shekhawat GS (1999) Potato late blight in India. Technical Bulletin No. 27 (revised) C.P.R.I., Shimla

Singh PH, Shekhawat GS (2000) Wart diseases of potato in Darjeeling hiils. Technical Bulletin No. 19, p 73

Singh BP, Nagaich BB, Saxena SK (1987a) Fungi associated with dry rot of potatoes, their frequency and distribution. Indian J Plant Pathol 5:142–145

Singh BP, Prasad B, Wallia MS (1987b) Effect of date of planting, crop age and environmental factors on potato leaf spots. Indian Phytopathol 40:70–75

Singh BP, Roy S, Bhatthcharyya SK (1994) Occurrence of A2 mating type of *Phytophthora infestans* in India. Potato Res 37:227–231

Singh BP, Ahmed I, Sharma VC, Shekhawat GS (2000) A computerized forecast of potato late blight in western Uttar Pradesh. Potato J 27:25–34

Sneh B, Burpee L, Ogoshi A (1991) Identification of *Rhizoctonia* species. American Phytopathological Society, St. Paul

Somani AK, Arora RK (2010) Field efficacy of *Trichoderma viride, Bacillus subtilis* and *Bacillus cereus* in consortium for control of *Rhizoctonia solani* causing black scurf disease of potato. Indian Phytopathol 63:23–25

Somani AK, Chohan RKS (1994) Herbicide paraquat spray as a predisposing factor for dry rot development in freshly harvested potatoes. In: Shekhawat GS, Paul Khurana SM, Pandey SK, Chandla VK (eds) Potato present and future. Indian Potato Association, Shimla, pp 219–220

Stevenson WR, Loria R, Franc GD, Weingartner DP (2001) Compendium of potato diseases, 2nd edn. The American Phytopathological Society, St. Paul

Thirumalachar MJ (1955) Incidence of charcoal rot of potato in Bihar (India). In relation to cultural conditions. Phytopathalogy 456:608–610

Thirumalachar MJ, O'Brien MJ (1977) Suppression of charcoal rot of potato with a bacterial antagonist. Plant Dis Rep 61:543–546

Tsror L, Perezt-Alon I (2005) The influence of inoculum source of *Rhizoctonia solani* on development of black scurf on potato. J Phytopathol 153:240–244

Tsror L, Balak R, Sneh B (2001) Biological control of black scruf on potato under organic amendment. Crop Prot 20:145–1509

Turkensteen LJ, Mulder A (1999) The potato disease *Phytophthora infestans*. Gewasbescherming 30:106–112

Turkensteen J, Flier WG, Wanningen R, Mulder A (2000) Production, survival and infectivity of oospores of *Phytophthora infestans*. Plant Pathol 49:688–696

Ulrich J, Schrodter H (1966) Nachrbl. dt. Pflaschutzdienst. Stuffa 18:33–40

Van Everdingen E (1926) Het berband tusschen der weersgesteld held on de aarda ppelziekte (*Phytophthora infestans*) (the relation between weather conditions and *Phytophthora infestans*). Tijdscher Pfziet 32:129–140

Verma M, Brar SK, Tyagi RD, Surampalli RY, Valéro JR (2007) Antagonistic fungi, *Trichoderma* spp.: panoply of biological control. Biochem Eng J 37:1–20

Virgen-Callerus G, Olalde-Portugal V, Carling DE (2000) Anastomosis group of *Rhizoctonia solani* on potato in central Mexico and potential for biological and chemical control. Am Potato J 77:214–219

Voral V (2005) Potato Sazava. Czech J Genet Plant Breed 41(4):175–176

van der Waals JE, Denner FDN, van Rij N, Korsten L (2003) Evaluation of PLANT-Plus a decision support system for control of early blight on potatoes in South Africa. Crop Prot 22(6): 821–828

Wale SJ (2000) Summary of the session on national potato production and the powdery scab situation. In: Merz U, Lees AK (eds) Proceedings of First European Powdery Scab Workshop, SAC, Aberdeen, Scotland, July 20–22, 2000. SCRI, Dundee, pp 3–9

Walker JC (1969) Plant pathology, 3rd edn. McGraw-Hill, New York

Walker JC (1983) *Synchytrium endobioticum*. CMI descriptions of pathogenic fungi and bacteria. No. 755. CAB International, Wallingford

Wallins JR (1962) Summary of recent progress in predicting late blight epidemics in United States and Canada. Am Potato J 39:306–312

Welsh RD, Callaghan MO (1996) Evaluation of fenpiclonil, as a potato seed treatment for the control of *Rhizoctonia solani* and *Helminthosporium solani*. In: Proceeding of 49th New Zealand plant protection conference. New Zealand, pp 152–156

Western JH (ed) (1971) Diseases of crop plants. Wiley, New York

Wharton P, Kirk W (2007) Fusarium dry rot. http://www.potatodiseases.org/contact.html

Wharton P, Fairchild K, Belcher A, Wood E (2012) First report of *in vitro* boscalid-resistant isolates of *Alternaria solani* causing early blight of potato in Idaho. Plant Dis 96(3):454–455

Wilson PS, Ahvenniemi PM, Lehtonen MJ, Kukkonen M, Rita H, Valkonen JPT (2008) Biological and chemical control and their combined use to control different stages of the *Rhizoctonia* disease complex on potato through the growing season. Ann Appl Biol 153:307–320

Winter W, Winiger FA (1983) Einflusverschiedener Fangpflenzensowie von Kalk und Kalkstickstoff auf die Bondenverseuchungmit Spongospora subterraneadem Erreger des Pulverschorfesbei Kartoffeln. Mitteilungenfur die Schweizerische Landwirtschaft 31:190–206

Yakhin IA, Ibragimov RI, Yakhin OI, Isaev RF, Vakhitov VA (1998) The induced effect of biopreparation stifun on the accumulation of trypsin inhibitors in potato tubers in storage. Russ Agr Sci 8:5–8

Zin-Woo K, Park BK, Hawang IY, Park CS (1998) Antifungal activity of root colonizing *Pseudomonas fluorescens* MCOT is responsible for its disease suppressing ability. Kor J Plant Pathol 14:606–611

Zwankhuizen MJ, Govers F, Zakoks JC (1998) Development of potato late blight epidemics: disease foci, disease gradients, and infection sources. Phytopathology 88:754–763

Zwankhuizen MJ, Govers F, Zadoks JC (2000) Inoculum sources and genotypic diversity of *Phytophthora infestans* in Southern Flevoland, The Netherlands. Eur J Plant Pathol 106: 667–680

Chapter 7
Host–Pathogen Interaction, Plant Diseases, Disease Management Strategies, and Future Challenges

Chakravarthula Manoharachary and Indra Kala Kunwar

7.1 Host–Pathogen Barriers: Some Issues

World population has been increasing enormously and is marching towards ten billion by 2050. The natural resources have been dwindling and getting exhausted due to over exploitation and anthropogenic activities. Diseases caused by fungi, bacteria, virus, mycoplasma, nematodes and other microbes on crop plants form the limiting factors in realizing the yield potential of crop plants thus resulting in about 10–25 % losses leading to famines, hunger, poverty and economic crisis. It is also known from literature that diseases of unknown etiology like Karnal bunt of wheat, sheath blight in rice, citrus die back, mango malformation, wilt of coconut, wilt of guava, brown blast of rubber and other such diseases are of great concern by virtue of their resurgence capabilities resulting in a change of cropping system, climate patterns, etc. Critical issues have become evident due to over and non-judicial use of chemicals to control diseases and pests including toxicity, residual effect and health problems besides resistance towards chemical by plant pathogens and insects. Successful management of diseases is the need of the hour to provide food security to the billions of people living around the world. Disease management practices must be safe, efficient, eco-friendly and cost effective.

Plant disease is defined as any disturbance that prevents normal development of plant and reduces its economic value besides having deviation in normal physiology and biochemical mechanisms that lowers the yields and reduces quality. Diseases are caused by pathogenic microorganisms, fungi, nematodes and insects. Parasitic plants and abiotic stresses are also known to cause diseases.

Disease production is not a simple process but involves many steps and critical issues. A susceptible host, virulent pathogen and congenial atmospheric conditions

C. Manoharachary, M.Sc., Ph.D., D.Sc. (✉) • I.K. Kunwar, M.Sc., D.Phil.
Department of Botany, Osmania University, Hyderabad, Telangana 500007, India
e-mail: cmchary@rediffmail.com

A. Goyal and C. Manoharachary (eds.), *Future Challenges in Crop Protection Against Fungal Pathogens*, Fungal Biology, DOI 10.1007/978-1-4939-1188-2_7,

are essential. The pathogen enters into the host through wounds, stomata, mechanical injuries, direct penetrations and also by degrading cell wall materials. The pathogenic fungi do produce several enzymes such as cutinases, cellulases, petinases, lignases and several other such biological weapons which degrade the complex cell wall materials of host and make successful entry into host tissues. However, there is a requirement of spore load for successful infection called inoculum potential. After the successful entry and establishment of pathogen, several toxins namely phytotoxins, vivotoxins and pathotoxins are released by the fungal pathogens into the host tissues which derail the host metabolism. Lycomarasmin is the example of phytotoxins and Fusaric acid is one of the toxins cited for vivotoxins. Pathotoxins include piricularin, tabotoxin, and victorin is the host specific toxin. Alteration of plant physiological and biochemical functions due to host–pathogen interaction results in disease syndrome formation. The sequential steps as mentioned above result in symptom production. Symptoms include leaf spots, cankers, galls, rusts, smuts, malformation, wilts, root and stem rots, blights, and others. Thus, various diseases on diversified crops result in heavy losses leading to epidemics and famines. In view of huge losses incurred in crop productivity, there is a necessity of plant disease control and management.

In the process of host–pathogen interaction several impediments and hurdles need to be faced both by fungal pathogens and host. The host plants defend themselves through defense structures that exist before infection. These include waxes, cuticles, multiple epidermis, closed stomata, formation of cork layers, tyloses formation, abscission layer formation, gum deposition, hyphal sheathing, etc.

Biochemical defense mechanism which offers resistance in host against pathogen being the release of inhibitors such as chlorogenic acid, phenolic compounds absence of nutrients, absence of common antigens and others. Phytoalexins are the fungicidal/fungistatic substances which serve the purpose of antibody. These are low molecular weight antimicrobial compounds that accumulate in plants as a result of infection or stress. Ipomeamarone and many other compounds play important role. In addition to phenolic substances produced in plant tissues in response to infection, induced synthesis of proteins and enzymes seems to play a key role in disease resistance. Pathogen related proteins are generally a group of plant proteins that are toxic to invading fungal pathogens or other pathogens. Signal transduction and several other genetic mechanisms have a greater role. There are also preformed chemicals (prohibitins) which prevent disease production.

Host–pathogen interaction is a struggle for survival between two organisms and there exists a state of balance and surviving plants and their parasites are capable of coexistence. Resistance and susceptibility are the major factors that play an important role in host–parasite interaction.

Interactions between a host and pathogen are medicated by environmental conditions, but are ultimately determined by genotype. Flor (1956) studied the genetics of flax cultivars and isolates of the linseed rust pathogen. The concept being that for each gene conditioning virulence or virulence in the pathogen, there is a corresponding gene conditioning resistance or susceptibility in the plant.

Genes for the resistance in the host are dominant (R) and those for susceptibility (lack of resistance) are recessive (r). Genes for virulence are dominant (A) and recessive (a). Gene for gene hypothesis operates therefore; four gene combinations are possible in host–pathogen interaction. These combinations are R-a, r-A, r-a, and R-A. Three combinations R-a, r-A, and r-a give rise to compatible reaction and infections are successful. One combination R-A results in an incompatible reaction and no infection occurs.

7.2 Plant Pathogen Diagnosis: Molecular Approach

Identification and diagnosis of plant pathogens in the process of early infection can help for the better management of the diseases. Martin et al. (2000) have discussed the molecular diagnosis of pathogen at length. Farmer is interested in the early and rapid detection of pathogen and disease diagnosis. It must be simple and cost effective. Symptoms and identification of plant pathogens based on morpho-toxonomic criteria have been used since ages. However, this kind of methodology is time consuming; hence DNA probes have been developed. RAPD, RLFP, ISSR, URP, EIISA, PCR, Q-PcR, DNA array technology membrane-based microarrays (Zhang et al. 2006) and others are developed for the detection of multiple pathogens present in the field samples. For understanding disease epidemics and their management, understanding variability in pathogen population is critical. Applications of DNA markers and probes have become very effective and less time consuming (Jalali 2008).

7.3 Host–Pathogen Interaction (Genetic Resistance and Biochemical Aspects)

Plants are known to posses preformed (constitutive) as well as inducible defense system to inhibit attack by pathogenic bacteria, fungi, viruses, nematodes and insects through metabolic, biochemical and molecular defenses, as well as by physical/structural barriers (Hammerschmidt 2007). Many studies also showed that chemicals or physical barriers are known to involve in preformed resistance. The exogenous applications of biotic or abiotic elicitors inducing resistance are categorized either as systemic acquired resistance (SAR) or induced systemic resistance (ISR) (Pieterse and Van Loon 2007). Induced resistance (SAR and ISR) involves in the activation of multiple genes and/or defense signaling pathways. Signal-transduction pathways are mediated by salicylic acid (SA), jasmonic acid (JA), and ethylene (ET) which are involved in regulating appropriate defense responses. The crosstalk between these different signal-transduction pathways allows the plant to fine-tune its defenses against different types of fungal microbial pathogens and insects (Pieterse and Van Loon 2007). Recently developed biotechnological tools are very helpful in recognizing the complex nature of plant defense mechanisms (Xing and Jordan 2000).

7.3.1 Passive Resistance

The cell wall is an important barrier monitoring its integrity and allows plants to quickly activate pathways to minimize pathogen entry and reduce the spread of disease (Hematy et al. 2009). Epidermal cell wall penetration and cell walls constituency are found to resist *Magnaporthe graminis* in resistant host varieties than susceptible host species. The alterations in cuticular structure lead to the release of fungitoxic substances and changes in gene expression that form a multifactorial defense response in *Arabidopsis thaliana* against *Botrytis cinerea* (Chassot et al. 2008). Variation in pectin composition of cell wall has been associated with stem rust (*Puccinia graminis*) resistance in wheat (Wietholter et al. 2003). In addition to these structural barriers, preformed antifungal compounds considered as prohibiting such as phenols and phenolic glycosides, unsaturated lactones, sulphur compounds, saponins, cyanogenic glycosides and glucosinolates (Khan et al. 2011) offer defense to plants against a variety of pathogenic microorganisms. In plants some defensins are known to be of antifungal or occasionally anti-bacterial activity. Salicylic acid synthesis pathway is shown to be responsible for accumulation of phenolics in plants after microbial infection in plants (Liu et al. 2007).

7.3.2 Induced Defense Responses

Many plants respond to pathogen attack by activating an array of inducible defense responses. Plant genotypes have specific R genes that are able to recognize pathogens to the corresponding *avr* genes. The *R* gene products are polymorphic and are coded by multigene families covering at least 150 loci in the *Arabidopsis* genome (The *Arabidopsis* Genome Initiative, 2000).

Many *R* genes confer resistance to various plant species against a range of pathogens and have been identified, isolated and cloned. *R* genes have been divided into five categories based on their structural features (Hammond-Kosack and Jones 1997). The first category encodes cytoplasmic serine/threonine protein kinases, suggesting their role in signaling mechanism via protein phosphorylation (Martin et al. 1993). The second category encodes cytoplasmic proteins with leucine-rich repeats (LRR), leucine zippers (LZ) and nucleotide binding sites (NBS) suggesting a model role for these molecules in intracellular signaling (Ori et al. 1997). The third category is very similar to the second class but in addition to LRR and NBS these *R* gene products contain N-terminal homology to Toll and the interleukin receptors (TIR) in *Drosophila* and mammals, respectively (Parker et al. 1997). Genes in the fourth category encode extracytoplasmic LRRs protein with membrane anchors and short C-terminal cytoplasmic tails. The fifth categories of R genes encode proteins with a clear extracellular LRR domain and an intracellular serine/threonine kinase domain (Song et al. 1995).

Avirulent (Avr) genes have dual functions namely pathogens containing Avr genes that are avirulent to plants carrying the matching R genes, while they are virulent in race, strain, pathovar or species-specific manner to plants without carrying the matching R genes. Selection pressure exerted by virulent pathogens on host plants might have resulted in the co-evolution of these *R* genes. Since *R* gene mediated resistance requires a corresponding *avr* gene, this resistance is also considered as gene-for-gene resistance. Triggering defense of responses by *R* genes occurs much faster than the responses to virulent pathogens and this leads to an effective expression of resistance. Several alterations in cellular metabolism occur as an outcome of R-avr binding, which finally results in the expression of resistance. The following are the alterations:

7.3.2.1 Cell Wall Fortification

Formation of new cell wall material adjacent to the site of infection in plants is an important defense response to pathogen penetration. These cell wall appositions (papillae) can occur even before the pathogen passes through the cell wall (Soylu et al. 2005). Callose offers defense against many pathogenic fungi and is manifested at the stage of penetration by the formation of papilla (Böhlenius et al. 2010). Silencing of callose synthase genes in *Arabidopsis* was to bring about a higher penetration rate of the pathogen *Hyaloperonospora parasitica*. Reduction in cellulose biosynthesis induces production of phytohormones, jasmonic acid (JA), salicylic acid, and ethylene which enhances pathogen resistance, and leads to changes in cell wall composition/structure, as well as causing ectopic lignin production (Hamann et al. 2009).

7.3.2.2 Phytoalexin Synthesis

Pathogen attack on specific host or diversified hosts that results in the synthesis of protective compounds (phytoalexins) which include a range of phenolics, terpenoids, hydroxycoumarins and hydroxycinnamate conjugates (Karou et al. 2005).

7.3.2.3 Oxidative Burst and Antioxidative Enzyme System

Studies related with the role of reactive oxygen species (ROS) during plant–pathogen interaction have been carried out in all kinds of interactions including in hemibiotrophic interaction (Shetty et al. 2007), necrotrophic interaction and biotrophic interaction (Romero et al. 2008).

During host–pathogen interaction production of ROS (mostly O_2, H_2O_2 and HO) at the cell surface called the 'oxidative burst' is the earliest event (Mendoza 2011) (Table 7.1). ROS plays a critical role in inducing tolerance by activating or inducing stress response related factors, such as mitogen-activated protein kinases (MAPKs), transcription factors, antioxidant enzymes, dehydrins, and low temperature induced shock, and pathogenesis-related proteins (Yoshioka et al. 2001). The primary

Table 7.1 Plant system generating ROS when challenged with various pathogens elicitors

Plant	Pathogen	AOS observed
Hordeum vulgare	*Blumeria graminis*	H_2O_2
Lycopersicum esculentum	*Fusarium oxysporum*	H_2O_2
Mimosa pudica	Ergosterol	O_2
Cicer arietinum	*Fusarium oxysporum*	H_2O_2
Arabidopsis thaliana	*Botrytis cinerea*	H_2O_2
Nicotiana tabacum	*Botrytis cinerea*	H_2O_2
Arabidopsis thaliana	*Fusarium oxysporum*	H_2O_2
Arabidopsis thaliana	*Colletotricum lindemuthianm*	H_2O_2
Phaseolus vulgaris	*C. lindemuthianum*	H_2O_2
Brassica spp.	*Alternaria brassicae*	H_2O_2
Bean	*C. lindemuthianum*	H_2O_2
Soybean	*Verticillium dahliae*	H_2O_2
Cotton	*Verticillium dahliae*	H_2O_2
Tomato	*Cladosporium fulvum*	H_2O_2 and O_2
Rice	N-Acetylchito-Oligosaccharides	H_2O_2, O_2 and OH
Tobacco	*Phytophthora crytogea*, pure cryptogein and capsicein protein elicitor	H_2O_2, O_2 and OH
Rose	*Phytophthora* sp., crude cellwall preparation	H_2O_2 and O_2
Cucumber	*Phytopthora sojae*	H_2O_2
Parsley	*Rhizosphaera kalkhoffii*	H_2O_2
Soybean	*Pseudomonas syringae*	H_2O_2
Arabidopsis thaliana	Protein elicitor harpin	H_2O_2 and O_2

response to elicitors is indicated by changes in membrane permeability and activation of specific ion channels leading to the influx of Ca^{2+} and H^+ and the efflux of Cl^- and K^+ (Cervone et al. 1997) and the production of ROS, sequentially followed by defense gene activation and phytoalexin accumulation (Jabs et al. 1996). Out of two oxidative bursts occurring in plant–pathogen interactions, the first phase, which occurs within a few minutes of encounter with the pathogen, is a general response observed in both compatible and incompatible plant–pathogen interactions. The second oxidative burst that is observed 3–6 h after the first one is characteristic of incompatible interactions and correlates with the expression of resistance. Cell wall located enzymes have been proved to be responsible for apoplastic ROS overexpression in sunflower or rape. They induced an increase in H_2O_2 levels following pathogen inoculation compared with wild-type plants (Dong et al. 2008). Pathogen induced oxidative burst is known to involve the expression of enzymes like NAD(P)H oxidase, peroxidase, oxalate oxidase and lipoxygenase. NAD(P)H oxidase brought about the generation of AOS in soybean suspension cultures treated with an elicitor from *Phytophthora sojae* (Mithofer et al. 1997).

Plants posses very efficient enzymatic (superoxide dismutase, SOD (EC 1.15.1.1); catalase, CAT (EC 1.11.1.6); ascorbate peroxidase, APX (EC 1.11.1.11); glutathione reductase, GR (EC 1.8.1.7); monodehydroascorbate reductase, MDHAR (EC 1.6.5.4);

dehydroascorbate reductase, DHAR (EC 1.8.5.1); glutathione peroxidase, GPx (EC 1.11.1.9); guaiacol peroxidase, GPX (1.11.1.7) and glutathione-S-transferase, GST (EC 2.51.18) and non-enzymatic systems such as ascorbic acid, ASH; glutathione, GSH; phenolic compounds, alkaloids, non-protein amino acids, and antioxidant defense systems which work to control the cascades of uncontrolled oxidation and protect plant cells from oxidative damage by scavenging of ROS (Shao et al. 2008). SOD plays an important role in first line of defense, it converts superoxide radicals into hydrogen peroxide and oxygen which is less toxic to plant cell whereas CAT, APX and POD convert these hydrogen peroxide into water to nullify the toxic effect of ROS.

Pathogen infection also leads to an increase in the antioxidant enzymes, SOD and guaiacol-dependent peroxidase (GDP) in *Brassica* variety HC1 (Joshi et al. 2010). *Botryosphaeria berengriana* f. sp. *piricola* (BBP) and *Monilinia fructigena* (MFH) inoculated pear calli showed significantly increased antioxidant enzymes SOD, POD and CAT and expression of their isozymes were also increased as compared to control (Zhao et al. 2012).

Researchers have indicated that the resistance inducing chemicals can improve resistance in all of the sunflower genotypes to downy mildew and increase enzyme activities of peroxidase and polyphenol oxidase, as well as accumulation of mRNAs of glutathione S-transferase, defensin and catalase. The activity of SOD and guaiacol peroxidase (GPX) increased in the *mungbean yellow mosaic India virus* (MYMIV) infected and SA treated *Vigna mungo* plants when compared with control (Kundu et al. 2012).

Tomato defense genes, such as PR-1, β-1, 3-glucanase (GLU), polyphenol oxidase (PPO), peroxidase (POD), and SOD, were rapidly and significantly up-regulated by exogenous ABA treatment. There was a significant increase in the activities of defense enzymes like phenylalanine ammonia lyase, peroxidase and β-1, 3-glucanase on two chemical elicitors (acibenzolar-*S*-methyl benzo-[1,2,3]–thiadiazole-7-carboxylic acid S-methyl ester) and salicylic acid treatments in tea leaves challenged with the blister blight disease caused by *Exobasidium vexans* pathogen than on unchallenged leaves (Ajay and Baby 2010).

Antioxidants have different types of isozymes and the purpose of isozymes is to allow fine adjustment of metabolism for the proper functioning during any type of stress. Numerous reports have demonstrated the relationship between the expression pattern of POD isozymes and their different antifungal abilities. High intensities of SOD, POD and PPO isozymes in RM (mixture of riboflavin and Methionine) pretreated and powdery mildew (*Sphaerotheca fuliginea*) infected cucumber plants clearly indicated an important host defensive mechanism against the infection (Nam 2008).

7.3.2.4 PR Protein Synthesis

The pathogenesis-related (PR) proteins are synthesized in response to pathogens. Currently 17 families of inducible pathogenesis-related proteins have been recognized (Van Loon et al. 2006) (Table 7.2). PR-2 family is identified as β-1, 3-endoglucanases and the PR-3,-4,-8 and -11 ad endochitinases which act against

Table 7.2 Families of pathogensis-related proteins in plants and their functions

Family	Type member	Function	Molecular mass range (kDa)
PR-1	Tobacco PR-1a	Unknown	15–17
PR-2	Tobacco PR-2	β-1,3-glucanase	30–41
PR-3	Tobacco P, Q	Chitinase type-I, II, IV, V, VI, VII	35–46
PR-4	Tobacco 'R'	Chitinase type I, II	13–14
PR-5	Tobacco 'S'	Thaumatin-like	16–26
PR-6	Tomato Inhibitor I	Proteinase-inhibitor	8–22
PR-7	Tomato P69	Endoproteinase	69
PR-8	Cucumber chitinase	Chitinase type III	30–35
PR-9	Tobacco 'lignin-forming peroxidase'	Peroxidase	50–70
PR-10	Parsley 'PRI'	Ribonuclease like	18–19
PR-11	Tobacco 'class V' chitinase	Chitinase, type I	40
PR-12	Radish Rs-AFP3	Defensin	5
PR-13	Arabidopsis THI2.1	Thionin	5–7
PR-14	Barley LTP4	Lipid-transfer protein	9
PR-15	Barley OxOa (germin)	Oxalate oxidase	22–25
PR-16	Barley OxOLP	Oxalate-oxidase-like	100 (Hexamer)
PR-17	Tobacco PRp27	Unknown	Not known

Source: Modified from Van Loon et al. (2006)

bacterial and fungal pathogens, respectively. The chitinases as well as proteinase inhibitors (PR-6) have been reported to target nematodes and herbivorous insects. Members of the PR-8 family show lysozyme activity and are directed against bacterial pathogens, whereas defensins (PR-12) and thionins (PR-13) have broad antibacterial and antifungal activities. Some lipid-transfer proteins (PR-14) have antifungal and antibacterial activities and members of PR-1 thaumatin-like PR-5 families have been associated with activity against oomycetes. PR-7 is an endoproteinase that is the most conspicuous PR in tomato. PR-9 is a specific peroxidase that is thought to act in cell wall reinforcement by catalyzing lignification. PR-10 shows homology to ribonucleases and is thought to play a role in resistance against viral pathogens PR-15 and -16, comprise families of germin like oxalate oxidases and SODs respectively. PR-17 proteins have been found in infected tobacco, wheat and barley and contain sequences resembling the active site of zinc-metalloproteinases. A putative novel family (PR-18), which comprises of fungus and SA-inducible carbohydrate oxidases has been reported from sunflower (Custers et al. 2004).

7.3.2.5 Signal Transduction Leading to Expression of Defense Response Genes

Interconnected signaling pathways operate within the infected plant cells and this leads to expression of defense genes and intermediates include molecules like SA, JA, and ET (Von Dahl and Baldwin 2007). Other plant hormones, including ABA, brassinosteroids and auxin, have been implicated in plant defense, but their significance is less studied. Genetic dissection of the signaling pathway using mutants showing alteration in defense responses has led to the identification of signaling components and their interactions.

7.3.2.6 SA Signaling Pathway

Salicyclic acid (SA) plays an important signaling role in the activation of plant defense response due to pathogenic invasion, including both systemic (termed systemic acquired resistance) and localized responses usually characterized by HR. The role of SA in signal transduction has been studied by genetic dissection of the pathway using mutants. Some mutants show constitutive expression of PR genes, enhanced SA accumulation (cpr1, cpr5, cpr6) and spontaneous lesion formation similar to those observed during expression of a hypersensitive response. Several potential components of the SA signaling pathway initially were identified and cloned in tobacco including the bZIP transcription factors (Zhou et al. 2000). The bZIP transcription factors bind to SA-responsive elements in the promoters of defense genes, primarily those of pathogenesis-related or PR genes. This coupled with increasing genetic and biochemical crosstalk between SA-, ethylene and JA-associated defence pathways, underlie the crucial significance of salicylic acid signalling in plant resistance (Pedley and Martin 2003).

7.3.2.7 JA Signaling Pathway

Jasmonic acid (JA) is a lipid-derived hormone that plays an important regulatory role in various features of plant development and defense. In tomato, gene coding for leucine aminopeptidase, coronatine-insensitive 1 (COII) and allene oxide cyclase (AOC) are crucial in the proper functioning of JA signaling. Leucine modulates immunity against herbivores by inducing wound responsive genes and acting downstream of JA (Fowler et al. 2009). However, induction of the gene can be observed upon local and systemic wounding or treating with JA and system in or glucose, suggesting temporal regulation of JA biosynthesis (Stenzel et al. 2008).

7.3.2.8 ET Signaling Pathway

Role of ethylene in defense can be said as dual since it has been implicated as both a signal molecule during plant resistance and a virulence factor that can lead to pathogenesis, symptom expression and plant susceptibility (Chagué et al. 2006). There, ethylene is thought to have differential effects during plant defense in different pathosystems. Regulation and control of events during ethylene signaling mostly happen at the level of ethylene biosynthesis. In tomato, key genes in ethylene biosynthesis have been cloned and described, including those encoding

1-amincyclopropane carboxylic acid (ACC) oxidase and ACC synthase. However, emerging data show that regulation occurring at the level of ethylene receptors is also important. Ethylene biosynthesis shows interplay with certain MAP kinase cascades (Kim et al. 2003b) demonstrates the undeniable role of ethylene during plant defense.

7.3.2.9 ABA Signaling Pathway

More recently abscisic acid (ABA) was shown to play role in pathogen-associated defense responses. Exogenous application of ABA was seen to increase susceptibility of plants to pathogens, suggesting that ABA is a negative regulator of defense responses. Mutants affected in ABA synthesis (aba1-2, aba2-1) or ABA signaling (abi1-1, abi2-1) displayed enhanced resistance to pathogen and also an expression of JA/ET responsive genes (PDF1.2). ABA signaling therefore appeared to antagonize the JA-ET signaling pathway. However, ABA was seen to bring about enhanced resistance to pathogens through rapid callose deposition during early stages of infection in *Arabidopsis–Alternaria* interactions (Ton and Mauch-Mani 2004).

7.3.2.10 Cross Talk

Cross talk between pathways provides a regulatory potential for activating multiple resistance mechanisms in varying combinations and may help the plant to prioritize the activation of particular defense pathways thereby providing optimal defense (Pieterse et al. 2001). Global expression-profiling studies provided ample evidence that SA, JA, and ET pathways interact, either positively or negatively (Mur et al. 2006). SA-mediated suppression of JA-inducible gene expression is blocked in mutant *npr1* plants, demonstrating a crucial role for NPRI in the cross talk between SA and JA signaling. WRKY transcription factors are important regulators of SA-dependent defense responses. ET was shown to enhance the response of *Arabidopsis* to SA, resulting in a potentiated expression of the SA-responsive marker gene PR-1 (De Vos et al. 2006).

7.3.2.11 Induced Resistance

Induced resistance is not based on direct defense activation by the inducing agent, but on a faster and stronger activation of inducible defense mechanism once the plant is exposed to the pathogen. This enhanced capacity to express basal defense mechanisms is called sensitization, potentiation, or priming. Elicitors of biotic (glucans, proteins, lipids, etc.) and abiotic origin, signaling intermediates like SA, JA, ET, ABA, as well as other chemicals like the non-protein amino acid β-aminobutyric acid (BABA) were also shown to induce resistance in many plants against various pathogens (Cohen 2002). Different types of induced resistance and their mechanism are as follows.

Pathogen-Induced Resistance. Induced resistance is often activated after primary infection with a necrotizing pathogen and renders the distant, uninfected plant parts more resistant towards a broad spectrum of virulent pathogens including viruses, bacteria and fungi (Durrant and Dong 2004; Heil and Ton 2008). This form of induced resistance is often referred to as SAR and has been demonstrated in many plant–pathogen interactions (Sticher et al. 1997) mediated through SA accumulation. SA has an important role in signaling pathogen leading to ISR. After infection, endogenous levels of SA increase locally and systemically. The identity of the long distance signals that travels from the site of primary infection to the remote parts of the plant to induce PR gene expression and SAR, however is still unclear (Grant and Lamb 2006). Over the past few years, several other signaling molecules have emerged as possible candidates for the endogenous long distance signal for SAR. These include methyl salicylate (MeSA), glycerolipids (Chaturvedi et al. 2008), azeleic acid, and glycerol-3-p (Chanda et al. 2011).

The exact nature of the systemic SAR signal in *Arabidopsis* (*Arabidopsis thaliana*) after localized infection by avirulent *Pseudomonas syringae* remains complex and has been a matter of debate. In tobacco, SAR activation by the primary pathogen resulted in a significant reduction of disease symptoms caused by the fungi *Phytophthora parasitica, Cercospora nicotianae* and *Peronospora tabacina.*

Resistance Induced by Non-pathogenic Organisms. ISR was first described in leaves of *Arabidopsis* plants that were subjected to prior inoculation of roots with non-pathogenic *Pseudomonas fluorescens* strain WCS 417r. Subsequently researchers have shown that specific strains of PGPR not only improve plant growth, but also induce ISR against a wide range of pathogens (Baker et al. 2003).

Interaction of plants with beneficial microorganisms other than those causing ISR can also result in systemic, broad-spectrum resistance. The symbiosis between barley roots and the endophytic basidiomycete *Piriformospora indica* confers systemic resistance to the necrotrophic root-rot fungus *Fusarium culmorum* and the biotrophic fungus *Blumeria graminis* f. sp. *hordei* (Waller et al. 2005). Systemic resistance induced by the endophytic fungus *Trichoderma asperellum T34* protected *Arabidopsis* against a wide range of pathogens through engagement of the same signaling components as used in *Pseudomonas fluorescens* (Strain WCS417r)-mediated ISR. Transcript profiling of the shoots of *Medicago truncatula* plants whose roots had been colonized by the arbuscular mycorrhizal fungus *Glomus intraradices* revealed both systemic expression of various defense associated genes and establishment of an IR response to the bacterium *Xanthomonas campestris* pv. *alfalfa* (Liu et al. 2007).

Resistance Induced by Exogenous Application of Signaling Intermediates and other Chemicals. Exogenous application of signaling intermediates like SA, JA and ET and their derivatives as well as other chemicals like the non-protein amino acid

BABA were seen to induce resistance in plants against a variety of pathogens. These chemicals did not show any antimicrobial activity in vitro, but activated plant defense responses which led to the expression of resistance.

7.3.2.12 SA and SA Analogues

Salicylic acid plays a critical signaling role in the activation of disease resistance in plants. Exogenous application of SA induced resistance in tomato against tomato stem canker disease. Also foliar application of low concentration of SA induced priming in lettuce (*Lactuca sativa)* plants against pill-bugs *(Armadillidium vulgare)* (Nancy et al. 2011). SAR-like state of enhanced disease resistance can be induced without an HR in plants by treatment with solutions of SA or certain other chemicals. SA could induce systemic protection against *Botrytis elliptica* in lily *cv.* Star Gazer, and that is one of the few cases of SA-induced disease resistance demonstrated in monocots. Exogenous application of 200 mM salicylic acid through root feeding and foliar spray induced resistance against *Fusarium oxysporum* f. sp. *lycopersici* (Fol) in tomato (Mandal et al. 2009). Similarly exogenous application of SA or its analogs on plants has been shown to induce SA-mediated plant defenses to a broad range of pathogens. The effect of two chemical elicitors acibenzolar-*S*-methyl benzo-(1,2,3)-thiadiazole-7-carboxylic acid S-methyl ester and salicylic acid showed inducing resistance in tea plants against blister blight disease caused by *Exobasidium vexans*.

SA analog ASM was reported as induces of SAR in many plant–pathogen interactions like, pepper-*Phytophthora capsici*, tomato-*Xanthomonas euvesicatoria*, cyclamen-*Fusarium oxysporum* f.sp. *cyclaminis* and tobacco-*Peronospora tabaci* (Perez et al. 2003). Application of BTH also induced resistance in a number of plants such as wheat, cotton, tomato, pea, tobacco and *Arabidopsis* (Kohler et al. 2002) against a variety of pathogens. BTH treatment protected melon seedlings against two soil-borne pathogens *Didymella bryoniae* and *Sclerotinia sclerotiorum.*

7.3.2.13 JA and Related Compounds

Genes in the biosynthetic pathway of JA namely *lox 1* and *lox 2* (encoding two liposygenases) were activated on exogenous application of JA. Treatments with methyl jasmonate, the methyl ester of jasmonic acid, induce the accumulation of alkaloid compounds in jaborandi leaves (*Pilocarpus microphyllus)* and phytoalexins in *Cupressus lusitanica* cell cultures (Zhao et al. 2004). In grapevine, MeJA has been shown to stimulate deposition of callose and accumulation of PR proteins in leaves. Belhadj et al. (2006) showed that MeJA treatment of grapevine in the vineyard induces the production of stilbene phytoalexins, the accumulation of PR protein related RNAs, and triggers enhanced resistance to *Erysiphe necator.* Moreover, jasmonic acid pathway was found to contribute to the sulfated laminarin-induced resistance against *Plasmopara viticola* (Trouvelot et al. 2008).

7.3.2.14 Ethylene

Ethylene is involved in the activation of plant defense-related processes such as the production of PR proteins, and phytoalexins. Diaz et al. (2002) demonstrated that ethylene pre-treatment of tomato plants resulted in an increased resistance to *Botrytis cinerea*. Recently Belhadj et al. (2008) showed that ethephon treatment of grapevine foliar cutting triggers enhanced resistance to *Erysiphe necator* by inducing expression of various PR protein genes and the production of stilbenes. In transgenic pepper seedlings overexpressing endogenous *P R 10* or esterase genes, which are induced by the ET treatment, completely resisted the infection, which corroborated the importance of these genes in the defense response (Nunez-Pastrana et al. 2011).

7.3.2.15 Abscisic Acid

Exogenous application of Abscisic acid (ABA) prior to inoculation increased the susceptibility of tomato, *Arabidopsis* or grapevine when inoculated by various pathogens indicates the role of ABA as a negative regulator of plant defense responses. Susceptibility of tomato plants to *Botrytis cinerea* was enhanced by ABA treatment (Audenaert et al. 2002) and a high ABA concentration was observed in grapevine shoots with Pierce's disease symptoms compared to healthy shoots. In contrast exogenous application of ABA prior to inoculation induced resistance against necrotrophic pathogens in *Arabidopsis*. Exogenous ABA also enhanced disease resistance against *A. solani* infection in tomato (Song et al. 2011).

7.3.2.16 Probenazole

Probenazole and its active metabolite 1, 2-benzisothiazole-1, 1-dioxide induces SAR by triggering signaling at a point upstream of SA accumulation. Probenazole (PBZ) is the active ingredient of Oryzemate, an agrochemical which is used for the protection of rice plants from *Magnaporthe grisea* (Iwai et al. 2007). Probenazole has been used as practical crop protectant in wet land rice cultivation to control blast fungus (Yoshioka et al. 2001).

7.3.2.17 β-Aminobutryic Acid

Various amino acids have been shown to induce resistance in plants (Kuc 2001). Among these, the non-protein amino acid β-aminobutyric acid (BABA) has received a lot of attention. The compound was shown to be a potent inducer of resistance against abiotic stress and microbial pathogens (Cohen 2002). BABA-protected potato plants against *Phytophthora infestans*, especially when applied early in crop development and also provided some protection in tubers against late blight (Altramiranda et al. 2008).

Table 7.3 Plants exhibiting β-aminobutryic acid (BABA) induced resistance against viral, bacterial, fungal and nematode pathogen

Plant	Pathogen
Potato	*Phytophthora infestans, Alternaria solani*
Tomato	*Phytophthora infestans, Botrytis cinerea, Fusarium oxysporum* f. sp. *lycopersici, Meloidogyne javanicum, Clavibacter michiganensis*
Tobacco	*Peronospora tabacina, Phytophthora parasitica* var. *nicotianae, Tobacco mosaic virus*
Pepper	*Phytophthora capsici, Colletotrichum coccodes*
Melon	*Pseudoperonospora cubensis*
Arabidopsis	*Peronospora parasitica, Botrytis cinerea, Pseudomonas syringae*
Cauliflower	*Peronospora parasitica*
Broccoli	*Alternaria brassicicola*
Sunflower	*Plasmopara halstedii*
Lettuce	*Bremia lactucae*
Maize	*Fusarium moniliforme*
Pearl millet	*Sclerospora graminicola*
Pea	*Aphanomyces euteiches*
Peanut	*Cercosporidium personatum*
Grapes	*Plasmopara viticola*
Cotton	*Verticillium dahliae*
Sweet orange fruit	*Penicillium* spp.

BABA action was seen to be mediated through signaling intermediates like SA and JA. BABA treatment led to an increase in SA levels and SA-induced PR proteins like PR-1 in pepper plants infected with *Phytophthora capsici* and in tobacco plants infected with TMV (Siegrist et al. 2000).

In BABA-treated apple plants priming is associated with the production of phenolics, peroxidase, callose, lignin and other defense reactions against pathogen (Table 7.3). Exogenous application of BABA primed a rapid increase in peroxidase activity within 24 h in Jute against *Macrophomina phaseolina* (Ray et al. 2011).

Various cellular defense responses are induced following attack by either pathogens or insects or in response to abiotic stress. Various natural and synthetic chemicals are involved to induce priming in plants. Microarray analysis revealed that 22 genes were significantly up-regulated in BABA-treated *Arabidopsis* at 22 h post-inoculation with *Pseudomonas syringae* pv. *tomato* DC3000. The primed state in *Arabidopsis* (*Arabidopsis thaliana*) was still functional in the next generation without additional treatment (Slaughter et al. 2012).

7.3.2.18 Transcription Factors

Elicitors or pathogen activated transcription factors play an important role in controlling defense gene expression and plant resistance responses. Five major families of plant transcription factors (bZIP, WRKY, MYB, EREBF, and homeodomain proteins) have been shown to participate in the regulation of plant defense responses

(Rushton and Somssich 1998). Several members of various transcription factor families such as TGA-bZIP, ERF, Myb, Whirly and WRKY, are shown to be linked with plant defense responses and specific gene regulation (Desveaux et al. 2005). Many transcription factors involved in JA and ET signal transduction are members of the AP2/ERF group and SA signal transduction involves mostly WRKY and bZIP members. Transcription of the genes encoding these transcription factors can be either up or down regulated by the treatments.

AP2/ERF Transcription Factors. AP2/ERF transcription factors are characterized by a 58- to 60-amino acid DNA-binding domain first identified in APETALA2 (AP2) and the ethylene-response factors (ERF). In *Arabidopsis* the subgroup of AP2/ERF transcription factors that are rapidly induced by JA is known as Octadecanoid-responsive *Arabidopsis* AP2/ERF (ORA). This indicates that *Arabidopsis* AP2/ERF transcription factors can be divided into a group that integrates JA and ET pathways to activate defense gene expression, a group that selectively represses JA-responsive genes, and a group that induces gene expression through ER only. Phosphorylation can be important for AP2/ERF transcription factor activity was also shown for the rice AP2/ERF transcription factor. Ethylene-responsive element binding protein (OsEREBP1) after phosphorylation showed an enhanced binding to GCC-boxes (Cheong et al. 2003).

MYB Transcription Factors. Plants have very large MYB (myeloblastosis) families; for example, *Arabidopsis* contains 125 MYB genes. Most plant MYB factors belong to the R2R3 group, which is divided in two types that can bind different DNA sequences. Type I binds the DNA sequence (T/C)AAC(T/G)G, while type II binds to G(G/T)T(A/T)G(G/T)T. In *Arabidopsis,* only a few R2R3-MYB proteins are involved in defense-related pathways. Defense responses regulated by MYB transcription factors seem to cover all signaling pathways and act against many types of pathogens. MYB transcription factors also play a role in the defense response against insects (De Vos et al. 2006).

bZIP Transcription Factors. bZIP transcription factors are characterized by their basic leucine zipper (bZIP) domain. Two of the ten groups of bZIP transcription factors in *Arabidopsis* have been implicated to play a role in plant innate immunity. *AtbZIP10* is controlled by Lesions simulating disease resistance 1 (LSD1), a plant-specific zinc-finger protein that negatively regulates cell death by inhibiting nuclear translocation of *AtbZIP10*. BZI-1 transcription is up-regulated in response to pathogen attack and pathogen-induced phosphorylation of BZI-1 related proteins has been described (Kuhlmann et al. 2003).

WRKY Transcription Factors. WRKY proteins are recently identified as a class of DNA-binding proteins that recognize the TTGAC(C/T) W-box elements found in the promoters of a large number of plant defense-related genes. In plants, many WRKY proteins are involved in the defense against attack by phytopathogens such as bacteria (Deslandes et al. 2002; Dong et al. 2003), and fungi (Kalde et al. 2003).

Plant resistance to pathogens can be classified into two main types from the breeding point of view. Vertical resistance or race-specific, pathotype-specific resistance is generally determined by major genes (R genes) and is characterized by pathotype-specificity. Race-specific resistance is usually expressed in a quantitative manner and is often highly effective against a given race. It is usually determined by one or a few genes, which makes it easier to work within a breeding programme. However it may have certain disadvantages from the epidemiological and durability point of view since the pathogen may evolve new pathogenicity genes which are not counteracted by the existing *R* genes.

Horizontal resistance on the other hand is generally controlled by polygenes and is pathotype-non-specific. Horizontal resistance does not prevent the development of symptoms of the disease, but it slows down the rate of spread of the disease in the population. It is mostly inherited polygenically and is more difficult and time consuming to incorporate into breeding programs. However the selection pressure leading to evolution of new races of pathogen is considerably reduced in breeding for improved horizontal resistance.

Induced resistance may be considered as a type of horizontal resistance. It can be induced in plants irrespective of their genetic background and the presence of resistance genes. The fact that pathogens cannot overcome horizontal resistance due to its polygenic nature makes it an attractive option to breeding for disease resistance.

Induced resistance not only diminishes the use of toxic chemicals for disease control but offers an alternative, long-lasting, non-biocidal and eco-friendly approach for plant protection and thus contributing to the development of sustainable agriculture.

7.4 Plant Diseases

Since times immemorial it is known that humans are dependent on plants for food, fiber, shelter, drugs and others. Further plants are also important because they utilize CO_2 in photosynthesis and release O_2. Numerous diseases are produced by bacteria, viruses, mycoplasma, fungi, nematode, insects and others. An example of the greater impact on human existence is by the late blight of potatoes caused by *Phytophthora infestans* in the year 1845 that lead to Irish famine in Ireland. Similarly coffee rust in Ceylon (Sri Lanka) and leaf spot of rice in India have caused huge losses in respective crop yields. This is followed by *Plasmopara viticola* which causes downy mildew of grapes and threatened the wine industry in France. Apple scab caused by *Venturia inaequalis* first reported in 1819 in Sweden has played havoc with apple cultivation in the Kashmir valley during 1973. Panama disease of banana, wilt disease of pigeon pea, castor, guava and also smut and rust of cereals have become challenges to plant pathologists. However, the chance discovery of Bordeaux mixture by PA Millardet in France made a beginning to the chemical control of plant diseases. In due course of time major diseases could be controlled employing fungicides, integrated disease management (IDM), biocontrol agents, breeding for resistance and several other methods. Unfortunately the diseases of minor nature not only have become major diseases but posing a threat to crop productivity. Newly emerging fungal, viral, mycoplasmal, bacterial, nemotode and insect pests are posing serious threats to crops and challenges to phytopathologists. The phytopathologists have to protect the environment and also have to ensure the safety and security of farmers in the field by making concerted and concentrated efforts to minimize crop losses due to fungi and other microbes. Several diseases listed in Table 7.4 clearly indicate that epidemics and plant diseases result in crop loss which in turn results in hunger, poverty and death of people besides dwindling of economy. Wheat rusts have caused heavy losses in wheat and posed serious threat to wheat production world over.

Table 7.4 Important crop diseases caused by pathogenic fungi

Crop	Disease	Pathogen
Rice	Blast	*Magnaporthe grisea*
	Leafspot	*Helminthosporium oryzae*
Wheat	Rusts	*Puccinia graminis* f. sp. *tritici, P. striiformis* f. sp. *tritici, P. recondita*
	Smuts	*Ustilago tritici*
	Karnal bunt	*Neovossia indica*
	Foliar blight	*Alternaria tritici*
Pearl millet	Downy mildew	*Sclerospora graminicola*
	Ergot	*Claviceps fusiformis*
	Smut	*Tolyposporium penicillariae*
Sorghum	Grain mold	*Sphacelotheca sorghi*
Corn	Stalk rot	*Macrophomina phaseolina, Fusarium graminearum*
	Head smut	*Sphacelotheca cruenta*
Pigeonpea	Wilt	*Fusarium udum*
	Blight	*Phytophthora drechsleri* f.sp. *cajani*
Chickpea	*Ascochyta* blight	*Ascochyta rabiei*
Pea	Powdery mildew	*Erysiphe polygoni*
Potato	Late blight	*Phytophthora infestans*
	Black wart	*Synchytrium endobioticum*
Groundnut	Leafspot	*Cercospora arachidicola, Cercosporidium personatum*
	Rust	*Puccinia arachidis*
	Root rot	*Aspergillus flavus*
Mustard	White rust	*Albugo candida*
Mango	Mango malformation	Complex disease with many pathogens, mainly *Fusarium moniliforme* var. *subglutinans*
Banana	Sigatoka disease	*Mycosphaerella musicola* (yellow sigatoka), *Mycosphaerella fijiensis* (black sigatoka)
Guava	Wilt	*Fusarium oxysporum* f. sp. *psidii*
Apple	Scab	*Venturia inaequalis*
Tea	Blister blight	*Exobasidium vexans*
Coconut	Budrot	*Phytophthora palmivora*
	Wilt	*Ganoderma lucidum*

(continued)

Table 7.4 (continued)

Crop	Disease	Pathogen
Sugarcane	Redrot	*Colletotrichum falcatum*
	Smut	*Ustilago scitaminea*
Cotton	Wilt	*Fusarium oxysporum* f. sp. *vasinfectum*
Tobacco	Blackleg	*Phytophthora nicotianae*
Ginger	Rhizome rot	*Pythium aphanidermatum*
Coffee	Rust	*Hemileia vastatrix*
Rubber	Leaf fall	*Phytophthora palmivora*
	Pink disease	*Pellicularia salmonicolor*
Jute	Stalk rot	*Macrophomina phaseolina*
Black pepper	Wilt	*Phytophthora* sp.
Grapes	Downy mildew	*Plasmopara viticola*
Flax	Rust	*Melanospora lini*
Dutch elm	Wilt	*Ophiostoma ulmi*
Chilly	Anthracnose	*Colletotrichum capsici*
Cabbage	Club root	*Plasmodiophora brassicae*

7.4.1 Plant Diseases and Their Impact

Millets are grown on 35.5 million hectares in the world resulting in the production of 28.5 million tonnes. Millets are important in the arid and semi-arid regions and offer some promise of food security to humans and livestock. Millets include pearl millet, sorghum, finger millet, foxtail millet, little millet, etc. However, pearl millet and sorghum are the most important and nutritionally valuable. Downy mildew, rusts, blast, smut, ergot, foliar diseases, soil/seed/root-borne diseases have resulted in huge losses both at field site, in the market and storage. The future challenges include effective monitoring of the pathogen, use of DNA markers in the characterization of virulence besides there is a need to develop near isogenic R-lines with different R-genes and also identification of new resistant genes.

Fruits form an important component in our daily diet as they provide sugars, vitamins, minerals and medicinally important compounds such as flavonoids which prevent cancer and heart related diseases. Fruits suffer from various diseases and contamination caused by fungi and other microbes both at pre-harvest and post-harvest stages. Unfortunately still we have yet to find new techniques and fungicide formulations to control several fruit diseases including bunch rot of grapes (*Botrytis cinerea*), apple scab (*Venturia inaequalis*), wilt of guava (*Fusarium solani*), Panama disease of banana (*Fusarium cubense*), mango malformation (*F. moniliforme*), blue mold of citrus (*Penicillium citrinum*), anthracnose of papaya (*Colletotrichum papayae*) and others. The overall post-harvest losses range from 5 to 15 % world over. Further there are no scientifically developed storage godowns/houses/containers in many countries.

The management practices of fungal pathogens of fruit crops include cultural practices, hot water treatment, hot vapour exposure, ionizing irradiation, UV illumination, wrapping in disease resistant paper impregnated with sodium orthophenyl butyrate and sodium metabisulphite, use of volatile compounds, coating with oil, waxes and colloidal solution, biocontrol agents, use of botanicals as antifungal agents, use of gels and latex derived from plants, inducing resistance, growing disease resistant varieties, host defense through gene silencing and others (Arya 2010; Tripathi 2005).

Wheat is an important cereal crop and is cultivated worldwide and requires 40 % increase to meet global food requirements. Various diseases associated with wheat are the important limiting factor in its productivity. Fungal diseases include *Fusarium* head blight, wheat rusts (stem rust, leaf rust, yellow rust), Karnal bunt and powdery mildew cause heavy losses in yield and quality. The development of resistant varieties is the only solution to overcome this problem. Chemical treatment with triazole, cultural practices, removal of alternate and collateral hosts and others are the methods used to control rust disease. Control of Karnal bunt involves cultural practices, sowing of disease-free seeds and use of resistant varieties etc. (Arya and Perello 2010).

Sugarcane is one of the cash crops grown worldwide and contributes 70 % of world sugar. This commercial crop is attacked by fungi, viruses, bacteria, mycoplasma, nematodes, insects etc. The most challenging fungal diseases which cause huge losses in the crop yield include whip smut (*Ustilago scitaminea*), rust (*Puccinia melanocephala*), red rot (*Glomerella tucumanensis, Colletotrichum falcatum*), eye spot (*Bipolaris sacchari*), Fusarial rot (*Fusarium moniliforme*) and others. The quantitative and qualitative losses due to above fungal diseases can be minimized using resistant varieties, fungicide (mancozeb, metalaxyl, carboxin), thermotherapy, biocontrol, maintaining soil health and minimum fertilizer application.

Pulses form an important source of protein since time immemorial besides being of multiple utility as human food, animal feed and soil health. Chickpea, pigeon pea, green gram, black gram, lentil and pea are the pulse crops grown in different parts of the world. The challenging diseases include wilt, root rot, *Ascochyta* blight, *Phytophthora* blight, leafspot, mildews, rust, anthracnose and others which limit yields. IDM is an eco-friendly approach which suppresses the pathogen and seems to be the right approach to mitigate the huge losses incurred due to these diseases. Besides the above, application of fungicides, crop rotation, inter-cropping, cultural practices, use of bio-agents etc. are some other alternate strategies employed for disease control.

Fungal pathogens in oil seed crops have been found to cause heavy economic losses. Oils form economically important household commodity required in day today affairs and some oils are also used as therapeutic agents. The losses of oil seed crops due to fungal and microbial diseases ranges from 5 to 20 %. Diseases of important oilseed crops are presented in Table 7.5.

Vegetable crops suffer from a number of diseases caused by fungi and other microbes. Major diseases include wilt of tomato, powdery mildew and anthracnose of pepper. Potato wart, *Phomopsis* blight of eggplant, powdery mildew of cucumbers, anthracnose and wilt of cucurbits, and *Fusarium* wilt of okra. Onion suffers from basal rot, purple blotch, white rot, anthracnose and *Stemphylum* blight.

Table 7.5 Diseases of important oilseed crops and their control

Crop	Symptom	Pathogen	Control
Groundnut	Leafspot	*Mycosphaerella arachidicola*	Tolerant varieties; inter-cropping with sorghum, pearl millet; foliar spray of Carbendazim etc.
		Phacoisariopsis personata	Resistant varieties; Carbendazim + Mancozeb; deep burrowing of crop residues and removal of volunteer plants
Sunflower	Leaf spot	*Alternaria helianthi*	Removal of plant debris
	Blight	*Alternaria alternata*	Early sowing resistant varieties
	Downy mildew	*Plasmidiophora halstedii*	Mancozeb spray Seed treatment with Metalaxyl Apron 35
	Rust	*Puccinia helianthi*	Altering the sowing date, deep summer ploughing, resistant varieties
Safflower	Leaf blight	*Alternaria carthami*	Triazoles, Hexaconazole, Thiram; hot water treatment; tolerant varieties
	Root rot	*Phytophthora drechsleri*	Avoidance of mono cropping; use of resistant varieties
	Rust	*Puccinia carthami*	Destruction of crop debris; collateral host; crop rotation; spraying of tridemorph; resistant varieties
Sesamum	Blight	*Phytophthora parasitica* var. *sesame*	Inter-cropping; application of farmyard manure; biocontrol; seed treatment with Captan
	Wilt	*Fusarium oxysporum* f. sp. *sesami*	Seed treatment with Benlate; biocontrol
Castor	Blight	*Alternaria carthami*	Mancozeb application; judicious use of fertilizer
	Grey rot	*Botrytis ricini*	Spacing; Carbendazim spray
	Root rot	*Macrophomina phaseolina*	Crop rotation; Thiram, Topsin M-70
	Wilt	*Fusarium oxysporum* f. sp. *ricini*	Healthy seeds; Biocontrol; Carbendazim, Thiram

Vegetable legumes are known to suffer from rust, powdery mildew, root rot and others. Carrot the common root crop mainly suffers from powdery mildew, *Cercospora* leaf spot and *Alternaria* blight. Cultural practices, IDM and breeding for resistance are some of the successful methods employed to control vegetable crop diseases.

The population is growing beyond expectations particularly in China and India besides improvement in life expectancy. Further, the world population has doubled from three billion in 1960 and by 2080 it will be varying from 8.1 to 14.0 billion certainly it will cross 9.0 billion by 2050. Land is basic to agriculture, is finite and fragile. Water is getting polluted despite scanty rain, pollution, human interference, degradation of land, soil fertility loss, erosion of soil, increase in waste land, productivity loss, dwindling economy, ground water depletion and biodiversity erosion. Seventy percent increase in grain yield would be required to feed increasing population by 2050. There is a necessity to control the diseases caused by fungi and other diseases as they have become stumbling block for crop productivity.

Securing food for everyone under the ever increasing population under shrinking resources without compromising the environment and health is the foremost important task for everybody. About 20 % food production is lost due to plant diseases at various stages. The famines in the history are consequences of either natural calamities or epiphytotic disease appearance. We need to be very attentive as the pathogens are evolving very fast and new diseases are developing rapidly. The misery is getting augmented further with the climate changes making the situation diverse for the crops and favorable for the diseases development.

7.4.2 Epidemiology and Forecasting

Epidemiology and forecasting of diseases, host–pathogen interaction in a given environment is an important aspect for disease development. Virulent pathogen, susceptible host and congenial environmental variables are responsible for disease production and disease advancement. Epidemiology deals with dynamics of plant pathogen infecting host population. Late blight disease of potato caused by *Phytophthora infestans* awakened the plant pathologists to look into reasons behind the same. Discovery of barberry as an alternate host of *Puccinia graminis* var. *tritici* and its eradication has led to the beginning of phyto-epidemiological approach for plant disease management (Mehta 1933). Waggoner and Horsfall (1969) simulated plant disease epidemic on computer followed by Van der Plank (1963) who considered that the epidemiology and forecasting need multidisciplinary approach. Plant disease forecasting is a management system used to predict the occurrence or change in severity of plant diseases. It is important to note that the potential yield and the yield realized at farmer's field are not the same. This yield gap has been due to severity of diseases, potential and abiotic stresses that affect crops. Crop prediction models forewarn these menaces.

There is a necessity to develop disease forecasting models which are nothing but set of formulae, algorithm pattern after a detailed study of biology of specific pathogen. Keeping in view the host and crop management practices, forecast models provide an alternate calendar spray schedule to bring need-based precision. Forecasts may be based on initial inoculums, meteorological parameters and their combinations. There are several disease forecasting networks available across globe for maize (EPICORN—Southern corn leaf blight), tomatoes (EPIDEM Tomcost), potatoes (BLITE CAST for early blights), apple (EPIVEN—Scab) followed by weather-based location specific forewarning models. An artificial neural network (ANNS) is another tool for disease forecasting. ICAR and ISRO have demonstrated identification of coconut wilt using aerial false colour photography; remote sensing has also been used to forecast the diseases. Forecasted information has to be simple and user friendly. Online decision support systems to forecast different diseases are in use across the globe viz. tan spot, *Septoria* leaf blotch, leaf rust, *Fusarium* head blight. Therefore, accurate information in yield losses due to occurrence of a disease is needed by growers or plant protection specialists to decide on cost-effective control measures.

7.4.3 Impact of Climate Change

Climate change due to variability in CO_2 emissions, rise in temperature, anthropogenic activities, natural calamities, rise in sea level and other related changes will have impact on cropping system, disease incidence, agricultural productivity and sustainability. Temperature, moisture and green house gases are considered as important variables. It is expected that global temperature may rise between 0.9 and 3.5 °C in the year 2100 (IPCC 2007), bringing changes in rainfall pattern and green house gases. All these will influence the management of plant diseases and epidemics caused by fungi, microbes and insects. The available data indicates that the climate change (temperature, moisture, increased CO_2, rainfall) will affect the geographical range of pathogen, its population, generations, loss of resistance in cultivation, changes in crop disease cycle, impact on pathogen interaction, affect the morphology, physiology and biomass of crops besides affecting the virulence of pathogens and their evolution.

Some of the examples include dry rot of chickpea caused by *Rhizoctonia bataticola* becoming severe in rain fed environment and also increased incidence of stem rot in soybean by *Sclerotinia sclerotiorum* with increased temperature and wetter conditions. Geographic information system (GIS) has been used to evaluate and model the spatial distribution of plant disease in relation to environmental factors. This information will be useful to the crop growers, scientists, extension agencies and others (Serge et al. 2011).

7.5 Plant Disease Management

Plant pathogens gained momentum since the Irish famine epidemic caused by *Phytophthora infestans* in Ireland during the 1840s which resulted in the death of a million people. This has awakened the scientific and farming community to begin search for plant protection measures. The discovery of Bordeaux mixture, the first generation fungicide by Millardet in 1885 to control the downy mildew of grapes is considered as the landmark in the history of chemical control of plant diseases. It is estimated that around 20 % losses are recorded in the yield of food and cash crops worldwide. Out of 100 thousand fungi described in the world, at least 20,000 are known to cause diseases on crop plants. Though the crop losses get reduced through disease resistant varieties, crop rotation and phytosanitation, fungicides offer better control of diseases and maximize crop yields.

Plant diseases have been playing an important role in agricultural productivity and influencing the world's economy. The annual world crop loss has been around 25,000 million dollars. In spite of having phytosanitation, biological control, cultural practices, growing disease resistant varieties, seed treatment with chemicals, solarization and other such ancient and modern methodologies employed in disease control, still chemical control holds the key.

Diagnosis of disease and identification of fungal pathogens have to be established accurately for plant disease management. The following are the five categories of plant disease management.

1. Avoidance of pathogen through exclusion of the pathogen from a geographical area, either voluntarily or by legislations and by evasion so as to prevent the pathogen from coming into contact with the host.
2. Eradication of the pathogen from the host, soil, other sources so as to reduce the inoculum rather than the total eradication of the pathogen.
3. Protection of plants from pathogens through physical and chemicals methods.
4. Use of disease resistant varieties.
5. Chemotherapy.

7.5.1 Fungicides and Fungicidal Resistance

Along with growing population which will be around 8.3 billion by 2030, it is essential to offer food security which is the greatest challenge posed. Therefore, it has become a necessary evil to use chemical fungicides for the control of plant diseases though the problems of residual effect, toxicity and fungicidal resistance among pathogens exist.

Millardet at Bordeaux in France used Bordeaux mixture for the control of downy mildew of grapes in 1882 and this copper fungicide has been considered as the oldest fungicide. However, some fungal pathogens have also developed fungicidal resistance. This fungicidal resistance has opened up new challenges to the plant pathologists.

The fungicides used for disease control have been classified as protectants, eradicants and therapeutants. A successful fungicide is the one which has consistency in action and constant characters. These include effective action, soluble enough to function, adhesiveness, non-phytotoxic, safe for handling, purity, good shelf life, with no or less residual effect and other parameters need to be satisfied. The fungicides which are in use belong to inorganic copper components, inorganic mercury compounds, sulphur sprays, organic sulphur compounds, quinine and phenolic fungicides, heterocyclic nitrogen compounds, benzene compounds, organomercurials, systemic fungicides, organophosphorus compounds, antibiotics and fumigants. All the above chemicals have proved to be effective at one or other concentration against a particular disease. Furthermore, lethal levels have been identified for each fungicide.

From 1934, the dithiocarbamates along with organotins, low soluble copper compounds, quinine, captan, chloronil, PCNB, surface protectants like inorganic fungicides and others made an entry as second generation fungicides which offered some hope of disease control. The third generation chemicals include 2-aminopyrimidene, benzimidazoles, carboxamides, organophosphorus compounds, triazoles and others which are able to penetrate the host tissue and kill the pathogen. Later strobilurins, phenylpyrazoles, quinooxyfen, oxazolidinediones, spiroxamine, valinamides, cyanoimidazoles, thiocarbamates, amidoximes and others which are eco-friendly, possess broad-spectrum activity at low dose rates and are promising have offered a promise to protect the crops from fungal diseases. Interestingly breeding for disease resistance offered a greater hope to protect plants from diseases and the fungicides are now considered to be second line of defense (Klittich 2008). Thind (2011) reported that around 52 fungicides are registered for use in India and a good number of fungicides are under evaluation (Table 7.6). He has also stated that Azoxystrobin and fenamidone have been registered for control of grape downy mildew and potato blight. Mandipropamid, iprovalicarb, benthiavalicarb, fluopicolide, famoxadone, cyazofamid, pyraclostrobin and picoxystrobin are some of the fungicides which are being tested against different diseases. A new thrust has been on the following areas:

1. Combinational chemistry
2. High throughput screening
3. Advanced formulation and
4. Molecular toxicology and environmental safety

Further, the discovery of crop protection chemicals like pyrethroids, ascomectins, spinosyns and others as natural products offered new avenue for the discovery of new chemical molecules offering plant protection against disease causing pathogen.

The fungicidal research has to focus on the following in future:

1. Environmental and public safety, impact on soil and air environments.
2. Role of fungicidal resistance among pathogens and impact on host health.

Table 7.6 Some plant diseases and their chemical control

Crop	Disease	Fungicide
Sorghum	Smut	Carboxin 2 g/kg seed dressing
Rice	Blast	Carbendazim 1 g/L water
	Sheath blight	Carbendazim 1 kg/100 L water
	Brown spot	$MnCo_3$ 2 kg/ha
Wheat	Loose smut	2.5 g Carboxycin/kg seed
	Karnal bunt	2.5 g Thiram and Bavistin/kg of seeds
	Rust	2 kg of Mancozeb/ha
Pearl millet	Rust	2 kg Mancozeb/ha in 800 L water
Sugarcane	Wilt	Carbendazim
Coconut	Root wilt	1 kg $MgSO_4$/ha
Banana	Panama disease	2 g Captan/L water
	Sigatoka	1 g Thiophanate methy/L water
Citrus	Dry root rot	0.1 % Carbendazim and 0.25 % Mancozeb
Mango	Anthracnose	Bordeaux mixture
	Powdery mildew	

3. Reduction of time and effective control of diseases.
4. Discovery of new antifungal compounds with high efficacy.
5. Preparation of disease diagnostic kits and host/pathogen specific fungicide.
6. Critical analysis of fungicidal residues and toxicity.
7. Purity and shelf life of fungicide.
8. Limiting the diseases levels.
9. Bringing out suitable laws to control adulteration and marketing of spurious fungicides.

Fungicides have come into existence since the discovery of Bordeaux mixture by Millardet in order to control the downy mildew of grapes in France. Fungicides have become important in modern agriculture for the management of various diseases in agricultural crops. In spite of having cultural practices, growing resistant varieties, use of botanicals and biocontrol agents, nonetheless farmers are showing keen interest for the use of fungicides as they result in the disease control within a short period of application. However, farmers believe in immediate gains and no realization prevails on them about the impact of chemicals on most plants. It is also important to note that all other alternate methods are time consuming and results obtained at field site are variable but for growing disease resistant varieties. Around 150 chemicals of different classes are employed to control different diseases in various countries. It is also known that a fungicide may fail to control a particular disease under certain circumstances which may be due to insufficient dosage, low effectiveness, improper timing of application, defective method of application, expiry of product, adulteration, at times excessive wash off in monsoon and many other reasons. Environmental conditions, crop cultivation, geographic location and chemical constitution of fungicide are crucial in the management of disease. The side effects like fungicidal residue in the host, toxicity, retention capacity, shelf life of fungicide, physiological and biochemical changes in host, potentiality of chemical and genetic constitution of host to accept the fungicide have become important. Fungal pathogens show resistance to fungicides which may be due to innate capacity of fungal pathogen to the chemical into simpler compounds or may be due to fungicidal inaction on pathogen as the fungicide not being specific (Brent 1995; Thind 2008). This kind of resistance was shown towards benomyl, dimethrimol and others. Mancozeb or copper fungicides are still in use to control diseases. However, strobilurins, phenylpyrroles, anilinopyrimidines, spiroxamines, quinolines and phenylpyridylamines have shown more potent action against diverse fungal pathogens at much lower rates. Interestingly sudden failures and inaction of dithiocarbamates and some other fungicides has been reported with reference to powdery mildews, apple scab, peanut leaf spot and *Botrytis* gray mould (Brent 1995). The fungicidal resistance has been interpreted as genetic adjustment by a fungus resulting in reduced sensitivity to a fungicide. The fungicide does not induce resistance but it only selects the resistant propagules already present in low frequency in a natural population of the pathogen. Fungicidal resistance evolves from a mutation in genes, although resistance can originate from mutation in mitochondrial genes of the pathogen as evidenced in strobilurins.

Early fungicides developed during 1950–1960 did not cause much damage to crops but increased yields at lower dosage. Site specific systemic fungicides like benzamidine, phenylamines and others were more effective than classical protectants. These fungicides being site specific have brought the problem of resistance development in target pathogens. New fungicides like azoles introduced during 1980–1990 have proved remarkable but with a caution that the modern synthetic chemistry presents a risk of resistance.

The buildup of resistant strain is caused by the frequent use of site specific fungicide which exerts a selection pressure on population. The fungicide selectively inhibits sensitive strains but allows the increase of resistant strains. The shift towards building resistance occurs at different rates depending on the number of genes conferring resistance. In some like benzimidazoles and phenylamides which are highly selective fungicides, development of resistance is often sudden and the process is called disruptive selection. This has been due to mutation of a single major gene in a pathogen that is not affected by fungicide. Such fungicide resistance is referred as qualitative resistance or as discrete or black and white resistance. Such pathogen populations can remain resistant for many years even if pathogen is withdrawn. It may take 2–7 years or a fungicide to lose its sensitivity of fungal population and cause decline in disease control (Brent and Hollomon 2000). The risk of resistance development depends greatly upon the chemical nature to which it belongs and the mode of action of member fungicides. In the last 30–40 years the problems of acquired resistance have affected performance of fungicides. Estimates of resistance risk in different chemical classes of fungicides are shown in Table 7.7.

Management of fungicide resistance can be achieved by:

1. Management of timing of fungicide application, spray intervals and area treated
2. Optimum use of fungicide

Table 7.7 Fungicides, their mode of action, target site and resistance risk

Fungicide	Mode of action	Target site	Resistance risk
Benzimidazoles	Mitosis and cell division	β-tubulin	High
Strobilurins	Respiration	Complex III cytochrome	High
Carboxamides	Nucleic acid synthesis	Complex II succinate dehydrogenase	Moderate
Morpholines	Sterol biosynthesis	D 8-7 dimethylase	Moderate
Copper and Sulphur compounds, Dithiocarbamates	Multisite disruption of cell	Reaction with enzymes	Low function

Table 7.8 Fungicide affecting the host and the pathogen

Fungicide	Host	Pathogen affected
Benomyl	Grapes	*Botrytis cinerea*
Carbendazim	Apples	*Venturia inaequalis*
Iprodione	Grapes	*Botrytis cinerea*
Edifenphos	Rice	*Magnaporthe grisea*
Dimethirimol	Cucumber	*Sphaerotheca fuliginea*
Myclobutanil	Barley	*Erysiphe graminis*
Metalaxyl	Potato	*Phytopthora infestans*

Source: Brent and Hollomon (1998) and Thind (2008)

3. Avoiding sole use of harmful fungicides
4. To alternate the application of risk fungicides with those of multisite contact fungicides
5. Integration with cultural practices
6. Avoiding post-symptom curative treatments
7. Reduction in number of applications

There is a necessity of risk assessment for newly developed fungicides before marketing. Further it is essential to understand the implications involved between fungal pathogen and fungicide interaction (Table 7.8). Molecular approaches may help in clearing certain unsolved problems.

7.5.2 Breeding Resistant Varieties

Plants are varied in their action towards fungal pathogens and this character is controlled by genes associated with different genera and species. Many plants showing natural resistance to disease have long been used by plant geneticists in their breeding programmes. Resistance and susceptibility of a host plant to a fungal pathogen are largely inherited characteristics. The use of hybrids and disease resistant varieties not only gives more yields, but it is free from disease incidence, eliminates fungicidal interference, pollution, health hazards, etc.

However, the performance of a disease resistant variety or hybrid is also dependent upon edaphic and meteorological factors. The breeding for disease resistance is based on the laws of inheritance. Some diseases got introduced into one or other country due to transport of crop or its seeds. Further, some important diseases have been imported into countries where there was no existence of such diseases. For example, the potato tubers from South America to England have introduced *Phytophthora infestans* in 1880s. The resistance gene of a cultivar may persist for a period of 5 or 10 years as documented in rust resistant variety of wheat by Borlaug (1965). Monogenic resistance is governed by single gene whereas oligogenic and polygenic resistance are governed by two or many genes.

Extra-chromosomal inheritance gene interaction, modifier genes and reversal dominance play an important role in offering diseases resistance. If a variety is resistant to some pathogenic races than other, such resistance is called vertical resistance whereas when resistance is uniformly spread against all races of pathogen is called horizontal resistance.

A breeder while working for disease resistance has to follow:

1. Selection from existing crops
2. Selection from crops that escape damage in infected fields
3. Pure line selection
4. Plant introduction
5. Hybridization
6. Selection from wild varieties
7. Induced mutations

7.5.3 Alternate Disease Management Strategies

Chemical control has become essential for the control of several diseases caused by fungal pathogens. The non-judicious and careless application of chemicals has led to damage of crops and created fungicidal resistance among fungal pathogens. Chemicals either kill or inhibit germination, growth and multiplication of the pathogen. However, the residual/toxic effects remain in some hosts. Further environmental pollution has increased due to toxic gaseous chemicals besides affecting soil health. Adulteration of fungicides and their application lead to ineffective management of diseases, natural resources of raw materials used for the manufacturing of fungicides/pesticides have depleted and labour has become costly. All such reasons made the scientists to think of alternate methods for disease control.

In addition to chemical methods and breeding for resistant varieties several alternate methods are used singly or in combination to control plant diseases, and one of the common methods being biocontrol. Von Tubeuf (1914) coined the term 'biological control' in relation to plant pathogens. Biocontrol of root diseases in plants was reported for the first time by Hartley (1921). According to Cook and Baker

(1983) biological control was defined as 'the reduction of inoculum or disease producing activity of a pathogen accomplished by one or more organisms other than man'. Baker (1987) defined that biological control is the decrease of pathogen activity accomplished by one or more organisms including host plant excluding humans. The biocontrol agents reported are *Trichoderma, Gliocladium, Aspergillus, Neurospora, Chaetomium, Dactylella, Arthrobotrys, Glomus, Penicillium* etc. Important biocontrol agents that serve as antagonistic agents against plant pathogenic fungi include *Ampelomyces, Aspergillus* spp., *Chaetomium globosum, Fusarium* sp., *Gliocladium virens, Penicillium citrinum, Peniophora gigantea, Trichoderma harzianum* and others.

The mechanisms of biological control of plant diseases being:

1. Competition: Competition being between micro-organisms for space or nutrition and this is applied for biocontrol agents.
2. Antibiosis: This process is the inhibition of an organism through a metabolic product from another organism. The root disease caused by *Heterobasidion annosum* can be controlled by *Peniophora gigantea.*
3. Hyperparasitism and mycoparasitism: Biocontrol can be achieved through direct parasitism. This involves the direct utilization of food of one organism by another organism. Hyperparasites are parasitic on other parasites. Few hyperparasites include *Darluca filum* parasitizing rust fungi. *Ampelomyces quisqualis* parasitizes powdery mildew fungi. *Tuberculina maxima* parasitize *Cronartium ribicola. Trichoderma viride* and other species are known to parasitize hyphae of *Rhizoctonia solani.*

7.5.3.1 *Trichoderma* as Biocontrol Agent

Trichoderma is the anamorphic state of *Hypocrea* and is represented by more than 110 species. It colonizes mostly soil, organic debris, litter and other such substrates. Various species are identified based on morpho-taxonomic criteria and molecular tools. The most successful antagonistic species being *Trichoderma harzianum* followed by *T. atroviride, T. asperellum, T. hamatum, T. longibrachiatum, T. reesei, T. virens* and others. The plant diseases caused by *Fusarium, Phytophthora, Pythium, Rhizoctonia, Sclerotium* and many other soil-borne and root-borne diseases have been controlled using *Trichoderma* spp. as biocontrol agents. The commercial products of *Trichoderma* spp. are listed in Table 7.9.

7.5.3.2 Hypovirulence

It is a term used to describe reduced virulence found in some strains of pathogens. This process has been reported in many pathogens like *Rhizoctonia solani, Gaeumannomyces graminis* var. *tritici* and *Ophiostoma ulmi,* but the transmissible elements responsible for hypovirulence or reduced vigour of fungi are subject to debate and may be due to dsRNA viruses or plasmids.

Table 7.9 Commercial products of *Trichoderma* spp., their use for control of various types of diseases and companies commercializing them

Product	Fungus	Disease controlled	Commercialization
Trichodex	*T. harzianum*	*Botrytis* on grape vine	Makhteshim chemical Works Ltd, USA
Trieco	*T. viride*	Soil/root-borne disease	Jeypee Biotechs, India
TY Tusal	*Trichoderma* spp.	Soil/root-borne disease	Mycocontrol Ltd, Israel, Spain
$T_{22}g$, t-22 HB	*T. harzianum*	Soil/root-borne disease	THT Inc, USA
Soil Gard, RUTOPIA	*Trichoderma* sp.	Soil/root-borne disease	NaEx Corp, USA Inc
Ecofit	*T. viride*	Soil/root-borne disease	Hoechst Schering Agro Evo Ltd, India
Bio-trek 22G	*T. harzianum*	Soil/root-borne disease	Bioworks Inc of Geneva, NY
Biofungus	*Trichoderma* sp.	Soil/root-borne disease	Grondortsmettingen De Cuester n. v. Belgium
Bioderma	*T. harzianum* *T. viride*	Soil/root-borne disease	Biotech International Ltd, India
Binab T	*T. harzianum* *T. polysporum*	Soil/root-borne disease	Bio-innovation AB, UK

7.5.3.3 PGIPs-Plant Immunity

In the process of host–pathogen interaction a number of changes, actions, and reactions do occur. These include defense reactions such as formation of antifungal proteins, enzymes, and other biochemical physiological alterations. Polygalacturonase inhibiting proteins (PGIPs) are the extracellular host-plant protein inhibitors and these are helpful in stopping the infection process of host-plant tissues by fungal pathogen. PGIPs are the better choice of developing hyperactive PGIPs. Further transformation of crop plants with these modified PGIPs or with more than two PGIPs will result in wide fungal resistances (Dangs and Jones 2001).

7.5.3.4 Endophytes as Biocontrol Agents

Endophytes are the microorganisms (bacteria, actinomycetes, algae, fungi) that reside in healthy plant tissue without causing any disease symptoms besides spending whole life or a period of their life cycle in plant tissue. Often plant tissues harbor one or more microorganisms. The harboring fungi and other microbes have a great potential to produce novel natural products which also include compounds useful as plant protectants from disease causing pathogens. Several antimicrobial compounds have been isolated from endophytic fungi (Strobel 2003) (Table 7.10).

Table 7.10 Antimycotic compounds of some endophytic fungi effective against some pathogenic fungi

Endophytic fungi	Compound	Pathogenic fungi
Fusarium oxysporum	Cyclosporine	*Sclerotinia sclerotiorum*
Cryptosporiopsis quercina	Cryptocandin	*Botrytis cinerea*
Cryptosporiopsis quercina	Cryptocin	*Magnaporthe grisea*
Colletotrichum gloeosporioides	Colletotric acid	*Helminthosporium sativum*
Oomycetes	Oocytin	Water moulds

Table 7.11 Commercially available products formulated from fungi for the biocontrol of plant pathogens

Biocontrol fungus	Commercial product
Ampelomyces quisqualis	AQ10
Coniothyrium minitans	Contans
Fusarium oxysporum	Fusaclean, Biofax C
Gliocladium sp.	Soilgard, primastorp
Myrothecium verrucaria	DiTera
Paecilomyces lilacinus	Paecil

Taxol has been a well-known compound isolated from endophytic fungus *Taxomyces andreanae* of *Taxus baccata*. One hundred and eighty-seven endophytic fungi isolated from woody plants were antifungal against *Phytophthora infestans* (Park et al. 2005). Duijff et al. (1998) have reported induced resistance against *Fusarium* wilt in tomato using endophytic fungus *Fusarium oxysporum. Phomopsis cassiae*, an endophytic fungus in *Cassia spectabilis* produces ethyl 2,4-dihydroxy-5, 6-dimethyl benzonate and phomopsilactone, which is antifungal to *Cladosporium cladosporioides* and *C. sphaerospermum.*

7.5.3.5 Induced Resistance and Cross Protection

There are many fungi, microbes and abiotic agents that induce resistance. The localized and systemic induced resistance can act against whole range of pathogens. Localized resistance occurs in many plants and systemic resistance is limited to some plants. The inoculation with avirulent strains of pathogen or either microbes, both inducing resistance and challenge pathogens occur on or within the protected tissue. Such process is called cross protection. For example, non-pathogenic Fusaria when inoculated into soils supporting fusarium wilt diseased plants, resulted in the control of wilts. Commercially available products formulated for the biocontrol of plant pathogen and inducting the defense in host are given in Table 7.11.

Table 7.12 Plant extracts/oils/products inhibitory to plant pathogens

Plant extracts/oils/products	Pathogens controlled
Extracts	
Datura innoxia	*Colletotrichum capsici*
Achyranthes aspera	*Alternaria pisi*
Catharanthus roseus	*Magnaporthe grisea*
Garlic	*Aspergillus flavus*
Neem	*Alternaria sesame*
Cabbage, alfalfa, garlic	*Phytophthora capsici*
Oils	
Emulsified rape seed oil	Apple powdery mildew
Rape seed oil	Grape powdery mildew
Marigold oil	Damping off of seedling
Eucalyptus oil	*Candida* infections
Products	
Nimin from Neem[a]	Clubroot of crucifers
Neemasse from Neem	Black arm of cotton
Triact 70 from Neem	Powdery mildew, rust of wheat
Fungastop from Mint	Soil-borne pathogen

[a]Neem = *Azadirachta indica*

7.5.3.6 Botanicals

Attention has been drawn for the exploitation of plant products as novel chemotherapeutants in plant protection. Plant products are non-phytotoxic, biodegradable and not harmful to host metabolism. The secondary metabolites of plants include phenols, flavonoids, quinines, tannins, essential oils, alkaloids, saponins and sterols (Varma and Dubey 1999). However, scientifically accurate data on botanicals is essential pertaining to pathogen specific action, purity of compound, shelf life and residual toxicity effect on the host plant. Some of the extracts, essential oils and other plant products used as plant protectants are given in the Table 7.12.

7.5.3.7 Organic Farming in Disease Control

Organic farming has received the needed attention of farmers as it helps in disease management and supports plant growth. In fact it reduces the dependence on chemical inputs besides being an eco-friendly approach. According to the FAO/WHO organic agriculture is a holistic approach for food production management systems, which promotes and enhances agro-ecosystem health, including biodiversity, biological cycles and soil biological activity. It includes the use of management practices in preference to the use of off farm inputs, taking into account the regional conditions required for locally adapted systems. This is accomplished by using wherever possible agronomic, biological and mechanical methods as opposed to

Table 7.13 Eco-friendly control measures for the management of diseases

Crop	Disease/pathogen	Treatment/amendment
Wheat	Loose smut—*Ustilago tritici*	Solar heat treatment
Bajra	Ergot disease—*Claviceps fusiformis*	Salt water treatment
Grape wine	Petri disease—*Phaeoacremonium* sp.	Hot water treatment at 51 °C
Wheat/barley	*Erysiphe graminis*	Compost added to soil
Beans	*Rhizoctonia* sp.	Compost added to soil
Cucumber	Powdery mildew	Compost mix
Pea	Damping off	Seeds soaked in different compost extracts and dried before sowing
Soybean	*Phytophthora* spp.	Application of 40 tons of compost per hectare
Banana	Wilt—*Fusarium oxysporum* f. sp. *cubense*	Sugarcane bagasse added to soil
Wheat	Take all—*Gaeumannomyces graminis*	Rape/pea greens added to soil
Crops	Wilts and rots—*Fusarium* spp., *Rhizoctonia* spp.	Oil cake (neem, groundnut etc.) added to soil
Potato	Black scurf—*Rhizoctonia solani*	Saw dust, oil cakes added to soil
Potato	Wilt—*Verticillium albo-atrum*	Wheat straw, barley straw added to soil

using synthetic materials, to fulfill any specific function within the system. Organic farming with reference to disease management includes:

1. Growing crops that are resistant to disease or proper selection of sowing times that prevents the disease incidence.
2. Hygiene/sanitation, clean cultivation improving soil health to resist soil pathogens and promote plant growth, application of naturally occurring biocontrol agents for control of diseases.
3. Site selection and appropriate cultivation.
4. Use of disease-free seed.
5. Application of suitable cropping system.
6. Input of organic matter, efficient water management, safe and minimum application of organically approved chemical protectants.
7. Use of botanicals, oils, etc.
8. Diseases controlled by alternate methods (solar, hot water, compost amendments) (Table 7.13).

Soil solarization is a novel and eco-friendly technique tested for treatment of soil to combat soil-borne diseases through solar heat. The moisturized soil is covered under transparent polythene film during summer to prevent dissipation of the trapped solar heat. The temperature of soil rises, at time 13 °C higher than outside temperature reaching 55 °C at 8 cm depth thus killing fungal propagules. Organically approved bicarbonate salts have demonstrated good activity against powdery mildew and other diseases. Thermophilic and other extremophile plant pathogens have to be controlled by other means.

Organic farming is the economically and ecologically viable method of disease management as a preventive process besides being cost-effective method for sustainable agriculture.

Table 7.14 Arbuscular mycorrhizal (AM) fungi as biocontrol agents against some plant diseases

AM fungi	Host	Disease	Pathogen
Glomus fasciculatum	Tomato	Wilt	*Fusarium oxysporum* f. sp. *lycopersici*
Glomus fasciculatum	Green gram, chick pea	Root rot	*Macrophomina phaseolina*
Glomus mosseae	Tomato	Blight	*Phytophthora parasitica*
Glomus mosseae	Tomato, egg plant	Wilt	*Verticillium dahliae*
Glomus intraradices	Common bean	Rot	*Fusarium solani*
Glomus geosporum	*Dalbergia sissoo*	Wilt	*Fusarium solani*
Glomus fasciculatum	Tea	Brown rot	*Fomes lamaoensis*

7.5.3.8 Arbuscular Mycorrhizal Fungi and Plant Protection

Arbuscular mycorrhiazal fungi (AMF) are cosmopolitan and ubiquitous microscopic living organisms associated symbiotically with 70–80 % of plants. The roots of plants get colonized by these soil fungi and produce mycelium, arbuscules and vesicles. Mycorrhiza forms an important link between the plant roots and soil. Arbuscular mycorrhizal (AM) fungi help in the mobilization of phosphorus and other nutrients from soil to root followed by their upward movement besides increasing plant growth offering resistance against pathogens and abiotic stress.

Further, AM fungi help in the alleviation of salt stress, water stress tolerance, maintaining plant health and soil fertility. Multiple signals and differential induction of gene expression mediate the complex interaction between AMF and plant cells. AMF prevent root infections by reducing the access sites. Concentrated efforts can be made for formulations, product development and stringent quality measures be adopted. Different theories have been proposed to explain protection by AM fungi against fungal pathogens (Table 7.14). These include:

1. Improvement of plant nutrition and root biomass in inoculated plants resulting in increased plant tolerance and compensates the root damage caused by the pathogen.
2. Change of root morphology.
3. Modification of mycorrhizosphere microflora.
4. Competition between AM fungi and pathogenic fungi besides inducting resistance.

7.5.3.9 Chitinase as an Agent for Disease Control

Chitinase elaborated by many fungi can serve as potential disease control agent of several fungal pathogens. Most probably chitinase application can serve as alternate source for disease control. Further the genes encoding chitinase can be incorporated into host plant either to resist pathogens from invasion or suppress the pathogen. Genetic transformation of existing biocontrol of fungi that are well adjusted to their environment is likely to enhance their biocontrol capability.

7.5.3.10 Role of Transgenics in Disease Management

Transgenic plants are those plants which contain, within its genome, a foreign DNA that does not belong to it and has been introduced artificially via genetic engineering. Resistance genes can also be introduced in that way from unrelated plant species. Desirable target genes are isolated from fungi or other plants and introduced into the plants. Powell et al. (1994) reported transgenic tomato that has been resisted to *Botrytis cinerea*. The fungal resistant gene Hml has been isolated from maize conferring resistance to *Helminthosporium carbonum*. DNA markers are used for assisting the transfer of genes from one variety to another. In the absence of resistance genes, it is necessary to search for resistance genes in other species followed by their identification and cloning. Such genes can be deployed in crop varieties by using transgenic approaches. The genes responsible for offering dosages or resistance through β-1,3-glucanases, chitinases, thaumatin-like proteins, ribosome inactivating protein (RIPs) and thionins are introduced through genetic engineering methods. The expression of such genes in transgenic plants render disease resistance to many fungal pathogens and this has been proved in wheat against powdery mildew and rust (Bieri et al. 2003; Oldach et al. 2001). Chitinase gene introduction resulted in increased protection in rice against sheath blight and blast (Kim et al. 2003a).

7.6 Nanotechnology and Its Application

Nanotechnology has been in use in recent times for the detection of plant pathogens and biosensor related disease management through nano-formulations of agrochemicals around the world (Gopal et al. 2011a). Silver nanoparticles got application in a number of fields including in the control of wilt pathogens *Fusarium culmorum*, *Rhizoctonia solani*, *Magnaporthe grisea* and others (Gopal et al. 2011b) (Table 7.15). Antifungal action of fluconazole, a triazole fungicide was improved by using biologically synthesized Ag NP for combating fungal pathogens like *Candida albicans, Phoma glomerata* and *Trichoderma* sp. Nano-Gro and Nano-5 were also released into the market to control grape mold, rice blast, early and late blight, powdery mildews (Gogoi et al. 2010). Gopal et al. (2011b) have developed nano-hexaconazole to control powdery mildew of vegetables and *Rhizoctonia solani.*

Table 7.15 Control of plant pathogens by metallic nanoparticles

Source	Metallic nanoparticle	Location	Size (nm)
Candida glabrata	Ag	Intercellular	20
Fusarium oxysporum	Ag	Extracellular	20–40
Aspergillus niger	Ag	Extracellular	5–25
Trichoderma asperellum	Ag	Extracellular	5–25
Phanerochaete chrysosporium	Ag	Extracellular	50–200

Source: Modified from Gopal et al. (2011b)

7.7 Transcriptomics

Genomics and proteomics are now applied in understanding plant–pathogen interactions and also to address the hypothesis in the context of biochemical pathways, phylogeny, gene network and others. This kind of approach is more applicable to the non-culturable pathogen. It is also known that molecular signaling between fungal pathogen and hosts plays a fundamental role both in pathogenesis and in the establishment of beneficial interactions between symbiotic or parasitic partners, respectively. These interactions have profound effect for designing of new strategies to combat diseases. It is evident that Flor's gene for gene hypothesis explained the genetics of disease phenotypes incited by plant pathogenic fungi and others as well as failures and success of certain plant-pathogen/insect interactions.

Transcriptomics is a revolutionary functional genomic tool for deciphering plant–pathogen interaction in the pathogenomics era. A total of 2,977 full genomic sequences have been published (Gold Statistics 2011). It is essential to analyze more than a single genome sequence of a pathogen species to understand the mechanisms of pathogenesis. The sequencing of hundreds of fungal and microbial genomes including that of pathogen initiated development of novel approaches for the study of functional genomics. Technologies such as gene expression, DNA micro-array analysis and others enable the analysis of gene expression profiles in a single experiment and provide deeper insight into host–pathogen interactions. The transcriptome is the set of all RNA molecules including mRNA, rRNA, tRNA and the non-coding RNA produced in a cell or a tissue. It can vary with external environmental conditions including the state of the pathogen and pathogenesis. Further the transcriptome reflects the genes that are active at any given time.

Transcriptome is also called expression profiling which examines the expression load of mRNAs in a given cell population. The next generation sequencing technology is also known as RNA-sequencing. Transcriptome data also elaborates about the activity of genes that change their expression pattern in response to a signal originated from the host plant or in the host tissue and may reveal mechanisms of pathogenesis as initiated by fungal/microbial pathogens. Global transcriptome analysis is also of much importance in order to understand the interaction between plants genome and its environment along with fungal–pathogen interaction. Microarray provided basis information for many transcriptomic studies besides the serial analysis of gene expression technique.

Future challenges in this area are:

1. Identification of genes and gene clusters involved in the host–pathogen interaction.
2. Exploitation of pathway analysis within fungal pathogen for culturing non-culturable pathogens.
3. Understanding of changes in the host due to fungal pathogen invasion at molecular level.
4. Unveil the intricacies of molecular cross talk between host and the pathogen.

7.8 Integrated Disease Management

The integrated disease management (IDM) is a system approach that combines a wide variety of crop production and protection methods to minimize the yield losses caused by fungal pathogens or other microbes. It covers dynamic monitoring of diseases and conservation of their natural enemies (biocontrol agents).

In IDM, various components of management are so tailored so as to get cost-effective and eco-friendly approach without any adverse effects on ecosystem. The IDM is a dynamic approach which includes host resistance, cultural practices, soil solarization, removal of infected plants, crop rotation, biocontrol agents, etc. (Table 7.16).

Further, non-lethal dosages of fungicides are also used for the control of specific fungal diseases. Following is the chart showing different components of Integrated Plant Diseases Management or IDM (Fig. 7.1).

Table 7.16 Integrated disease management of some plant diseases

Crop	Symptom	Pathogen	Control
Groundnut	Leafspot	*Mycosphaerella arachidicola*	Tolerant varieties; inter-cropping with sorghum, pearl millet; foliar spray of Carbendazim etc.
		Phacoisariopsis personata	Resistant varieties; Carbendazim + Mancozeb; deep burrowing of crop residues and removal of volunteer plants
Sunflower	Leaf spot	*Alternaria helianthi*	Removal of plant debris
	Blight	*Alternaria alternata*	Early sowing resistant varieties
	Downy mildew	*Plasmidiophora halstedii*	Mancozeb spray Seed treatment with Metalaxyl Apron 35
	Rust	*Puccinia helianthi*	Altering the sowing date, deep summer ploughing, resistant varieties
Safflower	Leaf blight	*Alternaria carthami*	Triazoles, Hexaconazole, Thiram; hot water treatment; tolerant varieties
	Root rot	*Phytophthora drechsleri*	Avoidance of mono cropping; use of resistant varieties
	Rust	*Puccinia carthami*	Destruction of crop debris; collateral host; crop rotation; spraying of tridemorph; resistant varieties
Sesamum	Blight	*Phytophthora parasitica* var. *sesami*	Inter-cropping; application of farmyard manure; biocontrol; seed treatment with Captan
	Wilt	*Fusarium oxysporum* f. sp. *sesami*	Seed treatment with Benlate; biocontrol
Castor	Blight	*Alternaria carthami*	Mancozeb application; judicious use of fertilizer
	Grey rot	*Botrytis ricini*	Spacing; Carbendazim spray
	Root rot	*Macrophomina phaseolina*	Crop rotaion; Thiram, Topsin M-70
	Wilt	*F. oxysporum* f.sp. *ricini*	Healthy seeds; Biocontrol; Carbendazim, Thiram

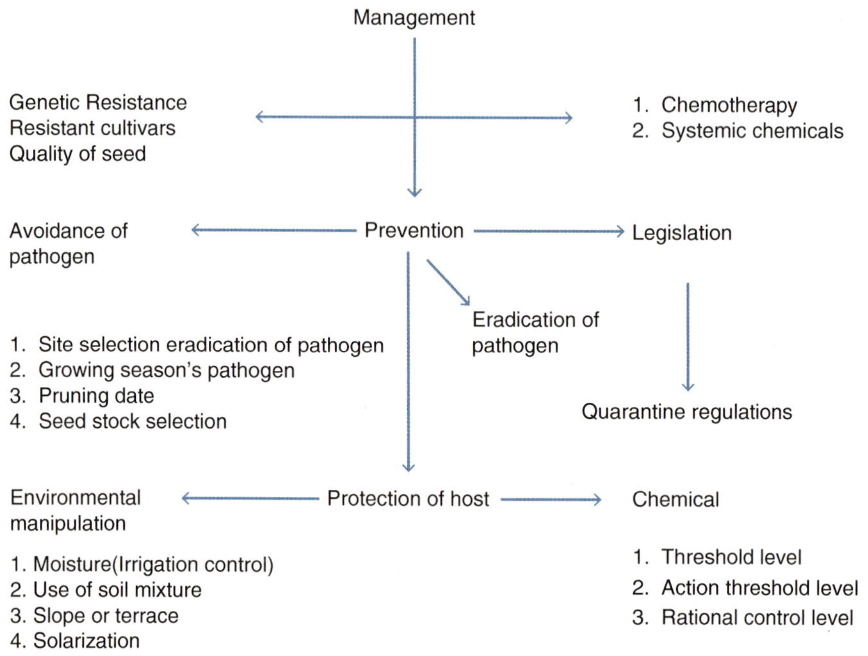

Fig. 7.1 Integrated plant disease management. (Modified from Smith et al. 1976)

7.9 Future Challenges

1. Exhaustive studies are necessary to be attempted on fungal pathogens and in particular from extreme habitats and different agro-climatic conditions.
2. Attention needs to be given to develop cost-effective, less time consuming and scientifically accurate disease diagnostic kits including molecular approaches.
3. Innovative methods have to be developed for cultivating non-culturable fungi and fungal culture collections are to be strengthened.
4. Use of reliable tools to forecast disease epidemics, application of IPM strategies and effective quarantine systems may become important in the future. Further strengthening of survey and surveillance system is needed.
5. Efforts to be continued to trace out and tag many other novel genes that contribute towards the disease resistance.
6. Emphasis has to be laid on development of efficient strains of bio-agents through biotechnological interventions which are cost-effective, efficient with reasonable shelf life, eco-friendly, non-toxic formulation and suitable to different agro-climatic conditions.
7. Scientifically accurate research is essential on the impact of climate change on disease scenario, succession severity and strategies to overcome the changing disease situation.

8. Advanced research has to be conducted on functional genomics in relation to virulence, resistant genes, multiple disease resistance breeding and introduction of disease resistance in transgenics.
9. Lack of user-friendly bio-informatics platform that supports the integration and use of available data on major plant pathogens.
10. Survey, surveillance, identification, disease diagnosis and control of fungal diseases which were earlier of minor importance but may become major, hence emphasis is needed.
11. The synergistic interaction of resistance genes with the IPM is least known and this holistic approach for plant protection has to be strengthened.
12. Risk assessment is critical for the newly developed fungicides before these are introduced for commercial use by the farmers. International organizations need to make more concerted efforts to prepare guidelines for resistance management strategies for different groups of fungicides. Further, it requires active co-operation from farmers, extension pathologists and advisors.
13. Proper understanding of host–pathogen interaction and population level for the deployment of best genes in different agro-climatic zones and also help in making pyramids of best gene combinations in agronomically superior cultivars.
14. Plant compounds that have antifungal activity need to be studied in depth.
15. Plant protection clinics are to be established.
16. Advances in nanotechnology and sensor design suggest that these challenges should be met in the near future.
17. Proteomics and genomics help in unraveling the mechanisms of fungal pathogenicity and this understanding needs to be strengthened as these processes help in development of novel disease tolerant varieties of agriculturally important crops.
18. The science of 'omics' including transcriptomics will be the functional genomics tools for deciphering plant–pathogen interactions at molecular level. Such studies will also trigger the process of some unknown host–pathogen interaction, hence proper understanding is necessary.
19. There is a necessity of bringing out an International Agricultural Biosecurity System for sustainable agriculture.

Acknowledgement One of the authors (CM) is grateful to NASI for financial support.

References

Ajay D, Baby UI (2010) Induction of systemic resistance to *Exobasidium vexans* in tea through SAR elicitors. Phytoparasitica 8:53–60

Altramıranda EAG, Andreu AB, Dalco GR, Olivieri FP (2008) Effect of β-aminobutyric acid (BABA) on protection against *Phytophthora infestans* throughout the potato crop cycle. Australas Plant Pathol 37:421–427

Arya A (2010) Recent advances in the management of fungal pathogens of fruit crops. In: Arya A, Analia E. Perello (eds) Management of fungal pathogens, pp 3–13. www.cabi.org.uk

Arya A, Perello AE (2010) Management of fungal plant pathogens. CABI, Wallingford, p 388
Audenaert K, Demeyer G, Hofte M (2002) Abscisic acid determines basal susceptibility of tomato to *Botrytis cinerea* and suppresses salicylic acid-dependent signaling mechanisms. Plant Physiol 128:491–501
Baker KF (1987) Evaluating concepts of biological control of plant pathogens. Annu Rev Phytopathol 25:67–85
Baker PAHM, Ran LX, Pieterse CMJ, VanLoon LC (2003) Understanding the involvement of rhizobacteria-mediated induction of systemic resistance in biocontrol of plant diseases. Can J Plant Pathol 25:5–9
Belhadj A, Saigne C, Telef N, Cluzet S, Bouscaut J, Corio-Costet F, Mérillon J (2006) Methyl jasmonate induces defense responses in grapevine and triggers protection against *Erysiphe necator*. J Agric Food Chem 54:9119–9125
Belhadj A, Telef N, Cluzet S, Bouscaut J, Coriocostet MF, Mérillon JM (2008) Ethephon elicits protection against *Erysiphe necator* in grapevine. J Agric Food Chem 56:5781–5787
Bieri S, Potrykus I, Futterer J (2003) Effects of combined expression of antifungal barley seed proteins in transgenic wheat on powdery mildew infection. Mol Breed 11:37–48
Böhlenius H, Mørch SM, Godfrey D, Nielsen ME, Thordal-Christensen H (2010) The multivesicular body-localized GTPase ARFA1b/1c is important for callose deposition and ROR2 syntaxin-dependent preinvasive basal defense in barley. Plant Cell 22:3831–3844
Borlaug NE (1965) Wheat, rust, and people. Phytopathology 55:1088–1098
Brent KJ (1995) Fungicide resistance in crop pathogens. How can it be managed? FRAC monograph no. 1. Global Crop Protection Federation, Brussels, p 48
Brent KJ, Hollomon DW (1998) Fungicide resistance: the assessment of risk. FRAC monograph II. Global Crop Protection Federation, Brussels
Brent KJ, Hollomon DW (2000) Fungicide resistance management. Plant Dis Res 15:1–13
Cervone F, Castoria R, Leckie F, De Lorenzo G (1997) Perception of fungal elicitors and signal transduction. In: Aducci P (ed) Signal transduction in plants. Birkhauser, Basel, pp 153–177
Chagué V, Danit LV, Siewers V, Gronover CS, Tudzynski P, Tudzynski B, Sharon A (2006) Ethylene sensing and gene activation in *Botrytis cinerea*: a missing link in ethylene regulation of fungus-plant interaction? Mol Plant Microbe Interact 19:33–42
Chanda B, Xia Y, Mandal MK, Yu K, Sekine K, Gao Q, Selote D, Hu Y, Stromberg A, Navarre D, Kachroo A, Kachroo P (2011) Glycerol-3-phosphate is a critical mobile inducer of systemic immunity in plants. Nat Genet 43(5):421–427. doi:10.1038/ng.798
Chassot C, Buchala A, Schoonbeek HJ, Métraux JP, Lamotte O (2008) Wounding of *Arabidopsis* leaves causes a powerful but transient protection against *Botrytis* infection. Plant J 55:555–567
Chaturvedi R, Krothapalli K, Makandar R, Nandi A, Sparks A, Roth M, Welti R, Shah J (2008) Plastid omega 3-fatty acid desaturase-dependent accumulation of a systemic acquired resistance inducing activity in petiole exudates of *Arabidopis thalliana* is independent of jasmonic acid. Plant J 54:106–117
Cheong YH, Moon BC, Kim JK, Kim CY, Kim MC, Kim IH, Park CY, Kim CJ, Park BO, Koo SC, Yoon HW, Chung WS, Lim CO, Lee SY, Cho MJ (2003) BWMK1, a rice mitogen-activated protein kinase, locates in the nucleus and mediates pathogenesis-related gene expression by activation of a transcription factor. Plant Physiol 132:1961–1972
Cohen Y (2002) β-Aminobutyric acid-induced resistance against plant pathogen. Plant Dis 86:448–457
Cook RJ, Baker RF (1983) The nature and practices of biological control of plant pathogens. APS Books, St Paul, p 530
Custers JH, Harrison SJ, Sela-Burrlage MB, van Deventer E, Lageweg PW et al (2004) Isolation and characterization of a class of carbohydrate oxidases from higher plants, with a role in active defence. Plant J 39:147–160
Dangs JL, Jones JDG (2001) Plant pathogens and integrated defence responses to infections. Nature 411:826–833
De Vos M, Van Zaanen W, Koornneef A, Korzelius JP, Dicke M, Van Loon LC, Pieterse CMJ (2006) Herbivore-induced resistance against microbial pathogen in *Arabidopsis*. Plant Physiol 142:352–363

Deslandes L, Olivier J, Theulieres T, Hirsch J, Feng D, Bittner-Eddy P, Beynon J, Marco Y (2002) Resistance to *Ralstonia solanacearum* in *Arabidopsis thaliana* is conferred by the recessive RRSI-R gene, a member of a novel family of resistance genes. Proc Natl Acad Sci U S A 99:2404–2409

Desveaux D, Maréchal A, Brisson N (2005) Whirly transcription factors: defense gene regulation and beyond. Trends Plant Sci 10:95–102, PubMed ID: 15708347

DeVos M, Van Zaanen W, Koornneer A, Korzelius J, Dicke M, Van Loon L, Pieterse C (2006) Herbivore-induced resistance against microbial pathogens in *Arabidopsis*. Plant Physiol 142:352–363

Diaz J, ten Have A, van Kan JL (2002) The role of ethylene and wound signaling in resistance of tomato to *Botrytis cinerea*. Plant Physiol 129:1341–1351

Dong J, Chen C, Chen Z (2003) Expression profiles of the *Arabidopsis* WRKY gene super family during plant defense response. Plant Mol Biol 51:21–37

Dong X, Ji R, Guo X, Foster SJ, Chen H, Dong C, Liu Y, Hu Q, Liu S (2008) Expressing a gene encoding wheat oxalate oxidase enhances resistance to *Sclerotinia sclerotiorum* in oilseed rape (*Brassica napus*). Planta 228:331–340

Duijff BJ, Poulain D, Olivain C, Alabouvette C, Kamancean C (1998) Implication of systemic induced resistance in the suppression of *Fusarium* wilt of tomato by *Pseudomonas fluorescens* WCS417r and by non-pathogenic *Fusarium oxysporum* Fo47. Eur J Plant Pathol 104:903–910

Durrant W, Dong X (2004) Systemic acquired resistance. Annu Rev Phytopathol 42:185–209

Flor HH (1956) The complementary genetic systems in flax and flax rust. Adv Genet 8:29–53

Fowler D, Freedman E, Scannella J (2009) Predatory functional morphology in raptor: interdigital variation in talon size is related to prey restraint and immobilisation technique. PLoS One 4:1–9

Gogoi R, Singh PK, Kumar R, Nair KK, Alam I, Srivastava C, Yadav S, Subramanyam B, Gopal M, Roy Choudhury S, Pradhan S, Goswami A (2010) Nano-sulphur offers better protection of vegetables against powdery mildew caused by *Erysiphe cichoracearum*. In: 63rd annual meeting national symposium perspective in the plant health management. Indian Phytopathological Society, 14–16 December 2010, p 41

Gold Statistics (2011) http://www.genomesonline.org/cgi-bin/GOLD/bin/gold. Cgi as accessed on 6 Nov 2011

Gopal M, Gogoi R, Srivastava C, Kumar R, Singh PK, Nair KK, Yadav S, Goswami A (2011a) Nanotechnology and its application in plant protection. In: Plant pathology in India: vision 2030, IPS, New Delhi, pp 224–232

Gopal M, Roy Choudhury S, Pradhan S, Srivastava, Gogoi R, Kumar R, Goswami A (2011b) Nanocapsulated hexaconazole: a novel fungicide and the process for making the same. Patent submitted to ITMU, IARI, Patent application No: 2021/Del/2011

Grant M, Lamb C (2006) Systemic immunity. Curr Opin Plant Biol 9:414–420

Hamann T, Bennet M, Mansfield J, Somerville CR (2009) Identification of cell-wall stress as a hexose-dependent and osmosensitive regulator of plant responses. Plant J 57:1015–1026

Hammerschmidt R (2007) Introduction: definitions and some history. In: Walters D, Newton A, Lyon G (eds) Induced resistance for plant defense: a sustainable approach to crop protection. Blackwell Publishing, Oxford, pp 1–8

Hammond-Kosack K, Jones J (1997) Plant disease resistance genes. Annu Rev Plant Physiol Plant Mol Biol 48:575–607

Hartley C (1921) Damping off in forest nurseries. U.S. Dept Agric Bull 934. pp 1-99

Heil M, Ton J (2008) Long-distance signaling in plant defense. Trends Plant Sci 13:264–272

Hematy K, Cherk C, Somerville C (2009) Host-pathogen warfare at the plant cell wall. Curr Opin Plant Biol 12:406–413

IPCC (2007) Climate change 2007. The physical science basis. Contribution on working group I to the fourth assessment report of the IPCC. In: Solomon SD, Qin D, Manning M, Chen Z, Marquis M, Averyt KB, Tignor M, Miller HL (eds) Report of the intergovernmental panel on climate change. Cambridge University Press, Cambridge, p 996

Iwai T, Seo S, Mitsuhara I, Ohashi Y (2007) Probenazole-induced accumulation of salicylic acid confers resistance to *Magnaporthe grisea* in adult rice plants. Plant Cell Physiol 48:915–924

Jabs T, Dietrich R, Dangl J (1996) Initiation of runaway cell death in an *Arabidopsis* mutant by extracellular superoxide. Science 273:1853–1856

Jalali BL (2008) Molecular plant pathology: where do we stand? J Mycol Plant Pathol 38: 419–429

Joshi S, Chavan V, Kamble A, Bhargava S (2010) Role of salicylic acid and oxidative burst in expression of resistance to *Alternaria* blight. Arch Phytopathol Plant Prot 43:1198–1207

Kalde M, Barth M, Somssich IE, Lippok B (2003) Members of *Arabidopsis* WRKY group III transcription factors are part of different plant defense signaling pathways. Mol Plant Microbe Interact 16:295–305

Karou D, Dicko M, Simpore J, Traore A (2005) Antioxidant and antibacterial activities of polyphenols from ethnomedicinal plants of Burkina Faso. Afr J Biotechnol 4:823–828

Khan A, Qureshi R, Gillani S, Ullah F (2011) Antimicrobial activity of selected medicinal plants of Margalla Hills, Islamabad. Pakistan J Med Plants Res 5:4665–4670

Kim JK, Jang IC, Wu R, Zuo WN, Bostan RS, Lee YH, Ahn IP, Nahm BH (2003a) Co-expression of a modified maize ribosome-inactivating protein and a rice basic chitinase gene in transgenic rice plants confers enhanced resistance to sheath blight. Transgenic Res 12:475–484

Kim C, Liu Y, Thorne E, Yang H, Fukushige H, Gassmann W, Hildebr D, Sharp R, Zhang S (2003b) Activation of a stress-responsive nitrogen-activated protein kinase cascade induces the biosynthesis of ethylene in plants. Plant Cell 15:2707–2718

Klittich CJ (2008) Milestones in fungicide discovery: chemistry that changed agriculture. Plant Health Progr. doi:10.1094/PHP-2008-0418-01-RV

Kohler A, Scchwindling S, Conrath U (2002) Benzothia diazole-induced priming for potentiated responses to pathogen infection, wounding and infiltration of water into leaves requires the NPRI/NIMI gene in *Arabidopsis*. Plant Physiol 128:1046–1056

Kuc J (2001) Concepts and direction of induced systemic resistance in plants and its application. Eur J Plant Pathol 107:7–12

Kuhlmann M, Horvay K, Strathmann A, Heinekamp T, Fischer U, Bottner S, Droge-Laser W (2003) The α-helical D1 domain of the tobacco bZIP transcription factor BZI 1 interacts with the Ankyrin-repeat protein ANK1 and 1S important for BZI-1 function, both in auxin signaling and pathogen response. J Biol Chem 278:8786–8794

Kundu S, Chakravarthy D, Pal A (2012) Salicylic acid ameliorates susceptible *Vigna mungo* cultivar to Mungbean yellow mosaic virus. Sci Cult 78:5–6

Liu J, Maldonado-Mendoza I, Lopez-Meyer M, Cheung F, Town C, Harisson M (2007) Arbuscular mycorrhizal symbiosis is accompanied by local and systemic alterations in gene expression and an increase in disease resistance in the shoots. Plant J 50:529–544

Mandal S, Mallick N, Mitra A (2009) Salicylic acid-induced resistance to *Fusarium oxysporum* f. sp. *lycopersici* in tomato. Plant Physiol Biochem 47:642–649

Martin G, Brommonschenkel S, Chunwongse J, Frary A, Ganal M, Spivey R, Wu T, Earle E, Tanksley S (1993) Map-based cloning of a protein kinase gene conferring disease resistance in tomato. Science 262(5138):1432–1436

Martin RR, James D, Levesgue CA (2000) Impact of molecular diagnostic technologies on plant disease management. Annu Rev Phytopathol 38:207–237

Mehta KC (1933) Rusts of wheat and barley in India—a study of their annual recurrences, life histories and physiologic forms. J Agric Sci 31:939–962

Mendoza M (2011) Oxidative burst in plant-pathogen interaction. Biotecnol Veg 11:67–75

Mithofer A, Daxberger A, Fromhold-Treu D, Ebel J (1997) Involvement of an NAD(P)H oxidase in the elicitor-inducible oxidative burst of soybean. Phytochem 45:1101–1107

Mur L, Kenton P, Atzorn R, Miersch O, Wsternack C (2006) The outcomes of concentration-specific interactions between salicylate and jasmonate signaling include synergy, antagonism, and oxidative stress leading to cell death. Plant Physiol 140:249–262

Nam JK (2008) Inhibition of powdery mildew development and activation of antioxidant enzymes by induction of oxidative stress with foliar application of a mixture of riboflavin and methionine in cucumber. Sci Hortic 118:181–188

Nancy T, Patricia S, Irineo T, Rosalia O, Enrique R, Sandra M, Ana F, Adan M, Marcela V, Genero S, Ramon G (2011) Effect of foliar salicylic acid and methyl jasmonate applications on protection against pill bugs in lettuce plants (*Lactuca sativa*). Phytoparasitica 39:137–144

Nunez-Pastrana R, Arcos-Ortega G, Souza-Perera R, Sanchez-Berges C, Nakazava-Ueji Y, Garcia-Villalobos F, Guzman-Antonio A, Zuniga- Agquilar J (2011) Ethylene, but not salicylic acid or methyl jasmonate, induces a resistant response against *Phytophthora capsici* in Habanero pepper. Eur J Plant Pathol 131:683–699

Oldach KH, Becker D, Lorz H (2001) Heterologous expression of genes mediating enhanced fungal resistance in transgenic wheat. Mol Plant Microbe Interact 14:832–883

Ori N, Eshed Y, Paran I, Presting G, Aviv D, Tanksley S, Zamir D, Fluhr R (1997) The 12C family from the wilt disease resistance locus 12 belongs to the nucleotide binding, leucine-rich repeat super family of plant resistance genes. Plant Cell 9:521–532

Park JH, Choi GJ, Lee HB, Kim KM, Jung HS, Lee SW, Jang KS, Cho KY et al (2005) Griseofulvin from *Xylaria* sp. strain F0010, an endophytic fungus of *Abies holophylla* and its antifungal activity against plant pathogenic fungi. J Microbiol Biotechnol 15:112–117

Parker J, Coleman V, Szabo V, Frost L, Schmidt R, van der Biezen A, Moores T, Dean C, Daniels M, Jones J (1997) The *Arabidopsis* downy mildew resistance gene RPP5 shares similarity to the toll and interleukin-1 receptors with N and L6. Plant Cell 9:879–894

Pedley K, Martin G (2003) Molecular basis of pro-mediated resistance to bacterial speck disease in tomato. Annu Rev Phytopathol 41:215–243

Perez L, Rodriguez M, Rodriguez F, Roson C (2003) Efficacy of acibenzolar-S-methyl, an inducer of systemic acquired resistance against tobacco blue mold caused by *Peronospora hyoscyami* f. sp. *tabacina*. Crop Prot 22:405–413

Pieterse C, Van Loon L (2007) Signaling cascades involved in induced resistance. In: Walter S, Newton A, Lyon G (eds) Induced resistance for plant disease control: a sustainable approach to crop protection. Blackwell Publishing, Oxford, pp 65–88

Pieterse C, Ton J, Van Loon L (2001) Cross-talk between plant defence signalling pathways: boost or burden? Ag Biotech Net 3:1–8

Powell ALT, Hallewin DG, Hall BD, Stotz H, Labavitch JM, Bennett AB (1994) Glycoprotein inhibitors of fungal polygalacturonases expression of pear PGIP improves resistance in transgenic tomatoes. Plant Physiol 105:159

Ray R, Ghosh A, Bera A, Dutta N, Chattopadhyay D, Chakrabarti K (2011) Analysis of differentially expressed transcripts in jute upon fungal infection and beta-amino butyric acid treatment. Physiol Mol Plant Pathol 76:59–66

Romero D, Rivera M, Cazorla F, Codina J, Fernandez-Ortuno D, Tores J, Perez-Garcia A, de Vicente (2008) Comparative histochemical analyses of oxidative burst cell wall reinforcement in compatible and incompatible melon-powdery mildew (*Podosphaera fusca*) interaction. J Plant Physiol 165:1895–1905

Rushton PJ, Somssich IE (1998) Transcriptional control of plant genes responsive to pathogens. Curr Opin Plant Biol 1:311–315

Serge S, Andrew N, Sparks A, Willocquet HL, Duveiller E, Madhuku G, Forbes G, Garrett K, Hudson D, Padgham J, Pande S, Sharma M, Yuen J, Djurie A (2011) International agricultural research tackling the effects of global and climate changes on plant diseases in the developing world. Plant Dis 59:1204–1216

Shao H, Chu L, Shao M, Jaleel C, Mi H (2008) Higher plant antioxidants and redox signaling under environmental stresses. C R Biol 331:433–441

Shetty N, Mehrabi R, Lutken H, Haldrup A, Kema G, Collinge D, Jorgensen H (2007) Role of hydrogen peroxide during the interaction between the hemibiotrophic fungal pathogen *Septoria tritici* and wheat. New Phytol 174:637–647

Siegrist J, Orober M, Buchenauer H (2000) α-Aminobutyric acid mediated enhancement of resistance in tobacco to tobacco mosaic virus depends on the accumulation of salicylic acid. Physiol Mol Plant Pathol 56:95–106

Slaughter A, Daniel X, Flors V, Luna E, Hohn B, Mauch-Mani B (2012) Descendants of primed *Arabidopsis* plants exhibit resistance to biotic stress. Plant Physiol 158:835–843

Smith RI, Apple JL, Bottrell DG (1976) Broad based integrated plant disease management. In: Apple JL, Smith RF (eds) The origins of integrated pest management concepts for agricultural crops integrated pest management. Plenum Press, New York, pp 1–16

Song W, Wang G, Chen L, Kim H, Pi L, Holsten T, Garden J, Wang B, Zhai W, Zhu L, Fauquet C, Ronald P (1995) A receptor kinase-like protein encoded by the rice disease resistance gene, Xa21. Science 270:1804–1806

Song W, Ma X, Tan H, Zhou J (2011) Abscisic acid enhances resistance to *Alternaria solani* in tomato seedlings. Plant Physiol Biochem 49:693–700

Soylu S, Brown I, Mansfield J (2005) Cellular reaction in *Arabidopsis* following challenge by strains of *Pseudomonas syringae* from basal resistance to compatibility. Physiol Mol Plant Pathol 66:232–243

Stenzel I, Hause B, Proels R, Miersch O, Oks M, Roitsch T, Wasternack C (2008) The AOC promoter of tomato is regulated by developmental and environmental stimuli. Phytochemistry 69:1859–1869

Sticher L, Mauch-Mani B, Metraux J (1997) Systemic acquired resistance. Annu Rev Phytopathol 35:235–270

Strobel GA (2003) Endophytes as sources of bioactive products. Microbes Infect 5:535–544

The Arabidopsis Genome Initiative (2000) Analysis of the genome sequence of the flowering plant *Arabidopsis thaliana*. Nature 408:796–815

Thind TS (2008) Fungicide resistance: a potential challenge in disease control. J Mycol Plant Pathol 38:407–418

Thind TS (2011) Agricultural fungicides: present scenario and future needs. In: Plant pathology in India: vision 2030, IPS, New Delhi, pp 206–211

Ton J, Mauch-Mani B (2004) β-aminobutyric acid-induced resistance against necrotrophic pathogen is based on ABA-dependent priming for callose. Plant J 38:119–130

Tripathi P (2005) Botanical pesticides in the management of post harvest fruit diseases. Daya Publishing House, New Delhi, p 174

Trouvelot S, Varnier AL, Allègre M, Mercier L, Baillieul F, Around C, Gianazzi-Pearson V, Klarzyynski O (2008) A β-1,3 glucan sulfate induces resistance in grapevine against *Plasmopara viticola* through priming of defense responses, including HR-like cell death. Mol Plant Microbe Interact 21:232–243

Van der Plank JE (1963) Plant disease epidemics and control. Academic, New York, p 349

Van Loon L, Rep M, Pieterse C (2006) Significance of inducible defense-related proteins in infected plants. Annu Rev Phytopathol 44:135–162

Varma J, Dubey ND (1999) Prospectives of botanical and microbial products as pesticides of tomorrow. Curr Sci 76:172–179

Von Dahl C, Baldwin I (2007) Deciphering the role of ethylene in plant herbivore interactions. J Plant Growth Regul 26:201–209

Von Tubeuf CF (1914) Biologische Bekämpfung von Pilzkrankheiten der Pflanzen. Naturw Z Forst-Landw 12:11–19

Waggoner PE, Horsfall IG (1969) Epidem—a simulator of plant disease written for a computer In: Bull Conn Agric Expt Stn No 698. Connecticut, p 80

Waller F, Achatz B, Baltruschat H, Fodor J, Becker K, Fischer M, Heier T, Hu Chelhoven R, Neumann C, Von Wettstein D, Franken P, Kogel K (2005) The endophytic fungus *Piriformospora indica* reprograms barley to salt-stress tolerance, disease resistance and higher yield. Proc Natl Acad Sci U S A 102:13386–13391

Wietholter N, Graebner B, Mierau M, Mort A, Moerschbacher B (2003) Differences in the methylester distribution of homogalacturonans from isogenic wheat lines resistant and susceptible to the wheat stem rust fungus. Mol Plant Microbe Interact 16:945–952

Xing T, Jordan M (2000) Genetic engineering of plant signal transduction mechanisms. Plant Mol Biol Rep 18:309–318

Yoshioka K, Nakashita H, Klessig D, Yamaguchi I (2001) Probenazole induces systemic acquired resistance in *Arabidopsis* with a novel type of action. Plant J 25:149–157

Zhang HW, Sarg YC, Tan RX (2006) Biology and chemistry of endophytes. Natl Prod Rep 23:753–771

Zhao J, Zheng SH, Fujita K, Sakai K (2004) Jasmonate and ethylene signaling and their interaction are integral parts of the elicitor signaling pathway leading to beta-thujaplicin biosynthesis in *Cupressus lusitanica* cell cultures. J Exp Bot 55:1003–1012

Zhao J, Wang Y, Zhang J, Han Y, Yang Z, Feng W (2012) Induction of defensive enzymes (isozymes) during defense against two different fungal pathogens in pear calli. Afr J Biotechnol 11:13670–13677

Zhou J, Trifa Y, Silva H, Pontier D, Lam E, Shah J, Klessig D (2000) *NPRI* differentially interacts with members of the TGA/OBF family of transcription factors that bind an element of the *PR-1* gene required for induction by salicylic acid. Mol Plant Microbe Interact 13(2):191–202

Chapter 8
Ug99-Future Challenges

Subhash Chander Bhardwaj, Mohinder Prashar, and Pramod Prasad

8.1 Introduction

Wheat, the second most important staple food crop after rice as a source of calories and first as a source of protein in the diets of developing country consumers, is grown on about 225 million ha worldwide from the equator to latitudes of 60°N and 44°S and at altitudes ranging from near sea level to more than 3,000 m. Approximately 600 million tons of wheat is produced annually, roughly half of which is in the developing countries (Aquino et al. 2002). Wheat provides 21 % of the food calories and 20 % of the protein to the tune of more than 4.5 billion people across 94 developing countries (Braun et al. 2010). The projected demand for wheat is estimated to increase by 60 % by 2050 in developing countries; at the same time, climate change-induced temperature increases, diseases and other pests are expected to reduce wheat production by more than 29 % in developing countries (Rosegrant et al. 1995). The major fungal diseases of wheat that are caused by biotrophs, include the three rusts, powdery mildew, bunts and smuts; whereas, those caused by hemi-biotrophs include blotches, foliar blight, tan spot and Fusarium head blight (scab).

Among the rusts, stem or black rust, caused by *Puccinia graminis* Pers. f. sp. *tritici* Eriks. & Henn., is infamous for causing severe losses to wheat (*Triticum aestivum* L. and *T. turgidum* var. *durum*) production. Historically, the most damaging disease of wheat has the capacity to turn a healthy looking crop, only weeks away from harvest, into nothing more than a tangle of black stems and shrivelled grains at

S.C. Bhardwaj, Ph.D. (✉) • P. Prasad, Ph.D.
Regional Station, Directorate of Wheat Research, Flowerdale,
Shimla, Himachal Pradesh 171002, India
e-mail: scbfdl@hotmail.com

M. Prashar, Ph.D.
Maharashtra Hybrid Seeds Company Limited, Aurangabad Road, At Dawalwadi,
Tq Badnapur, PO Box 76, Dit. JALNA 431203 Jalna, Maharashtra, India

A. Goyal and C. Manoharachary (eds.), *Future Challenges in Crop Protection Against Fungal Pathogens*, Fungal Biology, DOI 10.1007/978-1-4939-1188-2_8,

harvest. Under severe conditions, yield losses of 100 % are possible (Anonymous 1992). Stem rust is known for causing severe devastations periodically in all wheat-growing countries of the world.

The causal organism of stem rust was named *Puccinia graminis* in 1797 by Persoon and the first detailed reports about this pathogen were given independently by Italian Scientists Fontana (1932) and Tozetti and Alimurgia (1952). Stakman and Piemeisel (1917) showed that stem rust pathogen had various forms or races. Many devastating epidemics of stem rust of wheat are reported to have occurred in the past around the globe. The notable ones being the epidemics of North America in 1904 and 1916. In 1993 and 1994, the last major stem rust epidemic occurred in Ethiopia (Shank 1994), where Enkoy, a popular wheat variety, suffered major losses; however, the rest of the world has remained unscathed by stem rust for over four decades.

Puccinia graminis tritici (*Pgt*) is a basidiomycetous rust fungus, having a broad host range with an ability to infect about 365 species of plants (Anikster 1984). *P. graminis* has been broken into subspecies and *formae speciales* based on spore morphology, fertility crosses and host range. Urban (1967) separated the species into two subspecies based on morphology. Subspecies *graminicola* contained stem rusts found on non-cereal grasses while subspecies *graminis* contained the stem rusts found on cereal crops. Subspecies *graminis* was then further divided into variety *stakmanii*, found on barley, oat and rye and variety *graminis*, which is found mainly on wheat. The subspecies proposed by Urban are in disagreement with *formae speciales* designations based on fertility crosses. Crosses between *formae speciales tritici* and *secalis* and between *avenae* and *poae* were found to produce viable offspring (Johnson 1949). *Pgt* infects cereal crops as urediniospores during the warm months. Urediniospores are dikaryotic (n+n) and spread via agitation of the host plant such as from rain or wind. These spores can land on other cereal hosts and germinate to cause new infection. This process is the asexual stage of the fungus and can result in rapid increase of the organism. Near late summer when temperatures begin to drop teliospores are produced. These teliospores undergo karyogamy before overwintering (Boehm et al. 1992). In early spring the teliospores germinate producing four haploid basidiospores per spore cell. Basidiospores are carried by agitation to *Berberis* species where they directly penetrate the leaf tissue forming pycnia. Pycnia produce receptive hyphae, haploid pycniospores belonging to one of the two mating types (Johnson and Newton 1946). The pycniospores can be wind and water dispersed as well as transferred by insects that are attracted to the exudates (Leonard and Szabo 2005). When pycniospores of one mating type are transferred to receptive hyphae of a pycnium of the compatible mating type, aecia are formed below the pycnium (Johnson and Newton 1946). The aecia produce diploid aeciospores, which infect wheat and form uredia (Roelfs 1985). In the presence of water urediniospores (uredospores) germinate producing a germ tube (Leonard and Szabo 2005). The germ tube grows along the epidermis toward the stoma. If the germ tube comes in contact with a stoma, it will form an appressorium. From the base of the appressorium a penetration peg grows into the substomatal cavity and forms a vesicle. The substomatal vesicle then elongates and hyphae grow inside the plant tissue. Once a hypha comes in contact with a cell wall a haustorial mother cell is formed. The haustorial mother cell exudes enzymes in order to dissolve the plant cell wall. Upon invaginating the cell, a haustorium is produced. The haustoria are responsible for the exchange of nutrients and proteins with the plant and are required for sustaining the growth of the fungus. The genome of *Pgt* was determined to be made of 18 chromosomes (Boehm et al. 1992) and was estimated by genome sequence assembly to be 88.6 Mb (Duplessis et al. 2011). No evidence for whole genome duplication was found but the larger than expected genome seems to be a result of a large number of transposable elements, which composed 45 % of the genome (Duplessis et al. 2011). About 17,773 protein coding genes were predicted, of which only 35 % showed significant homology to known proteins (Duplessis et al. 2011).

An important finding came from the pioneering work of Dr. E.C. Stakman (Stakman and Piemeisel 1917) who showed that the stem rust pathogen had various forms or races. These races differ in their ability to infect different wheat varieties which were later found to carry distinct resistance genes or their combinations. At present wheat scientists use wheat lines that usually carry a single race-specific resistance gene to determine avirulence/virulence characteristics of a race. The races of *Pgt* are determined by phenotypic avirulence/virulence testing on a standard set of wheat differentials. Nomenclature used for designating *Pgt* races has evolved, starting with a chronological numbering system by Stakman (Stakman and Levine 1922) into the current system (Roelfs and Martens 1988). This system is based on the infection types produced on different isolates of *Pgt* lines of wheat. Infection types are classified as: 0—no infection; flecking, 1—small uredia often surrounded by necrosis; 2—small to medium size uredia surrounded by necrosis or chlorosis; 3—medium uredia associated with chlorosis; 4—large uredia without chlorosis or necrosis (Roelfs and Martens 1988). An infection type of 0, 1 or 2 was designated to be a resistant reaction where as 3 or 4 a susceptible reaction. In India, binomial system for designating the pathotypes of *Pgt* is followed (Bahadur et al. 1985) and which is being updated from time to time (Bhardwaj 2012).

8.2 Stem Rust Epidemiology in India

In India cyclonic movement that forms in the Bay of Bengal in November, crosses over to the Arabian Sea around 10°N, subsequently such deep depression recurves and hits the western coast enabling a long distance dispersal of *Pgt* urediniospores from Nilgiris to central India (Nagarajan et al. 1976). Nagarajan and Singh (1975) developed the Indian Stem Rust Rules (ISR) to define the climatic situation that enables long distance spread of *Pgt*. North Indian hills do not play much role in the recurrence of stem rust in India (Bhardwaj et al. 2012), as very weak and primitive type of pathotypes were observed in these areas, which do not occur elsewhere in India.

8.3 Emergence and distribution of Ug99

Many sources of resistance including the alien sources have been used for stem rust resistance. Introduction of rye (*Secale cereale* L.) gene (1B/1R translocation or substitution) into bread wheat (Mettin et al. 1973; Zeller 1973) which carries *Lr*26/*Sr*31/*Yr*9, completely linked resistance gene has not only contributed 12–20 % yield jump but also imparted resistance to major biotic and abiotic stresses (Cox et al. 1995). During many years, *Sr*31 and other stem rust resistance genes kept the stem rust fungus under control, thus it was believed that genetic resistance had overpowered this ancient plague of stem rust. Stem rust of wheat had became a disease of past, unfortunately a new race of *Pgt* was observed in Uganda in 1998 (Ug99) that has the ability to overcome the resistance imparted by majority of the resistance genes including *Sr*31, which have provided protection for the last 60 years (Singh et al. 2011). The Ug99 race of *Pgt* was discovered by William W. Wagoire in an experimental wheat field from the Buginyanya Zonal Agricultural Research and Development Centre in Uganda. Ravi Singh at CIMMYT quickly recognized Ug99's potential to devastate wheat crops (Fig. 8.1) and thereafter it was confirmed as a new strain of *P. graminis* by Zacharias A. Pretorius from the University of the Free State in Bloemfontein, South Africa. Initially the collections of Ug99 were designated as TTKS based on the North American nomenclature system and redesignated as TTKSK after adding a fifth set of differentials in the nomenclature system (Jin et al. 2008). The emergence of Ug99 is considered a highly significant event having far reaching consequences not only for India but also for global wheat production due to susceptibility of most of the wheat cultivars against Ug99 (Fig. 8.2).

It has been estimated by Singh et al. (2008) that the area under the risk of Ug99 amounts to around 50 Mha of wheat grown globally, i.e. about 25 % of the world's wheat area. Varieties of Indian subcontinent like PBW343, PBW373 and few others are susceptible to these races. The race is expected to move further in coming years to other areas, where wheat is one of the major food crops. Germplasm with resistance to Ug99 is available in many parts of the world including India (Singh et al. 2008).

Fig. 8.1 Stem rust infection on PBW343 in Ethiopia

Fig. 8.2 Shrivelled seeds produced in a wheat variety infected with Ug99 (Source Dr. Rabi Singh, CIMMYT, Mexico)

Being wind borne, the stem rust pathogen is capable of causing explosive epidemics. They are known to produce millions of urediniospores and these spores are transferred from one plant to other plant or from one area to the other. The long-distance dispersal (LDD) through the agency of air and may be unintended human interactions is well documented for wheat rusts. A major mode of dispersal for wheat rusts is stepwise range expansion through wind. It usually occurs over shorter distances, i.e. within country or region, and has a much higher probability of successful dispersal of the disease than other possible modes of dispersal. One interesting example of such type of dispersal mechanism is the spread of stripe rust by *Yr*9-virulent race of *P. striiformis* that evolved in Eastern Africa. This race migrated to South Asia through the Middle East and West Asia in a stepwise manner over about 10 years, and caused severe epidemics in its path (Singh et al. 2004).

Some of the important examples of LDD include the introduction of sugarcane rust into the America from Cameroon in 1978 and a wheat stem rust introduction into Australia from southern Africa in 1969. More recently, the arrival of Asian soybean rust into the USA in 2004 from North to South America/Caribbean was most likely carried by hurricane Ivan (Anonymous 2005). At present, assisted long-distance dispersal is also believed to occur by the clothing of travellers or infected plant material. There is strong evidence to support an accidental introduction of wheat yellow rust into Australia in 1979, probably on travellers clothing from Europe (Steele et al. 2001). Spread of stem rust race from Antarctic to New Zealand is well documented (McEvans 1969)

Ug99 has the similar story of its dispersal. According to Singh et al. (2008), the Ug99 race had been confirmed in Uganda, Kenya, Ethiopia, Sudan and Yemen up to 2007. By then, occurrence in Yemen was considered particularly significant, as it

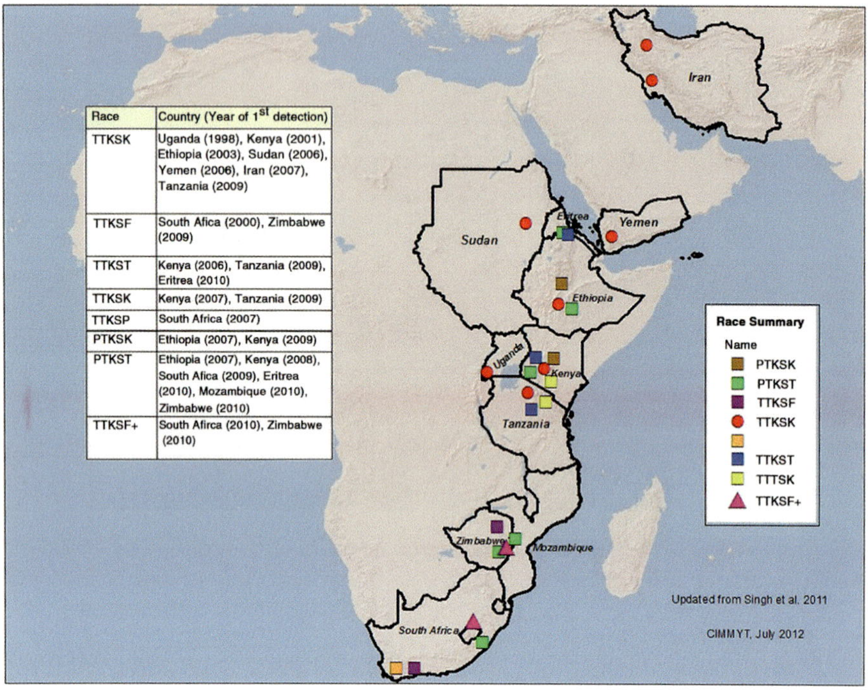

Fig. 8.3 Movement and distribution of different variants of Ug99 (Source Dr. Ravi Singh, CIMMYT, Mexico)

provided strong evidence that Ug99 was moving toward the important wheat areas of the Middle East and Asia (Singh et al. 2011). Subsequently, Ug99 race (TTKSK) was reported from Iran, by FAO (Food and Agriculture Organization) in March 2008 (FAO 2008). In 2009 Ug99 was reported from Khuzestan, one of the provinces in southern Iran. There were few reports of the occurrence of Ug99 in Pakistan during 2009, but pathotyping on differentials and DNA analysis of those samples convincingly indicated the absence of Ug99 (Mirza et al. 2010). So far it is not reported to occur in India also. The Ug99, is evolving very fast and till date eight variants have been documented as the progenies of this race (Singh et al. 2011) as given in Fig. 8.3. The highlands of east Africa have historically been a hot spot for the development of highly virulent races of *Puccinia* sp. Factors contributing to the development of highly virulent races of wheat rust include the year-round cultivation of susceptible wheat genotypes creating ideal conditions for disease development.

All the variants differ slightly in their avirulence/virulence profiles but genetically they are closely interrelated with nearly related DNA fingerprints. Simple sequence repeats marker similarity of Ug99 races have revealed that they evolved from a common ancestor. Race TTKSK (Ug99) is the only known pathotype of the lineage group, which is reported outside of Africa, but it is very likely that other members of the group may appear in the areas outside Africa. Race Ug99 has further evolved and the family has now eight members (Table 8.1).

Table 8.1 Variants in Ug99 over the years

Race	Common alias	Key virulence (+) or avirulence (−)[a]	Year of identification	Confirmed countries (year)
TTKSK	Ug99	+*Sr*31	1999	Uganda (1998/1999), Kenya (2001), Ethiopia (2003), Sudan (2006), Yemen (2006), Iran (2007), Tanzania (2009)
TTKSF	–	−*Sr*31	2000	South Africa (2000), Zimbabwe (2009)
TTKST	Ug99 + *Sr*24	+*Sr*31, +*Sr*24	2006	Kenya (2006), Tanzania (2009), Eritrea (2010)
TTTSK	Ug99 + *Sr*36	+*Sr*31, +*Sr*36	2007	Kenya (2007), Tanzania (2009)
TTKSP	–	−*Sr*31, +*Sr*24	2007	South Africa (2007)
PTKSK	–	+*Sr*31, −*Sr*21	2007	Uganda (1998), Ethiopia (2007), Kenya (2009)
PTKST	–	+*Sr*31, +*Sr*24, −*Sr*21	2008	Ethiopia (2007), Kenya (2008), South Africa (2009), Eritrea (2010), Mozambique (2010), Zimbabwe (2010)
TTKSF+	–	−*Sr*31	2012	South Africa (2010), Zimbabwe (2010)

[a]Characteristic features not the complete avirulence/virulence given

Till now Ug99 race migration has followed gradual stepwise range expansion, following the predominant West-East air flows. Moreover, there is well-recognized evidence connecting East Africa with West and South Asia as a single epidemiologic zone for migration of rust races, with East African origin (Singh et al. 2004). Keeping these facts in view, a group of wheat rust experts had forecasted in the year 2005 that within few years Ug99 shall reach across the Saudi Arabian peninsula and into the Middle East, South Asia, and eventually East Asia and the Americas (Anonymous 2005). Interestingly, recent evidences also support their forecast. The data obtained from GIS tools determines two potential air-borne migration routes for Ug99 to south Asia (Hodson et al. 2005). The first route matches the route described by Singh et al. (2004) for the *Yr*9-virulent race of *P. striiformis* and is considered the most likely route. The second route that connects East Africa directly with southern Pakistan/western India has no known precedence and is highly speculative and of much low probability (Joshi et al. 2008). So far, the movement of Ug99 has followed the first route and as mentioned above has reported to cross the Arabian Gulf in the 2006 crop season. Thus, predicting the exact route of the Ug99 movement for near future is not going to be that easy as the disease outcomes would depend on host susceptibility, prevailing environmental conditions and many other factors. Keeping in view the facts and figures, it may not be an issue in North Western Plain Zone of India (Nagarajan 2012).

More detailed analysis of further potential movements of Ug99 have been undertaken using the HYSPLIT (Hybrid Single-Particle Lagrangian Integrated Trajectory), an air-borne particle trajectory model developed by National Oceanic and Atmospheric Administration (NOAA), USA (Draxler and Rolph 2003), which supports the hypothesis that Yemen could be a staging post for onward movement into the Middle East and Asia. Expanding known range of Ug99 across continent and high mobility of people nationally and internationally, make us to think of the need for continuous monitoring and surveillance in all wheat-growing regions of the world.

8.4 History Repeats Itself

When *Lr*26 was incorporated through 1B/1R translocation for rust resistance, it not only conferred resistance to stem and stripe rust but also gave many advantages (Mettin et al. 1973; Zeller 1973). Few years of its introgression-virulent mutants were identified in Europe (Bartos et al. 1984), India (Nayar et al. 1987), North America (Kolmer 1991) and elsewhere, mostly these were independent mutations for gain in virulence for *Lr*26. In India first variant of *P. triticina* was identified in 1986 (Nayar et al. 1987) and now there are 19 pathotypes with virulence to *Lr*26 and 10 with combined virulence to *Lr*23 and *Lr*26 (Bhardwaj 2012). Likewise, virulence to *Sr*24 was identified in South Africa in 1984 (Roux and Rijkenberg 1987) whereas in India an independent mutation in pathotype 40A resulted in pathotypes 40-1 with virulence for *Sr*24 (Bhardwaj et al. 1990). Prior to that, *Sr*24 used to confer resistance for stem rust in India. Many incidents of independent mutations for virulence for one gene are well known in wheat rusts, example of *Yr*9 (McIntosh et al. 1995) is well documented. Nagarajan (2012) has classified North Pakistan and NW India zone (Indo-Gangetic Plains) comprising more than 14 Mha under bread wheat, as Epidemiological zone III. Referring to earlier publications (Mehta 1940, 1952; Joshi et al. 1971; Nagarajan and Joshi 1985) the proposed threat due to *Pgt* Ug99 (Singh et al. 2006, 2008, 2011 and Stokstady 2007) is not relevant to wheat production of NWPZ. Thus, keeping an eye on independent mutation and proneness of Peninsular and Central India and adjoining areas a long-term strategy is in place to combat this threat.

8.5 Ug99-Risk Mitigation Strategies

The estimated feasible area under the risk of Ug99 along its natural migration path in North Africa, Middle East and Asia (excluding China) might amount to 50 million ha of wheat, that is, about 25 % of the world's wheat area and accounting for an estimated 19 % of global production amounting to about 117 million tons (Reynolds and Borlaug 2006). If somehow the Ug99 reaches to these regions, it would affect estimated one billion people living in these parts of the world. To avoid such type of catastrophe, one of the best strategies is to identify and deploy resistant wheat genotypes that can prove suitable for the regions, which are really prone to Ug99. Extensive screening of wheat genotypes across the hotspots for Ug99 can be done to identify area-specific potential wheat genotypes with resistance to Ug99.

8.6 Borlaug Global Rust Initiative

Borlaug Global Rust Initiative (BGRI) (earlier Global Rust Initiative) was implemented on September 9, 2005 at Nairobi, Kenya with the objectives: to monitor the spread of wheat stem rust race Ug99, to screen the released varieties and germplasm for resistance to Ug99, to distribute the sources of resistance worldwide, breeding to incorporate diverse resistance genes and adult plant resistance gene into high-yielding adapted varieties. Under the framework of BGRI, the evolution and migration of the Ug99 group of races are being monitored carefully so as to provide early warning to the farmers and wheat rust researchers in case of an epidemic. It will help the farmers as well as researchers in decision making. India, one of the strong partners of BGRI, is actively participating in the germplasm testing in Kenya and Ethiopia along with that from CIMMYT, ICARDA and various other countries. The success of BGRI lies in a timely replacement of stem rust susceptible cultivars with resistant ones having equal or better yield potential and other necessary characteristics.

Global Cereal Rust Monitoring System (GCRMS) has been implemented under the umbrella of BGRI, Consultative Group on International Agricultural Research (CGIAR) centres, advanced research labs, national agricultural programmes and UN-FAO to integrate and disseminates up-to-date information on stem rust incidence, severity, as well as races. It has resulted in to emergence of strong, rapidly expanding, coordinated international rust surveillance network. So far more than 15 countries are reporting standardized field survey and surveillance data on wheat rust disease incidence and severity, and this number is expected to rise further in near future (Singh et al. 2011).

8.7 Durable Rust Resistance in Wheat Project

In April 2008 a collaborative effort began, the Durable Rust Resistance in Wheat (DRRW) Project led by Cornell University, seeks to mitigate the threat posed by Ug99 and other potential races causing wheat rust through the coordinated activities that will replace susceptible varieties with durably resistant varieties. The DRRW also aims to harness recent advances in genomics to introduce non-host resistance (immunity) into wheat. The major goal of the project is to improve international collaboration in wheat research to meet growing world food demand, which is estimated as 50 % production increase in wheat by 2020.

8.8 Screening and Deployment of Ug99 Resistant Global Wheat Varieties

As preparedness for combating the probable threat of Ug99, more than 250,000 wheat varieties, germplasm collections, and advanced breeding materials from wheat producing countries of Africa and Asia, were screened for resistance to stem rust race Ug99 and its derivatives at Njoro, Kenya and to a lesser extent, at Kulumsa and Debre Zeit, Ethiopia from 2005 to 2012. Out of total materials screened more than 85 % of materials from various countries were found to be susceptible to the Ug99 group of races. During 2005 and 2006, about 10 % wheat material from South America, Australia, USA and CIMMYT was documented as resistant to Ug99 race, possessing resistance gene *Sr*24, linked to leaf rust resistance gene *Lr*24. Unfortunately, during 2006 TTKST, another race in Ug99 lineage with virulence to *Sr*24 was detected in Kenya (Jin et al. 2008) and by 2007 it had built up sufficiently to cause an epidemic on the *Sr*24-carrying variety Mwamba, with about 30 % of the wheat area in Kenya at that time. TTKST is now the predominant stem rust race in Kenyan fields, and a majority of wheats carrying *Sr*24, including many hard red winter wheats that were resistant to the original Ug99, are now showing high susceptibility to TTKST in field screening nurseries (Singh et al. 2011). Similarly, in 2007 resistance rendered to Ug99 race by *Sr*36 gene for US wheats and for some spring wheats from Australia was knocked out by TTTSK, a new variant of Ug99 group. Interestingly, during the screening for Ug99 race resistance, it was observed that in contrast to bread wheat, a larger proportion of durum wheat varieties and germplasm were showing resistance to all the races of the Ug99 lineage. This phenomenon was attributed to the presence of stem rust resistance gene *Sr*13 in most of the durum wheat germplasm. Again the Ug99 race got mutated to produce two new races TRTTF and JRCQC in the Ug99 lineage from Ethiopia; both of them were able to breakdown the resistance to durum wheat provided by *Sr*13. TRTTF race, with virulence to *Sr*13, *Sr*Tmp, and *Sr*1A.1R has been reported from Africa, the Middle East, and Asia, thus reducing the focus in the utilization of these resistance genes in breeding programmes. It is predicted that most of the wheat-growing regions of the world will suffer more and more in the future because of existing favourable environmental conditions for stem rust build up and unavailability of suitable Ug99 race resistance wheat germplasm, which could lead to epidemic build up.

8.9 Exploring Available Ug99 Resistance Genes

To date, 58 genes have been designated for resistance to wheat stem rust (McIntosh et al. 2011; Mc Intosh personal comm). Over the last century, these genes have been identified within common wheat and wild relatives. Many *Sr* genes of common wheat origin have been deployed during major efforts to incorporate genetic resistance to stem rust in wheat cultivar development. Of the designated *Sr* genes many are single-locus major genes (McIntosh et al. 2011) conferring resistance at all stages of plant development, sometimes with varying effectiveness at the adult plant stage. Resistance can also be conferred by multiple minor genes that individually contribute small effects but together contribute significantly to the resistance phenotype. Stem rust resistance from two genes *Sr*2 and *Sr*55 are unique in a way that they confer quantitative adult plant resistance to stem rust and are pleiotropic (McIntosh et al. 1995); conferring resistance to diseases including leaf rust, stripe rust and powdery mildew. Major gene resistance to rust pathogens of wheat generally operates in a gene-for-gene manner where a single disease resistance gene corresponds to a single avirulence factor in the pathogen. Most stem rust seedling resistance genes confer a strong defence response involving chlorosis or necrosis that limit the formation and spread of fungal hyphae and uredia in host tissues. The type of defence response and the presence of either chlorosis or necrosis differ, sometimes greatly, between individual *Sr* genes and are used to classify the phenotypic expression of resistance (Stakman et al. 1962). *Sr*5, *Sr*17, *Sr*27, *Sr*35 and *Sr*36 all confer low, hypersensitive infection types whereas *Sr*22, *Sr*29 and *Sr*33 confer low chlorotic infection types (McIntosh et al. 1995).

In contrast, major gene resistance to stem rust that is race specific and observed at the seedling stage, stem rust resistance conferred by adult plant resistance (APR) genes is non-race specific and is expressed in adult plants. Most APR genes are minor genes acting as quantitative trait loci (QTL). The accumulation of multiple minor genes contributing to resistance has the effect of generating high levels of resistance in adult plants (Singh et al. 2011). One of the most widely utilized adult plant resistance genes is *Sr*2. The gene is most effective in combination with other resistance genes with small effects. The "*Sr*2-Complex" comprising *Sr*2 in combination with up to five additional stem rust resistance genes with small effects continues to be bases of adult plant resistance to stem rust in international breeding efforts (Singh et al. 2011). Although there are several genes showing considerable amount of resistance to Ug99 group of stem rust races yet, only *Sr*22, *Sr*26, *Sr*35 and *Sr*50 are known to be effective against all currently reported races of the group. *Sr*25 is known to confer high level of resistance only in some specific genetic backgrounds, especially when present with adult plant resistance gene; Misr 1 in Egypt and Muquawin 09 in Afghanistan are the example of such varieties with gene combination of *Sr*25 and *Sr*2. Jain and co-workers (2009) reported PKTSC, a race of stem rust from Nilgiri Hills of India in 2007, possessing virulence to *Sr*25 and several other resistance genes; it reduced the utility of *Sr25* in future breeding programmes. *Sr26* gene transferred to wheat chromosome 6AL from *Triticum elongatum* is likely to be utilized in many breeding programmes because of its availability in adapted genetic backgrounds. Screening of genes *Sr*29, *Sr*32, *Sr*37, *Sr*39, *Sr*40 and *Sr*44 have not been done that widely for their effectiveness to races in Ug99 lineage, which could be useful in breeding programme for stem rust resistance.

A resistance gene temporarily designated as *SrCad*, located on chromosome 6DS and present in Canadian wheat varieties AC Cadillac and Peace, confers moderate resistance to the Ug99 group of races when present alone and a high level of resistance when present together with slow rusting leaf rust resistance gene *Lr34* (Hiebert et al. 2010). Similarly, a few other temporarily designated genes (*SrSha7* derived from Chinese cultivar Shanghai#7, *SrND643* from ND643 and *SrHuw234* from HUW234) are present in improved wheat materials and, when present alone, confer moderate levels of resistance to the Ug99 group of races (Singh et al. 2011).

8.10 Breeding Approaches and Meeting Durable Resistance Against Ug99

Both major gene and minor gene resistance drive evolution of *Pgt* populations. Major genes, when deployed singly, have the effect of generating directional selection toward virulence resulting in boom and bust cycles. The result of continuous boom and bust cycles are a diminished gene pool of effective stem rust resistance genes. Minor genes also exert selection on *Pgt* populations. However, the result of selection is not a qualitative change as a single mutation from *Avr* to *avr* but as a quantitative increase in aggressiveness. Presumably, resistance from minor genes does not involve the recognition of *Avr* gene products to trigger resistance and will not induce selection pressure in pathogen populations for mutations in major gene targets. Both mechanisms of resistance show the potential to break down but quantitative resistance has shown to be more durable over time and space.

Single genes deployed over large acreages have short lifespans. Pyramids of stem rust resistance genes show promise in prolonged durability of major gene resistance. A genotype with multiple genes together, to which no virulence exists in the pathogen population, should in theory prolong the effectiveness of each individual gene. The virulence may not persist in the pathogen population as individuals with the acquired virulence cannot reproduce because of the presence of additional resistance genes to which they are avirulent. To cause disease on a cultivar carrying three major genes for resistance, a pathogen would have to acquire virulence to all three genes simultaneously which may be a very low probability. In order to assure that multiple modes of pathogen recognition are active, more than two major genes present together in a single genetic background should be sufficient to provide multiple modes of pathogen recognition and thereby prevent the simultaneous mutation to virulence. A mutation in a pathogen virulence component to overcome resistance could compromise pathogen fitness resulting in fitness cost (Bahri et al. 2009). This could potentially increase the durability of a resistance gene product that recognizes a virulence component that, if lost, confers a fitness penalty to the pathogen. Therefore, it is always advisable to introduce durable resistance in current wheat cultivars and recent wheat germplasm as a long-term strategy.

*Sr*2 gene, transferred from *Triticum turgidum* L. f. sp. *dicoccum* Schrank ex. Schübler (cv. Yaroslav) into hexaploid wheat by McFadden, one of the *Sr* gene deployed in modern wheat stem rust management, is reported to provide durable resistance to stem rust (McIntosh 1988). Hope cultivar of wheat with *Sr*2 gene was used in Mexico during the 1940s as the donor for developing the stem rust resistant wheat cultivar Yaqui 50, since then, the *Sr*2 gene has been employed widely by CIMMYT in Mexico and spread to many wheat production regions of the world. The gene has provided durable, broad-spectrum rust resistance effective against all isolates of *P. graminis* worldwide for more than 50 years. US wheat cultivar Chris, which is not known to carry *Sr*2 but possesses *Sr*7a (Singh and McIntosh 1987) also displayed high level of resistance and hence its adult plant resistance may involve interaction of moderately effective gene *Sr*7a and other unknown adult plant resistance genes. These observations, give researches an opportunity to look into possibility of deploying these sources of durable resistance in current day wheat cultivars available with them. The breeding strategies to combat the havoc of Ug99 for the foreseeable future would be to use race-specific resistance genes in combinations and not alone. However, by doing so we might have to compromise yield and other quality parameters (Singh et al. 2006). If two resistance genes are to be incorporated, a three-way cross strategy needs to be followed keeping adapted cultivar as the third parent (Singh et al. 2011). It is desirable to use large population size for increasing the chances of transgressive segregants. Molecular markers can be used in segregating generations to select superior segregants for resistance in the background of desired agronomic performance. Limited back crossing can also be followed to restore the characteristics of the recurrent parent (Singh et al. 2011).

8.11 Indian Preparedness for Ug99

Although the stem rust prone area in India is less than 25 % of the total area, the possible implications of entry of Ug99 race into the country or independent mutation for *Sr*31 cannot be ignored. In a study on diversity for stem rust resistance in Indian wheat, commendable diversity was observed (Bhardwaj et al. 2003). Seven different types of resistance to stem rust were observed in wheat lines evaluated during 2000–2001. Further we have preparedness to combat the threat named Ug99. We have already initiated activities in collaboration with CIMMYT and BGRI to identify and develop suitable resistant cultivars for rapid deployment in its different wheat zones.

So far more than 947 lines have been screened against Ug99 type of races in Kenya and Ethiopia (Table 8.2). During summer (off season) crop of 2005, a set of 19 Indian wheat varieties and 3 genetic stocks were screened under natural outbreak of Ug99 at Njoro, Kenya. Wheat variety HW 1085, developed by IARI Regional station, Wellington for South Hill zone and three genetic stocks, i.e. FLW 2 (PBW 343+*Sr*24), FLW 6 (HP 1633+*Sr*24) and FLW 8 (HI 1077+*Sr*25), developed at DWR Regional Station, Shimla, India, were found resistant against Ug99 race under natural conditions in Kenya. After the detection of TTKST, a new variant of Ug99 in 2006 from Kenya the usefulness of the gene *Sr*24 was reduced. However, it has been still reported effective against other races of Ug99 lineage.

Table 8.2 Number of Indian wheat entries screened in Ethiopia and Kenya

Year	Number of lines
2005–2006	19
2006–2007	248
2008–2009	250
2009–2010	241
2010–2011	189
Total	947

In the year 2006, out of 102 Indian lines along with lines from other neighbouring countries (Nepal, Pakistan and Bangladesh), screened at Kenya; GW273, GW322, HD2781, HI1500, MP4010, HI8498, MACS2846 and HD4672 showed satisfactory level of resistance. The cultivars UP2338 and HUW510, carrying *Sr*24, showed acceptable level of resistance in the year 2005 but lost usefulness next year due to the detection of TTKST. During testing in Kenya, it was observed that most of the varieties (PBW343, PBW373 and others carrying *Sr*31) grown in India are susceptible to Ug99.

Likewise, there is need to promote durum wheat cultivars (HI8498, MACS2846, HD4672) in the central and peninsular zones that showed good to moderate resistance against Ug99 (Joshi et al. 2008). Based on the studies conducted so far, rust resistance in durum wheat seems to be conditioned by genes different from the known ones, and hence, can provide the much needed diversity in gene deployment. A perusal of available data for Central and Peninsular India indicate that Lok1, HI9498, WH147, GW 322, HI 1531, HI 8627, HD 4672, DL 788-2 and MPO 1215 are in cultivation whereas in Peninsular Zone Lok 1, NI 5439, NIDW 295, MACS 2846, HI 8663, UAS 321 and UAS 431 are being cultivated. All of these are resistant to moderately resistant to Ug99 races. Lok 1 occupies at least 50 % area and others cover nearly 30 % area in two zones. Ug99 resistant varieties in the two zones work out to be 80 % of the ~7.0 million ha, i.e. 5.6 million ha.

8.12 Conclusion

Though Ug99 type of races have posed a threat for wheat cultivation throughout the world, however, with the collaboration of BGRI, Govt. of India through ICAR, New Delhi has taken proactive steps to meet this challenge. Evaluation in Kenya and Ethiopia has helped in identifying Ug99 resistant germplasm. Keeping into consideration the harsh cool weather in main wheat belt in India, Ug99 type of virulence may not be a threat in this area, however, our preparedness and pre-emptive efforts are in place. In much prone Central and Peninsular India, a sizeable area is under Ug99 resistant varieties.

References

Anikster Y (1984) The *formae speciales*. In: Bushnell WR, Roelfs AP (eds) Cereal rusts, vol 1. Academic Press, Orlando, FL, pp 115–130

Anonymous (1992) The wheat rust patrol—striking at a fast moving target. Partners in research for development. No. 5: 20–24. ACIAR, GPO Box1571, Canberra, ACT, 2001, Australia

Anonymous (2005) NCSRP. Soybean Asian rust—research update, 2005. http://www.planthealth.info/pdf_docs/AsianRust_screen.pdf. Accessed 1/4/07

Aquino P, Carrion F, Calvo R (2002) Selected wheat statistics. In: Ekboir J (ed) CIMMYT 2000–2001 world wheat overview and outlook: developing no-till packages for small scale farmers. CIMMYT, Mexico, DF, pp 52–62

Bahadur P, Nagarajan S, Nayar SK (1985) A proposed system of virulence analysis in India II *Puccinia graminis tritici*. Proc Indian Acad Sci 95:19–33

Bahri BO, Kaltz M, Leconte CD, Vallavieille-Pope C, Enjalbert J (2009) Tracking costs of virulence in natural populations of the wheat pathogen, *Puccinia striiformis* f. sp. *tritici*. BMC Evol Biol 12:1–12

Bartos P, Stuchlikova E, Kubova R (1984) Wheat leaf rust epidemics in Czechoslovakia in 1993. Cereal Rusts Bull 12:40–41

Bhardwaj SC (2012) Wheat rust pathotypes in Indian subcontinent then and now. In: Singh SS, Hanchinal RR, Singh G, Sharma RK, Tyagi BS, Saharan MS, Sharma I (eds) Wheat productivity enhancement under changing climate. Narosa Publishing House Pvt Ltd, New Delhi, pp 227–238

Bhardwaj SC, Nayar SK, Prashar M, Kumar J, Meven MK, Singh SB (1990) A pathotype of *Puccinia graminis* f. sp. *Tritici* on *Sr*24 in India. Cereal Rusts Powder Mild Bull 18:35–38

Bhardwaj SC, Nayar SK, Prashar M, Jain SK, Singh SB (2003) Diversity of resistance for *Puccinia graminis tritici* in wheat (*Triticum aestivum*) and triticale mutant. Indian J Agric Sci 73(12): 676–679

Bhardwaj SC, Gangwar OP, Singh SB, Saharan MS, Sharma S (2012) Rust situation and pathotypes of *Puccinia* species in Leh Ladakh in relation to recurrence of wheat rusts in India. Indian Phytopathol 65(3):230–232

Boehm EWA, Wenstrom JC, McLaughlin DJ, Szabo LJ, Roelfs AP, Bushnell WR (1992) An ultrastructural pachytene karyotype for *Puccinia graminis* f. sp. *tritici*. Can J Bot 70:410–413

Braun HJ, Atlin G, Payne T (2010) Multi-location testing as a tool to identify plant response to global climate change. In: Reynolds MP (ed) Climate change and crop production. CABI, London, pp 115–138

Cox TS, Gill BS, Sears RG (1995) Notice and release of KS94WGRC32 leaf rust resistant hard red winter wheat germplasm. Ann Wheat Newslett 41:241

Draxler RR, Rolph GD (2003) HYSPLIT (hybrid single-particle lagrangian integrated trajectory). http://www.arl.noaa.gov/ready/hysplit4.html. Accessed 29 November 2007. NOAA Air Resources Laboratory, Silver Spring, MD

Duplessis S, Cuomo CA, Lin Y, Aerts A, Tisserant E, Veneault-Fourrey C et al (2011) Obligate biotrophy features unraveled by the genomic analysis of rust fungi. Proc Natl Acad Sci U S A 108:9166–9171

FAO (2008) Wheat killer detected in Iran. http://www.fao.org/newsroom/en/news/2008/1000805/index.html

Fontana F (1932) Observations on the rust of grain. P.P. Pirone, Translated Phytopathologic Classics, No.2, Amer Phytopathol Soc, Washington, DC (Originally published in 1767)

Hiebert CW, Fetch TG, Zegeye T, Thomas JB, Somers DJ et al (2010) Genetics and mapping of seedling resistance to Ug99 stem rust in Canadian wheat cultivars 'Peace' and 'AC Cadillac'. Theor Appl Genet 121:1083–1091

Hodson DP, Singh RP, Dixon JM (2005) An initial assessment of the potential impact of stem rust (race Ug99) on wheat producing regions of Africa and Asia using GIS. In: Absts of the 7th international wheat conference, Mar del Plata, p 142

Jain SK, Prashar M, Bhardwaj SC, Singh SB, Sharma YP (2009) Emergence of virulence to *Sr*25 of *Puccinia graminis* f. sp. *tritici* on wheat in India. Plant Dis 93:840

Jin Y, Pretorius ZA, Singh RP, Fetch T Jr (2008) Detection of virulence to resistance gene *Sr*24 within race TTKS of *Puccinia graminis* f. sp. *tritici*. Plant Dis 92:923–926

Johnson T (1949) Inter varietal crosses in *Puccinia graminis*. Can J Res Sec C 27:45–65

Johnson T, Newton M (1946) Specialization, hybridization, and mutation in the cereal rust. Bot Rev 12:337–392

Joshi LM, Saari EE, Gera SD (1971) Epidemiology of wheat rusts in India. INSA Bull 373: 449–453

Joshi AK, Mishra B, Prashar M, Tomar SMS, Singh RP (2008) Ug99 race of stem rust pathogen: challenges and current status of research to sustain wheat production in India. Indian J Genet 68(3):231–241

Kolmer JA (1991) Physiologic specialization of *Puccinia recondite* f. sp. *triticina* in Canada in 1990. Can J Plant Pathol 13:371–373

Leonard K, Szabo LJ (2005) Stem rust of small grains and grasses caused by *Puccinia graminis*. Plant Mol Pathol 6:99–111

McEvans JM (1969) The races of stem rust infecting New Zealand wheat crops. N Z J Agric Res 19:536–541

McIntosh RA (1988) The role of specific genes in breeding for durable stem rust resistance in wheat and triticale. In: Simmonds NW, Rajaram S (eds) Breeding strategies for resistance to the rust of wheat. CIMMYT, DF, Mexico, p 1–9

McIntosh, Wellings RC, Park R (1995) Wheat rusts, an atlas of resistance genes. CSIRO, East Melbourne

McIntosh RA, Yamazaki Y, Dubcovsky J, Rogers J, Morris C, Somers DJ (2011) Catalogue of gene symbols for wheat: 2011 supplement. Annu Wheat Newslett 57:303–321

Mehta KC (1940) Further studies on cereal rusts in India. Part 1. Scientific monograph No. 15. Imperial Council of Agricultural Research, New Delhi, p 244

Mehta KC (1952) Further studies on cereal rusts in India. Part II. Scientific monograph No. 18. Indian Council of Agricultural Research, New Delhi, p 368

Mettin D, Bluthner WD, Schlegel R (1973) Additional evidence on spontaneous 1B/1R, wheat Rye substitutions and translocations. In: Sears ER, Sears LMS (eds) Proceedings of the 4th international wheat genetics symposium. Agricultural Experimental Station, University of Missouri, Columbia, MO, pp 179–184

Mirza JI, Rattu A, Khanzada KA, Ahmad I, Fetch T et al (2010) Race analysis of stem rust isolates collected from Pakistan in 2008–09. In: McIntosh R, Pretorius Z (eds) Proceedings of BGRI 2010 technical workshop, St Petersburg, May 30–31, p 5

Nagarajan S (2012) Is *Puccinia graminis* f. sp. *tritici* virulence Ug99 a threat to wheat production in North Western Plain Zone of India. Indian Phytopathol 65:219–226

Nagarajan S, Joshi LM (1985) Epidemiology in Indian subcontinent. Roelfs AP, Bushnell WR (eds) Cereal rust, vol II. Academic, New York, NY, pp 371–402

Nagarajan S, Singh H (1975) Indian stem rust rules an epidemiological concept on the spread of wheat stem rust. Plant Dis Rep 59:133–136

Nagarajan S, Joshi LM, Saari EC (1976) Meteorological conditions associated with long distance dissemination and deposition of *Puccinia graminis* f. sp. *tritici* urediospores in India. Phytopathology 66:198–203

Nayar SK, Nagarajan S, Bahadur P, Singh SB (1987) Occurrence of new virulent pathogenic forms in race 77 of *Puccinia recondita* f. sp. *tritici* in India. Curr Sci 56:844–845

Reynolds MP, Borlaug NE (2006) Applying innovations and new technologies from international collaborative wheat improvement. J Agric Sci 144:95–110

Roelfs AP (1985) Wheat and rye stem rust. In: Roelfs AP, Bushnell WR (eds) The cereal rusts, vol 2. Academic, Orlando, FL, pp 3–37

Roelfs AP, Martens JW (1988) An international system of nomenclature for *Puccinia graminis* f. sp. *tritici*. Phytopathology 78:526–533

Rosegrant MW, Agcaoili Sombilla M, Perez ND (1995) Global food projections to 2020: implications for investment. Food, Agriculture and the Environment Discussion, Paper 5. International Food Policy Research Institute, Washington, DC, p 54

Roux L, Rijkenberg FHJ (1987) Pathotypes of *Puccinia graminis* f. sp. *tritici* with increased virulence for *Sr*24. Plant Dis 77:1115–1119

Shank R (1994) Wheat stem rust and drought effects on bale agricultural production and future prospects. In: Report on February 17–28 assessment. United Nations Emergencies Unit for Ethiopia, Addis Ababa. http://www.africa.upenn.edu/eue_web/Bale_mar.txt

Singh RP, McIntosh RA (1987) Genetics of resistance to *Puccinia graminis tritici* in 'Chris' and 'W3746' wheats. Theor Appl Genet 3:846–855

Singh RP, William HM, Huerta-Espino J, Rosewarne G (2004) Wheat rust in Asia: meeting the challenges with old and new technologies. In: New directions for a diverse planet: Proceedings of the 4th International Crop Science Congress, Brisbane, Australia, 26 Sep–1 Oct 2004. 2004 ISBN 1 920842 20 9. http://www.cropscience.org.au/icsc2004/symposia/3/7/141singhrp.htm. Accessed 1/4/07

Singh RP, Hodson DP, Yue J, Huerta-Espino J, Kinyua MG, Wanyera R, Njau P, Ward RW (2006) Current status, likely migration and strategies to mitigate the threat to wheat production from race Ug99 (TTKS) of stem rust pathogen. CAB Rev Perspect Agric Vet Sci Nutr Nat Resour 1:1–13

Singh RP, Hodson DP, Huerta-Espino J, Jin Y, Njau P (2008) Will stem rust destroy the world's wheat crop? Adv Agron 98:271–309

Singh RP, Hodson DP, Huerta-Espino J, Jin Y, Bhavani S, Njau P, Herrera-Foessel S, Singh PK, Singh S, Govindan V (2011) The emergence of Ug99 races of the stem rust fungus is a threat to world wheat production. Annu Rev Phytopathol 49:465–481

Stakman EC, Levine, MN (1922) The determination of biologic forms of *Puccinia graminis* on *Triticum* spp. Minnesota Agricultural Experimental Station Technical Bulletin, Minneapolis, MN, p 8

Stakman EC, Piemeisel FJ (1917) A new strain of *Puccinia graminis*. Phytopathology 7:73

Stakman E, Steward D, Loegering W (1962) Identification of physiologic races of *Puccinia graminis* var. *tritici*. US Department of Agriculture, ARS (E-617), Washington, DC

Steele KA, Humphreys E, Wellings CR, Dickinson MJ (2001) Support for a stepwise mutation model for pathogen evolution in Australasian *Puccinia striiformis* f. sp. *tritici* by use of molecular markers. Plant Pathol 50:174–180

Stokstady E (2007) Deadly wheat fungus threatens world's breadbaskets. Science 315:1786–1787

Tozetti GT, Alimurgia V (1952) True nature, causes and sad effects of the rusts, the bunts, the smuts, and other maladies of wheat and oats in the field. In: Tehon LR (ed) Translated phytopathological classics no. 9. American Phytopathology Society, St Paul, MN (originally published 1767), p 139

Urban Z (1967) The taxonomy of some European graminicolous rusts. Ceska Mykologia 21:12–16

Zeller EJ (1973) 1B/1R wheat rye chromosome substitutions and translocations. In: Sears ER, Sears LMS (eds) Proceedings of the 4th International Wheat Genetics Symposium. Agricultural Experiment Station, University of Missouri, Columbia, MO, pp 209–211

Chapter 9
Increased Virulence of Wheat Rusts and the Threat to Global Crop Production

Thomas Fetch and Brent McCallum

9.1 Characteristics of Wheat Rust Diseases

Cereal rust diseases are caused by fungi in the genus *Puccinia*, and are distinct from most other fungi in that they are biotrophic and require a living plant host for their growth and reproduction, although they have been grown in axenic culture (Williams et al. 1967). Rust diseases are extremely dangerous because they cycle very rapidly on the plant (reproduce from infection to new spores in 7–14 days) depending on temperature (Roelfs 1985b). They produce prodigious numbers of spores (over two trillion stem rust spores per hectare at moderate infection levels (Rowell and Roelfs 1971), and are commonly transported hundreds of kilometers by wind (Roelfs 1985a). Rust fungi are highly specialized on the hosts they attack (known as *formae speciales* or f. sp.), which was first described by Eriksson in 1894 (Anikster 1984). For example, wheat leaf rust cannot infect corn and corn rust cannot infect wheat, but some species such as *P. graminis* have many f. sp. that can infect many host genera (Cummins 1971). Cereal rust fungi are also specialized in their ability to infect different genotypes within a host species. This was initially described by Stakman and Piemeisel (1917) as strains and later became known as physiological races, which vary in their ability to attack specific resistance genes and can attack some wheat cultivars but not others (Stakman et al. 1962).

The variability among strains or races to attack specific resistance genes was summarized by Flor (1971) in his gene-for-gene hypothesis where "For each gene that conditions reaction in the host, there is a corresponding gene in the parasite that conditions pathogenicity." A strain or race that is highly virulent can attack (produces a susceptible reaction) many specific resistance genes, whereas an avirulent race attacks only a few genes. Rust races have the ability to change in virulence, by both sexual and asexual mechanisms. Sexual recombination of virulence in the cereal

T. Fetch, Ph.D. (✉) • B. McCallum, Ph.D.
Agriculture and Agri-Food Canada, 101 Route 100, Morden, MB, Canada R6M 1Y5
e-mail: Tom.Fetch@agr.gc.ca

A. Goyal and C. Manoharachary (eds.), *Future Challenges in Crop Protection Against Fungal Pathogens*, Fungal Biology, DOI 10.1007/978-1-4939-1188-2_9,

rusts involves alternate host species (*Barberry* and *Mahonia* spp. for stem and stripe rust, *Thalictrum* spp. for leaf rust) to complete the life cycle (Jackson and Mains 1921; Craigie 1927; Jin et al. 2010). The barberry alternate host is highly significant in generating new stem rust races (Roelfs 1982) and causing major epidemics, resulting in a major eradication campaign in the USA and Canada (Campbell and Long 2001). However, mutation is the primary source of new variation in virulence (Groth 1984) and can account for most or all changes in virulence that occur (Samborski 1985b). While in many countries there have not been significant outbreaks of rust for many years, recently there has been an emergence of new highly virulent races that threaten the worldwide production of wheat. Discussion of the threat of new stem, leaf, and stripe rust races to wheat production will be presented.

9.2 Wheat Stem Rust (*Puccinia graminis* f. sp. *tritici*)

Stem rust of wheat (*Triticum aestivum* and *T. durum*), caused by the fungus *Puccinia graminis* f. sp. *tritici*, has historically been the most damaging disease of wheat worldwide. Stem rust epidemics date back to biblical times where it caused "mildewing and blasting of wheat" (Arthur 1929). One of the first published descriptions of black rust was by Fontana (1767) which described a devastating epidemic in Tuscany, Italy in 1766. Another excellent treatise of this epidemic was written by Targioni-Tozzetti, who wrote that "rust is the most serious and formidable malady of wheat" and described the physiology of infection of the spores through the plant stomata (Peterson 2001). Many scientists in several countries have studied the taxonomy, life cycle, cytology, epidemiology, host–plant resistance, physiology, and virulence in the cereal rusts and numerous reference books have been published since the 1800s (Schafer et al. 1984). More is known about stem rust epidemiology, physiology, histology, and virulence than all other plant diseases (Roelfs 1985b), due to the significant damage to crops that it can cause.

Stem rust can cause complete crop loss within a few weeks in a seemingly healthy-looking crop (Fetch et al. 2011), due to tremendous water loss from large pustules that rip open the epidermis on the stems (Fig. 9.1). Losses to stem rust occur in all continents where wheat is grown, and several major epidemics have been documented in Australia and North America (Saari and Prescott 1985). Stem rust caused numerous losses in wheat in Canada and the USA in the first half of the twentieth century (Roelfs 1978), with the most significant epidemics occurring in 1916, 1935, and 1953–1954 (Stakman and Harrar 1957). The epidemics from 1953 to 1954 due to the new race 15B were the most damaging in the history of North America, causing estimated losses of 2.5 and 2.1 million tons in the USA (Stakman and Harrar 1957) and 1.7 and 5.5 million tons in Canada (Peturson 1958), respectively. In Australia, the most severe epidemic in history occurred in 1973, causing an estimated $200–300 million loss, which led to the establishment of the Australian Cereal Rust Control Program (Park 2007). In Ethiopia, a new race of stem rust caused between 65 and 100 % yield loss in the cultivar Enkoy from 1993 to 1994 (Dubin and Brennan 2009), and in 2013 near total losses were reported on 10,000 ha.

Fig. 9.1 Ug99 stem rust infection on wheat in Kenya

of wheat due to regional epidemics of race TKTT (D. Hodson, pers. comm.). In South America, stem rust caused a severe epidemic in 1950 by a variant of race 15, and widespread epidemics occurred during 1975–1976 (Germán et al. 2007). In India, an epidemic during 1946–1947 caused 20 % losses estimated at 2 million tons (Dubin and Brennan 2009).

Epidemics of cereal rusts occur when there is a combination of a widely grown susceptible host, virulent race(s), early infection of the crop, and favorable environment (Roelfs 1985a). An environment that favors rust infection and spread combines high humidity for spore germination and frequent winds that can rapidly disperse urediniospores over hundreds of kilometers. The movement of urediniospores on prevailing wind currents from maturing crops to green hosts is known as the "Puccinia path" (Stakman 1934), and has been described in Australia (Luig 1985), India (Nagarajan and Joshi 1985), and North America (Fetch et al. 2011). A "Puccinia pathway" can also extend across countries and even oceans, with published reports of stem rust migration from Africa to western Australia (Watson and de Sousa 1982). There also appears to exist a "Puccinia pathway" from west Africa to South America (Saari and Prescott 1985), and the frequent wind movement from South Africa to South America (Isard and Russo 2011) may vector the dangerous stem rust strain Ug99 into the western hemisphere. This has serious implications for the threat of stem rust to worldwide wheat production, as we shall see later.

Changes in virulence of *Puccinia graminis* can occur both by sexual and asexual mechanisms. The sexual mechanism requires a functional alternate host (*Berberis* and *Mahonia* spp.) to complete the life cycle (Craigie 1927). In Australia, native barberry does not exist and the sexual cycle is unimportant (Watson and Luig 1958). In contrast, common barberry was imported into North America by European colonists and contributed significantly in the development of new virulent races and early season infection of wheat, which led to reoccurring epidemics of stem rust in the late 1800s and early 1900s. This led to the development of barberry eradication programs in both Canada and the USA in 1916 (Campbell and Long 2001). Currently, the role of barberry in the rust life cycle has been almost entirely eliminated in North America (Roelfs 1982). While importation and movement of barberry is still regulated in Canada, the US government terminated their program in 1978 and turned over regulatory responsibilities to state agencies. It has been speculated that it is a matter of time before the sexual stage of *P. graminis* f. sp. *tritici* will revive and again produce new strains of wheat stem rust in North America (Leonard 2000). Eradication of ornamental barberry was critical because it resulted in: (1) elimination of initial spore inoculum coming from the barberry bushes that can directly infect emerging wheat crops; and (2) reduced the ability of the fungus to recombine into new virulent races. Species of barberry are present in Russia (Skolotneva et al. 2010), Central Asia, and Turkey (Mert et al. 2012) near wheat production areas, which may be a factor in developing new races of stem rust.

The other mechanism in the development of new stem rust races is by mutation in asexual populations. Mutation in asexual rust populations has traditionally been thought of as occurring is a stepwise fashion, by accumulation of one additional virulence factor at a time (Muller 1932; Roelfs and Groth 1980). However, in large asexual populations two mutations can occur more quickly than anticipated because of recurrent rather than unique events (Maynard 1968). The deployment of resistant wheat varieties in many countries has been a huge factor in reducing the overall amount of stem rust inoculum, thus limiting asexual population size and mutations in virulence. This can be seen in the overall reduction in variability in asexual populations of *P. graminis tritici* in North America since resistant cultivars began to be deployed in the 1950s, which have been documented in numerous virulence surveys.

Johnson and Newton (1946) stated that "new physiologic races with different pathogenic traits different from those of known races may arise at any time" and that "the price of security from rust damage is a continuous vigilance in the form of surveys to detect any pathogenic changes in the economically important rusts and unrelenting efforts to develop cereal varieties resistant to any virulent races that may arise." Annual surveys of virulence in *P. graminis tritici* were initiated in 1919 in Australia, Canada, and the USA (Kolmer 2000; Park 2007) and have continued without interruption to date. Surveys of virulence in *P. graminis tritici* have been conducted sporadically in South Africa (Pretorius et al. 2007), South America (Germán et al. 2007), and many other countries. In North America, the variability in *P. graminis tritici* has dropped from an average of about 20 races per year in the 1920s to few races in recent years (Roelfs and Groth 1980; Fetch et al. 2011), attributed primarily to the eradication of barberry (Roelfs 1982). In Australia, pathotype diversity increased from six races in 1919 to over 50 races in the 1950–1960s, but has declined both in inoculum level and diversity since the last major epidemic in 1973 (Park 2007). Changes in virulence in other countries are difficult to document since long-term surveys have not been conducted. The last international survey of virulence in *P. graminis tritici* was conducted in 18 countries from 1969 to 1971 (Luig 1983), but comparison of races was difficult because of differences in the nomenclature and rust differential lines that were used. New global rust surveillance protocols for stem rust have been proposed (Park et al. 2011) using 20 standard differential lines, the letter-code system of nomenclature (Fetch et al. 2009), and DNA profiling. This was an outcome of the identification of the highly virulent strain known as Ug99.

A novel stem rust race identified in 1999 and originating in Uganda, now known as Ug99 (Pretorius et al. 2000), is currently a major threat to wheat production worldwide (Singh et al. 2008). Ug99 is virulent on most deployed wheat stem rust resistance (*Sr*) genes, and an estimated 85–95 % of wheat lines worldwide are susceptible to Ug99 (Jin et al. 2007; Singh et al. 2011). With the exception of the 1993–1994 and 2013 epidemics in Ethiopia, stem rust has not been a disease problem for decades (Singh 2006). This resulted in a significant decline in stem rust research and surveillance, with a concomitant reduction in worldwide facilities and abilities to conduct race analysis (Park et al. 2011). However, the broad virulence spectrum of Ug99 has raised a worldwide alarm to this threat, resulting in the establishment of a global rust initiative (now Borlaug global rust initiative, or BGRI) to monitor the spread and detect changes in virulence of rusts that threaten wheat, including Ug99.

Currently, Ug99 stem rust is mutating and migrating. The original isolate of Ug99 from Uganda was identified as race TTKSK (Jin et al. 2007) on the international standard set of 20 wheat differential lines (Fetch et al. 2009). The exact origin of Ug99 is unknown, but may have evolved via step-wise mutation from an ancestral race that appears to have been present since the early 1970s (Park et al. 2011). Alternatively, new strains could have arisen on barberry that is present near Mt. Kenya (Jin unpublished data). Six additional variants of Ug99 (races TTKSF, TTKSP, TTKST, TTTSK, PTKSK, PTKST) are now known and their evolutionary relationships have been described (Park et al. 2011). Ug99 has spread from Uganda to Kenya (2001), Ethiopia (2003), Sudan and Yemen (2006), Iran (2007), Tanzania, Zimbabwe, South Africa (2009), and Eritrea (2010), primarily on regional winds (Hodson 2011; Singh et al. 2011; Wolday et al. 2011). Rust fungi have been documented to travel hundreds or even thousands of kilometers by wind (Brown and Hovmoller 2002). Stem rust has previously been reported to migrate from central Africa to Australia (Watson and de Sousa 1982), and coffee rust migrated from west Africa to Brazil in 1970, resulting in the removal of all coffee trees in a 50 × 800 km area in an effort to prevent establishment of the disease (Bowden et al. 1971). Since Ug99 stem rust is now prevalent across east Africa and into South Africa and Iran, spore movement by prevailing winds threatens wheat production in Australia, South America, and southern Asia (Pakistan and India). If Ug99 was vectored by wind to South America, it could cause significant epidemics as current wheat cultivars rely mainly on genes *Sr24* and *Sr31* (Germán et al. 2007), which are ineffective to race PTKST that is present in South Africa (Pretorius et al. 2010). Further movement of Ug99 from South America to North America could occur quickly as is exemplified by the movement of soybean rust from its introduction to Paraguay in 2001 to introduction into the southern USA in 2004 (Pan et al. 2006). Additionally, rust spores can be carried on clothing of tourists and establish exotic introductions of virulent races (see section on stripe rust). With the millions of tourists that travel to countries known to harbor Ug99 spores, introduction to other countries via this mechanism could happen at any time and is unpredictable.

In addition to Ug99, there are other dangerous races of stem rust that exist and could threaten wheat production if they migrated. A highly virulent race of stem rust detected in Pakistan from 2005 to 2009 was initially thought to be Ug99, but was subsequently pathotyped to race RRTTF (Iqbal et al. 2010). This race is also present in Ethiopia, Eritrea, and Yemen and is highly virulent to most Canadian wheat cultivars (Fetch et al. 2012). In Turkey, several races with high virulence (e.g., TKTTC, RTTTF) were identified in surveys conducted in 2007 and 2008 (Mert et al. 2012). Since the advent of the BGRI in 2005, there has been renewed interest in studying virulence and movement of cereal rusts worldwide. A new web-based system called "Rust SPORE" is now available (http://www.fao.org/agriculture/crops/rust/stem/rust-report/en/) that provides pathotype information and maps to locate races of stem rust worldwide, based on virulence data provided by international collaborators (Hodson 2011). This system in tandem with increased worldwide survey data of stem rust races would be highly useful in documenting new races with high virulence that threaten global wheat production.

9.3 Wheat Leaf Rust (*Puccinia triticina*)

Wheat leaf rust, caused by *Puccinia triticina* Eriks., is the most common and widely distributed of the wheat rusts (Huerta-Espino et al. 2011). Typical symptoms of leaf rust infection are small round orange pustules on plant leaves (Fig. 9.2). Although damage caused by leaf rust is not as dramatic as stem or stripe rust, the annual losses are higher because of the wide distribution and frequent presence of the disease. In order to identify effective leaf rust resistance (*Lr*) genes to develop resistant wheat cultivars, many countries initiated virulence surveys that are either conducted annually, at regular intervals, or sporadically. Virulence surveys around the world have revealed that *P. triticina* is a highly variable pathogen (Huerta-Espino et al. 2011). This variation can result from a number of different sources including introduction or movement of *P. triticina*, mutation, sexual recombination, and parasexual or somatic recombination followed by selection due to resistance in the host (Johnson 1961).

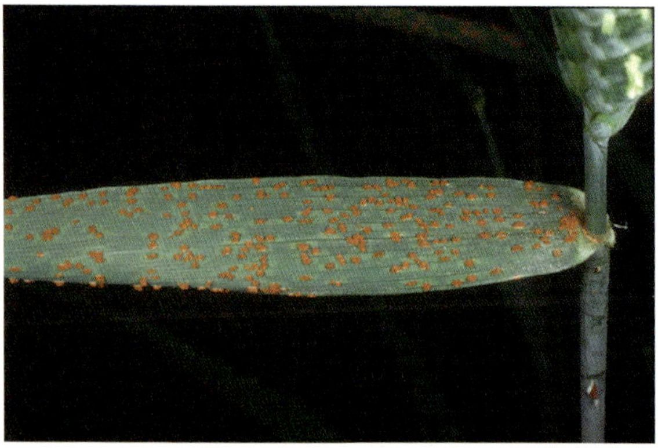

Fig. 9.2 Leaf rust infection on flag leaf of wheat

Sexual recombination is relatively rare for *P. triticina* in most parts of the world since the alternate or sexual host, *Thalictrum speciosissimum* L., is native to southern Europe and southwest Asia, but is absent from most wheat growing areas (Kolmer 2013). Therefore, *P. triticina* is confined mostly to asexual or clonal reproduction, with the relatively rare occurrence of parasexual recombination (Park and Wellings 2012). Where *P. triticina* populations have been analyzed in detail with virulence and molecular markers, they are usually described as a collection of clonal groups (Singh et al. 2004a; Ordonez and Kolmer 2009; Kolmer et al. 2011, 2012a). These groups are distinct from each other, and members within a group have more similarity to each other than to members of other groups (Wang et al. 2010; Kolmer 2013). However, there can be significant diversity within groups, due to mutation and selection. Saari and Prescott (1985) and Huerta Espino (1992) divided the world into 11 major epidemiological zones including South Africa, North Africa and Western Europe, West Asia (including the middle east), South Asia, South East Asia, Far East, North America, Mexico, South America, Australia and New Zealand, and Eastern Europe and Egypt. These regions were defined by the wind patterns which determine the spread of rust pathogens. Within each region, subdivision can occur due to the crop species and cultivars grown, or to physical barriers to movement such as mountain ranges.

Virulence changes in *P. triticina* populations are most evident in countries that have conducted annual surveys over a long period of time such as Australia, Canada, and the USA. One of the earliest studies on physiologic specialization in *P. triticina* was done in the USA by Mains and Jackson (1926), which described 12 different races of the fungus differentiated by their virulence on a set of 11 standard differential lines. Johnston et al. (1968) summarized annual race surveys conducted in the USA between 1926 and 1960 and found 100 different races described over that period using eight of the differentials described by Mains and Jackson. There were a small number of relatively common races that were found repeatedly over a number of years, but most races were relatively rare and were only found either once or a few times. They also noted that the various regions of the country differed in the prevalent races that were found. By 1961 there were 183 different races reported worldwide (Johnston 1961). Virulence analysis in the USA and other countries subsequently moved away from using differential cultivars, which often contained more than one resistance gene, to near-isogenic lines that only had a single *Lr* gene in a susceptible genetic background (Long et al. 1985). Changes in the pathogen population could now be tracked by changes in virulence frequency to each *Lr* gene as well as changes in frequency of races. An expanded set of 12 near-isogenic lines were used to distinguish races in North America after 1986, along with the development of a letter-code nomenclature system (Long and Kolmer 1989). This system is currently used in Canada and the USA, with the addition of one more set of four differential lines. Although hundreds of unique races have been identified in annual virulence surveys, they were found to form six distinct groups or race clusters (Ordonez and Kolmer 2009). While there is considerable variation within a group, the differences between races within the same group were smaller than those between groups. The groups tend to stay isolated because of asexual reproduction, and to some extent because of geographic isolation. Variation within each race group is thought to be primarily due to mutation and selection.

Virulence surveys of *P. triticina* were initiated in Canada in 1931 (Johnson 1956) and have been conducted annually since that time. In reviewing the virulence spectrum found between 1931 and 1955, Johnson (1956) reported 51 different races. Most were relatively rare and different races tended to predominate in eastern Canada compared to western Canada. From 1956 to 1987, there was selection for virulence in the Canadian *P. triticina* population over time on *Lr* genes used in North American cultivars, although virulence also existed for *Lr* genes that were not used in commercial wheat production (Kolmer 1989). From 1987 to 1997, the eastern and western populations in Canada were clearly distinct based on virulence differences. The effects of host selection were more apparent in western Canada because of the higher use of *Lr* genes within the wheat cultivars of the Great Plains (Kolmer 1999). The Canadian population from 1997 to 2007 resembled the population in the USA with six race clusters based on virulence phenotype, but was less clearly distinguished using molecular markers (Wang et al. 2010).

The 11 different global epidemiological zones tend to have leaf rust populations that are geographically isolated from each other (Huerta-Espino et al. 2011). However, an important component to changes in *P. triticina* populations is the movement of the pathogen between zones. *Puccinia triticina* co-evolved with wheat in Middle East and spread throughout the world along with wheat production. The introduction of both wheat and *P. triticina* into areas such as North and South America, Australia, and New Zealand has been relatively recent. The race groups in South America were found to closely resemble those in North America, because both populations were likely derived from European introductions (Ordonez et al. 2010).

In North America, the MBDS race cluster first appeared in 1996 and was rapidly selected due to virulence on *Lr17*, which was present in many of the USA wheat cultivars (Ordonez and Kolmer 2009). Race MBDS became prevalent in both the USA and Canada between 1996 and 2004. This race cluster was thought to be an exotic introduction (Kolmer 2001) because it is unique from the other previously existing race groups for a number of virulence and molecular markers (Ordonez and Kolmer 2009; Wang et al. 2010). This race group appeared in South America in 1999 (Ordonez et al. 2010), and it is speculated that this race group could have originated from Mexico and spread to both North and South America.

Virulence surveys for leaf rust have been conducted in Australia and New Zealand since the 1920s (Park et al. 1995). Exotic introductions have occurred several times. Race 53-1,(6),(7),10,11 appeared in New Zealand in 1981, and differed from previous races for many virulence genes and also had a unique isozyme allele pattern (Luig et al. 1985). In 1984, race 104-2,3,(6),(7),11 was detected, which differed from previous races in at least nine virulence genes and two isozyme loci (Park et al. 1995). The original race also diversified rapidly through mutation to virulence on additional resistance genes and selection to form a group of related races.

In many areas of the world durum wheat is highly resistant to the resident *P. triticina* populations. In Mexico, durum was grown for over 25 years without losses to leaf rust. However, in 2001 a new *P. triticina* race (BBG/BN) started to cause serious leaf rust epidemics (Singh et al. 2004a). This race was highly virulent on Mexican durum cultivars and also on those from many other countries, but was very avirulent on bread wheat. Increased virulence was also noticed in other durum growing countries. When *P. triticina* isolates found on durum from various regions around the world were compared for virulence they were very similar, although very different from *P. triticina* isolated from bread wheat in the same regions, suggesting a common origin (Ordonez and Kolmer 2007).

Changes in the virulence of *P. triticina* and other rust fungi are most commonly attributed to mutation that allows a formerly avirulent race to become virulent and infect specific wheat cultivars. Mutation to virulence may be a relatively rare event, but the effective enormous size of the pathogen population increases the occurrence (Samborski 1985a). New mutant virulent races increase rapidly in frequency because of host selection. The evolution from avirulence to virulence is often a two-step process because of the dikaryotic nature of *P. triticina*. Races that have mutated at one of the two loci controlling avirulence to a particular resistance gene often have an intermediate phenotype and a partial level of virulence (Samborski 1963; Kolmer and Dyck 1994) because of partial dominance. A number of genetic studies on *P. triticina* have found that avirulence loci are heterozygous, but if avirulence is fully dominant then the phenotypic change from avirulence to virulence is not observed until both loci have mutated (Samborski 1985a).

A series of mutational virulence changes from a single progenitor can lead to the development of race clusters typically found in many regions of the world. This process is illustrated by Park et al. (1995) who postulated seven different mutational changes for virulence to create seven new races from one originally introduced race. The appearance of a damaging *Lr24* virulent race in Australia in 2000 was attributed to mutation, since the new race was very similar to a previously existing race with the addition of *Lr24* virulence (Park et al. 2002). In North America, races virulent to *Lr21* recently appeared after many years of complete effectiveness of this gene (McCallum et al. 2011b; Kolmer et al. 2012b). These races most likely are mutants from preexisting races because of their close similarity apart from *Lr21* virulence. The evolution of *Lr21* virulence is a major problem for wheat breeders in the north central USA and Canadian prairies because of extensive use of this resistance gene in wheat cultivars. A number of other virulence changes in *P. triticina* populations have been attributed to this step-wise mutation process (Kolmer et al. 2009).

Parasexual or somatic hybridization is another mechanism of variation in *P. triticina* and other rust fungi (Park and Wellings 2012). The frequency of this mechanism of recombination in nature is unknown. However, a strong case for the occurrence of somatic hybridization involves the appearance of race 64-(6),(7),(10),11 in Australia in 1990 (Park and Wellings 2012). This race combined a number of different virulence features and isozyme alleles that were previously detected in two different races. DNA marker analysis strengthened the hypothesis that race 64-(6),(7),(10),11 was a somatic hybrid from the two previously existing races (Park and Wellings 2012). The actual mechanism of somatic hybridization and recombination in rust fungi is not fully understood, and attempts to replicate this process under controlled conditions have had mixed success (Park and Wellings 2012).

However, hyphae of different *P. triticina* races are capable of vegetative fusion, and nuclei from the paired strains come into very close association suggesting a nuclear fusion process (Wang and McCallum 2009). It is unknown how common somatic hybridization is in *P. triticina*, but in the absence of the alternate host it represents a powerful mechanism for rapid recombination of the genomes of reproductively isolated races.

Sexual recombination in *P. triticina* is only possible in those areas of the world with an alternate host, primarily *Thalictrum speciosissimum* L., although a small number of other species have been reported to function as alternate hosts (Samborski 1985a). *Thalictrum speciosissimum* is native to southern Europe and southwest Asia. However, when populations from Europe (Kolmer et al. 2012a) and Central Asia (Kolmer and Ordonez 2007) were analyzed for molecular markers and virulence, there was little evidence of sexual reproduction. Sexual reproduction in *P. triticina* populations has been shown to increase diversity, generate a more random distribution of virulence, and reduce linkage disequilibrium under controlled conditions (Kolmer 1992; Liu and Kolmer 1998). Sexual reproduction does not appear to be functional in most *P. triticina* populations analyzed to date because populations were characterized by high levels of linkage disequilibrium, heterozygosity, and clonally reproducing groups of related races.

Genetic changes in *P. triticina* populations around the world have resulted in virulence on wheat cultivars that were formerly resistant, leading to epidemic yield losses. The mechanisms for this genetic change include migration or introduction, mutation, parasexual recombination, sexual recombination, and selection. The ability for a resistance gene or a cultivar to remain resistant over a long period of time is termed durability. Most race-specific resistance genes and the cultivars that have only race-specific resistance genes have been overcome by genetic changes in the *P. triticina* population leading to susceptibility (Kolmer et al. 2009), and therefore have not been durable. However, the non-race-specific gene *Lr34* has been deployed globally for many years and has been durable in Canadian cultivars since the 1970s (McCallum et al. 2011a). The wheat cultivar Pasqua (Dyck 1993) has *Lr34* combined with four race-specific resistance genes and has been essentially immune to leaf rust in Canada since its release in 1991 (McCallum and Thomas 2011). While it is difficult to control or slow the rate of genetic change in *P. triticina*, effective resistance strategies based on non-race-specific genes like *Lr34*, *Lr46*, and *Lr67* (Hiebert et al. 2010) are the best defense against losses to wheat leaf rust.

9.4 Wheat Stripe Rust (*Puccinia striiformis* f. sp. *tritici*)

Stripe rust, caused by *Puccinia striiformis* Westend. f. sp. *tritici* Erikss., is currently the most damaging of the cereal rust pathogens worldwide (Singh 2004). Symptoms of stripe rust infection are long linear chains of pustules constricted by veins on leaf tissue (Fig. 9.3), but can also infect wheat heads. It was first described in 1777 and its center of origin is presumed to be in the Middle East (Stubbs 1985), but has

Fig. 9.3 Stripe rust infection on leaves of wheat

spread to all countries where wheat is grown. Although stripe rust has been said to lack the "killing power" of stem rust, numerous epidemics have occurred sporadically across the world and losses of up to 80 % have been reported (Wellings 2011). Stripe rust is the most important disease on wheat in China, where 80 % of the current varieties are susceptible to the predominant races, and the last epidemic in 2002 caused losses of 1.3 million metric tons (Wan et al. 2007). Severe epidemics of stripe rust have occurred in Australia since 2003 due to an exotic introduction of a new race, resulting in fungicide applications estimated between AUS$40–90 annually (Wellings 2007). Stripe rust is endemic in the Pacific Northwest region of the USA, with the most severe loss of 0.8 million metric tons occurring in 1976 (Chen 2007). Losses are often associated with new races that develop virulence for host resistance genes in cultivars that have recently been deployed.

Virulence variation in *P. striiformis* is accomplished by both sexual and asexual mechanisms. The alternate host for *P. striiformis* was previously unknown, but recently has been identified as species of *Berberis* (Jin et al. 2010), thus the life cycle is similar to *P. graminis*. Generation of new virulent races of stripe rust by sexual means is most likely occurring in areas where species of *Berberis* are abundant (Jin 2011), particularly in western China (Chen et al. 2009) and in the Pacific Northwest region of the USA. Prior to the discovery of the sexual cycle, changes in virulence were thought to be strictly by asexual means of mutation or somatic recombination (Stubbs 1985). The high mutation rate in stripe rust may be explained by exposure to UV irradiation during spore movement (Johnson et al. 1978), and has been estimated at three times higher compared to stem rust urediniospores (Maddison and Manners 1972). Stepwise mutation is the most common method of genetic variation that leads to development of new races (Wellings 2011). This was documented in Australia where 20 new pathotypes were derived from an exotic introduction (Wellings 2007), and in the USA where 62 new races have been identified since 2000 (Chen 2007).

Work on identifying races of stripe rust has been conducted worldwide since the discovery of strains by Allison and Isenbeck (1930). One significant problem in race identification is the lack of an agreed-upon system in naming races of *P. striiformis*. The first international nomenclature system was suggested by Johnson et al. (1972) and used a binary notation based on nine "International" differential lines. This has been modified by the use of a second set of eight lines denoted by the letter "E" for the "European" differentials, plus the reaction on cultivar "Avocet" in Australia (McIntosh et al. 1995). A similar system is in India, using seven of the "International" set of lines and the eight "European" lines, plus two supplemental lines (Prashar et al. 2007). In China, a numerical system with the prefix CYR (Chinese Yellow Rust) uses 17 differential lines to determine the race (Wan et al. 2004). The numerical system is also used in the USA, with a prefix of PST and 20 differential lines (Chen 2007). Comparison of virulence across the various nomenclature systems is difficult since many differential lines are wheat cultivars that contain more than one resistance gene (*Yr* gene). Recently, a number of near-isogenic lines each containing a single *Yr* gene in the "Avocet" background has been developed (Wellings 2009). The use of selected Yr genes in the Avocet NIL background and the letter-code nomenclature system similar to that used in stem and leaf rust would enable the direct comparison of stripe rust virulence on a global basis.

Regardless of the systems used to characterize the races of stripe rust, new races are migrating worldwide. Stripe rust is primarily dispersed and spread on prevailing winds like the other cereal rusts, but also have several documented exotic introductions via human-mediated travel. Stripe rust was introduced into Australia in 1979 on the clothing of a tourist that returned from Europe (O'Brien et al. 1980). Shortly thereafter, it spread to New Zealand on prevailing winds (Beresford 1982). A second exotic introduction of race 64E0A– occurred in 1999, followed by a third in 2002 of race 134E16A+ in western Australia (Wellings 2007). Race 134E16A+ is nearly identical to race PST-78, an exotic introduction into the USA in 2000, and indicates an association of international travel in the migration of dangerous races of stripe rust (Wellings 2011). Stripe rust also was introduced into South Africa in 1996 (Pretorius et al. 2007). Once new races are introduced or develop by mutation, they can be rapidly spread by wind. Virulence for gene *Yr9* occurred initially in East Africa and spread rapidly across the Middle East to Pakistan and India in the 1990s (Singh et al. 2004b). Races 134E16A+ and PST-78 spread rapidly across Australia and the USA, respectively, shortly after their introduction. Virulence to *Yr27*, a gene present in many wheat cultivars in the Central and Western Asia and Northern Africa (CWANA) regions, initially was detected in India in 2001 but has spread to several countries in the CWANA region resulting in severe epidemics in 2010 (Sharma-Poudyal et al. 2013).

Most epidemics of stripe rust occurring recently are due to the increased virulence seen globally in populations of *P. striiformis*. In China, races found recently have a wider virulence spectrum than those found earlier (Wan et al. 2007). Races collected in 1999 in Syria and Lebanon had greater virulence compared to those collected only 5 years earlier (Yahyaoui et al. 2002). New stripe rust races have been detected in India since 1995 and are evolving frequently (Prashar et al. 2007). Numerous pathotypes with increased virulence have been detected in Australia since the three exotic introductions beginning in 1979 (Wellings 2007). In the USA, virulence has increased since the exotic introduction of PST-78 in 2000 (Chen 2007). Additionally, new variants of PST-78 are more aggressive and can tolerate higher environmental temperatures (Milus et al. 2006). Wellings (2011) surveyed

25 scientists worldwide and found that regions in the USA, East Asia, South Asia, Australia, and Kenya were the most "at risk" for stripe rust epidemics based on incidence and severity scores. Clearly the continued evolution of new races and their subsequent migration makes stripe rust the most damaging of all the cereal rusts at this time on a global scale.

9.5 Conclusions

Wheat is the primary source of nutrition for about 85 % of the world's population, which is anticipated to grow to nine billion people by 2050 (Chaves et al. 2013). Thus, it is critical to maximize production of this crop. The most important constraints to wheat production are the rust diseases, which are very explosive and can cause severe epidemics with substantial losses in a short period of time. Rust diseases are controlled mainly by deployment of resistance genes, but can be overcome by new races with increased virulence that overcome host resistance. Although numerous wheat cultivars have been developed with good levels of rust resistance both before and subsequent to the Green Revolution, rusts are "shifty enemies" (Stakman 1947) that constantly change in virulence in response to deployment of new resistance genes. New races of stem, leaf, and stripe rust with increased virulence have been recently described worldwide and present a significant threat to global wheat production. Dr. Norman Borlaug said "Rust never sleeps," and it is critical that new rust scientists are trained and virulence surveys are conducted worldwide on an annual basis. As earlier stated by Johnson and Newton (1946), "the price of food security from rust damage is continued vigilance in the form of surveys to detect any pathogenic changes … and unrelenting efforts to develop cereal varieties resistant to virulent races as they arise." Thus, we must not become complacent after decades of research and control of rust diseases in many countries, because we may have won the battle but have not yet won the war.

References

Allison CC, Isenbeck K (1930) Biologische Spezialisierung von *Puccinia glumarum tritici*, Erikss. und Henn. Phytopathol Z (n.d.);2:87–98

Anikster Y (1984) The formae speciales. In: Bushnell WR, Roelfs AP (eds) The cereal rusts. Academic, Orlando, pp 115–130

Arthur JC (1929) The plant rusts (Uredinales). Wiley, New York

Beresford RM (1982) Stripe rust (Puccinia striiformis), a new disease of wheat in New Zealand. Cer Rust Bull 10:35

Bowden J, Gregory PH, Johnson CG (1971) Possible wind transport of coffee leaf rust across the Atlantic Ocean. Nature 229(5285):500–501. doi:10.1038/229500b0

Brown JKM, Hovmoller MS (2002) Aerial dispersal of pathogens on the global and continental scales and its impact on plant disease. Science 297:537–541

Campbell CL, Long DL (2001) The campaign to eradicate the common *Barberry* in the United States. In: Peterson PD (ed) Stem rust of wheat: from ancient enemy to modern foe. APS Press, St. Paul, pp 16–50
Chaves M, Martinelli J, Wesp-Guterres C, Graichen F, Brammer S, Scagliusi S et al (2013) The importance for food security of maintaining rust resistance in wheat. Food Secur 5(2): 157–176
Chen XM (2007) Challenges and solutions for stripe rust control in the United States. Aust J Agr Res 58(6):648–655
Chen WQ, Wu LR, Liu TG, Xu SC, Jin SL, Peng YL et al (2009) Race dynamics, diversity, and virulence evolution in Puccinia striiformis f. sp. tritici, the causal agent of wheat stripe rust in China from 2003 to 2007. Plant Dis 93(11):1093–1101
Craigie JH (1927) Discovery of the function of the pycnia of the rust fungi. Nature 120:1–7
Cummins GB (ed) (1971) The rust fungi of cereals, grasses, and bamboos. Springer, Berlin
Dubin HJ, Brennan JP (2009) Combating stem and leaf rust of wheat. IFPRI Discussion paper 910 [serial on the Internet]. http://www.ifpri.org/publication/combating-stem-and-leaf-rust-wheat
Dyck PL (1993) The inheritance of leaf rust resistance in the wheat cultivar Pasqua. Can J Plant Sci 73:903–906
Fetch T, Jin Y, Nazari K, Park R, Prashar M, Pretorius Z (2009) Race nomenclature systems: can we speak the same language? In: McIntosh R (ed) Proc. BGRI 2009 Technical Workshop, Obregon, Mexico, pp 61–64. http://www.globalrust.org/sites/default/files/2009%20BGRI%20 Plenary%20papers.pdf
Fetch T, McCallum B, Menzies J, Rashid K, Tenuta A (2011) Rust diseases in Canada. Prairies Soils Crops J [serial on the Internet] 4:86–96. http://www.prairiesoilsandcrops.ca/
Fetch T, Zegeye T, Singh D, Wanyera R, Penner M, Rashid K (2012) Virulence of Ug99 (race TTKSK) and race RRTTF on Canadian wheat cultivars. Can J Plant Sci 92(3):602
Flor HH (1971) Current status of the gene-for-gene concept. Annu Rev Phytopathol 9:275–296
Fontana F (1767) Observations on the rust of grain. In: Phytopathological classics. American Phytopathological Society, Washington, DC, pp 1–38
Germán S, Barcellos A, Chaves M, Kohli M, Campos P, De Viedma L (2007) The situation of common wheat rusts in the Southern Cone of America and perspectives for control. Aust J Agr Res 58(6):620–630
Groth JV (1984) Virulence frequency dynamics of cereal rust fungi. In: Bushnell WR, Roelfs AP (eds) The cereal rusts. Academic, Orlando, pp 231–252
Hiebert CW, Thomas JB, McCallum BD, Humphreys DG, DePauw RM, Hayden MJ et al (2010) An introgression on wheat chromosome 4DL in RL6077 (Thatcher*6/PI 250413) confers adult plant resistance to stripe rust and leaf rust (*Lr67*). Theor Appl Genet 121:1083–1091
Hodson DP (2011) Shifting boundaries: challenges for rust monitoring. Euphytica 179(1): 93–104
Huerta Espino J (1992) Analysis of wheat leaf and stem rust virulence on a worldwide basis [PhD thesis]. University of Minnesota, St. Paul
Huerta-Espino J, Singh RP, Germán S, McCallum BD, Park RF, Chen WQ et al (2011) Global status of wheat leaf rust caused by *Puccinia triticina*. Euphytica 179(1):143–160
Iqbal MJ, Ahmad I, Khanzada KA, Ahmad N, Rattu AUR, Fayyaz M et al (2010) Local stem rust virulence in Pakistan and future breeding strategy. Pakistan J Bot 42(3):1999–2009
Isard SA, Russo JM (2011) Risk assessment of aerial transport of rust pathogens to the western hemisphere and within North America. In: McIntosh R (ed) Proc BGRI 2011 Tech Workshop. St. Paul, MN, pp 25–34. http://www.globalrust.org/db/attachments/knowledge/134/2/2011%20 BGRI%20proceedings%20PAPERS-pp1-53.pdf
Jackson HS, Mains EB (1921) Aecial stage of the orange leaf rust of wheat, Puccinia triticina Erikss. J Agric Res 22:151–172
Jin Y (2011) Role of Berberis spp. as alternate hosts in generating new races of Puccinia graminis and P. striiformis. Euphytica 179(1):105–108
Jin Y, Singh RP, Ward RW, Wanyera R, Kinyua M, Njau P et al (2007) Characterization of seedling infection types and adult plant infection responses of monogenic Sr gene lines to race TTKS of Puccinia graminis f. sp. tritici. Plant Dis 91(9):1096–1099

Jin Y, Szabo LJ, Carson M (2010) Century-old mystery of *Puccinia striiformis* life history solved with the identification of berberis as an alternate host. Phytopathology 100(5):432–435

Johnson T (1956) Physiologic races of leaf rust of wheat in Canada 1931 to 1955. Can J Agric Sci 36:371–379

Johnson T (1961) Man-guided evolution in plant rusts. Science 133:357–362

Johnson T, Newton M (1946) Specialization, hybridization, and mutation in the cereal rusts. Bot Rev 12:337–392

Johnson R, Stubbs RW, Fuchs E, Chamberlain NH (1972) Nomenclature for physiologic races of *Puccinia striiformis* infecting wheat. Trans Br Mycol Soc 58:475–480

Johnson R, Priestley RH, Taylor EC (1978) Occurrence of virulence in *Puccinia striiformis* for Compair wheat in England. Cer Rusts Bull 6:11–13

Johnston CO (1961) Sixth revision of the international register of physiologic races of *Puccinia recondita* rob. ex desm. (formerly *P. rubigo-vera tritici*). U.S. Department of Agriculture, Washington, DC, pp 1–15

Johnston CO, Caldwell RM, Compton LE, Browder LE (1968) Physiologic races of *Puccinia recondita* f. sp. *tritici* in the United States from 1926 through 1960. US Dept Agric Tech Bull 1393:1–22

Kolmer JA (1989) Virulence and race dynamics of *Puccinia recondita* f. sp. *tritici* in Canada during 1956-1987. Phytopathology 79:349–356

Kolmer JA (1992) Effect of sexual recombination in two populations of the wheat leaf rust fungus, *Puccinia recondita*. Can J Bot 70:359–363

Kolmer JA (1999) Virulence dynamics, phenotypic diversity, and virulence complexity in two populations of *Puccinia triticina* in Canada from 1987 to 1997. Can J Bot 77:333–338

Kolmer J (2000) Early research on the genetics of *Puccinia graminis* and stem rust resistance in wheat in Canada and the United States. In: Peterson PD (ed) Stem rust of wheat: from ancient enemy to modern foe. APS, St. Paul, pp 51–82

Kolmer JA (2001) Molecular polymorphism and virulence phenotypes of the wheat leaf rust fungus *Puccinia triticina* in Canada. Can J Bot 79(8):917–926

Kolmer J (2013) Leaf rust of wheat: pathogen biology, variation and host resistance. Forests 4:70–84

Kolmer JA, Dyck PL (1994) Gene expression in the *Triticum aestivum-Puccinia recondita* f. sp. *tritici* gene-for-gene system. Phytopathology 84:437–440

Kolmer JA, Ordonez ME (2007) Genetic differentiation of *Puccinia triticina* populations in central Asia and the Caucasus. Phytopathology 97:1141–1149

Kolmer J, Chen X, Jin Y (2009) Diseases which challenge global wheat production—the wheat rusts. In: Carver BF (ed) Wheat science and trade. Wiley-Blackwell, Ames, pp 89–124

Kolmer JA, Ordonez ME, Manisterski J, Anikster Y (2011) Genetic differentiation of *Puccinia triticina* populations in the Middle East and genetic similarity with populations in Central Asia. Phytopathology 101(7):870–877

Kolmer JA, Hanzalova A, Goyeau H, Bayles R, Morgounov A (2012a) Genetic differentiation of the wheat leaf rust fungus Puccinia triticina in Europe. Plant Pathol 62:21–31

Kolmer JA, Long DL, Hughes ME (2012b) Physiologic specialization of *Puccinia triticina* on wheat in the United States in 2010. Plant Dis 96(8):1216–1221

Leonard KJ (2000) Stem rust- Future enemy? In: Peterson PD (ed) Stem rust of wheat: from ancient enemy to modern foe. APS, St. Paul, pp 119–142

Liu JQ, Kolmer JA (1998) Molecular and virulence diversity and linkage disequilibria in asexual and sexual populations of the wheat leaf rust fungus, Puccinia recondita. Genome 41: 832–840

Long DL, Kolmer JA (1989) A North American system of nomenclature for *Puccinia recondita* f. sp *tritici*. Phytopathology 79:525–529

Long DL, Schafer JF, Roelfs AP (1985) Specific virulence of *Puccinia recondita* f. sp. *tritici* in the United States from 1978 through 1983. Plant Dis 69:343–347

Luig NH (1983) editor. A survey of virulence genes in wheat stem rust, *Puccinia graminis* f. sp. *tritici*. Adv. Plant Breed, Verlag Paul Parey, Berlin Vol. 11

Luig NH (1985) Epidemiology in Australia and New Zealand. In: Roelfs AP, Bushnell WR (eds) The cereal rusts. Academic, Orlando, pp 301–328
Luig NH, Burdon JJ, Hawthorn WM (1985) An exotic strain of Puccinia recondita tritici in New Zealand. Can J Plant Pathol 7(2):173–176
Maddison AC, Manners JG (1972) Sunlight and the viability of cereal rust uredospores. Trans Br Mycol Soc 59:429–443
Mains EB, Jackson HS (1926) Physiologic specialization in the leaf rust of wheat; *Puccinia triticina* Erikss. Phytopathology 16:89–120
Maynard SJ (1968) Evolution in asexual and sexual populations. Am Nat 102:469–473
McCallum BD, Thomas J (2011) Effectiveness of wheat leaf rust gene combinations derived from the cultivar Pasqua. In: Plant and animal genome conference XIX, San Diego, Jan 15–19
McCallum BD, Humphreys DG, Somers DJ, Dakouri A, Cloutier S (2011a) Allelic variation for the rust resistance gene *Lr34/Yr18* in Canadian wheat cultivars. Euphytica 183(2):261–274
McCallum BD, Seto-Goh P, Xue A (2011b) The detection of virulence to *Lr21* in Canada during 2011. Can J Plant Sci 92:605
McIntosh RA, Wellings CR, Park RF (1995) Wheat rusts—an atlas of resistance genes. In: McIntosh RA (ed). CSIRO, East Melbourne
Mert Z, Karakaya A, Düşünceli F, Akan K, Çetin L (2012) Determination of *Puccinia graminis* f. sp. *tritici* races of wheat in Turkey. Turk J Agric For 36(1):107–120
Milus EA, Seyran E, McNew R (2006) Aggressiveness of Puccinia striiformis f. sp. tritici isolates in the south-central United States. Plant Dis 90(7):847–852
Muller HJ (1932) Some genetic aspects of sex. Am Nat 8:118–138
Nagarajan S, Joshi LM (1985) Epidemiology in the Indian subcontinent. In: Roelfs AP, Bushnell WR (eds) The cereal rusts. Academic, Orlando, pp 371–402
O'Brien L, Brown JS, Young RM, Pascoe T (1980) Occurrence and distribution of wheat stripe rust in Victoria and susceptibility of commercial wheat cultivars. Australas Plant Pathol 9:14
Ordonez ME, Kolmer JA (2007) Virulence phenotypes of a worldwide collection of *Puccinia triticina* from durum wheat. Phytopathology 97:344–351
Ordonez ME, Kolmer JA (2009) Differentiation of molecular genotypes and virulence phenotypes of *Puccinia triticina* from common wheat in North America. Phytopathology 99:750–758
Ordonez ME, Germán SE, Kolmer JA (2010) Genetic differentiation within the Puccinia triticina population in South America and comparison with the North American population suggests common ancestry and intercontinental migration. Phytopathology 100(4):376–383
Pan Z, Yang XB, Pivonia S, Xue L, Pasken R, Roads J (2006) Long-term prediction of soybean rust entry into the continental United States. Plant Dis 90(7):840–846
Park RF (2007) Stem rust of wheat in Australia. Aust J Agr Res 58(6):558–566
Park RF, Wellings CR (2012) Somatic hybridization in the Uredinales. Annu Rev Phytopathol 50(1):219–239
Park RF, Burdon JJ, McIntosh RA (1995) Studies on the origin, spread, and evolution of an important group of *Puccinia recondita* f. sp. *tritici* pathotypes in Australasia. Eur J Plant Pathol 101(6): 613–622
Park RF, Bariana HS, Wellings CR, Wallwork H (2002) Detection and occurence of a new pathotype of *Puccinia triticina* with virulence for *Lr24* in Australia. Aust J Agr Res 53:1069–1076
Park R, Fetch T, Hodson D, Jin Y, Nazari K, Prashar M et al (2011) International surveillance of wheat rust pathogens: progress and challenges. Euphytica 179(1):109–117
Peterson PD (2001) Stem rust of wheat: exploring the concepts. In: Peterson PD (ed) Stem rust of wheat: from ancient enemy to modern foe. APS, St. Paul, pp 1–16
Peturson B (1958) Wheat rust epidemics in Western Canada in 1953, 1954, and 1955. Can J Plant Sci 38:16–28
Prashar M, Bhardwaj SC, Jain SK, Datta D (2007) Pathotypic evolution in *Puccinia striiformis* in India during 1995–2004. Aust J Agr Res 58(6):602–604
Pretorius ZA, Singh RP, Wagoire WW, Payne TS (2000) Detection of virulence to wheat stem rust resistance gene Sr31 in *Puccinia graminis* f. sp. *tritici* in Uganda. Plant Dis 84(2):203

Pretorius ZA, Pakendorf KW, Marais GF, Prins R, Komen JS (2007) Challenges for sustainable cereal rust control in South Africa. Aust J Agr Res 58(6):593–601

Pretorius ZA, Bender CM, Visser B, Terefe T (2010) First report of a Puccinia graminis f. sp. tritici race virulent to the Sr24 and Sr31 wheat stem rust resistance genes in South Africa. Plant Dis 94(6):784

Roelfs AP (1978) Estimated losses caused by rust in small grain cereals in the United States—1918 to 1976. USDA Misc Publ No 1363. U.S. Department of Agriculture, Washington, DC

Roelfs AP (1982) Effects of barberry eradication on stem rust in the United States. Plant Dis 72:177–181

Roelfs AP (1985a) Epidemiology in North America. In: Roelfs AP, Bushnell WR (eds) The cereal rusts. Academic, Orlando, pp 403–434

Roelfs AP (1985b) Wheat and rye stem rust. In: Roelfs AP, Bushnell WR (eds) The cereal rusts. Academic, Orlando, pp 3–37

Roelfs AP, Groth JV (1980) A comparison of virulence phenotypes in wheat stem rust *Puccinia graminis* f. sp. *tritici* populations reproducing sexually and asexually. Phytopathology 70:855–862

Rowell JB, Roelfs AP (1971) Evidence for an unrecognized source of overwintering wheat stem rust in the United States. Plant Dis Rep 55:990–992

Saari EE, Prescott JM (1985) World distribution in relation to economic losses. In: Roelfs AP, Bushnell WR (eds) The cereal rusts, vol 2. Academic, Orlando, pp 259–298

Samborski DJ (1963) A mutation in *Puccinia recondita* Rob. ex. Desm. f. sp. *tritici* to virulence on transfer, Chinese spring×*Aegilops umbrellulata* Zhuk. Can J Bot 41:475–479

Samborski DJ (1985) Wheat leaf rust. In: Roelfs AP, Bushnell WR (eds) The cereal rusts, vol II, Diseases, distribution, epidemiology, and control. Academic, Orlando, pp 39–59

Schafer JF, Roelfs AP, Bushnell WR (1984) Contributions of early scientists to knowledge of cereal rusts. In: Bushnell WR, Roelfs AP (eds) The cereal rusts: origins, specificity, structure, and physiology. Academic, Orlando, pp 3–38

Sharma-Poudyal D, Chen XM, Wan AM, Zhan GM, Kang ZS, Cao SQ et al (2013) Virulence characterization of international collections of the wheat stripe rust pathogen, Puccinia striiformis f. sp. tritici. Plant Dis 97(3):379–386

Singh RP (ed) (2004) The cost to agriculture of recent changes in cereal rusts. In: Proc 11th Intl cereal rusts and powdery mildews conf, 22–27 Aug, John Innes Centre, Norwich

Singh RP (2006) Current status, likely migration and strategies to mitigate the threat to wheat production from race Ug99 (TTKS) of stem rust pathogen. CAB Rev 1:54

Singh RP, Huerta-Espino J, Pfeiffer W, Figueroa-Lopez P (2004a) Occurrence and impact of a new leaf rust race on durum wheat in Northwestern Mexico from 2001 to 2003. Plant Dis 88(7):703–708

Singh RP, William HM, Huerta-Espino J, Rosewarne GM (eds) (2004b) Wheat rust in Asia: meeting the challenges with old and new technologies. 4th Intl Crop Sci Congr, 26 September—1 October, Brisbane, Australia. http://www.cropscience.org.au/icsc2004/symposia/3/7/141_singhrp.htm

Singh RP, Hodson DP, Huerta-Espino J, Jin Y, Njau P, Wanyera R et al (2008) Will stem rust destroy the world's wheat crop? In: Sparks D (ed) Advances in agronomy. Academic, Burlington, pp 271–309

Singh RP, Hodson DP, Huerta-Espino J, Jin Y, Bhavani S, Njau P et al (2011) The emergence of Ug99 races of the stem rust fungus is a threat to world wheat production. Annu Rev of Phytopathol 49:465–481

Skolotneva ES, Volkova VT, Zaitseva LG, Lekomtseva SN (2010) The virulence of wheat stem rust pathogen in the central region of Russia. Mikol Fitopatol 44(4):367–372

Stakman EC (1934). Epidemiology of cereal rusts. In: Proc 5th Pac Sci Congr, vol 4, pp 3177–3184

Stakman EC (1947) Plant diseases are shifty enemies. Am Sci 35:321–350

Stakman EC, Harrar JG (1957) Principles of plant pathology. Ronald, New York

Stakman EC, Piemeisel FJ (1917) Biologic forms of *Puccinia graminis* on cereals and grasses. J Agric Res (Washington, DC) 10:429–495
Stakman EC, Stewart DM, Loegering WQ (1962) Identification of physiologic races of *Puccinia graminis* var. *tritici*. US Dept Agric Res Serv Bull E617 (revised 1962), pp 1–53
Stubbs RW (1985) Stripe rust. In: Roelfs AP, Bushnell WR (eds) The cereal rusts. Academic, Orlando, pp 61–101
Wan A, Zhao Z, Chen X, He Z, Jin S, Jia Q et al (2004) Wheat stripe rust epidemic and virulence of Puccinia striiformis f. sp. tritici in China in 2002. Plant Dis 88(8):896–904
Wan AM, Chen XM, He ZH (2007) Wheat stripe rust in China. Aust J Agr Res 58(6):605–619
Wang X, McCallum B (2009) Fusion body formation, germ tube anastomosis, and nuclear migration during the germination of urediniospores of the wheat leaf rust fungus *Puccinia triticina*. Phytopathology 99(12):1355–1364
Wang X, Bakkeren G, McCallum B (2010) Virulence and molecular polymorphisms of the wheat leaf rust fungus *Puccinia triticina* in Canada from 1997 to 2007. Botany 88:575–589
Watson IA, de Sousa CNA (1982) Long distance transport of spores of *Puccinia graminis tritici* in the southern hemisphere. Proc Linn Soc NSW 106:311–321
Watson IA, Luig NH (1958) Widespread natural infection of barberry by *Puccinia graminis* in Tasmania. Proc Linn Soc NSW 83:181–186
Wellings CR (2007) *Puccinia striiformis* in Australia: a review of the incursion, evolution, and adaptation of stripe rust in the period 1979-2006. Aust J Agr Res 58(6):567–575
Wellings CR (2009) The development and application of near—isogenic lines for monitoring cereal rust pathogens. Proc Borlaug Global Rust Initiative Technical Workshop, BGRI Cd Obregon, Mexico, p 77
Wellings CR (2011) Global status of stripe rust: a review of historical and current threats. Euphytica 179(1):129–141
Williams PG, Scott KJ, Kuhl JL, Maclean DJ (1967) Sporulation and pathogenicity of *Puccinia graminis* f. sp. *tritici* grown on an artificial medium. Phytopathology 57:326–327
Wolday A, Fetch T, Hodson DP, Cao W, Briere S (2011) First report of Puccinia graminis f. sp. tritici races with virulence to wheat stem rust resistance genes Sr31 and Sr24 in Eritrea. Plant Dis 95(12):1591
Yahyaoui AH, Hakim MS, El Naimi M, Rbeiz N (2002) Evolution of physiologic races and virulence of *Puccinia striiformis* on wheat in Syria and Lebanon. Plant Dis 86:499–504

Chapter 10
Fusarium Diseases of Canadian Grain Crops: Impact and Disease Management Strategies

Nora A. Foroud, Syama Chatterton, Lana M. Reid, T. Kelly Turkington, Sheryl A. Tittlemier, and Tom Gräfenhan

10.1 *Fusarium* Species Classification and Genetics

Species of the genus *Fusarium* are ascomycetes that are characterized by their typical conidia, which are often fusiform to sickle-shaped with an elongated apical cell and pedicellate basal (foot) cell. Several important *Fusarium* species, including *F. avenaceum*, *F. graminearum* and *F. pseudograminearum*, are known to produce a teleomorph state that was formerly classified in the genus *Gibberella*. In other species, such as *F. culmorum*, *F. oxysporum* and *F. sporotrichioides*, no teleomorph has been reported, so far. Geiser et al. (2013) proposed to recognize the genus *Fusarium* as the sole name for a group that includes virtually all important saprophytic, plant pathogenic, and mycotoxigenic species. *Fusarium* spp. can be identified by morphological features and also by genetic analysis. Morphological species identification can be based on microscopic and/or macroscopic characters, such as conidia, phialides, chlamydospores, ascospores and colony characteristics of pure cultures (see Figs. 10.1 and 10.2). Macroconidia, a form of asexual spores most often aggregated to sporodochia, are usually 'banana'-shaped in *Fusarium* species, which can also produce microconidia in the aerial mycelium and/or chlamydospores in hyphae. The size, shape and number of septa in the macroconidia are often used to differentiate

N.A. Foroud, Ph.D. (✉) • S. Chatterton, Ph.D.
Lethbridge Research Centre, Agriculture and Agri-Food Canada,
5403 1st Ave. S, Lethbridge, AB, Canada T1J 4B1
e-mail: nora.foroud@agr.gc.ca

L.M. Reid, Ph.D.
Eastern Cereal and Oilseed Research Centre, Agriculture and Agri-Food Canada,
Central Experimental Farm, Ottawa, ON, Canada

T.K. Turkington, Ph.D.
Lacombe Research Centre, Agriculture and Agri-Food Canada, Lacombe, AB, Canada

S.A. Tittlemier, Ph.D. • T. Gräfenhan, Ph.D.
Grain Research Laboratory, Canadian Grain Commission, Winnipeg, MB, Canada

A. Goyal and C. Manoharachary (eds.), *Future Challenges in Crop Protection Against Fungal Pathogens*, Fungal Biology, DOI 10.1007/978-1-4939-1188-2_10,

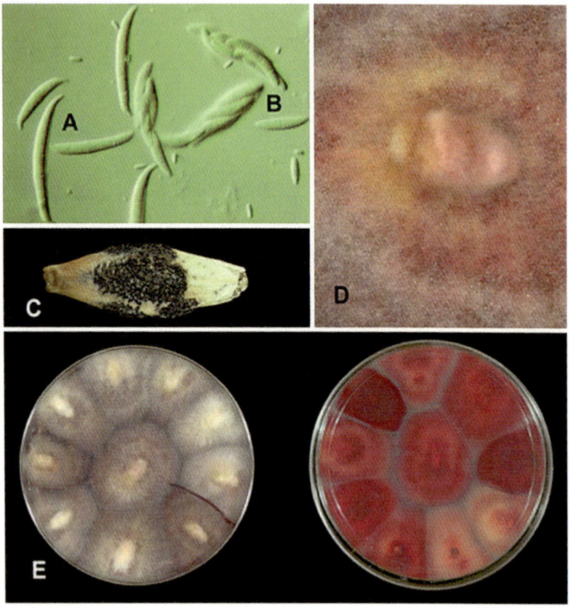

Fig. 10.1 *Fusarium graminearum*: (**a**) Conidia of a culture grown on potato dextrose agar (PDA); (**b**) asci and ascospores of the teleomorph; (**c**) purplish-black perithecia (teleomorph) on barley seed; (**d**) close-up of typically red mycelia with a yellow tint; (**e**) colonies growing from wheat seed as seen from above (*left*) and below (*right*)

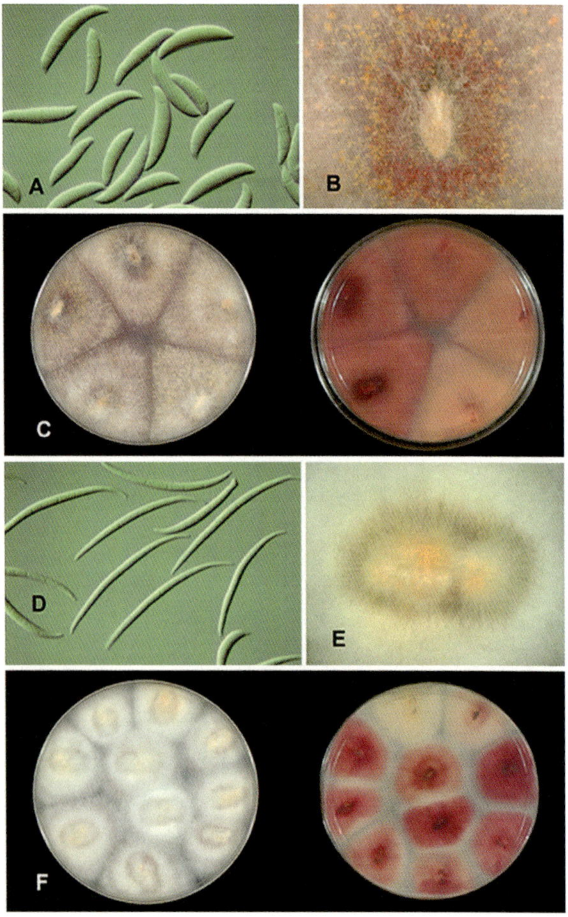

Fig. 10.2 *Fusarium culmorum*: (**a**) short and broad conidia of a culture grown on PDA; (**b**) close-up of loose mycelium with abundant orange and red sporodochia; (**c**) fast growing colonies from wheat seed as seen from above (*left*) and below (*right*). *Fusarium avenaceum*: (**d**) long and slender conidia of a culture grown on PDA; (**e**) close-up of dense white mycelium predominantly without sporodochia; (**f**) colonies often with white margin growing from durum seed as seen from above (*left*) and below (*right*)

between species (Leslie and Summerell 2006). The teleomorph state is characterized by black, flask-shaped perithecia with a single ostiole (narrow opening) at the top, from which the asci (containing the ascospores) are released. The asci of *F. graminearum* (formerly *Gibberella zeae*) have been shown to be forcibly ejected through the ostiole under conditions of high humidity (Trail et al. 2002).

Genetic analysis is a more accurate approach for species identification and is enhanced by an increase in available *Fusarium* genome sequences. The genomes of several important *Fusarium* species such as *F. graminearum*, *F. oxysporum*, *F. pseudograminearum*, *F. solani* and *F. verticillioides* are publicly available through various data portals (e.g. Broad Institute of Harvard and MIT; DOE Joint Genome Institute; NCBI GenBank). The *F. graminearum* strain Ph1 was the first complete *Fusarium* genome to be published and annotated (Cuomo et al. 2007). The *F. graminearum* genome is 36 megabases (Mb) with over 13,000 genes. A comparison of the Ph1 strain with the partially completed genome of *F. graminearum* strain GZ3639 revealed over 10,000 single-nucleotide polymorphisms found along all four chromosomes (Cuomo et al. 2007). The complete genome sequence and assembly of *F. graminearum* GZ3639 and seven other *F. graminearum* strains were recently prepared by R. Subramaniam and colleagues (personal communications). *F. graminearum* has a small number of chromosomes compared to other *Fusarium* spp., and this is believed to be a result of ancestral chromosome fusion (Cuomo et al. 2007). *F. verticillioides* strain 7600 and *F. oxysporum* f. sp. *lycopersici* strain 4287 were sequenced and compared to *F. graminearum* Ph1 by Ma et al. (2010). *F. verticillioides* is 42 Mb with over 14,000 genes found on 11 chromosomes. *F. oxysporum* f. sp. *lycopersici* genome is 60 Mb, the largest of the three genomes, with over 17,000 genes encoded on a total of 15 chromosomes. Four of the 15 chromosomes are lineage specific and are composed primarily of transposable elements. Ma et al. (2010) also observed that the lineage specific regions differ in sequence among different formae speciales of *F. oxysporum*, and some of these regions encode virulence factors specific to the host range of a given strain. They were able to demonstrate that chromosome 14, which encodes host-specific virulence factors, is able to undergo horizontal transfer between different *F. oxysporum* f. sp., and that this transfer leads to a change in host-specificity. *F. solani* (*N. haematococca* MPVI isolate 77-13-4) has also been sequenced and has 17 chromosomes with a genome size of 54.43 Mb. Of the 17 chromosomes, three (14, 15 and 17) are non-essential, and at least one of these (chromosome 14) is involved in host-specificity (Coleman et al. 2009).

A subset of *Fusarium* spp. produces a series of mycotoxins, including trichothecenes and fumonisins. Both of these classes of toxins are secondary metabolites and their production is controlled by a specific set of genes found in clusters on the genome. Fumonisins are produced by a number of *Fusarium* spp., including *F. verticillioides*, an important pathogen of maize. Over 28 fumonisins have been identified to date, and are divided into four groups (Rheeder et al. 2002): A-, B-, C- and P-series. Fumonisins are composed of a 19- to 20-carbon aminopolyhydroxyalkyl backbone, similar in structure to sphingosine (Shier 1992). The B-series is the most abundant of the four series (where fumonisin B_1 is the major fumonisin found in *Fusarium*-infected grain) and forms the basic fumonisin structure.

The A-series is characterized by the acetylation of the amino group (Abbax et al. 1993); members of the C-series do not have a terminal methyl group (Branham and Plattner 1993); members of the P-series have a 3-hydroxypyridinium in place of the carbon-2 amine (Musser et al. 1996). Fumonisin toxicity is related to its ability to disrupt sphingolipid metabolism through inhibition of ceramide synthase activity (Voss et al. 2007). These toxins have been shown to affect mitochondrial respiration (Domijan and Abramov 2011) and have been associated with various cancers in humans and animals (Gelderblom et al. 1988; Müller et al. 2012). The fumonisin biosynthetic genes are referred to as FUM genes and are found in the FUM cluster (Proctor et al. 2003, 2006).

Trichothecenes are potent inhibitors of eukaryotic protein biosynthesis and are expressed by *Fusarium* pathogens that affect cereal crops, including *F. culmorum*, *F. graminearum* and *F. sporotrichioides*. Over 200 trichothecenes have been identified from a variety of fungal species (Cole and Cox 1981; Schollenberger et al. 2007). The trichothecenes are divided into four groups based on specific structural features (reviewed in Shank et al. 2011): Types A, B, C and D. Trichothecene-producing *Fusarium* spp. produce either Type A trichothecenes (such as T-2 toxin and HT-2 toxin), or Type B trichothecene (such as nivalenol (NIV), 4-deoxynivalenol (DON; also known as vomitoxin) and acetylated derivatives). The genes encoding trichothecene biosynthesis and metabolism (*TRI* genes) are mainly found in the TRI cluster (Hohn et al. 1993; Ward et al. 2002; Brown et al. 2004), and the specific trichothecenes produced by a given species are determined by the sequences of the TRI genes within this cluster. A summary of TRI genes and their functions are reviewed in Foroud and Eudes (2009). Sequence differences among specific TRI genes, which define their trichothecene genotype (Desjardins 2008), have been used to predict the trichothecene chemotype of a given *Fusarium* strain (Lee et al. 2001; Ward et al. 2008; Alexander et al. 2011; Boutigny et al. 2011; Reynoso et al. 2011). For example, NIV chemotypes are determined by the presence of functional sequences of the Tri13 and Tri7 genes for NIV and 4-acetylnivalenol production, respectively (Lee et al. 2002; Kim et al. 2003). DON producers, which do not express functional Tri13/Tri7 genes, are divided into two chemotypes (3-acetyldeoxynivalenol (3-ADON) and 15-acetyldeoxynivalenol (15-ADON)) determined by the esterase specificity encoded in the *TRI8* gene (Alexander et al. 2011).

From the *Fusarium* TRI cluster, DNA sequences of the TRI12 gene encoding a trichothecene efflux pump have been used to develop real-time PCR assays for quantitative diagnostics of the 3ADON, 15ADON and NIV genotypes of *F. graminearum* and *F. culmorum* (Kulik 2011; Nielsen et al. 2012). These assays utilize genetic markers directly involved in the production of trichothecenes and provide a powerful, cost effective tool to monitor the genotype composition and shifts in pathogen populations of mycotoxigenic *Fusarium* species. Over the past decade, a number of quantitative assays for the detection and quantification of other toxin producing *Fusarium* species have been published (reviewed by Morcia et al. 2013). Applications of these quantitative diagnostic tools included traceability studies of different *Fusarium* species on small cereals (Waalwijk et al. 2004, 2008; Yli-Mattila et al. 2006; Fredlund et al. 2010, 2013; Lindblad et al. 2013), grain dust (Halstensen et al. 2006), and along wheat processing chains (Terzi et al. 2007; Tittlemier et al. 2014).

For the development of species-specific detection assays, DNA sequences of barcode regions, including genes encoding the translation elongation factor 1α, β-tubulin, or mating type (MAT) were successfully used for the detection and quantification of *Fusarium* species (Nicolaisen et al. 2009; Demeke et al. 2010). In *Fusarium*, however, universal barcode markers, such as the internal transcribed spacer (ITS) regions of the ribosomal DNA and the cytochrome oxidase 1 (COX1) gene, commonly used for identification of other groups of fungi have been reported to be non-orthologous and paralogous, respectively (O'Donnell and Cigelnik 1997; Gilmore et al. 2009). Multiple copies of these barcode regions are present in *Fusarium* genomes, showing a rather low degree of divergence among homologous sequences with a number of closely related species sharing identical sequences. In metagenomic profiling of microbial communities employing next generation sequencing (NGS), universal barcode markers often provide insufficient resolution at or below species level when used for identification and semi-quantification of plant-associated *Fusarium* species. Other universal barcodes, such as the protein coding chaperonin-60 (*cpn*60) gene have been reported to be a robust target for species-level characterization in bacteria (Links et al. 2012). The potential of *cpn*60 to identify and differentiate species of *Fusarium* is currently being explored by the authors.

Studies on population dynamics based on single-nucleotide polymorphisms (SNP) have proven to provide sufficient resolution for population- and individual-level analyses of *Fusarium* species. They are adaptable to high-throughput DNA chip-based methods, but finding markers to characterize and distinguish populations is often problematic. Multilocus genotyping (MLGT) assays are a powerful tool that can facilitate accurate identification of species and trichothecene chemotype for large numbers of *Fusarium* isolates. Population dynamics behind adaptive shifts observed for mycotoxin chemotypes and newly emerging pathotypes in *Fusarium graminearum* were studied employing multiplex PCR assays to enable simultaneous determination of species identity and trichothecene chemotype (Ward et al. 2008; Gale et al. 2011). For other population studies (Suga et al. 2004), the genome of *Fusarium graminearum* was mined for repeat sequences to analyze genotypes based on variable number of tandem repeats (VNTR). A number of VNTR markers were selected based on length polymorphisms and used to analyze population genetics in *F. graminearum* and closely related taxa (Ward et al. 2008; Gale et al. 2011; Zhang et al. 2012).

10.2 Fusarium Diseases of Pulse Crops

10.2.1 Pathogens

Canada is a major world exporter of pulse crops. In particular, lentil (*Lens culinaris*) and field pea (*Pisum sativum*) production in Canada has been rising due to the benefits of crop diversification, nitrogen inputs into the soil, and increased worldwide demand for pulse crops (Graham and Vance 2003). In 2012, 2.5 million hectares of pulse crops were planted in Canada, with the majority planted to lentil (1.0 million hectares), and peas (1.35 million hectares) (Statistics Canada). Dry beans (*Phaseolus vulgaris*) and chickpeas (*Cicer arietinum*) comprise the remainder of pulse crops with 120,000 and 80,000 hectares planted, respectively, in 2012. The goal of the Canadian pulse industry is to realize an increase of pulse acreage to 15 % of total planted area. Currently, pulse acreage accounts for an average of 5–10 % total crop area over the three main pulse-growing provinces of Manitoba, Saskatchewan and Alberta.

As pulse acreage increases, the prevalence and incidence of root rots, caused by *Fusarium* spp., have also been increasing. Most pulse crops are subject to root rot pathogens that build up in the soil over several years and reduce plant stands and yields (Persson et al. 1997; Bailey et al. 2003; Infantino et al. 2006; Naseri and Marefat 2011). Root rot is a general term that describes disease symptoms which include reddish-brown-black lesions on the hypocotyl and tap root, often accompanied by vascular discolouration, foliar chlorosis and wilt (Agrios 1997; Bailey et al. 2003; Infantino et al. 2006). Yield losses of 10–30 % are commonly observed in pulse crops affected by moderate to severe root rot, but yield loss potential can be even higher under favourable environmental conditions (Oyarzun 1993; Schneider et al. 2001; Schwartz et al. 2005; Cichy et al. 2007). Root rots, causing wilt and death of mature plants, are reported throughout pulse-growing regions in Canada, and experienced growers are increasingly challenged with yield loss due to stand death. Annual disease surveys indicate that root rot incidence is now widespread in most pulse-growing regions of Canada.

Root rots can be caused by a number of fungi, including *Pythium*, *Rhizoctonia* and/or *Fusarium* spp. (Bailey et al. 2003). However, Fusarium root rot is considered the most prevalent root disease in field peas, dry beans and lentils (Henriquez et al. 2012b; McLaren et al. 2012; Miller et al. 2012). In recent years, 80–100 % of pea fields surveyed in Saskatchewan and Manitoba had plants with root rot symptoms, with severity usually occurring at a moderate level, or 30–40 % of roots and lower stem with symptoms (McLaren et al. 2010, 2011, 2012; Dokken-Bouchard et al. 2011). *Fusarium solani* f. sp. *pisi* and *F. avenaceum* (teleomorph formerly *Giberella avenacea*) are the most prevalent pathogens in field pea crops, with *F. avenaceum* becoming the predominant species isolated from rotted roots in recent years (McLaren et al. 2012). *Fusarium avenaceum* is a common soil saprophyte in temperate regions, and has traditionally been associated with crown rot and Fusarium head blight (FHB) of cereals (Leslie and Summerell 2006; Fernandez 2009). However, it is also very aggressive on all pulse crops, including lentils, dry bean and field peas, and is now the principal species associated with root rot of field pea across the prairie region (Feng et al. 2010; Chittem et al. 2012). The '*F. solani* species complex' comprises over 50 phylogenetic species, of which many members are common soil-dwelling fungi, and act as saprophytes and/or plant pathogens (Coleman et al. 2009). Pathogenic isolates in this group are further characterized by formae speciales to indicate the specific host plants to which they are restricted, such that *F. solani* f. sp. *pisi* is only pathogenic to pea, and *F. solani* f. sp. *phaseoli* is only pathogenic to beans (Oyarzun et al. 1993).

F. solani f. sp. *phaseoli* is generally considered to be the most common pathogen causing dry bean root rots, but recent surveys conducted in Manitoba indicate that there is a complex of *Fusarium* spp. associated with root rot symptoms (Henriquez et al. 2012a). Lentil root rot can also be caused by a number of different pathogenic species including *F. avenaceum*, *F. acuminatum* and *F. redolens* (Hwang et al. 1994; Bailey et al. 2000; Esmaeili Taheri et al. 2011). The vascular wilt pathogens *F. oxysporum* ff. spp. *pisi* and f. sp. *phaseoli* are frequently found in pea and dry bean fields, respectively, but are generally associated with low disease severity indexes (Henriquez et al. 2012b; McLaren et al. 2012). *Fusarium graminearum*, the primary causal agent of FHB in Canada, has also been associated with root rot of legumes (Chongo et al. 2001; Goswami et al. 2008; Bilgi et al. 2011; Esmaeili Taheri et al. 2011; Henriquez et al. 2012a). *Fusarium redolens*, a *Fusarium* species closely related to *F. oxysporum* (Bogale et al. 2007), is also frequently isolated from diseased roots of pulse crops, including field pea, lentil and chickpea (Esmaeili Taheri et al. 2011; Jiménez-Fernández et al. 2011). *Fusarium redolens* has a broad-host range, and has also been shown to induce root rot symptoms in durum wheat in Saskatchewan (Esmaeili Taheri et al. 2011).

10.2.2 Infection Pathways and Symptoms

Most of the *Fusarium* spp. capable of causing root rots on pulse crops produce the same disease symptoms, making it difficult to distinguish between causal agents (Hwang et al. 1995, 2000; Bailey et al. 2000; Bilgi et al. 2008; Feng et al. 2010). Symptoms first appear as small reddish-brown lesions at the base of the hypocotyl and taproot (Fig. 10.3a) (Stahl et al. 1994; Schwartz et al. 2005). As the disease advances, lesions coalesce to form large necrotic areas which encircle the stem and expand vertically (Fig. 10.3c) (Bailey et al. 2003). A reduction in root mass also becomes evident at this stage. In the final stages of root rots, root mass will be reduced by 80–100 %, the hypocotyl becomes pithy and lesions can extend vertically upwards of 2 cm (Bilgi et al. 2008). At this point, the plant is functionally dead with obvious signs of yellowing, wilting and collapse. Infection with *F. solani* and *F. avenaceum* also causes red or brown streaking of the vascular system, indicating that these pathogens can enter the xylem (Fig. 10.3b) (Bailey et al. 2000, 2003; Feng et al. 2010). A major impact of root-rotting fungi on pulse crops is the reduction in the number of nodules on the roots, primarily because secondary root growth is severely impacted (Hwang et al. 1994, 1995). This then results in a reduction in nitrogen fixation, thus reducing the benefit of pulses to subsequent crops in a rotation.

Infection pathways of *F. solani* f. sp. *pisi* on field pea and *F. solani* f. sp. *phaseoli* on dry bean have been well characterized. *Fusarium solani* survives in crop residues and in soil as chlamydospores, which serve as the primary inoculum source (Leslie and Summerell 2006). Chlamydospores are produced in infected tissues of host crops at the end of the growing season, and can survive in the soil for extended periods of time (Bailey et al. 2003; Schwartz et al. 2005). Germination of chlamydospores are stimulated by the presence of nutrients exuded from germinating

Fig. 10.3 Symptoms of root rot on field pea: (**a**) early symptoms with small brown lesions at point of seed attachment; (**b**) red streaking of the vascular system characteristic of *Fusarium* infection; (**c**) extended brown/black lesions on tap root, and loss of secondary root mass and root nodules

seeds of host crops, and thus use of rotations with non-host crops is essential to reduce survival of chlamydospores (Mondal et al. 1996; Oyarzun et al. 1998). Chlamydospores germinate to produce hyphae which can directly infect the developing hypocotyl and epicotyl of seedlings, or the hyphae produces macroconidia which then infect the seedling (Nelson 2004). In vitro studies have shown that macroconidia of *F. solani* invade root tissues by primarily colonizing the zone of elongation. The remainder of the root zones appear to be resistant to primary infection, even in the presence of large numbers of macroconidia and fungal mycelia (Gunawardena et al. 2005). However, Stahl et al. (1994) describe direct penetration of the epidermis of the epicotyl. Production of cutinases by *F. solani* f. sp. *pisi* has also been implicated in initial infection and penetration of the cuticle barrier (Li et al. 2002; Hadwiger 2008). After penetrating the epidermis, mycelium then advances through the cortex both inter- and intra-cellularly until it reaches the Casparian strips present in the endodermis of epicotyl stems. At this point, degradation of the vascular parenchyma is visible in advance of the invading hyphae, suggesting that cell wall degrading enzymes aid in breaking down the barrier to the vascular system, and resulting in colonization of the vascular bundles. Studies of *F. solani* on peas have shown that this pathogen will exclusively colonize the xylem stem tissues beyond the epicotyl, while external lesions on the stem abruptly stop on the epicotyl 1–2 cm above ground (Stahl et al. 1994). It is unknown whether the other root-rotting fungi, such as *F. avenaceum* and *F. graminearum*, colonize and infect tissues of all host pulse crops in a similar manner.

Unlike the other Fusarium root rot pathogens of pulses, *F. avenaceum* is unable to produce chlamydospores, and thus survives in crop residues of susceptible host crops (Leslie and Summerell 2006). Modern agronomic practices, such as reduced tillage, increased glyphosate use and crop rotation with susceptible hosts, have likely allowed pathogenic *Fusarium* spp. to accumulate to damaging levels (Fernandez et al. 2008, 2009, 2011). The increasing prevalence of *F. avenaceum* associated with both broad-leaf pulse and cereal crops in Saskatchewan suggests that *Fusarium* inoculum is being maintained or even increasing on residues of these host crops (Bailey et al. 2001; Fernandez 2007; Abdellatif et al. 2010; Feng et al. 2010). *Fusarium avenaceum* isolates display genetic and ecological plasticity, allowing this fungus to occupy several ecological niches, such as root tissues of pulses, head and root tissues of cereals and residues of host crops (Abdellatif et al. 2010). *Fusarium avenaceum* survived in colonized stem bases of winter wheat over a period of 10 months in the Netherlands (Köhl et al. 2007) with DNA levels decreasing by only 50 % over the winter months. *F. graminearum* survived in standing wheat stubble for up to 20 months, and provided sufficient inoculum levels to serve as a primary inoculum for subsequent crops (Hogg et al. 2010).

10.2.3 Resistance Mechanisms

No commercial cultivars of field pea, lentil or dry beans are completely resistant to Fusarium root rot (Bailey et al. 2003; Grünwald et al. 2003; Xue 2003). Partial resistance to Fusarium root rot caused by both *F. solani* and *F. avenaceum* has been reported in one commercial cultivar, 'Franklin' (Chittem et al. 2012). The mechanism of resistance is not known, although generally genotypes of dry beans and field peas with large, robust root systems show better resistance than those with small root systems (Kraft and Boge 2001; Cichy et al. 2007). Partial resistance is present in several field pea accessions, but these have not yet been transferred into lines with desirable commercial attributes (Grünwald et al. 2003). Quantitative trait locus/loci (QTL) that confer partial resistance to *F. solani* and *F. avenaceum* have been described in field pea, however these QTL did not account for 60 % and 80 %, respectively, of the observed phenotypic variation in root rot resistance (Feng et al. 2011; Li et al. 2012). This suggests that additional resistance genes or QTL are associated with root rot resistance (Feng et al. 2011; Li et al. 2012).

In general, large-seeded Andean dry beans (e.g. kidney beans) tend to be more susceptible to Fusarium root rot than the small-seeded Mesoamerican type beans (e.g. black beans) (Bilgi et al. 2008). Cultivars with partial resistance to *F. solani* f. sp. *phaseoli* also appear to have resistance to other Fusarium root rot pathogens, such as *F. graminearum* (Bilgi et al. 2011). Similar to the situation in field peas, QTL have been identified in dry bean from bean lines with different root rot resistance sources, but these QTL generally account for a small proportion of root rot variation (Schneider et al. 2001; Román-Avilés and Kelly 2005; Ronquillo-López et al. 2010). Most of these QTL are present in regions of the bean genome where resistance genes, such as pathogenesis-related proteins (*PVPR-2*), polygalacturonase-inhibiting protein (*Pgip*) and chalcone synthase (*ChS*) are located (Schneider et al. 2001). This would indicate that partial physiological resistance to Fusarium root rot is associated with generalized host defense responses that are induced upon host attack

(Schneider et al. 2001; Román-Avilés and Kelly 2005). However, markers associated with field root rot resistance often do not correlate with greenhouse root rot screening experiments (Román-Avilés and Kelly 2005). This lack of association suggests that environmental variation is the most important factor contributing to disease development and resistance responses to Fusarium root rot. As a result, breeding for resistance and elucidation of genetic resistance to Fusarium root rots has been challenging, and limited progress has been made in identifying sources of genetic resistance (Singh and Schwartz 2010).

The pea-*F. solani* interaction has been studied as a model system to understand the biochemical and molecular components of non-host resistance by comparing the difference in responses of pea to infection with *F. solani* ff. spp. *pisi* and *phaseoli*. This topic has been reviewed extensively in a recent article (Hadwiger 2008), and thus will not be reviewed again here.

10.3 *Fusarium* Diseases of Cereals

10.3.1 Pathogens and Associated Mycotoxins

The two main *Fusarium* diseases of cereal crops are FHB and Fusarium crown rot (FCR), both of which have been observed in wheat, barley, rye, oats and triticale. FHB is reviewed here in greater detail since it is the main *Fusarium* disease of cereal crops in Canada and worldwide. *Fusarium culmorum*, *F. graminearum* and *F. pseudograminearum* (teleomorph aka *Gibberella coronicola*) are the major species responsible for FHB and/or FCR (O'Donnell et al. 2000; Liddell 2003; Backhouse et al. 2004; Smiley et al. 2005; Tóth et al. 2005); although, *F. pseudograminearum*, primarily responsible for FCR, is only found occasionally in Canada, as this species prefers warmer and drier climates. The majority of *Fusarium* spp. involved in FHB and FCR produce mycotoxins belonging to the trichothecene class, although other *Fusarium* mycotoxins including fumonisins, moniliformin (MON) and zearalenone (ZEA), have also been found in FHB-infected cereals worldwide (Golrnski et al. 1996; Palacios et al. 2011). Trichothecenes contaminate the kernels of FHB-infected spikes, and can also accumulate in the kernels of FCR-infected cereals when the fungus moves up the stem and into the spike (Mudge et al. 2006). Trichothecenes are harmful for human and animal consumers (Eriksen and Pettersson 2004; Godfray et al. 2010), and also interfere with downstream processing including malting (Wolf-Hall 2007). Various cytotoxic effects of trichothecenes have been observed in mammalian and plant systems (Ueno 1983; Rocha et al. 2005; Pestka 2010; Arunachalam and Doohan 2013), although inhibition of eukaryotic protein synthesis machinery is the main mechanism of toxicity (Ueno et al. 1968; McLaughlin et al. 1977). Consumption of contaminated grain can lead to a condition known as alimentary toxic aleukia (ATA), where symptoms of ingestion include gastroenteritis, abdominal and oesophageal pain, ataxia, dyspenia, and subcutaneous haemorrhaging (Lutsky et al. 1978; Peraica et al. 1999). The main potential source of trichothecene contamination of food is from FHB-infected cereals, and

DON is the main trichothecene detected in grain. For this reason, limits are in place to manage DON in food and feed, as described in Sect. 10.6.

The *Fusarium* spp. belonging to the *F. graminearum* (*Fg*) complex, responsible for FHB in North America, are Type B trichothecene producers (Ward et al. 2002; Starkey et al. 2007). Historically, *Fg* populations dominated by 15-ADON chemotypes were responsible for FHB in North America. Some 3-ADON producers were identified more frequently on the continent over 10 years ago, and have since been replacing the 15-ADON populations (Ward et al. 2008). The 3-ADON producers were shown to be the predominant genotype representing more than 90 % of the *Fg* populations in the Canadian Maritimes (R. Clear, unpublished). Since the 1990s, the 3-ADON populations have been moving from the Red River valley in Manitoba to eastern Saskatchewan, and currently represent up to 60 % of the *Fg* population in central Alberta. The 3-ADON producers tend to be more aggressive (Foroud et al. 2012a) and produce higher levels of toxins both in culture and in planta (Ward et al. 2008; Puri and Zhong 2010; von der Ohe et al. 2010; Yli-Mattila and Gagkaeva 2010; Foroud et al. 2012a; Clear et al. 2013). The other important Type B trichothecene producing species, *F. culmorum* (*Fc*), can be associated with FHB and FCR of cereals. Its distribution in western Canada appears to depend partly on environmental factors. Especially in cooler and wet years, *F. culmorum* can be more frequently detected (Fig. 10.4) and contribute significantly to DON contamination in cereal grains (Clear et al. 1993). Similarly to *F. graminearum*, the 3-ADON chemotype of *F. culmorum* is reported to be the more aggressive and toxigenic genotype (Miedaner et al. 2004). In Canada, the 3-ADON genotype represents 100 % of the *Fc* populations found on cereals.

While DON producers are the main species in North America, other chemotypes have also been identified in cereal crops. NIV producers, for example, encompass 79 % of the *F. graminearum* strains identified in Louisiana (Gale et al. 2011), although, this is not representative of the entire population in the United States. In Canada, the NIV chemotypes represent less than 1 % of the *F. graminearum* population (Gräfenhan, unpublished). NIV has been shown to be less phytotoxic than DON, and accumulates in lower quantities in the kernels of infected spikes in cereal crops (Muthomi et al. 2000; Miedaner et al. 2001; Foroud et al. 2012a). However, to animal and human health NIV is more acutely toxic than DON, with one tenth the emetic potential compared to that of DON (Wu et al. 2013). In Canada, occasional contamination of barley with NIV is often caused by infections with *F. poae*. The Type A trichothecenes, produced by species such as *F. sporotrichioides* and *F. poae*, tend to be more toxic in mammalian systems than Type B trichothecenes and have also been identified in FHB-infected crops. In western Canada, trichothecene Type A producing *Fusarium* species are more frequently recovered from FHB-diseased oat and barley seeds (Gilbert and Tekauz 2011). On durum wheat, the predominant *Fusarium* species found in Canada are *F. graminearum* and *F. avenaceum* (Clear et al. 2005; Gräfenhan et al. 2013; Tittlemier et al. 2013b). Depending on the year, the latter species can cause significant damage on heads and seed of durum, especially in the main growing areas of western Canada (Fig. 10.5). Tittlemier et al. (2013b) demonstrated that *F. avenaceum* is the main producer of emerging mycotoxins, including MON and enniatins (ENNs), on durum wheat in Canada.

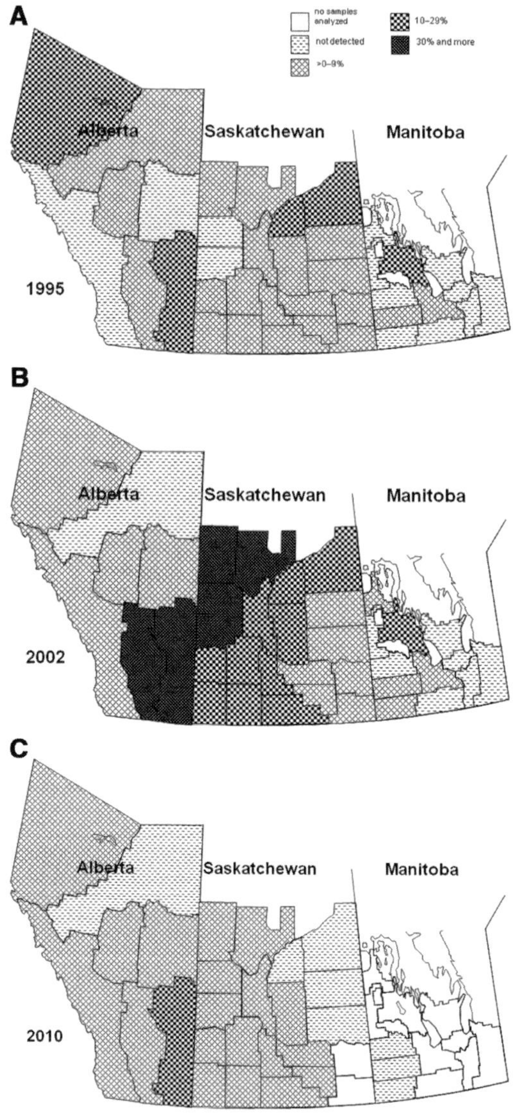

Fig. 10.4 Frequency of occurrence of *Fusarium culmorum* on Fusarium damaged kernels (FDK) of wheat from western Canada in the years 1995 (**a**), 2002 (**b**), and 2010 (**c**)

10.3.2 Infection Pathways and Symptoms

Fusarium infection of cereals is caused by inoculum build up in the soil on crop residues. Initial FCR infection, which can be caused by *Fusarium* mycelium or spores, occurs on emerging shoots, or at the crown or stem base of cereals (Burgess et al. 2001). It has been shown that the trichothecene biosynthesis is initiated during early stages of infection, and while trichothecene accumulation is not necessary for symptoms to develop, a higher infection rate is observed in the presence of the toxin

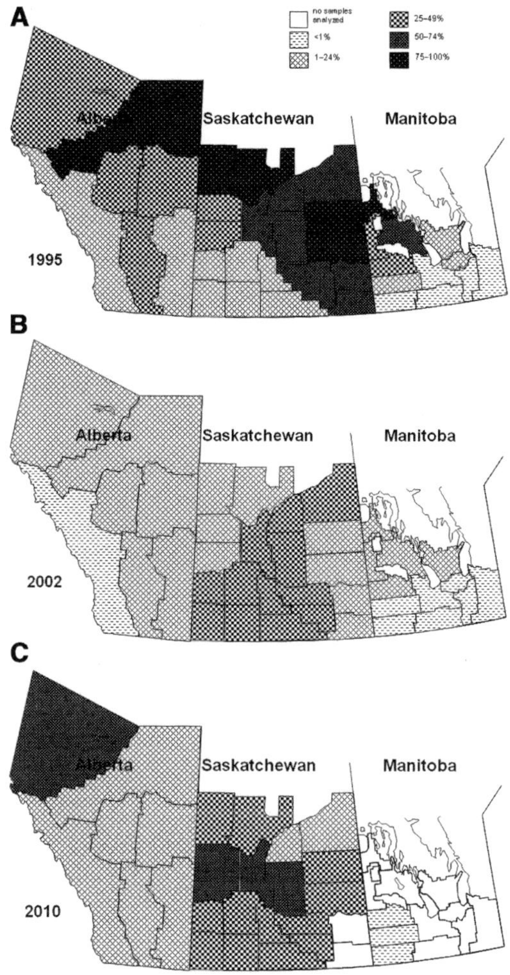

Fig. 10.5 Frequency of occurrence of *Fusarium avenaceum* on Fusarium damaged kernels (FDK) of wheat from western Canada in the years 1995 (**a**), 2002 (**b**), and 2010 (**c**)

(Mudge et al. 2006). The fungus can be isolated from the stem base, the flag leaf node, mature heads and kernels of FCR-infected plants (Mudge et al. 2006). FCR has been observed in cereal crops worldwide, including Canada and the United States (Smiley et al. 2005; Fernandez et al. 2011), and tends to be a major problem in Australia (Backhouse et al. 2004; Obanor et al. 2013).

While FHB can be caused by macroconidia or chlamydospores, ascospores released from the perithecium under humid conditions (reviewed in Bai and Shaner 1994; Parry et al. 1995; Gilbert and Haber 2013) provide the main source of inoculum under field conditions (Sutton 1982; Fernando et al. 2000; Markell and Francl 2003). Infection of cereal inflorescence occurs during anthesis and grain development. Initial symptoms appear as brown water spots on individual spikelets, typically near the base of the glume. As infection progresses, the whole spikelet shows signs of necrosis or premature senescence (Fig. 10.6), and sometimes white or

Fig. 10.6 Symptoms of FHB: (**a**) browning/discoloration of infected spikelets in wheat; (**b**) premature senescence of FHB-infected wheat spike; (**c**) browning/discoloration of infected spikelets in barley and (**d**) oat

pinkish mycelium is visible on the surface of the spikelet (Parry et al. 1995). Detailed studies of *Fusarium* invasion of the wheat spike have been conducted using sophisticated microscopy techniques (Kang and Buchenauer 1999, 2000a, b, c, 2002a, b, 2003; Siranidou et al. 2002; Wanjiru et al. 2002; Jansen et al. 2005; Kang et al. 2005). *Fusarium* can gain access to the host cell through the stomata; however, the primary mode of invasion is by direct penetration of the adaxial epidermal cell walls of the spikelet (Kang and Buchenauer 2000a; Pritsch et al. 2000). Penetration could be facilitated by cutinases and lipases which may lead to cuticle degradation, and expression of a *Fusarium* gene encoding the latter has been implicated in FHB aggressiveness (Voigt et al. 2005). While cuticle degradation has not been experimentally observed, degradation of cell wall components has been observed during *Fusarium* infection of wheat (Kang and Buchenauer 2000b; Wanjiru et al. 2002; Kang et al. 2005). Furthermore, expression or accumulation of cell wall degrading enzymes, including cellulases and pectate esterases, has been observed in the *F. graminearum* secretome and upon exposure to plants or cell wall components (Phalip et al. 2005; Paper et al. 2007; Carapito et al. 2013; Rampitsch et al. 2013). Once established within the spikelet, hyphae can spread to other spikelets within the head through the rachis (Parry et al. 1995). Disease spread typically occurs below the infected spikelets, and premature senescence, or wilt, is sometimes observed above the infected spikelets. Trichothecene biosynthesis is induced upon colonization of the developing kernel, and again at the rachis node (Ilgen et al. 2009). DON accumulates ahead of the growing hyphae, and by 4–6 days after inoculation the hyphae can be found at the rachis both inside and outside of the vascular bundles (Kang and Buchenauer 1999). The production of trichothecenes has been shown to be necessary for *Fusarium* disease spread in *Triticeae* (Proctor et al. 1995; Eudes et al. 2001; Bai et al. 2002; Langevin et al. 2004; Jansen et al. 2005; Maier et al. 2006). By contrast trichothecenes are not required for the establishment of initial infection (Bai et al. 2002; Jansen et al. 2005). FHB infection of the wheat spike leads to yield losses when infection occurs during anthesis and early stages of kernel development, and the kernels that do develop are often contaminated with mycotoxins (Bushnell et al. 2003; Steffenson 2003; Del Ponte et al. 2007).

10.3.3 Physiological Mechanisms of Resistance

The most effective means to prevent *Fusarium*-related damage is to cultivate crops with high levels of resistance (Foroud and Eudes 2009). The mechanisms of FCR resistance are not well characterized. FHB resistance mechanisms are well described, with two major forms of resistance initially defined by Schroeder and Christensen (1963): Type I, resistance to initial infection; and Type II, resistance to disease spread within an infected spike. Other forms of resistance, as summarized by Mesterházy (2003a), include: Type III, resistance to kernel infection (Mesterházy 1995); Type IV, tolerance to FHB and trichothecenes (Mesterházy 1995); and Type V, resistance to trichothecene accumulation (Miller et al. 1985). Type V resistance can be further subdivided into two classes based on the method of resistance, as defined by Boutigny et al. (2008): Type V class 1 is defined as resistance to trichothecene accumulation by chemical modification, and Type V class 2 is defined as resistance to trichothecene accumulation by inhibition of its biosynthesis.

Cell wall lignification or thickening of the rachis node, accompanied with delayed hyphal colonization of the rachis, has been implicated in Type II resistance in wheat (Kang and Buchenauer 2000c). Jansen et al. (2005) also observed cell wall thickening at the rachis node of susceptible wheat inoculated with a trichothecene non-producing mutant (Proctor et al. 1995) of *F. graminearum* that is unable to spread in otherwise susceptible cultivars. As previously mentioned, trichothecene biosynthesis is induced when the hyphae reaches the rachis node (Ilgen et al. 2009), and trichothecene production is required for disease spread to occur (Proctor et al. 1995). Thus, it is likely that accumulation of trichothecenes is involved in weakening the barrier at the rachis node, and that Type II resistant genotypes are able to prevent and/or slow this process through enhanced cell wall thickening compared with susceptible genotypes.

Among the major cereal crops, wheat is the most susceptible and the most heavily FHB-affected crop, where tetraploid (AABB) durum wheat is more susceptible than hexaploid (AABBDD) bread wheat (Langevin et al. 2004). Barley is the second most affected cereal crop—although 6-row barley is nearly as susceptible as wheat, whereas 2-row barley is more resistant. Barley has inherent Type II resistance, and while unconventional mycelial spread by external routes has been observed (Langevin et al. 2004), disease spread does not occur through the rachis (Langevin et al. 2004; Jansen et al. 2005). Rye and oats are the most FHB resistant among the major cereals (Langevin et al. 2004), although disease symptoms are not as clearly discernible in standing oats as they are in other cereals (Fig. 10.6) (Tekauz et al. 2004, 2008). Furthermore, oats tend to accumulate more DON and T-2 toxin than wheat (Langseth and Rundberget 1999; Tekauz et al. 2004). Triticale, a hybrid of wheat and rye which has generally been shown to have higher disease resistances than wheat, is shown to have similar FHB-susceptibility as hexaploid wheat (Langevin et al. 2004).

10.3.4 Genetics of Resistance

FHB resistance is polygenic and also tends to be associated with poor agronomics, making it challenging for breeders to incorporate high levels of resistance into favourable cultivars. Over 100 QTL have been identified in FHB resistance of hexaploid wheat, and 22 of these have been reported in multiple mapping populations (reviewed in Bürstmayr et al. 2009). One of the most widely used and best characterized sources of resistance is 'Sumai3', a Chinese cultivar with very strong Type II resistance. Three major QTL have been identified in 'Sumai3': 3BS (also known as Fhb1), which is the best source of Type II resistance; 5A, which is associated with Type I resistance and is found in different germplasm from different regions worldwide; and 6BS (Fhb2) (Bürstmayr et al. 2009). It has been proposed that the 3BS QTL encodes or regulates expression of a UDP-glycosyltransferase (Lemmens et al. 2005) or a pectin methyl esterase inhibitor (Zhuang et al. 2012). UDP-glycosyltransferases can detoxify DON through condensation of glucose with the C-3 hydroxyl group (Poppenberger et al. 2003). Glycosylated-DON and derivatives thereof have been observed in *Fusarium*-infected cereals (Berthiller et al. 2005; Dall'Asta et al. 2005; Lemmens et al. 2005). Pectin methylesterase inhibitors interfere with the activity of pectin methylesterase, a key enzyme involved in pectin biosynthesis. Pectin is a major component of the plant cell wall, and its degradation has been observed during *Fusarium* infection of wheat (Kang and Buchenauer 2000b). Transgenic expression of an *Actinidia chinensis* pectin methylesterase inhibitor has been shown to improve resistance to fungal diseases (including FHB) in durum wheat (Volpi et al. 2011).

Despite the higher susceptibility and lower genome complexity of durum wheat, only five QTL mapping studies on FHB resistance have been reported in tetraploid wheat, and some of these were conducted in wild relatives of *Triticum turgidum* subsp. *durum* (Ban and Watanabe 2001; Stack et al. 2002; Somers et al. 2006; Bürstmayr et al. 2012). Several FHB-resistance QTL identified in tetraploid wheat correspond to genomic regions of resistance QTL from hexaploid wheat, including 3B from *T. turgidum* subsp. *durum* and 6B from *T. turgidum* subsp. *dicoccum*, corresponding to the hexaploid 3BS and 6BS QTL, respectively (Bürstmayr et al. 2012).

In barley, FHB-resistance QTL have been identified on all seven chromosomes (de la Pena et al. 1999; Zhu et al. 1999; Ma et al. 2000; Yu et al. 2010). The Vrs1 locus, which confers row-type, is associated with a QTL that confers the higher FHB resistance observed in 2-row barley compared with 6-row. It is not known whether this resistance is result of a pleiotropy or if it is directly linked to Vrs1 (reviewed in Massman et al. 2011).

In addition to QTL studies, a series of functional genomic experiments have been conducted in wheat and barley to identify genes and/or molecular pathways involved in mediating FHB resistance (Pritsch et al. 2000, 2001; Wang et al. 2005; Zhou et al. 2005, 2006; Boddu et al. 2006, 2007; Bernardo et al. 2007; Golkari et al. 2007, 2009; Geddes et al. 2008; Li and Yen 2008; Jia et al. 2009; Steiner et al. 2009; Cho et al. 2012; Foroud et al. 2012b). In these studies, *Fusarium*-induced up-regulation of pathogenesis-related (PR) proteins and antioxidants has been observed, and in some cases this up-regulation was higher and/or sooner in resistant lines compared with susceptible ones (Pritsch et al. 2000; Geddes et al. 2008; Golkari et al. 2009; Foroud et al. 2012b). Changes in expression of genes involved in regulating plant hormone biosynthesis and responses have also been observed. In microarray studies, complemented by hormone treatment experiments, Li and Yen (2008) reported that the hormones jasmonic acid (JA) and ethylene (ET) are involved in mediating FHB resistance. Similarly, Desmond et al. (2005) reported a role for JA signalling in FCR resistance in wheat; although, it should be noted that, a different set of host genes is believed to be responsible for FHB and FCR resistances (Li et al. 2010). Virus-induced gene silencing experiments in wheat, where suppression of the ET signalling pathway leads to increased FHB-susceptibility (Gillespie et al. 2012), support results presented by Li and Yen (2008). By contrast, genetic silencing of *ET-INSENSITIVE 2* (*EIN2*; involved in ET signalling) by RNA-interference led to reduced FHB-susceptibility in wheat cv. 'Bobwhite' (Chen et al. 2009). Furthermore, exogenous applications of an ethylene precursor or inhibitor demonstrated that ET signalling can enhance FHB-susceptibility in wheat and barley (Chen et al. 2009). This discrepancy in results was also observed in a separate hormone silencing experiment, where *EIN2* was silenced by RNA-interference in three wheat genotypes (Foroud 2011). In this study, ET silencing led to increased susceptibility in the susceptible genotype, had no impact on the Type I resistant genotype, and led to increased resistance in the Type II resistant genotype. Different outcomes were also observed in different genetic backgrounds silenced in the JA- and salicylic acid (SA)-signalling pathways (Foroud 2011). In the dicot plant *Arabidopsis*, Makandar et al. (2010) observed crosstalk between SA and JA signalling pathways in *Fusarium* resistance, and proposed that the timing of SA and JA signalling is critical in differentiating between resistant and susceptible outcomes. Together, these studies suggest that the role of plant hormones in mediating disease outcomes is genotype-dependent, and may be dependent on crosstalk among different signalling pathways.

10.4 Fusarium Diseases of Maize

10.4.1 Pathogens and Associated Mycotoxins

Several species of *Fusarium* infect maize with infection of the ear and the stalk being the most commonly found diseases. The predominant species causing ear and stalk rot in Canada is *F. graminearum* (Koehler 1957, 1959; Sutton 1982; Reid 1996). A less predominant species is *F. verticillioides* (previously referred to as *F. moniliforme*, Seifert et al. 2003). A comprehensive list of other *Fusarium* species responsible for ear and stalk rots is provided by Mesterházy et al. (2012). The optimum temperature for *F. graminearum* development is 26–28 °C, while *F. verticillioides* tends to grow best at higher temperatures (Reid et al. 1999); therefore, *F. graminearum* is predominantly found in northern regions worldwide and *F. verticillioides* in the southern regions or in dryer years in a northern area (Reid et al. 1999). A single ear or grain can occasionally be infected by different *Fusarium* spp. (Logrieco et al. 2002). Pathogenicity between *Fusarium* spp. and aggressiveness within a species is quite variable and highly dependent on the environmental conditions in a given field season (Reid et al. 2002; Garcia et al. 2009; Iglesias et al. 2010; Miedaner et al. 2010).

Fusarium spp. produce a large number of chemically very different mycotoxins (Logrieco et al. 2002). *Fusarium graminearum* infected ears are usually contaminated with the trichothecene toxin DON before harvest and ZEA during storage. If contaminated grain is fed to livestock, especially swine, DON results in vomiting, feed refusal, decreased weight gain and reproductive problems (Vesonder et al. 1981; Prelusky et al. 1994). This toxin is also an immunosuppressant and thus predisposes animals to other diseases and masks underlying toxicoses (Pestka and Bondy 1994). ZEA causes reproductive problems including reduced litter size, swine estrogenic syndrome and male infertility (Prelusky et al. 1994). Grain contaminated with the polyketide fumonisin mycotoxins produced by *F. verticillioides* can result in equine leukoencephalomalacia (Kellerman et al. 1990), porcine pulmonary edema (Harrison et al. 1990), liver cancer in rats (Gelderblom et al. 1988) and neural tube defects in mice (Voss et al. 2006). Fumonisins have also been associated with human esophageal cancer (International Agency for Research on Cancer (ICARC) 1993). These fungal contaminations cause both direct and indirect economic losses to the maize and livestock industry but they also affect the health of grain handlers and processors.

10.4.2 Infection Pathways and Symptoms

There are many infection pathways by which *Fusarium* spp. can enter maize plants. Stalk rot is often initiated from root infection, through stalk nodes or through holes in the stalk often created by insects and sometimes mechanical damage from cultural practices after planting. There are three potential fungal entry points for ear infection: (1) by fungal spores landing on the silks of the flowering ears, germinating and then the fungal mycelia grow down the silks to infect the kernels and cob (rachis) (Koehler 1942); (2) through wounds created by insects, hail, or birds on the ear (Sutton 1982); and (3) from systemic stalk infections of *F. verticillioides* (Foley 1959; Munkvold et al. 1997b). Which infection pathway is more important depends on the *Fusarium* spp. that is predominant, the insect pressures in a given geographical location and the environmental conditions. For example, Munkvold et al. (1997a) reported less ear rot on maize hybrids with the Bt trait which considerably lowers European corn borer

Fig. 10.7 Symptoms of Fusarium ear blight in maize

populations. Larger populations of thrips, especially on ears with looser husks, were correlated to ear rot (Farrar and Davis 1991; Parsons and Munkvold 2010a, b).

The symptoms of ear rot are depicted in Fig. 10.7. Ear rot caused by *F. graminearum* is characterized by a pinkish coloured mold (White 1999). Infection from the silk commonly begins as white mycelium moving down from the ear tip. This mycelium later turns reddish-pink on infected kernels. In some cases, pinkish fungal growth can be found on the exterior husk leaves and in severe infections it is impossible to separate the husks from the kernels as the entire ear becomes a tightly bound mass of fungal and plant tissue that appears 'mummified'. When infection occurs through kernel wounds, a similar fungal growth pattern is seen but it starts from the initial wound site and tends to spread to the tip of the ear faster than to the butt of the ear (Reid and Sinha 1998). Once the kernels reach 22–23 % moisture it is difficult for the fungus to further infect (Christensen and Kaufmann 1969; Xiang et al. 2010a); however cob (rachis) moisture can be 15–25 % higher than kernel moisture, so the infection may spread in the cobs and can enter younger kernels via the pedicel (Reid and Sinha 1998). In some cases the ear may appear to be symptomless but when squeezed by hand it will feel quite spongy and the cob will be wet and often pink/red in colour. Symptoms of *F. verticillioides* infection on maize ears are quite different from that of *F. graminearum*. Depending on the mode of fungal entry, the symptoms often occur on individual kernels or on a limited area of the ear (White 1999). Infected kernels develop a cottony white growth or may develop white streaks on the pericarp and fungal growth on the cob. How fast symptoms develop in a given year is highly dependent on the environment which, not only influences ear development and subsequent kernel drydown, but also fungal growth. Infection through the silks cannot proceed once the silks have dried out (Reid et al. 1992a; Reid and Sinha 1998) and there is a relationship between kernel drydown rates and ear rot severity symptoms (Xiang et al. 2010a).

Maize plants with *Fusarium* stalk infections often wilt and the leaves may change from a light to a dull green colour while the lower stalk becomes dry and the pith tissue disintegrates to a shredded appearance. For *F. graminearum*, distinctive symptoms are a tan to dark brown discolouration of the lower internodes and pink to reddish discolouration of the pith tissue. Bluish-black coloured perithecia or reddish-white asexual spores may form on the stalk surface. For *F. verticillioides*, brown streaks appear on the lower internodes and the rotted pith tissue may be whitish-pink to salmon in colour. For both pathogens, symptoms usually appear late in the season and plants may lodge if infection is severe. Plants that are stressed, such as from an early frost, are more susceptible to stalk rot.

For *F. graminearum* ear rot, visual symptoms are highly correlated to DON levels (Reid et al. 1996; Perkowski et al. 1997; Reid and Sinha 1998; Bolduan et al. 2009). Correlations between symptoms and fumonisin levels are less reliable for *F. verticillioides* infections possibly in part due to systemic infections from the stalk leading to more asymptomatic infections (Pascale et al. 1997; Murillo-Williams and Munkvold 2008).

10.4.3 Resistance Mechanisms

There is variability within the maize gene pool for levels of resistance to Fusarium ear and stalk rots and breeders have successfully developed genotypes with high levels of resistance to some of these diseases (Reid et al. 2001a, b, 2003); however, it is not clear what the mechanism of this resistance is. Phenotypically, two forms of resistance have been described in maize that are somewhat similar to resistance to initial infection and disease spread, respectively, in cereals: (1) 'silk resistance', where the fungus does not penetrate the silk channel, and thus does not infect the kernels (Reid et al. 1992b); and (2) 'kernel resistance', where the fungus does not penetrate the cob, and thus does not spread from kernel to kernel (Chungu et al. 1996). Studies have indicated that the resistance mechanisms may be associated with flavone content in the silks, stalks and kernels (Reid et al. 1992a; Sekhon et al. 2006; Santiago et al. 2007), (E)-ferulic acid content and dehydrodimers of ferulic acid in kernels (Assabgui et al. 1993; Bily et al. 2003), and 4-acetylbenzoxazolin-2-one (4-ABOA) in kernels (Miller et al. 1997). Recently, Cao et al. (2011) researched the role of hydroxycinnamic acids and reported that several changes in cell wall bound compounds of silk tissues were observed after inoculation with *F. graminearum*. It has been postulated that the *An2* gene which encodes an ent-copalyl synthase gene which has a role in gibberellin synthesis might play a role in silk resistance as this gene is strongly up-regulated after maize silk is inoculated with *F. graminearum* (Harris et al. 2005).

Hoenisch and Davis (1994) observed a correlation between higher pericarp thickness and resistance to *F. verticillioides*. The thicker pericarp may inhibit fungal growth as well as act as a barrier to insect feeding. Sampietro et al. (2009) identified various properties of the pericarp and its wax layer as resistance factors.

Sweet corn, which has been bred to have a thin pericarp, is extremely susceptible to both *F. graminearum* and *F. verticillioides* (Reid et al. 2000). Long chain alkanes on the surface of maize silks have also been implicated in resistance to *F. graminearum* (Miller et al. 2003).

Genotypes developed with selection for *F. graminearum* ear rot resistance also exhibit high levels of resistance to *F. verticillioides* and common smut (*Ustilago zeae*) in inoculated trials (Reid et al. 2009) indicating that there may be an associated resistance mechanism to multiple ear diseases. Resistance to ear rot and stalk rot do not correlate (Mesterházy and Kovács 1988).

10.4.4 Genetics of Resistance

The inheritance of resistance to *Fusarium* spp. in maize is complex and maize genotypes possess different resistance levels as regards to kernel and silk channel resistance (Lemmens et al. 2005). Resistance to *F. graminearum* ear rot through kernel infection is under both simple (additive and dominance) and digenic (dominance x dominance) effects (Chungu et al. 1996). Estimates of the number of factors affecting kernel resistance ranged from 4.6 to 13.7. For *F. verticillioides*, Boling and Grogan (1965) estimated several additive, dominant and additive x dominant digenic epistatic gene effects. They estimated an average dominance of approximately 0.5 and the number of participating genes was estimated at 1.47. Eller et al. (2008) established that resistance to *F. verticillioides* ear rot is determined by polygenes. Maternal effects for both species have also been reported (Headrick and Pataky 1991; Kovács et al. 1994).

Several studies have found QTL associated with resistance to *Fusarium* in maize. Robertson-Hoyt et al. (2006) found 7 QTL that explained 47 % of the phenotypic variation for *F. verticillioides* ear rot and nine were found for fumonisin content explaining 67 % of the variation. Working with two maize populations, they found that three QTL for ear rot and two for fumonisin were mapped in similar positions. Two QTL, localized on chromosome 4 and 5, appeared to be consistent in both populations. Ding et al. (2008) reported two QTL on chromosome 3. Pérez-Brito et al. (2001) identified nine and seven QTL in two populations, three of which were co-located. Recently, Martin et al. (2011) identified co-localized QTL for both *F. graminearum* ear rot resistance and reduced levels of DON in different mapping populations. Reinprecht et al. (2008) identified about 100 genes behind the QTL, among them chitinase and protein kinase. A meta-analysis of QTL associated with ear rot resistance (Xiang et al. 2010b) from the data of 14 studies representing *F. graminearum*, *F. verticillioides* and *Aspergillus flavus* QTL studies found that resistance QTL against the three fungi were clustered on the same chromosomes. These data seem to support the idea of common resistance. Various other studies have reported the identification of possible genes and genetic resistance mechanisms related to ear rot resistance (Jenczmionka and Schäfer 2005; Igawa et al. 2007; Yuan et al. 2008; Lanubile et al. 2010; Zhang et al. 2011).

10.5 Management of Fusarium Head Blight Caused by *Fusarium graminearum*

The occurrence of plant disease depends on the interaction of three factors that 'have often been visualized as a triangle…' (Agrios 1988); a virulent pathogen, a susceptible host and a favourable environment are needed for disease to occur. Variation in any one factor will influence the ultimate level and severity of disease. For example, disease severity may be low if the host has some resistance even though the environment is conducive and a pathogen is present at sufficient levels. Disease severity may also be low if weather conditions are too hot, too cold or too dry, even with a susceptible host and a source of disease inoculum. Where a virulent pathogen is not present or is at low levels, disease may either not occur or be at low levels even when the host is susceptible and there is a favourable environment. Disease management strategies employed by farmers rely on manipulation of one or more components of the disease triangle. The ultimate goal is to create cropping conditions that do not favour pathogen survival and/or disease development. Unfortunately, effective management of FHB, while limiting its impact, cannot be achieved by simply manipulating a single component of the disease triangle (e.g. host resistance). As McMullen et al. (2008, 2012) suggests, effective management of FHB and its impacts on crop production and quality require the use of a combination of strategies.

10.5.1 Crop Rotation

Fusarium graminearum overwinters mainly on infected crop residue, but can also be seed-borne (Wiese 1987; Mathre 1997; Gilbert and Tekauz 2000). Survival in crop residue is highest in plant tissues that are resistant to decay, especially the node tissues of small grain cereals (Burgess and Griffin 1968; Sutton 1982). Gilbert and Tekauz (2000) suggested that *F. graminearum* was unlikely to survive in soil without crop residues. Sutton (1982) also indicated that soil is not likely a 'major inoculum source' and referred to work by Gordon (1954, 1956) in Canada where *F. graminearum* was not isolated from soil samples collected from cereal fields. In Australia, Wearing and Burgess (1977) were able to isolate *F. graminearum* from soil, but it was mainly associated with small pieces of debris.

Given the key role of infested crop residues as a source of inoculum, crop rotation to nongramineous hosts and avoiding corn in rotations, or in close proximity, have been suggested as methods of reducing the risk from FHB or ear/stalk rot in corn (Seaman 1982; Wiese 1987; Parry et al. 1995; Mathre 1997; White 1999; Gilbert and Tekauz 2000; Stack 2000). Corn is an important host of *F. graminearum* (White 1999) and can support extensive colonization of not only infected ears, but also of stalks (Windels and Kommedahl 1984; Kommedahl and Windels 1985; Windels et al. 1988). Although Wiese (1987) and Mathre (1997) recommended at least 1 year between grass or cereal production, rotations with at least 2 years between susceptible crops are needed to reduce the risk of FHB (Burgess and Griffin 1968; Warren and Kommedahl 1973; Khonga and Sutton 1988).

For example, Khonga and Sutton (1988) placed infested corn and wheat residue in the field for up to 3 years and found production of perithecia and ascospores by *Gibberella zeae* (perfect state of *F. graminearum*) occurred primarily in the first and second years. Inch and Gilbert (1999) also found that *F. graminearum* could survive in infected seed for up to 2 years regardless of whether it was on the soil surface or buried up to 10 cm deep in the soil.

Inclusion of highly susceptible crop types either directly in the rotation or in adjacent fields can exacerbate FHB issues. Mathre (1997) reported that barley production in the first half of the 1900s was more or less eliminated when corn was grown in rotation with barley in the eastern and central corn belt of the United States, because the level of FHB became so severe. Other research has highlighted the risk of FHB associated with corn in rotation with small grain cereals. The first report of significant levels of FHB and DON contamination in Manitoba wheat, caused by *F. graminearum*, was associated with two fields that were previously cropped to corn (Clear and Abramson 1986). In Ontario, Teich and Nelson (1984) and Teich and Hamilton (1985) found that FHB levels were lower in wheat when it was not sown after corn. More recently, Schaafsma et al. (2001) conducted a survey of hand-harvested grain from commercial wheat fields in Ontario under a range of agronomic practices. They found that in 2 of 4 years (1996–1999), DON levels in wheat were significantly higher when planted on corn residue compared with wheat or soybean residue. In 1996, Schaafsma et al. (2001) found that levels of DON were similar when corn or wheat had been planted 2 years previously and were significantly higher than when soybean was the previous crop. In a Minnesota trial, Dill-Macky and Jones (2000) found that FHB and DON were higher when wheat followed corn, lowest when wheat followed soybean and intermediate with wheat on wheat. In contrast, Yi et al. (2001) found that FHB and DON levels were similar when winter wheat was grown after maize or spring wheat, whether it had been harvested for grain or silage. However, Yi et al. (2001) stated that inoculation of the pre-crop treatments with infested oat grain may have precluded treatment differences. A spore trapping study by Francl et al. (1999) found that inoculum of *G. zeae* was significantly higher on wheat spikes exposed in fields with corn residues than wheat residues. Wheat heads exposed in fields with corn residue had an average number of colony forming units (CFU) of *G. zeae* per wheat spike (head) per day of 126 versus 13 CFU for wheat heads sampled next to wheat residue. Khonga and Sutton (1988) found that corn residue, including kernels and stalk pieces, tended to be more abundant producers of both conidiospores and ascospores than wheat stems, but not wheat kernels or spikelets.

10.5.2 Tillage

Tillage is a traditional strategy that has been recommended for managing FHB, while conservation tillage has often been implicated as a risk factor for FHB caused by *F. graminearum*, as crop residues are the most important source of inoculum (Parry et al. 1995; McMullen et al. 1997; Stack 2000). Teich and Hamilton (1985) found that FHB was lower in ploughed fields than fields with 'light tillage', but in an earlier study, Teich and Nelson (1984) found that FHB levels were similar with or without ploughing. In a separate study, FHB levels were found to be lower in ploughed treatments versus treatments that had been 'disc-cultivated' (Teich unpublished) (Teich 1989). Dill-Macky and Jones (2000) found that tillage regime did have a significant effect on FHB and DON levels in wheat. When averaged over tillage systems, the incidence of FHB was 63.5 %, 71.8 % and 70.8 % for moldboard ploughing, chisel plough, and no-till, respectively. Small, but significant differences in DON level were observed for the tillage treatments, with moldboard ploughing having 8.1 ppm, compared to chisel plough (10.6 ppm), and no-till (11.1 ppm) which were not significantly different. However, other work indicates that FHB and DON may not always be reduced with tillage. Clear and Abramson (1986) found that the initial appearance of significant levels of Fusarium damaged kernel (FDK) and DON in Manitoba occurred in two wheat fields that had been disced in the previous fall and in the following spring, a tillage regime that would be considered to be conventional and fairly aggressive in western Canada. In a subsequent survey, Gilbert and Tekauz (1993) found no difference between tillage practices during the 1993 FHB epidemic in Manitoba. In Ontario, Miller et al. (1998) found that tillage system (moldboard ploughing versus no-till) did not have a significant influence on the level of FHB or kernel infection. The authors suggested that under weather conditions favourable for disease, other factors such as variety resistance, rotation and previous history of disease would likely be more critical for FHB than the tillage system used. Schaafsma et al. (2001) found similar results from a survey of hand-harvested grain from commercial fields in Ontario from 1996 to 1999. DON levels tended to be slightly higher under minimum tillage versus no-till or conventional, which had similar levels. Overall, Schaafsma et al. (2001) found that tillage system accounted for very little of the variation in DON levels from 1996 to 1999. Other factors such as year, cultivar, and rotation accounted for more variation in DON compared with tillage system. Fernandez et al. (2001) found zero tillage did not result in more FHB compared with conventional tillage in eastern Saskatchewan. FHB severity tended to be highest under minimum tillage, but was lower under both zero and conventional tillage. Khonga and Sutton (1988) suggested that complete burial of infested residue by moldboard ploughing may help to prevent spore production, if residues are not brought back to the soil surface by subsequent tillage. Earthworm activity, which is enhanced under conservation tillage practices (House and Parmelee 1985; Wardle 1995; Kladivko et al. 1997; Chan 2001; Chan and Heenan 2006; Eriksen-Hamel et al. 2009), may help to reduce the amount of *F. graminearum*-infested crop residue under direct seeding (Oldenburg et al. 2008; Schrader et al. 2009; Wolfarth et al. 2011) and perhaps this has contributed to the variable effect of conservation tillage in relation to FHB. In areas where *F. graminearum* is commonly found on crop residues, a general background level of inoculum may preclude any differences in disease risk among tillage systems. Ascospore dispersal from one field to another would introduce the pathogen into fields where infested residues were not present either as a result of burial by tillage or extended crop rotation to non-host crops.

10.5.3 Field Location

Head infections in wheat typically arise from wind-borne ascospores released from fruiting bodies (perithecia) produced by the sexual stage of *F. graminearum*, *G. zeae*, and are formed on old crop residue and infected seed left on the soil surface. Although production of perithecia typically occurs in the spring, these fruiting structures can also be found on harvested grain, especially barley, and can be produced in the fall depending on the location (Paulitz 1996; Mathre 1997). Old crop residues including vegetative and reproductive plant tissues and infected seed are the main sources of inoculum (Sutton 1982).

Dispersal of ascospores appears to occur over relatively short distances. Gilbert and Tekauz (2000) have suggested that the appearance of FHB in eastern Saskatchewan is not likely the result of long-distance (300 km) transport of ascospores, based on results from Fernando et al. (1997). Fernando et al. (1997) demonstrated gradients of head and seed infection resulting from ascospores of *G. zeae*, over distances of at least 22 m. Gilbert and Tekauz (2000) cited reports by Stack (1997) who suggested, based on analysis of spore dispersal gradients, that ascospores could be dispersed and result in head blight symptoms in fields up to 1 mile away from the source of inoculum. Francl et al. (1999) suggested that dispersal of ascospores produced by *G. zeae*, may occur over 'kilometers to tens of kilometers or more....' Based on current research, immediately adjacent fields or areas would be most at risk from air-borne ascospores. Maldonado-Ramirez et al. (2005) and Schmale III and Bergstrom (2007) demonstrated the presence of viable ascospores in the planetary boundary layer suggesting the occurrence of long-distance transport of *G. zeae* ascospores. Recent work using clonal sources of *G. zeae* inoculum identified using microsatellite markers demonstrated dispersal of a released clone up to 750 m, with the majority being collected within 100–250 m of the source (Prussin 2013). Keller et al. (2010) also used clones to study inoculum dispersal of *G. zeae* and found that head infections resulting from a local source of inoculum decreased by 90 % within 6 m of the source. Overall, research suggests that an FHB epidemic within an individual field would largely originate from inoculum produced within the field itself or in adjacent fields. However, as Schmale III and Bergstrom (2007) suggest, long-distant transport of viable ascospores of *G. zeae* may result in the introduction of novel strains into regions where they were not previously present.

Long-distance transport of ascospores into Alberta from eastern Saskatchewan and Manitoba is unlikely. Moreover, there would be a greater potential for a significant reduction in ascospore viability during long-distance dispersal as the ascospores would be exposed to greater periods of ultraviolet (UV) radiation (Waggoner et al. 1983; Rotem and Aust 1991). Radiation has been shown to influence spore survival for many fungi (Leach and Anderson 1982; Caesar and Pearson 1983; Boland 1984; Rotem et al. 1985). Caesar and Pearson (1983) found that average ascospore survival for *Sclerotinia sclerotiorum* was 51 and 22 % after 2 and 4 days field exposure on the upper leaves of a bean canopy. Survival rates of <10 % were observed after 6 days exposure. Ascospore survival also decreased rapidly at relative humidities of >35 % and temperatures of ≥25 °C. Boland (1984) also demonstrated decreased *S. sclerotiorum* ascospore viability, with average survival rates of <50 % after 2 days and <1 % after 3 days field exposure of ascospores on Millipore filter paper. Higher ascospore survival was observed by Boland (1984) and Caesar and Pearson (1983) when ascospores were shielded from UV radiation. Caesar and Pearson (1983) also found that survival was increased on shaded leaves in the lower part of a bean canopy. Rotem and Aust (1991) found that exposure to UV radiation reduced spore viability from up to several days to less than 50 min for various pathogens including *Aspergillus macrospora*, *A. niger* and *Mycosphaerella pinodes*.

10.5.3.1 Integration of Strategies to Limit Inoculum Availability and Host Infection

Development of less susceptible, and eventually more FHB-resistant cultivars, has been a key focus of Canadian cereal breeding programmes. However, unlike resistance to many of the cereal rusts, high levels of resistance to FHB have been elusive, although substantial improvements in reducing the level of susceptibility and moving towards FHB resistance have been made since the early 1990s in western Canada. Extensive reviews of the topic of host resistance have been published by numerous authors (Parry et al. 1995; Gilbert and Tekauz 2000; Tekauz et al. 2000; Mesterházy 2003a; Steffenson 2003; McMullen et al. 2012).

Like host resistance, fungicides have not provided high levels of FHB control and DON suppression, but depending on the level of host resistance can provide moderate reductions in FHB severity, Fusarium damaged kernel levels, and DON contamination (Mesterházy 2003b; Paul et al. 2008; McMullen et al. 2012). The other major approach to limiting inoculum availability is crop rotation, which if sufficient time is given between host crops, substantial reductions in pathogen viability and inoculum availability can be achieved. However, given the ability of the pathogen to produce wind-borne ascospores, which readily move to adjacent fields and the potential for regional epidemics of FHB to occur as consequence of inoculum dispersal over tens of kilometres, crop rotation in itself may not provide a high level of FHB management where the pathogen is well established on crop residues. McMullen et al. (2008, 2012) emphasized that effective FHB management cannot rely on individual strategies, but rather an integration of multiple disease management strategies that limit inoculum availability and host infection. The combination of growing small grain cereals on residue of non-host crops, use of a moderately resistance host genotype, and application of effective fungicides has been found to greatly reduce the level of disease and DON contamination, while significantly increasing crop yield (McMullen et al. 2008). The combination of host resistance, rotation and fungicide represents a foundation on which other strategies can be added to further reduce inoculum availability and disease development. For example, producers growing small grain cereals under irrigation may be able to reduce the risk of head and seed infection by careful water management (McLaren et al. 2003) In Washington State, FHB or scab, caused by various *Fusarium* spp. including

F. graminearum, was found in irrigated fields, but not in dryland wheat fields (Strausbaugh and Maloy 1986). More recently in Idaho (Marshall et al. 2012) and southern Alberta (Turkington et al. 2005, 2006) irrigation was an important contributing factor to FHB outbreaks in these areas. However, the most difficult aspect of irrigation management for FHB control in the irrigated dry regions such as southern Alberta will be trying to balance the water requirements of the crop during flowering versus the need to reduce the risk of FHB. Efetha (2008) has produced a set of recommendations to help producers meet the water needs of their cereal crops, but at the same time reduce the risk of FHB and potential DON contamination of harvested grain.

Harvest management can be an important consideration when dealing with an infected crop. In areas where the disease is severe, producers are advised to adjust their combines to blow out scabby wheat kernels, FDK, (which are lighter than the other seeds) and infected chaff as a way of improving the grade and reducing toxin levels in harvested grain (Tkachuk et al. 1991; Anonymous 1996; Gilbert and Tekauz 2000; Salgado et al. 2011; McMullen et al. 2012). However, this will not completely eliminate problems in wheat, especially when wet harvest conditions allow for continued fungal growth on the maturing crop and potential DON contamination issues even though FHB and FDK levels appear to be low. Removing severely infected kernels during harvesting is not very effective with barley and oat, although removing the hull in hulless barley is an effective way of reducing DON levels (Clear et al. 1997). The downside to harvest management is that it will typically return highly infected wheat kernels and chaff back into the field where this material can act as a source of inoculum in future growing seasons.

McMullen et al. (2012) also suggest that effective chopping and distribution of straw may help to encourage decomposition of infested residue, thereby reducing the availability of inoculum for subsequent epidemics. Chopping of crop residues into smaller pieces, which exposes a greater surface area to microbial activity increases the rate of decomposition of crop residues (Sims and Frederick 1970; Bremer et al. 1991; Angers and Recous 1997; Jensen and Ambus 1998; Gunnar 2001), thereby removing a potential source of FHB inoculum. Moreover, retention of crop residues under conservation tillage can enhance soil flora and fauna activity (House and Parmelee 1985; Chan 2001; Chan and Heenan 2006), which can result in enhanced residue decomposition, especially where residues are chopped into smaller pieces (Boström and Lofs-Holmin 1986; Lowe and Butt 2003). Ultimately, enhanced activity of soil fauna such as earthworms may help to reduce FHB inoculum availability (Schrader et al. 2009; Wolfarth et al. 2011).

Integration of irrigation and residue management with the combination of rotation, host resistance and fungicide may help to further reduce the impact of FHB. Moreover, strategies such as irrigation management may also help to reduce the amount of infested residue, thereby reducing inoculum availability. The use of more resistance crop varieties has been shown to reduce the amount of infested residue thereby reducing the amount of inoculum available to initiate subsequent epidemics (Salas and Dill-Macky 2003, 2004, 2005). Fungicide application may also help to reduce the amount of infested residue and thus the level of inoculum.

There may be a synergistic effect of using crop rotation, host resistance, residue and irrigation management, and fungicides in relation to the availability of inoculum to initiate FHB. Further research is needed to study these interactive effects and whether they have the potential to further reduce the impact of FHB. Ultimately, for FHB management strategies employed by farmers to be more effective, they must incorporate practices that influence all components of the disease triangle, with the goal to create cropping conditions that do not favour pathogen survival, inoculum production and/or disease development.

10.6 Modern Detection Methods for *Fusarium*-Related Mycotoxins

Many jurisdictions, including Canada, have established regulations and guidelines for the presence of *Fusarium* mycotoxins in grains that are used in the production of food and feed in order to protect consumers. Health Canada has set maximum limits of 1.0 and 2.0 mg/kg for DON in soft wheat used in baby foods and non-staple foods, respectively. These limits are currently under review by Health Canada (2011). The Canadian Food Inspection Agency (CFIA) has guidelines and recommended tolerances for a wider range of *Fusarium* mycotoxins in feed, including diacetoxyscirpenol, T-2 and HT-2 toxins, ZEA, and DON. These values range from 0.025 mg/kg for T-2 toxin in diets for dairy animals up to 5 mg/kg in diets for cattle and poultry (CFIA 2012b). The Canadian regulatory limits for food and guidance and recommended levels for feed are consistent with those in other countries.

In Canada, the analysis of grain and grain products is performed along the grain handling and processing chains in order to demonstrate compliance with established regulatory and guidance levels. Domestic and export shipments of bulk grain are monitored by the Canadian Grain Commission (Tittlemier et al. 2014), feed components are monitored by the Canadian Food Inspection Agency, and grain-based foods are monitored by the Canadian Food Inspection Agency (CFIA 2012a) and Health Canada (Scott 1997).

10.6.1 Sampling and Sample Preparation

The determination of any analyte in a given material involves the following general steps: sampling of the material, processing of the sample and a chemical test to detect and quantify the analyte. The initial sampling step is the basis of the entire analysis—without proper sampling, the final analytical result will be meaningless if it does not relate back to the original material of interest.

Proper sampling is especially important for the analysis of mycotoxins in particulate material such as grains because mycotoxins are heterogeneously distributed in this type of material. Consequently, small portions of the larger sample can contain very different concentrations of mycotoxins such as DON (Biselli et al. 2008) because kernel to kernel concentrations can vary over two orders of magnitude (Sinha and Savard 1997). This effect of sampling is magnified for samples composed of larger particles such as maize kernels. Concentrations of mycotoxins in cereal kernels can also significantly differ from concentrations in chaff and other non-kernel segments. DON concentrations were approximately 2–10 times higher in chaff, peduncle and rachis tissues from wheat heads (Sinha and Savard 1997); ZEA was similarly elevated in chaff versus *Fusarium* damaged kernels (Golinski et al. 2010).

Sample preparation can help to minimize the heterogeneous distribution of mycotoxins in materials and reduce the variability in analytical results. For example, grinding of whole grain samples reduces the variability of DON and NIV measurements (Champeil et al. 2004). Increasing the size of the sample analyzed also reduces the variance of the entire analysis (Whitaker et al. 2002).

10.6.2 Screening Methods for the Detection and Quantification of Mycotoxins

There are a number of methods in use to detect and quantify *Fusarium*-related mycotoxins based on a variety of technologies; Shephard et al. (2012) provide an overview of recent advances. Methods can be classified and organized based on the different technologies they are based upon, however a more user-friendly way to classify methods is to place them along a continuum from screening to confirmatory methods.

Screening methods generally emphasize ease of use, speed and an overall reduced cost of analysis. Development of many screening methods is aimed towards use in settings outside of the traditional laboratory, such as in-field assessment or monitoring of incoming deliveries at processing facilities. However, screening methods can still be useful in laboratory settings where large numbers of samples need analysis. Screening methods often incorporate quick sample clean-up and rapid detection, but depending on their scope, they may still require access to fume hoods and other laboratory safety equipment due to the use of solvents for extraction.

The most basic screening method is visual inspection. Due to the physical damage that can be produced by *Fusarium* infection, visual inspection of wheat may be used as a screening method in order to estimate *Fusarium* mycotoxin concentrations. Such visual inspection is feasible for wheat, since FDK can be distinguished from healthy kernels due their shriveled and discoloured appearance. However, visual inspection is not feasible for other grains because the physical damage is not as easily discerned. It has been shown that FDK can serve as an estimate of DON (Miedaner et al. 2001; Mesterházy 2002) in wheat, and that FDK are associated with MON in durum wheat (Tittlemier et al. 2014). In order to manage DON concentrations in wheat, FDK is used as a grading factor in Canada where tolerances for FDK in various wheat classes have been established by the Canadian Grain Commission.

In addition to a restriction to wheat, visual inspection of *Fusarium* damage has other limitations that confine its use to a screening tool. Asymptomatic wheat kernels can still contain mycotoxins (Sinha and Savard 1997). As well, later stage infection can also affect the presence of visual *Fusarium* damage. Reduced damage can be observed from spikes infected past the soft dough stages of kernel development (Del Ponte et al. 2007).

Commercially available chemistry-based screening methods are available for a limited number of *Fusarium*-related mycotoxins in grain. The majority of kits are for the analysis of DON, but some are available for ZEA and T-2/HT-2 (Meneely et al. 2011). Technologies currently available are predominantly immuno-based and include lateral flow devices, enzyme linked immunosorbent assays (ELISA) (Meneely et al. 2011), and planar waveguide-based methods (Tittlemier et al. 2013a).

The performance of commercially available screening methods has been recently reviewed by different groups. The Grain Inspection, Packers & Stockyards Administration (GIPSA) of the United States Department of Agriculture evaluates submitted screening methods against criteria for the quantitative determination of mycotoxins in grains, oilseeds and processed-grain products. GIPSA has evaluated a number of quantitative and qualitative screening methods for the analysis of DON or ZEA, and has posted the results on their website (http://www.gipsa.usda.gov/fgis/insp_weigh/raptestkit.html). Aamot et al. (2012) and Tangni et al. (2011) also report on the performance of commercially available ELISA and lateral flow devices for the analysis of DON in oats and wheat.

Screening methods are useful tools to gauge mycotoxin concentrations in grain outside of a laboratory, or to process a large number of grain samples without requiring complex laboratory equipment. However, there are limitations to screening methods. One very important limitation is the potential for cross reactivity in immuno-based methods. Methods that are based upon the recognition of the mycotoxin analyte by an antibody or other receptor can return inaccurate results when another molecule interacts with the antibody (Tangni et al. 2010). For example, mean results from ELISA tests for DON submitted to an interlaboratory study were approximately 2–3 times higher than those obtained from confirmatory methods that had analyzed the same test material. The difference was attributed to the cross reactivity of the ELISA tests towards 15-ADON, a metabolite of DON (Josephs et al. 2001). There is no confirmation of analyte identity in these screening methods that would avoid issues caused by cross reactivity.

Until recently, screening methods have also focused on single analytes, thus additional tests would need to be run to obtain data on additional analytes. However, multi-mycotoxin screening methods are being developed and commercialized for use in laboratory (Dorokhin et al. 2011; Meneely et al. 2012; Tittlemier et al. 2013a) and non-laboratory settings (He et al. 2012; Lattanzio et al. 2012).

10.6.3 Confirmatory Methods for the Detection and Quantification of Mycotoxins

Confirmatory methods are more complex than screening methods, and require expensive instrumentation and associated technical expertise to operate and maintain the instrumentation. These methods mainly employ chromatography with mass spectrometry, and with advances in fast chromatography, can generate results as quickly as screening methods.

The strength of modern confirmatory methods lies with their ability to confirm analyte identity as well as perform a simultaneous sensitive analysis of many mycotoxins. For example, methods that use liquid chromatography with tandem mass spectrometry have been developed to analyze over 150 various fungal and bacterial metabolites, including many *Fusarium*-related compounds (Vishwanath et al. 2009). The co-occurrence of multiple mycotoxins drives the need for multi-mycotoxin methods, and almost all new methods incorporate multiple mycotoxin analytes (Shephard et al. 2012).

Advances in confirmatory methods include a move towards 'dilute and shoot' methods that minimize sample clean-up in order to decrease analysis time and increase sample throughput (van der Fels-Klerx et al. 2012; Warth et al. 2012). Simple clean-up of sample extracts may also be used to minimize matrix effects and increase sensitivity further. Methods incorporating simple clean-up predominantly use solid phase extraction in cartridge (Schenzel et al. 2012) or dispersive format (Rubert et al. 2012). Sorbents used for the clean-up of sample extracts containing *Fusarium*-related mycotoxins include immunoaffinity materials (Tang et al. 2012) as well as conventional abiotic materials such as silica or C_{18} (Rubert et al. 2012; Schenzel et al. 2012).

10.6.4 General Considerations for the Detection of Fusarium-Related Mycotoxins

Many chemical test methods are available for mycotoxin analysis, and as with any tool, users need to ensure they are properly used in order to generate accurate and precise data. As described above, proper sampling and sample preparation must be performed for the results of the chemical test method to meaningfully relate back to the original sample of interest.

Consumers of chemical test method data must also ensure the methods used follow general proper analytical chemistry practices. All methods used to generate data must be validated so that accurate and precise data are obtained. Validation should be performed for different matrices of interest, because a method that works for one matrix may not work for another (Malachova et al. 2012). There must be routine monitoring of method performance as well. There are many commercially available certified reference materials containing characterized amounts of the more commonly analyzed *Fusarium*-related mycotoxins (DON, HT-2, T-2), as well as many proficiency tests that have been established for laboratories to verify and monitor their performance.

References

Aamot HU, Hofgaard IS, Brodal G, Elen O, Jestoi M, Klemsdal SS (2012) Evaluation of rapid test kits for quantification of deoxynivalenol in naturally contaminated oats and wheat. World Mycotoxin J 5:339–350

Abbax HK, Gelderblom WCA, Cawood ME, Shier WT (1993) Biological activities of fumonisins, mycotoxins from Fusarium moniliforme, in jimsonweed (Datura stramonium L.) and mammalian cell cultures. Toxicon 31:345–353

Abdellatif L, Fernandez MR, Bouzid S, Vujanovic V (2010) Characterization of virulence and PCR-DGGE profiles of *Fusarium avenaceum* from western Canadian Prairie Ecozone of Saskatchewan. Can J Plant Pathol 32:468–480

Agrios GN (1988) Plant pathology, 3rd edn. Academic, San Diego

Agrios GN (1997) Plant pathology, 4th edn. Academic, San Diego, CA

Alexander NJ, McCormick SP, Waalwijk C, van der Lee T, Proctor RH (2011) The genetic basis for 3-ADON and 15-ADON trichothecene chemotypes in *Fusarium*. Fungal Genet Biol 48:485–495

Angers D, Recous S (1997) Decomposition of wheat straw and rye residues as affected by particle size. Plant and Soil 189:197–203

Anonymous (1996) Sustainable agriculture facts—growing for tomorrow: dealing with Fusarium head blight. Canada-Manitoba Agreement on Agricultural Sustainability

Arunachalam C, Doohan FM (2013) Trichothecene toxicity in eukaryotes: cellular and molecular mechanisms in plants and animals. Toxicol Lett 217:149–158

Assabgui RA, Reid LM, Hamilton RI, Arnason JT (1993) Correlation of kernel (E)-ferulic acid content of maize with resistance to *Fusarium graminearum*. Phytopathology 83:949–953

Backhouse D, Abubakar AA, Burgess LW, Dennis JI, Hollaway GJ, Wildermuth GB et al (2004) Survey of *Fusarium* species associated with crown rot of wheat and barley in eastern Australia. Australas Plant Pathol 33:255–261

Bai G, Shaner G (1994) Scab of wheat: prospects for control. Plant Dis 78:760–766

Bai G-H, Desjardins AE, Plattner RD (2002) Deoxynivalenol-nonproducing *Fusarium graminearum* causes initial infection, but does not cause disease spread in wheat spikes. Mycopathologia 153:91–98

Bailey KL, Gossen BD, Derksen DA, Watson PR (2000) Impact of agronomic practices and environment on diseases of wheat and lentil in southeastern Saskatchewan. Can J Plant Sci 80:917–927

Bailey KL, Gossen BD, Lafond GP, Watson PR, Derksen DA (2001) Effect of tillage and crop rotation on root and foliar diseases of wheat and pea in Saskatchewan from 1991 to 1998: univariate and multivariate analyses. Can J Plant Sci 81:789–803

Bailey KL, Gossen BD, Gugel RK, Morrall RAA (2003) Diseases of field crops in Canada, 3rd edn. University Extension Press, University of Saskatchewan, Saskatoon, SK

Ban T, Watanabe N (2001) The effects of chromosomes 3A and 3B on resistance to Fusarium head blight in tetraploid wheat. Hereditas 135:95–99

Bernardo A, Bai G, Guo P, Xiao K, Guenzi AC, Ayoubi P (2007) *Fusarium graminearum*-induced changes in gene expression between Fusarium head blight-resistant and susceptible wheat cultivars. Funct Integr Genomics 7:69–77

Berthiller F, Dall'Asta C, Schuhmacher R, Lemmens M, Adam G, Krska R (2005) Masked mycotoxins: determination of a deoxynivalenol glucoside in artificially and naturally contaminated wheat by liquid chromatography-tandem mass spectrometry. J Agric Food Chem 53:3421–3425

Bilgi VN, Bradley CA, Khot SD, Grafton KF, Rasmussen JB (2008) Response of dry bean genotypes to Fusarium root rot, caused by Fusarium solani f. sp. phaseoli, under field and controlled conditions. Plant Dis 92:1197–1200

Bilgi VN, Bradley CA, Mathew FM, Ali S, Rasmussen JB (2011) Root rot of dry edible bean caused by *Fusarium graminearum*. Plant Health Prog. doi:10.1094/PHP-2011-0425-01-RS

Bily AC, Reid LM, Taylor JH, Johnston D, Malouin C, Burt AJ et al (2003) Dehydrodimers of ferulic acid in maize grain pericarp and aleurone: resistance factors to *Fusarium graminearum*. Phytopathology 93:712–719

Biselli S, Persin C, Syben M (2008) Investigation of the distribution of deoxynivalenol and ochratoxin A contamination within a 26 t truckload of wheat kernels. Mycotoxin Res 24:98–104

Boddu J, Cho S, Kruger WM, Muehlbauer GJ (2006) Transcriptome analysis of the barley-Fusarium *graminearum* interaction. Mol Plant Microbe Interact 19:407–417

Boddu J, Cho S, Muehlbauer GJ (2007) Transcriptome analysis of trichothecene-induced gene expression in barley. Mol Plant Microbe Interact 20:1364–1375

Bogale M, Wingfield BD, Wingfield MJ, Steenkamp ET (2007) Species-specific primers for Fusarium redolens and a PCR-RFLP technique to distinguish among three clades of *Fusarium oxysporum*. FEMS Microbiol Lett 271:27–32

Boland GL (1984) Epidemiology and management of diseases caused by *Sclerotinia sclerotiorum* in white bean and soybean [Thesis]. Dissertation, University of Guelph, Guelph, ON

Bolduan C, Miedaner T, Schipprack W, Dhillon BS, Melchinger AE (2009) Genetic variation for resistance to ear rots and mycotoxins contamination in early European maize inbred lines. Crop Sci 49:2019–2028

Boling MB, Grogan CO (1965) Gene action affecting host resistance to *Fusarium* ear rot of maize. Crop Sci 5:305–307

Boström U, Lofs-Holmin A (1986) Growth of earthworms (*Allolobophora caliginosa*) fed shoots and roots of barley, meadow fescue and lucerne. Studies in relation to particle size, protein, crude fiber content and toxicity. Pedobiologia 29:1–12

Boutigny AL, Richard FF, Barreau C (2008) Natural mechanisms for cereal resistance to the accumulation of *Fusarium* trichothecenes. Eur J Plant Pathol 121:411–423

Boutigny A-L, Ward TJ, Van Coller GJ, Flett B, Lamprecht SC, O'Donnell K et al (2011) Analysis of the *Fusarium graminearum* species complex from wheat, barley and maize in South Africa provides evidence of species-specific differences in host preference. Fungal Genet Biol 48:914–920

Branham BE, Plattner RD (1993) Isolation and characterization of a new fumonisin from liquid cultures of Fusarium moniliforme. J Nat Prod 56:1630–1633

Bremer E, Houtum W, Kessel C (1991) Carbon dioxide evolution from wheat and lentil residues as affected by grinding, added nitrogen, and the absence of soil. Biol Fertil Soils 11:221–227

Brown DW, Dyer RB, McCormick SP, Kendra DF, Plattner RD (2004) Functional demarcation of the *Fusarium* core trichothecene gene cluster. Fungal Genet Biol 41:454–462

Burgess LW, Griffin DM (1968) The recovery of *Gibberella zeae* from wheat straws. Aust J Exp Agric Anim Hus 8:364–370

Burgess LW, Backhouse D, Summerell BA, Swan LJ (2001) Crown rot of wheat. In: Summerell B, Leslie J, Backhouse D, Bryden W, Burgess L (eds) Fusarium: Paul E. Nelson memorial symposium. APS Press, St. Paul, MN, pp 271–294

Bürstmayr H, Ban T, Anderson AJ (2009) QTL mapping and marker-assisted selection for Fusarium head blight resistance in wheat: a review. Plant Breed 128:1–26

Bürstmayr M, Huber K, Heckmann J, Steiner B, Nelson J, Bürstmayr H (2012) Mapping of QTL for Fusarium head blight resistance and morphological and developmental traits in three backcross populations derived from *Triticum dicoccum* × *Triticum durum*. Theor Appl Genet 125:1751–1765

Bushnell WR, Hazen BE, Pritsch C (2003) Histology and physiology of Fusarium head blight. In: Leonard KJ, Bushnell WR (eds) Fusarium head blight of wheat and barley. The American Phytopathological Society, St. Paul, MN, pp 44–84

Caesar AJ, Pearson RC (1983) Environmental factors affecting survival of ascospores of *Sclerotinia sclerotiorum*. Phytopathology 73:1024–1030

Canadian Food Inspection Agency (2012a) RG-8 regulatory guidance: contaminants in feed. http://www.inspection.gc.ca/animals/feeds/regulatory-guidance/rg-8/eng/1347383943203/1347384015909?chap=1#s1c1. Accessed 5 April 2013

Canadian Food Inspection Agency (2012b) 2009–2010 Ochratoxin A and deoxynivalenol in selected foods. http://www.inspection.gc.ca/food/chemical-residues-microbiology/chemical-residues/ochratoxin-a-and-deoxynivalenol/eng/1348258196979/1348258304536. Accessed 5 April 2013

Cao A, Reid LM, Butrón A, Malvar RA, Souto XC, Santiago R (2011) Role of hydroxycinnamic acids in the infection of maize silks by *Fusarium graminearum* Schwabe. Mol Plant Microbe Interact 24:1020–1026

Carapito R, Vorwerk S, Jeltsch J-M, Phalip V (2013) Genome-wide transcriptional responses of *Fusarium graminearum* to plant cell wall substrates. FEMS Microbiol Lett 340:129–134

Champeil A, Fourbet J-F, Doré T (2004) Effects of grain sampling procedures on *Fusarium* mycotoxin assays in wheat grains. J Agric Food Chem 52:6049–6054

Chan KY (2001) An overview of some tillage impacts on earthworm population abundance and diversity—implications for functioning in soils. Soil Tillage Res 57:179–191

Chan KY, Heenan DP (2006) Earthworm population dynamics under conservation tillage systems in south-eastern Australia. Soil Res 44:425–431

Chen X, Steed A, Travella S, Keller B, Nicholson P (2009) *Fusarium graminearum* exploits ethylene signalling to colonize dicotyledonous and monocotyledonous plants. New Phytol 182: 975–983

Chittem K, Porter L, McPhee K, Khan M, Goswami RS (2012) *Fusarium avenaceum* as causal agent of root rot in field peas and its control. Phytopathology 100:S25

Cho S-H, Lee J, Jung K-H, Lee Y-W, Park J-C, Paek N-C (2012) Genome-wide analysis of genes induced by *Fusarium graminearum* infection in resistant and susceptible wheat cultivars. J Plant Biol 55:64–72

Chongo G, Gossen BD, Kutcher HR, Gilbert J, Turkington TK, Fernandez MR et al (2001) Reaction of seedling roots of 14 crop species to *Fusarium graminearum* from wheat heads. Can J Plant Pathol 23:132–137

Christensen CM, Kaufmann HH (1969) Grain storage: the role of fungi in quality loss. University of Minnesota Press, Minneapolis

Chungu C, Mather DE, Reid LM, Hamilton RI (1996) Inheritance of kernel resistance to *Fusarium graminearum* in maize. J Heredity 87:382–385

Cichy K, Snapp S, Kirk W (2007) Fusarium root rot incidence and root system architecture in grafted common bean lines. Plant and Soil 300:233–244

Clear RM, Abramson D (1986) Occurrence of Fusarium head blight and deoxynivalenol (vomitoxin) in two samples of Manitoba wheat in 1984. Can Plant Dis Surv 66:9–11

Clear RM, Babb J, Bullock P (1993) Fusarium head blight on the prairies: past, present, and future. Agriforum 1993:1–4

Clear RM, Patrick SK, Nowicki T, Gaba D, Edney M, Babb JC (1997) The effect of hull removal and pearling on *Fusarium* species and trichothecenes in hulless barley. Can J Plant Sci 77:161–166

Clear RM, Patrick SK, Gaba D, Abramson D, Smith DM (2005) Prevalence of fungi and fusariotoxins on hard red spring and amber durum wheat seed from western Canada, 2000 to 2002. Can J Plant Pathol 27:528–540

Clear RM, Tucker JR, Gaba D, Patrick SK, Lee SJ, Demeke T et al (2013) Deoxynivalenol levels and chemotype frequency in barley cultivars inoculated with two chemotypes of *Fusarium graminearum*. Can J Plant Pathol 35:37–45

Cole RJ, Cox RH (1981) The trichothecenes. In: Cole RA, Cox RH (eds) Handbook of toxic fungal metabolites. Academic Press, Toronto, ON, pp 152–263

Coleman JJ, Rounsley SD, Rodriguez-Carres M, Kuo A, Wasmann CC, Grimwood J et al (2009) The genome of *Nectria haematococca*: contribution of supernumerary chromosomes to gene expansion. PLoS Genet 5:e1000618

Cuomo CA, Gcoldener U, Xu JR, Trail F, Turgeon BG, Di Pietro A et al (2007) The *Fusarium graminearum* genome reveals a link between localized polymorphism and pathogen specialization. Science 317:1400–1402

Dall'Asta C, Berthiller F, Schuhmacher R, Adam G, Lemmens M, Krska R (2005) DON-glycosides: characterisation of synthesis products and screening for their occurrence in DON-treated wheat samples. Mycotoxin Res 21:123–127

de la Pena RC, Smith KP, Capettini F, Muehlbauer GJ, Gallo MM, Dill MR et al (1999) Quantitative trait loci associated with resistance to Fusarium head blight and kernel discoloration in barley. Theor Appl Genet 99:561–569

Del Ponte EM, Fernandes JMC, Bergstrom GC (2007) Influence of growth stage on Fusarium head blight and deoxynivalenol production in wheat. J Phytopathol 155:577–581

Demeke T, Gräfenhan T, Clear RM, Phan A, Ratnayaka I, Chapados J et al (2010) Development of a specific TaqMan real-time PCR assay for quantification of Fusarium graminearum clade 7 and comparison of fungal biomass determined by PCR with deoxynivalenol content in wheat and barley. Int J Food Microbiol 141:45–50

Desjardins AE (2008) Natural product chemistry meets genetics: when is a genotype a chemotype? J Agric Food Chem 56:7587–7592

Desmond OJ, Edgar CI, Manners JM, Maclean DJ, Schenk PM, Kazan K (2005) Methyl jasmonate induced gene expression in wheat delays symptom development by the crown rot pathogen *Fusarium pseudograminearum*. Physiol Mol Plant Pathol 67:171–179

Dill-Macky R, Jones RK (2000) The effect of previous crop residues and tillage on Fusarium head blight of wheat. Plant Dis 84:71–76

Ding J-Q, Wang X-M, Chander S, Yan J-B, Li J-S (2008) QTL mapping of resistance to Fusarium ear rot using a RIL population in maize. Mol Breed 22:395–403

Dokken-Bouchard FL, Banniza S, Bassendowski KA, Chant S, Cruise D, Gross G et al (2011) Survey of field pea diseases in Saskatchewan, 2010. Can Plant Dis Surv 91:133–135

Domijan A-M, Abramov AY (2011) Fumonisin B_1 inhibits mitochondrial respiration and deregulates calcium homeostasis—implication to mechanism of cell toxicity. Int J Biochem Cell Biol 43:897–904

Dorokhin D, Haasnoot W, Franssen MR, Zuilhof H, Nielen MF (2011) Imaging surface plasmon resonance for multiplex microassay sensing of mycotoxins. Anal Bioanal Chem 400: 3005–3011

Efetha A (2008) FHB and irrigation management—frequently asked questions. Alberta Agriculture, Food and Rural Development, Lethbridge. http://www1.agric.gov.ab.ca/$department/deptdocs.nsf/all/faq6824?opendocument

Eller MS, Robertson-Hoyt LA, Payne GA, Holland JB (2008) Grain yield and Fusarium ear rot of maize hybrids developed from lines with varying levels of resistance. Maydica 53:231–237

Eriksen GS, Pettersson H (2004) Toxicological evaluation of trichothecenes in animal feed. Anim Feed Sci Technol 114:205–239

Eriksen-Hamel NS, Speratti AB, Whalen JK, Légère A, Madramootoo CA (2009) Earthworm populations and growth rates related to long-term crop residue and tillage management. Soil Tillage Res 104:311–316

Esmaeili Taheri A, Hamel C, Gan Y, Vujanovic V (2011) First report of *Fusarium redolens* from Saskatchewan and its comparative pathogenicity. Can J Plant Pathol 33:559–564

Eudes F, Comeau A, Rioux S, Collin J (2001) Impact of trichothecenes on Fusarium head blight (*Fusarium graminearum*) development in spring wheat (*Triticum aestivum*). Can J Plant Pathol 23:318–322

Farrar JJ, Davis RM (1991) Relationships among ear morphology, western flower thrips, and Fusarium ear rot of corn. Phytopathology 81:661–666

Feng J, Hwang R, Chang KF, Hwang SF, Strelkov SE, Gossen BD et al (2010) Genetic variation in *Fusarium avenaceum* causing root rot on field pea. Plant Pathol 59:845–852

Feng J, Hwang R, Chang KF, Conner RL, Hwang SF, Strelkov SE et al (2011) Identification of microsatellite markers linked to quantitative trait loci controlling resistance to Fusarium root rot in field pea. Can J Plant Sci 91:199–204

Fernandez MR (2007) *Fusarium* populations in roots of oilseed and pulse crops grown in eastern Saskatchewan. Can J Plant Sci 87:945–952

Fernandez MR (2009) Common root rot of barley in Saskatchewan and north—central Alberta. Can J Plant Pathol 31:96

Fernandez MR, Stolhandske-Dale S, Zentner RP, Pearse P (2001) Progress in management of Fusarium head blight in Saskatchewan. In: Fedak G, Choo A (eds) Proceedings Canadian workshop on Fusarium head blight, Ottawa, ON

Fernandez MR, Huber D, Basnyat P, Zentner RP (2008) Impact of agronomic practices on populations of *Fusarium* and other fungi in cereal and noncereal crop residues on the Canadian Prairies. Soil Tillage Res 100:60–71

Fernandez MR, Zentner RP, Basnyat P, Gehl D, Selles F, Huber D (2009) Glyphosate associations with cereal diseases caused by *Fusarium* spp. in the Canadian Prairies. Eur J Agron 31: 133–143

Fernandez MR, Ulrich D, Brandt SA, Zentner RP, Wang H, Thomas AG et al (2011) Crop management effects on root and crown rot of wheat in west-central Saskatchewan, Canada. Agron J 103:756–765

Fernando WGD, Paulitz TC, Seaman WL, Dutilleul P, Miller JD (1997) Head blight gradients caused by *Gibberella zeae* from area sources of inoculum in wheat field plots. Phytopathology 87:414–421

Fernando WGD, Miller JD, Seaman WL, Seifert K, Paulitz TC (2000) Daily and seasonal dynamics of airborne spores of Fusarium graminearum and other Fusarium species sampled over wheat plots. Can J Bot 78:497–505

Foley DC (1959) Systemic infection of corn by *Fusarium moniliforme*. Phytopathology 49:538

Foroud NA (2011) Investigating the molecular mechanisms of Fusarium head blight resistance in wheat [Thesis]. Dissertation, The University of British Columbia, Vancouver, BC

Foroud NA, Eudes F (2009) Trichothecenes in cereal grains. Int J Mol Sci 10:147–173

Foroud NA, McCormick SP, MacMillan T, Badea A, Kendra DF, Ellis BE et al (2012a) Greenhouse studies reveal increased aggressiveness of emergent Canadian *Fusarium graminearum* chemotypes in wheat. Plant Dis 96:1271–1279

Foroud NA, Ouellet T, Laroche A, Oosterveen B, Jordan MC, Ellis BE et al (2012b) Differential transcriptome analyses of three wheat genotypes reveal different host response pathways associated with Fusarium head blight and trichothecene resistance. Plant Pathol 61:296–314

Francl L, Bergstrom G, Gilbert J, Pedersen W, Dill-Macky R, Sweets L et al (1999) Daily inoculum levels of *Gibberella zeae* on wheat spikes. Plant Dis 83:662–666

Fredlund E, Gidlund A, Pettersson H, Olsen M, Börjesson T (2010) Real-time PCR detection of *Fusarium* species in Swedish oats and correlation to T-2 and HT-2 toxin content. World Mycotoxin J 3:77–88

Fredlund E, Gidlund A, Sulyok M, Borjesson T, Krska R, Olsen M et al (2013) Deoxynivalenol and other selected *Fusarium* toxins in Swedish oats—occurrence and correlation to specific *Fusarium* species. Int J Food Microbiol 167:276–283

Gale LR, Harrison SA, Ward TJ, O'Donnell K, Milus EA, Gale SW et al (2011) Nivalenol-type populations of *Fusarium graminearum* and *F. asiaticum* are prevalent on wheat in Southern Louisiana. Phytopathology 101:124–134

Garcia D, Ramos AJ, Sanchis V, Marín S (2009) Predicting mycotoxins in foods: a review. Food Microbiol 26:757–769

Geddes J, Eudes F, Laroche A, Selinger LB (2008) Differential expression of proteins in response to the interaction between the pathogen *Fusarium graminearum* and its host, *Hordeum vulgare*. Proteomics 8:545–554

Geiser DM, Aoki T, Bacon CW, Baker SE, Bhattacharyya MK, Brandt ME et al (2013) One fungus, one name: defining the genus *Fusarium* in a scientifically robust way that preserves longstanding use. Phytopathology 103:400–408

Gelderblom WC, Jaskiewicz K, Marasas WF, Thiel PG, Horak RM, Vleggaar R et al (1988) Fumonisins—novel mycotoxins with cancer-promoting activity produced by *Fusarium moniliforme*. Appl Environ Microbiol 54:1806–1811

Gilbert J, Haber S (2013) Overview of some recent research developments in Fusarium head blight of wheat. Can J Plant Pathol 35:149–174

Gilbert J, Tekauz A (1993) Effects of delayed harvest on Manitoba wheat. In: Proceedings of the 1993 regional scab forum, Moorhead, MN

Gilbert J, Tekauz A (2000) Review: recent developments in research on Fusarium head blight of wheat in Canada. Can J Plant Pathol 22:1–8

Gilbert J, Tekauz A (2011) Strategies for management of Fusarium head blight (FHB) in cereals. Prairie Soils Crops J 4:97–104

Gillespie ME, Brandt AS, Scofield SR (2012) Ethylene-signaling is essential for basal resistance to Fusarium head blight in wheat. In: Proceedings of the national Fusarium head blight forum, Orlando, FL

Gilmore SR, Grafenhan T, Louis-Seize G, Seifert KA (2009) Multiple copies of cytochrome oxidase 1 in species of the fungal genus Fusarium. Mol Ecol Resour 9(suppl s1):90–98

Godfray HCJ, Beddington JR, Crute IR, Haddad L, Lawrence D, Muir JF et al (2010) Food security: the challenge of feeding 9 billion people. Science 327:812–818

Golinski P, Waskiewicz A, Wisniewska H, Kiecana I, Mielniczuk E, Gromadzka K et al (2010) Reaction of winter wheat (*Triticum aestivum* L.) cultivars to infection with *Fusarium* spp.: mycotoxin contamination in grain and chaff. Food Addit Contam Part A 27:1015–1024

Golkari S, Gilbert J, Prashar S, Procunier JD (2007) Microarray analysis of *Fusarium graminearum-induced* wheat genes: identification of organ-specific and differentially expressed genes. Plant Biotechnol J 5:38–49

Golkari S, Gilbert J, Ban T, Procunier JD (2009) QTL-specific microarray gene expression analysis of wheat resistance to Fusarium head blight in Sumai-3 and two susceptible NILs. Genome 52:409–418

Golrnski P, Kosteckt M, Lasocka I, Wisniewska H, Chelkowski J, Kaczmarek Z (1996) Moniliformin accumulation and other effects of *Fusarium avenaceum* (Fr.) Sacc. on kernels of winter wheat cultivars. J Phytopathol 144:495–499

Gordon WL (1954) The occurrence of Fusarium species in Canada. IV Taxonomy and prevalence of Fusarium species in the soil of cereal plots. Can J Bot 32:622–629

Gordon WL (1956) The occurrence of *Fusarium* species in Canada. V Taxonomy and geographic distribution of Fusarium species in soil. Can J Bot 34:833–846

Goswami RS, Matthew F, Gambhir A, Barasubiye T, Lamppa RS, Bradley C et al (2008) *Fusarium* species associated with root rots of pulse crops in North Dakota. In: The proceedings of the seventh Canadian pulse research workshop, Winnipeg, MB, 5–7 Nov 2008

Gräfenhan T, Patrick SK, Roscoe M, Trelka R, Gaba D, Chan JM et al (2013) *Fusarium* damage in cereal grains from Western Canada. 1. Phylogenetic analysis of moniliformin-producing *Fusarium* species and their natural occurrence in mycotoxin-contaminated wheat, oats, and rye. J Agric Food Chem 61:5425–5437

Graham PH, Vance CP (2003) Legumes: importance and constraints to greater use. Plant Physiol 131:872–877

Grünwald NJ, Coffman VA, Kraft JM (2003) Sources of partial resistance to Fusarium root rot in the *Pisum* core collection. Plant Dis 87:1197–1200

Gunawardena U, Rodriguez M, Straney D, Romeo JT, VanEtten HD, Hawes MC (2005) Tissue-specific localization of pea root infection by *Nectria haematococca*. Mechanisms and consequences. Plant Physiol 137:1363–1374

Gunnar L (2001) Cutting length of straw by combine harvesting—technical possibilities and biological effects. In: Reports Agriculture and Industry, TI—Swedish Institute of Agricultural and Environmental Engineering. Report 282. http://www.jti.se/index.php?page=publicationinfo&publicationid=229

Hadwiger LA (2008) Pea-*Fusarium solani* interactions contributions of a system toward understanding disease resistance. Phytopathology 98:372–379

Halstensen AS, Nordby KC, Eduard W, Klemsdal SS (2006) Real-time PCR detection of toxigenic Fusarium in airborne and settled grain dust and associations with trichothecene mycotoxins. J Environ Monit 8:1235–1241

Harris LJ, Saparno A, Johnston A, Prisic S, Xu M, Allard S et al (2005) The maize An2 gene is induced by *Fusarium* attack and encodes an ent-copalyl diphosphate synthase. Plant Mol Biol 59:881–894

Harrison LR, Colvin BM, Greene JT, Newman LE, Cole JR (1990) Pulmonary edema and hydrothorax in swine produced by fumonisin B1, a toxic metabolite of *Fusarium moniliforme*. J Vet Diagn Invest 2:217–221

He QH, Xu Y, Wang D, Kang M, Huang ZB, Li YP (2012) Simultaneous multiresidue determination of mycotoxins in cereal samples by polyvinylidene fluoride membrane based dot immunoassay. Food Chem 134:507–512

Headrick JM, Pataky JK (1991) Maternal influence of the resistance of sweet corn lines to kernels infection by *Fusarium moniliforme*. Phytopathology 81:268–274

Health Canada (2011) Canadian standards (maximum levels) for various chemical contaminants in food. www.hc-sc.gc.ca/fn-an/securit/chem-chim/contaminants-guidelines-directives-eng.php. Accessed 18 March 2013

Henriquez MA, McLaren DL, Conner RL, Balasubramanian PB, Chang KF (2012a) The proceedings of the ninth Canadian pulse research workshop, Niagara Falls, ON, 6–9 Nov 2012

Henriquez MA, McLaren DL, Conner RL, Penner WC, Kerley T (2012b) Diseases of field bean in Manitoba in 2011. Can Plant Dis Surv 92:120–121

Hoenisch RW, Davis RM (1994) Relationship between kernel pericarp thickness and susceptibility to Fusarium ear rot in field corn. Plant Dis 78:517–519

Hogg AC, Johnston RH, Johnston JA, Klouser L, Kephart KD, Dyer AT (2010) Monitoring Fusarium crown rot populations in spring wheat residues using quantitative real-time polymerase chain reaction. Phytopathology 100:49–57

Hohn TM, McCormick SP, Desjardins AE (1993) Evidence for a gene cluster involving trichothecene-pathway biosynthetic genes in *Fusarium sporotrichioides*. Curr Genet 24:291–295

House GJ, Parmelee RW (1985) Comparison of soil arthropods and earthworms from conventional and no-tillage agroecosystems. Soil Tillage Res 5:351–360

Hwang SF, Howard RJ, Chang KF, Park B, Burnett PA (1994) Etiology and severity of Fusarium root rot of lentil in Alberta. Can J Plant Pathol 16:295–303

Hwang SF, Howard RJ, Chang KF, Park B, Lopetinsky K, McAndrew DW (1995) Screening of field pea cultivars for resistance to Fusarium root rot under field conditions in Alberta. Can Plant Dis Surv 75:51–56

Hwang SF, Gossen BD, Turnbull GD, Chang KF, Howard RJ, Thomas AG (2000) Effect of temperature, seeding date, fungicide seed treatment and inoculation with *Fusarium avenaceum* on seedling survival, root rot severity and yield of lentil. Can J Plant Sci 80:899–907

Igawa T, Takahashi-Ando N, Ochiai N, Ohsato S, Shimizu T, Kudo T et al (2007) Reduced contamination by the *Fusarium* mycotoxin zearalenone in maize kernels through genetic modification with a detoxification gene. Appl Environ Microbiol 73:1622–1629

Iglesias J, Presello DA, Botta G, Lori GA, Fauguel CM (2010) Aggressiveness of *Fusarium* section *Liseola* isolates causing maize ear rot in Argentina. J Plant Pathol 92:205–211

Ilgen P, Hadeler B, Maier FJ, Schäfer W (2009) Developing kernel and rachis node induce the trichothecene pathway of *Fusarium graminearum* during wheat head infection. Mol Plant Microbe Interact 22:899–908

Inch S, Gilbert J (1999) Survival of Fusarium graminearum on Fusarium damaged kernels. In: Clear R (ed) Proceedings Canadian workshop on Fusarium head blight, Winnipeg, MB

Infantino A, Kharrat M, Riccioni L, Coyne CJ, McPhee KE, Grünwald NJ (2006) Screening techniques and sources of resistance to root diseases in cool season food legumes. Euphytica 147:201–221

International Agency for Research on Cancer (ICARC) (1993) Monograph 56: some naturally occurring substances: some food items and constituents heterocyclic amines and mycotoxins. IARC, Lyon, France

Jansen C, von Wettstein D, Schaefer W, Kogel KH, Felk A, Maier FJ (2005) Infection patterns in barley and wheat spikes inoculated with wild-type and trichodiene synthase gene disrupted *Fusarium graminearum*. Proc Natl Acad Sci U S A 102:16892–16897

Jenczmionka NJ, Schäfer W (2005) The Gpmk1 MAP kinase of *Fusarium graminearum* regulates the induction of specific secreted enzymes. Curr Genet 47:29–36

Jensen ES, Ambus P (1998) Plant litter particle size—effects on decomposition and nitrogen dynamics in soil. In: Proceedings of the 16th International Soil Science Society Congress, CD-ROM, Montpellier, France, August 1998. http://natres.psu.ac.th/Link/SoilCongress/bdd/symp41/482-t.pdf

Jia H, Cho S, Muehlbauer GJ (2009) Transcriptome analysis of a wheat near-isogenic line pair carrying Fusarium head blight-resistant and -susceptible alleles. Mol Plant Microbe Interact 22:1366–1378

Jiménez-Fernández D, Navas-Cortés JA, Montes-Borrego M, Jiménez-Díaz RM, Landa BB (2011) Molecular and pathogenic characterization of *Fusarium redolens*, a new causal agent of Fusarium yellows in Chickpea. Plant Dis 95:860–870

Josephs RD, Schuhmacher R, Krska R (2001) International interlaboratory study for the determination of the *Fusarium* mycotoxins zearalenone and deoxynivalenol in agricultural commodities. Food Addit Contam 18:417–430

Kang Z, Buchenauer H (1999) Immunocytochemical localization of Fusarium toxins in infected wheat spikes by *Fusarium culmorum*. Physiol Mol Plant Pathol 55:275–288

Kang Z, Buchenauer H (2000a) Cytology and ultrastructure of the infection of wheat spikes by *Fusarium culmorum*. Mycol Res 104:1083–1093

Kang Z, Buchenauer H (2000b) Ultrastructural and cytochemical studies on cellulose, xylan and pectin degradation in wheat spikes infected by *Fusarium culmorum*. J Phytopathol 148:263–275

Kang Z, Buchenauer H (2000c) Ultrastructural and immunocytochemical investigation of pathogen development and host responses in resistant and susceptible wheat spikes infected by *Fusarium culmorum*. Physiol Mol Plant Pathol 57:255–268

Kang Z, Buchenauer H (2002a) Immunocytochemical localization of β-1,3-glucanase and chitinase in *Fusarium culmorum*-infected wheat spikes. Physiol Mol Plant Pathol 60:141–153

Kang Z, Buchenauer H (2002b) Studies on the infection process of *Fusarium culmorum* in wheat spikes: degradation of host cell wall components and localization of trichothecene toxins in infected tissue. Eur J Plant Pathol 108:653–660

Kang Z, Buchenauer H (2003) Immunocytochemical localization of cell wall-bound thionins and hydroxyproline-rich glycoproteins in *Fusarium culmorum* infected wheat spikes. J Phytopathol 151:120–129

Kang Z, Zingen-Sell I, Buchenauer H (2005) Infection of wheat spikes by *Fusarium avenaceum* and alterations of cell wall components in the infected tissue. Eur J Plant Pathol 111:19–28

Keller MD, Waxman KD, Bergstrom GC, Schmale DG III (2010) Local distance of wheat spike infection by released clones of *Gibberella zeae* disseminated from infested corn residue. Plant Dis 94:1151–1155

Kellerman TS, Marasas WF, Thiel PG, Gelderblom WC, Cawood M, Coetzer JA (1990) Leukoencephalomalacia in two horses induced by oral dosing of fumonisin B_1. Onderstepoort J Vet Res 57:269–275

Khonga EB, Sutton JC (1988) Inoculum production and survival of *Gibberella zeae* in maize and wheat residues. Can J Plant Pathol 10:232–239

Kim H-S, Lee T, Dawlatana M, Yun S-H, Lee Y-W (2003) Polymorphism of trichothecene biosynthesis genes in deoxynivalenol-and nivalenol-producing *Fusarium graminearum* isolates. Mycol Res 107:190–197

Kladivko EJ, Akhouri NM, Weesies G (1997) Earthworm populations and species distributions under no-till and conventional tillage in Indiana and Illinois. Soil Biol Biochem 29:613–615

Koehler B (1942) Natural mode of entrance of fungi into Corn ears and some symptoms that indicate infection. J Agric Res 64:421–442

Koehler B (1957) Prevalence in dent corn and relation to seedling blights. Univ Illinois Ill Agric Exp Stn Bull 617:1–72

Koehler B (1959) Corn ear rots in Illinois. Univ Illinois Ill Agric Exp Stn Bull 639:1–72

Köhl J, de Haas BH, Kastelein P, Burgers SLGE, Waalwijk C (2007) Population dynamics of *Fusarium* spp. and *Microdochium nivale* in crops and crop residues of winter wheat. Phytopathology 97:971–978

Kommedahl T, Windels CE (1985) The changing nature of stalk rot of maize caused by *Gibberella zeae*. In: Parker CA, Rovira AD, Moore KJ, Wong PTW, Kollmorgen JF (eds) Ecology and management of soilborne plant pathogens. American Phytopathological Society, St. Paul, MN

Kovács K, Kovács G Jr, Mesterházy Á (1994) Expression of resistance to fusarial ear blight in corn inbreds and their hybrids. Maydica 39:187–190

Kraft JM, Boge W (2001) Root characteristics in pea in relation to compaction and Fusarium root rot. Plant Dis 85:936–940

Kulik T (2011) Development of TaqMan assays for 3ADON, 15ADON and NIV *Fusarium* genotypes based on *Tri12* gene. Cereal Res Commun 39:200–214

Langevin F, Eudes F, Comeau A (2004) Effect of trichothecenes produced by *Fusarium graminearum* during Fusarium head blight development in six cereal species. Eur J Plant Pathol 110:735–746

Langseth W, Rundberget T (1999) The occurrence of HT-2 toxin and other trichothecenes in Norwegian cereals. Mycopathologia 147:157–165

Lanubile A, Pasini L, Marocco A (2010) Differential gene expression in kernels and silks of maize lines with contrasting levels of ear rot resistance after *Fusarium verticillioides* infection. J Plant Physiol 167:1398–1406

Lattanzio VMT, Nivarlet N, Lippolis V, Gatta SD, Huet AC, Delahaut P et al (2012) Multiplex dipstick immunoassay for semi-quantitative determination of *Fusarium* mycotoxins in cereals. Anal Chim Acta 718:99–108

Leach CM, Anderson AJ (1982) Radiation quality and plant diseases. In: Hatfield J, Thomason IJ (eds) Proceedings of a conference on biometeorology in integrated pest management. Academic, New York

Lee T, Oh D-W, Kim H-S, Lee J, Kim Y-H, Yun S-H et al (2001) Identification of deoxynivalenol- and nivalenol-producing chemotypes of *Gibberella zeae* by using PCR. Appl Environ Microbiol 67:2966–2972

Lee T, Han YK, Kim KH, Yun SH, Lee YW (2002) *Tri13* and *Tri7* determine deoxynivalenol- and nivalenol-producing chemotypes of *Gibberella zeae*. Appl Environ Microbiol 68:2148–2154

Lemmens M, Scholz U, Berthiller F, Dall'Asta C, Koutnik A, Schuhmacher R et al (2005) The ability to detoxify the mycotoxin deoxynivalenol colocalizes with a major quantitative trait locus for Fusarium head blight resistance in wheat. Mol Plant Microbe Interact 18:1318–1324

Leslie JF, Summerell BA (2006) The Fusarium laboratory manual. Blackwell, Ames, IA

Li G, Yen Y (2008) Jasmonate and ethylene signaling pathway may mediate Fusarium head blight resistance in wheat. Crop Sci 48:1888–1896

Li D, Sirakova T, Rogers L, Ettinger WF, Kolattukudy PE (2002) Regulation of constitutively expressed and induced cutinase genes by different zinc finger transcription factors in *Fusarium solani* f. sp. *pisi* (*Nectria haematococca*). J Biol Chem 277:7905–7912

Li H, Xie G, Ma J, Liu G, Wen S, Ban T et al (2010) Genetic relationships between resistances to Fusarium head blight and crown rot in bread wheat (Triticum aestivum L.). Theor Appl Genet 121:941–950

Li WJ, Feng J, Chang KF, Conner RL, Hwang SF, Strelkov SE et al (2012) Microsatellite DNA markers indicate quantitative trait loci controlling resistance to pea root rot caused by *Fusarium avenaceum* (Corda ex Fries) Sacc. Plant Pathol J 11:114–119

Liddell CM (2003) Systematics of *Fusarium* species and allies associated with Fusarium head blight. In: Leonard KJ, Bushnell WR (eds) Fusarium head blight of wheat and barley. The American Phytopathological Society, St. Paul, MN, pp 35–43

Lindblad M, Gidlund A, Sulyok M, Borjesson T, Krska R, Olsen M et al (2013) Deoxynivalenol and other selected *Fusarium* toxins in Swedish wheat—occurrence and correlation to specific *Fusarium* species. Int J Food Microbiol 167:284–291

Links MG, Dumonceaux TJ, Hemmingsen SM, Hill JE (2012) The chaperonin-60 universal target is a barcode for bacteria that enables de novo assembly of metagenomic sequence data. PLoS One 7:e49755

Logrieco A, Mulè G, Moretti A, Bottalico A (2002) Toxigenic *Fusarium* species and mycotoxins associated with maize ear rot in Europe. Eur J Plant Pathol 108:597–609

Lowe CN, Butt KR (2003) Influence of food particle size on inter- and intra-specific interactions of *Allolobophora chlorotica* (Savigny) and *Lumbricus terrestris* (L.). Pedobiologia 47:574–577

Lutsky I, Mor N, Yagen B, Joffe AZ (1978) The role of T-2 toxin in experimental alimentary toxic aleukia: a toxicity study in cats. Toxicol Appl Pharmacol 43:111–124

Ma Z, Steffenson BJ, Prom LK, Lapitan NL (2000) Mapping of quantitative trait loci for Fusarium head blight resistance in barley. Phytopathology 90:1079–1088

Ma L-J, van der Does HC, Borkovich KA, Coleman JJ, Daboussi M-J, Di Pietro A et al (2010) Comparative genomics reveals mobile pathogenicity chromosomes in *Fusarium*. Nature 464:367–373

Maier FJ, Miedaner T, Hadeler B, Felk A, Salomon S, Lemmens M et al (2006) Involvement of trichothecenes in fusarioses of wheat, barley and maize evaluated by gene disruption of the trichodiene synthase (Tri5) gene in three field isolates of different chemotype and virulence. Mol Plant Pathol 7:449–461

Makandar R, Nalam V, Chaturvedi R, Jeannotte R, Sparks AA, Shah J (2010) Involvement of salicylate and jasmonate signaling pathways in *Arabidopsis* interaction with *Fusarium graminearum*. Mol Plant Microbe Interact 23:861–870

Malachova A, Varga E, Schwartz H, Krska R, Berthiller F (2012) Development, validation and application of an LC-MS/MS based method for the determination of deoxynivalenol and its conjugates in different types of beer. World Mycotoxin J 5:261–270

Maldonado-Ramirez SL, Schmale Iii DG, Shields EJ, Bergstrom GC (2005) The relative abundance of viable spores of *Gibberella zeae* in the planetary boundary layer suggests the role of long-distance transport in regional epidemics of Fusarium head blight. Agr Forest Meteorol 132:20–27

Markell SG, Francl LJ (2003) Fusarium head blight inoculum: species prevalence and Gibberella zeae spore type. Plant Dis 87:814–820

Marshall J, Bissonnette K, Chen J (2012) The re-emergence of Fusarium head blight in Idaho. Impact: University of Idaho, Extension, Idaho Falls Research and Extension Center, Idaho Falls, ID. http://www.extension.uidaho.edu/impacts/Pdf_12/30-12jmarshall-fusarium.pdf

Martin M, Miedaner T, Dhillon BS, Ufermann U, Kessel B, Ouzunova M et al (2011) Colocalization of QTL for Gibberella ear rot resistance and low mycotoxin contamination in early European maize. Crop Sci 51:1935–1945

Massman J, Cooper B, Horsley R, Neate S, Dill-Macky R, Chao S et al (2011) Genome-wide association mapping of Fusarium head blight resistance in contemporary barley breeding germplasm. Mol Breed 27:439–454

Mathre DE (1997) Compendium of barley diseases, 2nd edn. APS Press, St. Paul, MN

McLaren DL, Irvine RB, Turkington TK (2003) Irrigation management strategies to reduce Fusarium head blight and deoxynivalenol. Can J Plant Pathol 25:117–118 (Abstr.)

McLaren DL, Conner RL, Hausermann DJ, Henderson TL, Penner WC, Kerley TJ (2010) Field pea diseases in Manitoba in 2009. Can Plant Dis Surv 90:148–149

McLaren DL, Hausermann DJ, Henderson TL, Kerley TJ (2011) Field pea diseases in Manitoba in 2010. Can Plant Dis Surv 91:140–141

McLaren DL, Hausermann DJ, Henriquez MA, Kerley TJ, Chang KF (2012) Field pea disease in Manitoba in 2011. Can Plant Dis Surv 92:150–151

McLaughlin CS, Vaughn MH, Campbell JM, Wei CM, Stafford ME (1977) Inhibition of protein synthesis by trichothecenes. In: Rodricks JV, Hesseltine CW, Mehlman MA (eds) Mycotoxins in human and health. Pathotox Publishers, Park Forest South, IL, pp 263–275

McMullen M, Jones R, Gallenberg D (1997) Scab of wheat and barley: a re-emerging disease of devasting impact. Plant Dis 81:1340–1348

McMullen M, Halley S, Schatz B, Meyer S, Jordahl J, Ransom J (2008) Integrated strategies for Fusarium head blight management in the United States. Cereal Res Commun 36:563–568

McMullen MP, Bergstrom GC, De Wolf E, Dill-Macky R, Hershman DE, Shaner G et al (2012) A unified effort to fight an enemy of wheat and barley: Fusarium head blight. Plant Dis 96:1712–1728

Meneely JP, Ricci F, van Egmond HP, Elliott CT (2011) Current methods of analysis for the determination of trichothecene mycotoxins in food. Trends Analyt Chem 30:192–203

Meneely JP, Quinn JG, Flood EM, Hajšlová J, Elliott CT (2012) Simultaneous screening for T-2/HT-2 and deoxynivalenol in cereals using a surface plasmon resonance immunoassay. World Mycotoxin J 5:117–126

Mesterházy Á (1995) Types and components of resistance to Fusarium head blight of wheat. Plant Breed 114:377–386

Mesterházy Á (2002) Role of deoxynivalenol in aggressiveness of *Fusarium graminearum* and *F. culmorum* and in resistance to Fusarium head blight. Eur J Plant Pathol 108:675–684

Mesterházy Á (2003a) Breeding wheat for Fusarium head blight resistance in Europe. In: Leonard KJ, Bushnell WR (eds) Fusarium head blight of wheat and barley. The American Phytopathological Society, St. Paul, MN, pp 211–240

Mesterházy Á (2003b) Control of Fusarium head blight of wheat by fungicides. In: Leonard KJ, Bushnell WR (eds) Fusarium head blight of wheat and barley. The American Phytopathological Society, St. Paul, MN, pp 363–380

Mesterházy Á, Kovács K (1988) Breeding corn against fusarial stalk rot, ear rot and seedling blight. Acta Phytopathol Acad Sci Hung 21:231–249

Mesterházy Á, Lemmens M, Reid LM (2012) Breeding for resistance to ear rots caused by *Fusarium* spp. in maize—a review. Plant Breed 131:1–19

Miedaner T, Reinbrecht C, Lauber U, Schollenberger M, Geiger HH (2001) Effects of genotype and genotype-environment interaction on deoxynivalenol accumulation and resistance to Fusarium head blight in rye, triticale, and wheat. Plant Breed 120:97–105

Miedaner T, Schilling A, Geiger HH (2004) Competition effects among isolates of Fusarium culmorum differing in aggressiveness and mycotoxin production on heads of winter rye. Eur J Plant Pathol 110:63–70

Miedaner T, Bolduan C, Melchinger AE (2010) Aggressiveness and mycotoxin production of eight isolates each of *Fusarium graminearum* and *Fusarium verticillioides* for ear rot on susceptible and resistant early maize inbred lines. Eur J Plant Pathol 127:113–123

Miller JD, Young JC, Sampson DR (1985) Deoxynivalenol and Fusarium head blight resistance in spring cereals. Phytopathol Z 113:359–367

Miller JD, Miles M, Fielder DA (1997) Kernel concentrations of 4-acetylbenzoxazolin-2-one and diferuloylputrescine in maize genotypes and Gibberella ear rot. J Agric Food Chem 45: 4456–4459

Miller JD, Culley J, Fraser K, Hubbard S, Meloche F, Ouellet T et al (1998) Effect of tillage practice on fusarium head blight of wheat. Can J Plant Pathol 20:95–103

Miller SS, Reid LM, Butler G, Winter SP, McGoldrick NJ (2003) Long chain alkanes in silk extracts of maize genotypes with varying resistance to *Fusarium graminearum*. J Agric Food Chem 51:6702–6708

Miller SG, Dokken-Bouchard FL, Cruise D, Ippolito J, Schemenauer I (2012) Survey of lentil diseases in Saskatchewan, 2010 and 2011. Can Plant Dis Surv 92:136–137

Mondal SN, Kageyama K, Hyakumachi M (1996) Chlamydospore formation in *Fusarium solani* f. sp. *phaseoli* in root extracts of crop plants and their virulence to kidney beans (*Phaseolus vulgaris* L.). Soil Biol Biochem 28:539–543

Morcia C, Rattotti E, Stanca AM, Tumino G, Rossi V, Ravaglia S et al (2013) *Fusarium* genetic traceability: role for mycotoxin control in small grain cereals agro-food chains. J Cereal Sci 57:175–182

Mudge AM, Dill-Macky R, Dong Y, Gardiner DM, White RG, Manners JM (2006) A role for the mycotoxin deoxynivalenol in stem colonisation during crown rot disease of wheat caused by *Fusarium graminearum* and *Fusarium pseudograminearum*. Physiol Mol Plant Pathol 69:73–85

Müller S, Dekant W, Mally A (2012) Fumonisin B_1 and the kidney: modes of action for renal tumor formation by fumonisin B_1 in rodents. Food Chem Toxicol 50:3833–3846

Munkvold GP, Hellmich RL, Showers WB (1997a) Reduced Fusarium ear rot and symptomless infection in kernels of maize genetically engineered for European corn borer resistance. Phytopathology 87:1071–1077

Munkvold GP, McGee DC, Carlton WM (1997b) Importance of different pathways for maize kernel infection by *Fusarium moniliforme*. Phytopathology 87:209–217

Murillo-Williams A, Munkvold GP (2008) Systemic infection by *Fusarium verticillioides* in maize plants grown under three temperature regimes. Plant Dis 92:1695–1700

Musser SM, Gay ML, Mazzola EP, Plattner RD (1996) Identification of a new series of fumonisins containing 3-hydroxypyridine. J Nat Prod 59:970–972

Muthomi JM, Schuetze A, Dehne HW, Mutitu EW, Oerke EC (2000) Characterization of *Fusarium culmorum* isolates by mycotoxin production and aggressiveness to winter wheat. J Plant Dis Prot 107:113–123

Naseri B, Marefat A (2011) Large-scale assessment of agricultural practices affecting Fusarium root rot and common bean yield. Eur J Plant Pathol 131:179–195

Nelson EB (2004) Microbial dynamics and interactions in the spermosphere. Annu Rev Phytopathol 42:271–309

Nicolaisen M, Supronienė S, Nielsen LK, Lazzaro I, Spliid NH, Justesen AF (2009) Real-time PCR for quantification of eleven individual *Fusarium* species in cereals. J Microbiol Methods 76:234–240

Nielsen LK, Jensen JD, Rodriguez A, Jorgensen LN, Justesen AF (2012) TRI12 based quantitative real-time PCR assays reveal the distribution of trichothecene genotypes of *F. graminearum* and *F. culmorum* isolates in Danish small grain cereals. Int J Food Microbiol 157:384–392

O'Donnell K, Cigelnik E (1997) Two divergent intragenomic rDNA ITS2 types within a monophyletic lineage of the fungus Fusarium are nonorthologous. Mol Phylogenet Evol 7:103–116

O'Donnell K, Kistler HC, Tacke BK, Casper HH (2000) Gene genealogies reveal global phylogeographic structure and reproductive isolation among lineages of *Fusarium graminearum*, the fungus causing wheat scab. Proc Natl Acad Sci U S A 97:7905–7910

Obanor F, Neate S, Simpfendorfer S, Sabburg R, Wilson P, Chakraborty S (2013) Fusarium graminearum and Fusarium pseudograminearum caused the 2010 head blight epidemics in Australia. Plant Pathol 62:79–91

Oldenburg E, Kramer S, Schrader S, Weinert J (2008) Impact of the earthworm *Lumbricus terrestris* on the degradation of *Fusarium*-infected and deoxynivalenol-contaminated wheat straw. Soil Biol Biochem 40:3049–3053

Oyarzun PJ (1993) Bioassay to assess root rot in pea and effect of root rot on yield. Netherlands J Plant Pathol 99:61–75

Oyarzun P, Gerlagh M, Hoogland AE (1993) Pathogenic fungi involved in root rot of peas in the Netherlands and their physiological specialization. Netherlands J Plant Pathol 99:23–33

Oyarzun PJ, Gerlagh M, Zadoks JC (1998) Factors associated with soil receptivity to some fungal root rot pathogens of peas. Appl Soil Ecol 10:151–169

Palacios SA, Ramirez ML, Cabrera Zalazar M, Farnochi MC, Zappacosta D, Chiacchiera SM et al (2011) Occurrence of *Fusarium* spp. and fumonisin in durum wheat grains. J Agric Food Chem 59:12264–12269

Paper JM, Scott-Craig JS, Adhikari ND, Cuomo CA, Walton JD (2007) Comparative proteomics of extracellular proteins *in vitro* and *in planta* from the pathogenic fungus *Fusarium graminearum*. Proteomics 7:3171–3183

Parry DW, Jenkinson P, McLeod L (1995) *Fusarium* ear blight (scab) in small grain cereals—a review. Plant Pathol 44:207–238

Parsons MW, Munkvold GP (2010a) Associations of planting date, drought stress, and insects with Fusarium ear rot and fumonisin B1 contamination in California maize. Food Addit Contam Part A 27:591–607

Parsons MW, Munkvold GP (2010b) Relationships of immature and adult thrips with silk-cut, Fusarium ear rot and fumonisin B_1 contamination of maize in California and Hawaii. Plant Pathol 59:1099–1106

Pascale M, Visconti A, Prończuk M, Wiśniewska H, Chelkowski J (1997) Accumulation of fumonisins in maize hybrids inoculated under field conditions with *Fusarium moniliforme* Sheldon. J Sci Food Agric 74:1–6

Paul PA, Lipps PE, Hershman DE, McMullen MP, Draper MA, Madden LV (2008) Efficacy of triazole-based fungicides for Fusarium head blight and deoxynivalenol control in wheat: a multivariate meta-analysis. Phytopathology 98:999–1011

Paulitz TC (1996) Diurnal release of ascospores by Gibberella zeae in inoculated wheat plots. Plant Dis 80:674–678

Peraica M, Radić B, Lucić A, Pavlović M (1999) Toxic effects of mycotoxins in humans. Bull World Health Organ 77:754–766

Pérez-Brito D, Jeffers D, González-de-León D, Khairallah M, Cortés-Cruz M, Velázquez-Cardelas G et al (2001) QTL mapping of Fusarium moniliforme ear-rot resistance in highland maize, Mexico. Agrosciencia 35:181–196

Perkowski J, Prończuk M, Chelkowski J (1997) Deoxynivalenol and acetyldeoxynivalenol accumulation in field maize inoculated with *Fusarium graminearum*. J Phytopathol 145:113–116

Persson L, Bødker L, Larsson-Wikström M (1997) Prevalence and pathogenicity of foot and root rot pathogens of pea in southern Scandinavia. Plant Dis 81:171–174

Pestka J (2010) Toxicological mechanisms and potential health effects of deoxynivalenol and nivalenol. World Mycotoxin J 3:323–347

Pestka JJ, Bondy GS (1994) Immunotoxic effects of mycotoxins. In: Miller JD, Trenholm HL (eds) Mycotoxins in grain: compounds other than aflatoxin. American Phytopathological Society, St. Paul, MN

Phalip V, Delalande F, Carapito C, Goubet F, Hatsch D, Leize-Wagner E et al (2005) Diversity of the exoproteome of *Fusarium graminearum* grown on plant cell wall. Curr Genet 48:366–379

Poppenberger B, Berthiller F, Lucyshyn D, Sieberer T, Schuhmacher R, Krska R et al (2003) Detoxification of the *Fusarium* mycotoxin deoxynivalenol by a UDP-glucosyltransferase from *Arabidopsis thaliana*. J Biol Chem 278:47905–47914

Prelusky DB, Rotter BA, Rotter RG (1994) Toxicology of mycotoxins. In: Miller JD, Trenholm HL (eds) Mycotoxins in grain: compounds other than aflatoxin. American Phytopathological Society, St. Paul, MN

Pritsch C, Muehlbauer GJ, Bushnell WR, Somers DA, Vance CP (2000) Fungal development and induction of defense response genes during early infection of wheat spikes by *Fusarium graminearum*. Mol Plant Microbe Interact 13:159–169

Pritsch C, Vance CP, Bushnell WR, Somers DA, Hohn TM, Muehlbauer GJ (2001) Systemic expression of defense response genes in wheat spikes as a response to *Fusarium graminearum* infection. Physiol Mol Plant Pathol 58:1–12

Proctor RH, Hohn TM, McCormick SP (1995) Reduced virulence of *Gibberella zeae* caused by disruption of a trichothecene toxin biosynthetic gene. Mol Plant Microbe Interact 8:593–601

Proctor RH, Brown DW, Plattner RD, Desjardins AE (2003) Co-expression of 15 contiguous genes delineates a fumonisin biosynthetic gene cluster in Gibberella moniliformis. Fungal Genet Biol 38:237–249

Proctor RH, Plattner RD, Desjardins AE, Busman M, Butchko RA (2006) Fumonisin production in the maize pathogen *Fusarium verticillioides*: genetic basis of naturally occurring chemical variation. J Agric Food Chem 54:2424–2430

Prussin AJ II (2013) Monitoring and predicting the long distance transport of *Fusarium graminearum*, causal agent of Fusarium head blight in wheat and barley [Thesis]. Virginia Polytechnic Institute and State University, Blacksburg, VA

Puri KD, Zhong S (2010) The 3ADON population of *Fusarium graminearum* found in North Dakota is more aggressive and produces a higher level of DON than the prevalent 15ADON population in spring wheat. Phytopathology 100:1007–1014

Rampitsch C, Day J, Subramaniam R, Walkowiak S (2013) Comparative secretome analysis of *Fusarium graminearum* and two of its non-pathogenic mutants upon deoxynivalenol induction *in vitro*. Proteomics 13:1913–1921

Reid LM (1996) Screening maize for resistance to Gibberella ear rot. Agric Agri Food Can Tech Bull Publ 196-5E:1–40

Reid LM, Sinha RC (1998) Maize maturity and the development of Giberrella ear rot symptoms and deoxynivalenol after inoculation. Eur J Plant Pathol 104:147–154

Reid LM, Bolton AT, Hamilton RI, Woldemariam T, Mather DE (1992a) Effect of silk age on resistance of maize to *Fusarium graminearum*. Can J Plant Pathol 14:293–298
Reid LM, Mather DE, Hamilton RI, Bolton AT (1992b) Genotypic differences in the resistance of maize silk to Fusarium graminearum. Can J Plant Pathol 14:211–214
Reid LM, Stewart DW, Hamilton RI (1996) A 4-year study of the association between Gibberella ear rot severity and deoxynivalenol concentration. J Phytopathol 144:431–436
Reid LM, Nicol RW, Ouellet T, Savard M, Miller JD, Young JC et al (1999) Interaction of *Fusarium graminearum* and *F. moniliforme* in maize ears: disease progress, fungal biomass, and mycotoxin accumulation. Phytopathology 89:1028–1037
Reid LM, Zhu X, Savard ME, Sinha RC, Vigier B (2000) Pre-harvest accumulation of deoxynivalenol in sweet corn ears inoculated with *Fusarium graminearum*. Food Addit Contam 17:689–701
Reid LM, McDiarmid G, Parker AJ, Woldemariam T, Hamilton RI (2001a) CO388 and CO389 corn inbred lines. Can J Plant Sci 81:457–459
Reid LM, McDiarmid G, Parker AJ, Woldemariam T, Hamilton RI (2001b) CO430, CO431 and CO432 corn inbred lines. Can J Plant Sci 81:283–284
Reid LM, Woldemariam T, Zhu X, Stewart DW, Schaafsma AW (2002) Effect of inoculation time and point of entry on disease severity in *Fusarium graminearum*, *Fusarium verticillioides*, or *Fusarium subglutinans* inoculated maize ears. Can J Plant Pathol 24:162–167
Reid LM, McDiarmid G, Parker AJ, Woldemariam T (2003) CO441 corn inbred line. Can J Plant Sci 83:79–80
Reid LM, Zhu X, Parker A, Yan W (2009) Increased resistance to *Ustilago zeae* and *Fusarium verticillioides* in maize inbred lines bred for *Fusarium graminearum* resistance. Euphytica 165:567–578
Reinprecht Y, Wu X, Yan S, Labey L, Dasilva E, Martin J, Pauls K (2008) A microarray-based approach for identifying genes for resistance to *Fusarium graminearum* in maize (*Zea Mays* L.). Cereal Res Commun 36:253–259
Reynoso MM, Ramirez ML, Torres AM, Chulze SN (2011) Trichothecene genotypes and chemotypes in *Fusarium graminearum* strains isolated from wheat in Argentina. Int J Food Microbiol 145:444–448
Rheeder JP, Marasas WFO, Vismer HF (2002) Production of fumonisin analogs by *Fusarium* species. Appl Environ Microbiol 68:2101–2105
Robertson-Hoyt LA, Jines MP, Balint-Kurti PJ, Kleinschmidt CE, White DG, Payne GA et al (2006) QTL mapping for Fusarium ear rot and fumonisin contamination resistance in two maize populations. Crop Sci 46:1734–1743
Rocha O, Ansari K, Doohan FM (2005) Effects of trichothecene mycotoxins on eukaryotic cells: a review. Food Addit Contam 22:369–378
Román-Avilés B, Kelly JD (2005) Identification of quantitative trait loci conditioning resistance to Fusarium root rot in common bean. Crop Sci 45:1881–1890
Ronquillo-López MG, Grau CR, Nienhuis J (2010) Variation in reaction to *Fusarium* spp. identified in a common bean (*Phaseolus vulgaris* L.) population developed for field-based resistance to root rot and wilt. Crop Sci 50:2303–2309
Rotem J, Aust HJ (1991) The effect of ultraviolet and solar radiation and temperature on survival of fungal propagules. J Phytopathol 133:76–84
Rotem J, Wooding B, Aylor DE (1985) The role of solar radiation, especially ultraviolet, in the mortality of fungal spores. Phytopathology 75:510–514
Rubert J, James KJ, Mañes J, Soler C (2012) Study of mycotoxin calibration approaches on the example of trichothecenes analysis from flour. Food Chem Toxicol 50:2034–2041
Salas B, Dill-Macky R (2003) Previous crop affecting soil populations of Fusarium head blight pathogens in Minnesota. In: Canty SM, Lewis J, Ward RW (eds) Proceedings of 2003 national Fusarium head blight forum, Bloomington, MN, p 171
Salas B, Dill-Macky R (2004) Incidence of *Fusarium graminearum* in pre-harvest and overwintered residues of wheat cultivars differing in Fusarium head blight resistance. In: Canty SM, Boring T, Versdahl K, Wardwell J, Ward RW (eds) Proceedings of second international sympo-

sium on Fusarium head blight, incorporating the eighth European Fusarium seminar, Orlando, FL, pp 502–503

Salas B, Dill-Macky R (2005) Effect of residue management and host resistance on the epidemiology of Fusarium head blight. Phytopathology 95:952 (Abstr.)

Salgado JD, Wallhead M, Madden LV, Paul PA (2011) Grain harvesting strategies to minimize grain quality losses due to Fusarium head blight in wheat. Plant Dis 95:1448–1457

Sampietro DA, Vattuone MA, Presello DA, Fauguel CM, Catalán CAN (2009) The pericarp and its surface wax layer in maize kernels as resistance factors to fumonisin accumulation by *Fusarium verticillioides*. Crop Prot 28:196–200

Santiago R, Reid LM, Arnason JT, Zhu X, Martinez N, Malvar RA (2007) Phenolics in maize genotypes differing in susceptibility to *Gibberella* stalk rot (*Fusarium graminearum* Schwabe). J Agric Food Chem 55:5186–5193

Schaafsma AW, Ilinic LT, Miller JD, Hooker DC (2001) Agronomic considerations for reducing deoxynivalenol in wheat grain. Can J Plant Pathol 23:279–285

Schenzel J, Forrer H-R, Vogelgsang S, Bucheli T (2012) Development, validation and application of a multi-mycotoxin method for the analysis of whole wheat plants. Mycotoxin Res 28:135–147

Schmale DG III, Bergstrom GC (2007) Aerobiology of *Gibberella zeae*: Whence come the spores for Fusarium head blight? In: 2007 national Fusarium head blight forum, Kansas City, MO, p 70

Schneider KA, Grafton KF, Kelly JD (2001) QTL analysis of resistance to Fusarium root rot in bean. Crop Sci 41:535–542

Schollenberger M, Drochner W, Müller H-M (2007) *Fusarium* toxins of the scirpentriol subgroup: a review. Mycopathologia 164:101–118

Schrader S, Kramer S, Oldenburg E, Weinert J (2009) Uptake of deoxynivalenol by earthworms from *Fusarium*-infected wheat straw. Mycotoxin Res 25:53–58

Schroeder HW, Christensen JJ (1963) Factors affecting resistance of wheat to scab by *Gibberella zeae*. Phytopathology 53:831–838

Schwartz HF, Steadman JR, Hall R, Forster RL (2005) Compendium of bean disease, 2nd edn. APS Press, St. Paul, MN

Scott PM (1997) Multi-year monitoring of Canadian grains and grain-based foods for trichothecenes and zearalenone. Food Addit Contam 14:333–339

Seaman WL (1982) Epidemiology and control of mycotoxigenic fusaria on cereal grains. Can J Plant Pathol 4:187–190

Seifert KA, Aoki T, Baayen RP, Brayford D, Burgess LW, Chulze S et al (2003) The name *Fusarium moniliforme* should no longer be used. Mycol Res 107:643–644

Sekhon RS, Kuldau G, Mansfield M, Chopra S (2006) Characterization of *Fusarium*-induced expression of flavonoids and PR genes in maize. Physiol Mol Plant Pathol 69:109–117

Shank RA, Foroud NA, Hazendonk P, Eudes F, Blackwell BA (2011) Current and future experimental strategies for structural analysis of trichothecene mycotoxins—a prospectus. Toxins 3:1518–1553

Shephard GS, Berthiller F, Burdaspal PA, Crews C, Jonker MA, Krska R et al (2012) Developments in mycotoxin analysis: an update for 2010–2011. World Mycotoxin J 5:3–30

Shier WT (1992) Sphingosine analogs: an emerging new class of toxins that includes the fumonisins. Toxin Rev 11:241–257

Sims JL, Frederick LR (1970) Nitrogen immobilization and decomposition of corn residue in soil and sand as affected by residue particle size. Soil Sci 109:355–361

Singh SP, Schwartz HF (2010) Breeding common bean for resistance to diseases: a review. Crop Sci 50:2199–2223

Sinha RC, Savard ME (1997) Concentration of deoxynivalenol in single kernels and various tissues of wheat heads. Can J Plant Pathol 19:8–12

Siranidou E, Kang Z, Buchenauer H (2002) Studies on symptom development, phenolic compounds and morphological defence responses in wheat cultivars differing in resistance to Fusarium head blight. J Phytopathol 150:200–208

Smiley RW, Gourlie JA, Easley SA, Patterson L-M, Whittaker RG (2005) Crop damage estimates for crown rot of wheat and barley in the Pacific Northwest. Plant Dis 89:595–604

Somers DJ, Fedak G, Clarke J, Cao W (2006) Mapping of FHB resistance QTLs in tetraploid wheat. Genome 49:1586–1593

Stack RW (1997) Gradients of Fusarium head blight in wheat along transects away from a concentrated source of *Gibberella zeae* ascospore inoculum. In: Dill-Macky R, Jones RK (eds) National Fusarium head blight forum, St. Paul, MN

Stack RW (2000) Return of an old problem: Fusarium head blight of small grains. APSnet Plant Health Reviews. http://www.apsnet.org/publications/apsnetfeatures/Pages/headblight.aspx

Stack RW, Elias EM, Fetch JM, Miller JD, Joppa LR (2002) Fusarium head blight reaction of Langdon durum-*Triticum dicoccoides* chromosome substitution lines. Crop Sci 42:637–642

Stahl DJ, Theuerkauf A, Heitefuss R, Schäfer W (1994) Cutinase of *Nectria haematococca* (*Fusarium solani* f. sp. *pisi*) is not required for fungal virulence or organ specificity on pea. Mol Plant Microbe Interact 7:713–725

Starkey DE, Ward TJ, Aoki T, Gale LR, Kistler HC, Geiser DM et al (2007) Global molecular surveillance reveals novel Fusarium head blight species and trichothecene toxin diversity. Fungal Genet Biol 44:1191–1204

Statistics Canada. Table 001-0010 - Estimated areas, yield, production and average farm price of principal field crops, in metric units, annual, CANSIM (database). (accessed: 2014-07-02)

Steffenson BJ (2003) Fusarium head blight of barley: impact, epidemics, management, and strategies for identifying and utilizating genetic resistance. In: Leonard KJ, Bushnell WR (eds) Fusarium head blight of wheat and barley. The American Phytopathological Society, St. Paul, Minnesota, pp 241–295

Steiner B, Kurz H, Lemmens M, Buerstmayr H (2009) Differential gene expression of related wheat lines with contrasting levels of head blight resistance after *Fusarium graminearum* inoculation. Theor Appl Genet 118:753–764

Strausbaugh CA, Maloy OC (1986) *Fusarium* scab of irrigated wheat in central Washington. Plant Dis 70:1104–1106

Suga H, Gale LR, Kistler HC (2004) Development of VNTR markers for two *Fusarium graminearum* clade species. Mol Ecol Notes 4:468–470

Sutton JC (1982) Epidemiology of wheat head blight and maize ear rot caused by *Fusarium graminearum*. Can J Plant Pathol 4:195–209

Tang Y, Lin H, Chen Y, Su W, Wang S, Chiueh L et al (2012) Development of a quantitative multimycotoxin method in rice, maize, wheat and peanut using UPLC-MS/MS. Food Anal Methods 6:727–736

Tangni EK, Motte J-C, Callebaut A, Pussemier L (2010) Cross-reactivity of antibodies in some commercial deoxynivalenol test kits against some fusariotoxins. J Agric Food Chem 58: 12625–12633

Tangni E, Motte J-C, Callebaut A, Chandelier A, Schrijver M, Pussemier L (2011) Deoxynivalenol loads in matched pair wheat samples in Belgium: comparison of ELISA VERATOX kit against liquid chromatography. Mycotoxin Res 27:105–113

Teich AH (1989) Epidemiology of wheat (*Triticum aestivum* L.) scab caused by *Fusarium* spp. In: Chelkowski J (ed) Topics in secondary metabolism 2: Fusarium—mycotoxins, taxonomy, and pathogenicity. Elsevier Science, New York, NY, pp 269–282

Teich AH, Hamilton JR (1985) Effect of cultural practices, soil phosphorus, potassium, and pH on the incidence of Fusarium head blight and deoxynivalenol levels in wheat. Appl Environ Microbiol 49:1429–1431

Teich AH, Nelson K (1984) Survey of Fusarium head blight and possible effects of cultural practices in wheat fields in Lambton County in 1983. Can Plant Dis Surv 64:11–13

Tekauz A, McCallum B, Gilbert J (2000) Review: Fusarium head blight of barley in western Canada. Can J Plant Pathol 22:9–16

Tekauz A, McCallum B, Ames N, Fetch JM (2004) Fusarium head blight of oat—current status in western Canada. Can J Plant Pathol 26:473–479

Tekauz A, Mitchell Fetch JW, Rossnagel BG, Savard ME (2008) Progress in assessing the impact of Fusarium head blight on oat in western Canada and screening of avena germplasm for resistance. Cereal Res Commun 36:49–56

Terzi V, Morcia C, Faccioli P, Faccini N, Rossi V, Cigolini M et al (2007) *Fusarium* DNA traceability along the bread production chain. Int J Food Sci Technol 42:1390–1396

Tittlemier SA, Roscoe M, Drul D, Blagden R, Kobialka C, Chan J et al (2013a) Single laboratory evaluation of a planar waveguide-based system for a simple simultaneous analysis of four mycotoxins in wheat. Mycotoxin Res 29:55–62

Tittlemier SA, Roscoe M, Trelka M, Gaba D, Chan J, Patrick S et al (2013b) *Fusarium* damage in cereal grains from western Canada. Part II: occurrence of mycotoxins and their source organisms in *Fusarium* damaged durum wheat harvested in 2010. J Agric Food Chem 61: 5438–5448

Tittlemier SA, Roscoe M, Trelka R, Patrick SK, Bamforth JM, Gräfenhan T et al (2014) Fate of moniliformin during milling of Canadian durum wheat, processing, and cooking of spaghetti. Can J Plant Sci 94:555–563

Tkachuk R, Dexter JE, Tipples KH, Nowicki W (1991) Removal by specific gravity table of tombstone kernels and associated trichothecenes from wheat infected with Fusarium head blight. Cereal Chem 68:428–431

Tóth B, Mesterházy Á, Horvath Z, Bartók T, Varga M, Varga J (2005) Genetic variability of central European isolates of the *Fusarium graminearum* species complex. Eur J Plant Pathol 113:35–45

Trail F, Xu H, Loranger R, Gadoury D (2002) Physiological and environmental aspects of ascospore discharge in *Gibberella zeae* (anamorph *Fusarium graminearum*). Mycologia 94:181–189

Turkington TK, Clear RM, Patrick SK (2005) Characteristics of wheat fields positive for Fusarium graminearum in Alberta, in 2002 and 2003. Can J Plant Pathol 27:479 (Abstr.)

Turkington TK, Kuzyk A, Dunn R, McLaren D, Irvine B, Clear RM et al. (2006) Irrigation and plant disease management. In: Proceedings of the irrigated crop production update: 2006 conference, Lethbridge, AB. http://www1.agric.gov.ab.ca/$department/deptdocs.nsf/all/ind10759

Ueno Y (1983) Trichothecenes: chemical, biological and toxicological aspects. Elsevier Scientific, Amsterdam

Ueno Y, Hosoya M, Morita Y, Ueno I, Tatsuno T (1968) Inhibition of the protein synthesis in rabbit reticulocyte by nivalenol, a toxic principle isolated from *Fusarium nivale*-growing rice. J Biochem 64:479–485

van der Fels-Klerx HJ, de Rijk TC, Booij CJH, Goedhart PW, Boers EAM, Zhao C et al (2012) Occurrence of Fusarium head blight species and *Fusarium* mycotoxins in winter wheat in the Netherlands in 2009. Food Addit Contam Part A 29:1716–1726

Vesonder RF, Ellis JJ, Rohwedder WK (1981) Elaboration of vomitoxin and zearalenone by *Fusarium* isolates and the biological activity of *Fusarium*-produced toxins. Appl Environ Microbiol 42:1132–1134

Vishwanath V, Sulyok M, Labuda R, Bicker W, Krska R (2009) Simultaneous determination of 186 fungal and bacterial metabolites in indoor matrices by liquid chromatography/tandem mass spectrometry. Anal Bioanal Chem 395:1355–1372

Voigt CA, Schafer W, Salomon S (2005) A secreted lipase of *Fusarium graminearum* is a virulence factor required for infection of cereals. Plant J 42:364–375

Volpi C, Janni M, Lionetti V, Bellincampi D, Favaron F, D'Ovidio R (2011) The ectopic expression of a pectin methyl esterase inhibitor increases pectin methyl esterification and limits fungal diseases in wheat. Mol Plant Microbe Interact 24:1012–1019

von der Ohe C, Gauthier V, Tamburic-Ilincic L, Brule-Babel A, Fernando W, Clear R et al (2010) A comparison of aggressiveness and deoxynivalenol production between Canadian *Fusarium graminearum* isolates with 3-acetyl and 15-acetyldeoxynivalenol chemotypes in field-grown spring wheat. Eur J Plant Pathol 127:407–417

Voss KA, Gelineau-van Waes JB, Riley RT (2006) Fumonisins: current research trends in developmental toxicology. Mycotoxin Res 22:61–69

Voss KA, Smith GW, Haschek WM (2007) Fumonisins: toxicokinetics, mechanism of action and toxicity. Anim Feed Sci Technol 137:299–325

Waalwijk C, van der Heide R, de Vries I, van der Lee T, Schoen C, Costrel-de Corainville G et al (2004) Quantitative detection of *Fusarium* species in wheat using TaqMan. Eur J Plant Pathol 110:481–494

Waalwijk C, Koch S, Ncube E, Allwood J, Flett B, de Vries I et al (2008) Quantitative detection of *Fusarium* spp. and its correlation with fumonisin content in maize from South African subsistence farmers. World Mycotoxin J 1:39–47

Waggoner PE, Green JSA, Smith FB (1983) The aerial dispersal of the pathogens of plant disease. Philos Trans R Soc Lond B Biol Sci 302:451–462

Wang Y, Yang L, Xu H, Li Q, Ma Z, Chu C (2005) Differential proteomic analysis of proteins in wheat spikes induced by *Fusarium graminearum*. Proteomics 5:4496–4503

Wanjiru MW, Kang Z, Buchenauer H (2002) Importance of cell wall degrading enzymes produced by *Fusarium graminearum* during infection of wheat heads. Eur J Plant Pathol 108:803–810

Ward TJ, Bielawski JP, Kistler HC, Sullivan E, O'Donnell K (2002) Ancestral polymorphism and adaptive evolution in the trichothecene mycotoxin gene cluster of phytopathogenic *Fusarium*. Proc Natl Acad Sci U S A 99:9278–9283

Ward TJ, Clear RM, Rooney AP, O'Donnell K, Gaba D, Patrick S et al (2008) An adaptive evolutionary shift in Fusarium head blight pathogen populations is driving the rapid spread of more toxigenic *Fusarium graminearum* in North America. Fungal Genet Biol 45:473–484

Wardle DA (1995) Impacts of disturbance on detritus food webs in agro-ecosystems of contrasting tillage and weed management practices. In: Begon M, Fitter AH (eds) Advances in ecological research, vol 26. Academic, New York, pp 105–185

Warren HL, Kommedahl T (1973) Fertilization and wheat refuse effects on *Fusarium* species associated with wheat roots in Minnesota. Phytopathology 63:103–108

Warth B, Parich A, Atehnkeng J, Bandyopadhyay R, Schuhmacher R, Sulyok M et al (2012) Quantitation of mycotoxins in food and feed from Burkina Faso and Mozambique using a modern LC-MS/MS multitoxin method. J Agric Food Chem 60:9352–9363

Wearing AH, Burgess LW (1977) Distribution of *Fusarium roseum* 'Graminearum' Group 1 and its mode of survival in eastern Australian wheat belt soils. Trans Br Mycol Soc 69:429–442

Whitaker TB, Hagler WM Jr, Giesbrecht FG, Johannson AS (2002) Sampling wheat for deoxynivalenol. In: DeVries JW, Trucksess M, Jackson LS (eds) Mycotoxins and food safety. Kluwer, New York, NY, pp 73–83

White DG (1999) Compendium of corn diseases, 3rd edn. APS Press, St. Paul, MN

Wiese MV (1987) Compendium of wheat diseases, 2nd edn. APS Press, St. Paul, MN

Windels CE, Kommedahl T (1984) Late-season colonization and survival of *Fusarium graminearum* Group II in cornstalks in Minnesota. Plant Dis 68:791–793

Windels CE, Kommedahl T, Stienstra WC, Burnes PM (1988) Occurrence of Fusarium species in symptom-free and overwintered cornstalks in northwestern Minnesota. Plant Dis 72:990–993

Wolfarth F, Schrader S, Oldenburg E, Weinert J, Brunotte J (2011) Earthworms promote the reduction of *Fusarium* biomass and deoxynivalenol content in wheat straw under field conditions. Soil Biol Biochem 43:1858–1865

Wolf-Hall CE (2007) Mold and mycotoxin problems encountered during malting and brewing. Int J Food Microbiol 119:89–94

Wu W, Bates MA, Bursian SJ, Link JE, Flannery BM, Sugita-Konishi Y et al (2013) Comparison of emetic potencies of the 8-ketotrichothecenes deoxynivalenol, 15-acetyldeoxynivalenol, 3-acetyldeoxynivalenol, fusarenon X, and nivalenol. Toxicol Sci 131:279–291

Xiang K, Reid LM, Zhu X (2010a) Relationship among kernel drydown rates, environmental factors and resistance to Gibberella ear rot, Fusarium ear rot and common smut of corn. In: Joint annual meeting of the Canadian Phytopathological Society and the Pacific Division of the American Phytopathological Society, Vancouver, BC

Xiang K, Zhang ZM, Reid LM, Zhu XY, Yuan GS, Pan GT (2010b) A meta-analysis of QTL associated with ear rot resistance in maize. Maydica 55:281–290

Xue AG (2003) Efficacy of *Clonostachys rosea* strain ACM941 and fungicide seed treatments for controlling the root rot complex of field pea. Can J Plant Sci 83:519–524

Yi C, Kaul HP, Kübler E, Schwadorf K, Aufhammer W (2001) Head blight (*Fusarium graminearum*) and deoxynivalenol concentration in winter wheat as affected by pre-crop, soil tillage and nitrogen fertilization. J Plant Dis Protect 108:217–230

Yli-Mattila T, Gagkaeva T (2010) Molecular chemotyping of *Fusarium graminearum*, *F. culmorum*, and *F. cerealis* isolates from Finland and Russia. In: Gherbawy Y, Voigt K (eds) Molecular identification of fungi. Springer, Berlin, pp 159–177

Yli-Mattila T, Paavanen-Huhtala S, Parikka P, Jestoi M, Klemsdal SS, Rizzo A (2006) Genetic variation, real-time PCR, metabolites and mycotoxins of *Fusarium avenaceum* and related species. Mycotoxin Res 22:79–86

Yu GT, Franckowiak JD, Neate SM, Zhang B, Horsley RD (2010) A native QTL for Fusarium head blight resistance in North American barley (*Hordeum vulgare* L.) independent of height, maturity, and spike type loci. Genome 53:111–118

Yuan J, Ali ML, Taylor J, Liu J, Sun G, Liu W et al (2008) A guanylyl cyclase-like gene is associated with Gibberella ear rot resistance in maize (*Zea mays* L.). Theor Appl Genet 116:465–479

Zhang Y, Choi Y-E, Zou X, Xu J-R (2011) The FvMK1 mitogen-activated protein kinase gene regulates conidiation, pathogenesis, and fumonisin production in *Fusarium verticillioides*. Fungal Genet Biol 48:71–79

Zhang H, Van der Lee T, Waalwijk C, Chen W, Xu J, Xu J et al (2012) Population analysis of the *Fusarium graminearum* species complex from wheat in China show a shift to more aggressive isolates. PLoS One 7:e31722

Zhou W, Kolb FL, Riechers DE (2005) Identification of proteins induced or upregulated by Fusarium head blight infection in the spikes of hexaploid wheat (*Triticum aestivum*). Genome 48:770–780

Zhou W, Eudes F, Laroche A (2006) Identification of differentially regulated proteins in response to a compatible interaction between the pathogen *Fusarium graminearum* and its host, *Triticum aestivum*. Proteomics 6:4599–4609

Zhu H, Gilchrist L, Hayes P, Kleinhofs A, Kudrna D, Liu Z et al (1999) Does function follow form? Principal QTLs for Fusarium head blight (FHB) resistance are coincident with QTLs for inflorescence traits and plant height in a doubled-haploid population of barley. Theor Appl Genet 99:1221–1232

Zhuang Y, Gala A, Yen Y (2013) Identification of functional genic components of major *Fusarium* head blight resistance quantitative trait loci in wheat cultivar Sumai 3. Mol Plant Microbe Interact 26:442–450

Chapter 11
Pseudomonas fluorescens: A Potential Biocontrol Agent for Management of Fungal Diseases of Crop Plants

D. Majumder, J.D. Kongbrailatpam, E.G. Suting, B. Kangjam, and D. Lyngdoh

11.1 Introduction

Rhizosphere region of the soil is inhabited by the large number of microorganisms having potential to promote plant growth, yield, and also to provide protection against diseases by suppressing pathogenic microorganisms. Several biologically important processes and interactions take place in the region which is primarily due to the influx of mineral nutrients from accumulation of plant root exudates. Rhizospheric microorganisms have the ability to solubilize the insoluble phosphates and maintain the soil health and quality. Interest in biological control of plant pathogens has been stimulated in recent years. Several bacteria belonging to the genera *Bacillus* and *Pseudomonas* have been intensively investigated as biocontrol agents primarily due to their ability to produce antimicrobial metabolites and ecological fitness in soil (Shanahan et al. 1992; Nielsen et al. 2000). Fluorescent pseudomonads are often predominant among plant rhizosphere-associated bacteria (Glick et al. 1995), make up a diverse group that can be visually distinguish from other *Pseudomonas* by their ability to produce a water-soluble yellow green pigment and has been considered as an important group due to their biofertilizing and biocontrol properties. The rhizobacteria are found in large numbers in all the major natural environments viz. terrestrial, freshwater, and marine and they also form intimate associations with plants and animals. This universal distribution suggests a remarkable degree of physiological and genetic adaptability (Spiers et al. 2005). They comprise *Pseudomonas aeruginosa,* the type species of the genus, *P. aureofaciens, P. chlororaphis, P. fluorescens, P. putida*, and the plant pathogenic species *P. cichorii* and *P. syringae* (Dwivedi and Johri 2003). *P. fluorescens* becomes a promising biocontrol agent for its noble

D. Majumder, Ph.D. (✉) • J.D. Kongbrailatpam, M.Sc. • E.G. Suting, M.Sc.
B. Kangjam, M.Sc. • D. Lyngdoh, M.Sc.
School of Crop Protection, College of Post Graduate Studies, Central Agricultural University, Umroi Road, Umiam, Shillong, Meghalaya 793 103, India
e-mail: dipali_assam@yahoo.co.in

A. Goyal and C. Manoharachary (eds.), *Future Challenges in Crop Protection Against Fungal Pathogens*, Fungal Biology, DOI 10.1007/978-1-4939-1188-2_11,

antagonistic property against a wide variety of phytopathogenic fungi and bacteria. Beneficial effect of *P. fluorescens* on plant also account for plant growth promotion. Selective strains of fluorescent pseudomonads have also been reported for biodegradation of agricultural pollutants and for weed control in agricultural fields as bioinoculants (Ramamoorthy et al. 2001; Sunish et al. 2005). In recent years, fluorescent pseudomonads isolated from rhizosphere of several crops have drawn attention worldwide owing to the ability of production of secondary metabolites such as antibiotics, volatile compounds, hydrogen cyanide (HCN), siderophores, cell wall-degrading enzymes, and phytohormones (Bakker and Schippers 1987; O'Sullivan and O'Gara 1992; Nielsen et al. 2000). Pseudomonads are aggressive colonizers of the rhizosphere of various crop plants, and have a broad spectrum antagonistic activity against different group of plant pathogens. Plant pathologists have been enthralled by the idea that antagonistic microorganisms could be used as environment friendly biocontrol agents, both in the field and in greenhouses. However, as noted by Garrett et al. (1965) that there were no shortcuts to biological control.

11.2 Genus *Pseudomonas*

Pseudomonads are rod-shaped Gram-negative bacteria that are characterized by metabolic versatility, are oxidase positive with aerobic respiration (some strains also have anaerobic respiration with nitrate as the terminal electron acceptor and/or arginine fermentation), and are motile owing to one or several polar flagella. They are ubiquitous in nature and belong to the Pseudomonadaceae family (subclass: γ-Proteobacteria; order: Pseudomonadales). Pseudomonads have simple nutritional requirements, and this is reflected by the relative abundance of these organisms in nature. They are found in soils, foliage, fresh water, sediments, and sea water. The genus *Pseudomonas* includes mostly fluorescent pseudomonads as well as a few non-fluorescent species. Fluorescent *Pseudomonas* group represents: (1) phytopathogenic cytochrome c oxidase-positive species, such as *P. cichorii*, *P. marginalis*, and *P. tolaasii*; (2) non-phytopathogenic, non-necrogenic strains, such as *P. fluorescens*, *P. putida*, *P. chlororaphis*, *P. aureofaciens*, and *P. aeruginosa* type species; (3) phytopathogenic necrogenic fluorescent *Pseudomonas* spp. without cytochrome c oxidase: *P. syringae* and *P. viridiflava*. Non-fluorescent *Pseudomonas* group constitutes *P. stutzeri*, *P. mendocina*, *P. alcaligenes*, and *P. pseudoalcaligenes* (Palleroni 1984; Holt et al. 1994; Bossis et al. 2000).

Fluorescent pseudomonads are the most studied group within the genus *Pseudomonas* that can generally be visually distinguished from other pseudomonads by their ability to produce a water-soluble yellow-green pigment in culture media. All fluorescent pseudomonads fall into one of the five rRNA group (Palleroni et al. 1973). The Guanine + Cytosine (G + C) content of their DNA ranges from 58 to 68 % (Palleroni 1975). *Pseudomonas* is characterized by their ability to grow in simple media at the expense of a great variety of simple organic compounds, without needing organic growth factors. King's medium B (KMB) is an optimal medium for isolation of most species of *Pseudomonas* (King et al. 1954). The optimum temperature for growth is between 25 and 30° C. As a group, the fluorescent pseudomonads are of primary significance in diverse areas such as animal pathogenicity, plant pathogenicity, food spoilage, and biological control.

11.3 Mechanisms of Biocontrol by *P. fluorescens*

Fluorescent pseudomonads have emerged as the largest and potentially the most promising group of biocontrol agent as well as PGPR with their rapid growth, simple nutritional requirements, ability to utilize diverse organic substrates, and mobility. Several basic mechanisms of the bacterial-induced biocontrol, particularly concerning the *Pseudomonas* genus are antibiosis, fungistasis, competition for nutrients, modification of the biophysical root environment, active exclusion of pathogenic organisms from the rhizosphere, detoxification of pathogen virulence factors, and the induction of plant disease resistance (Bakker and Schippers 1987; O'Sullivan and O'Gara 1992; Nielsen et al. 2000 and Daval et al. 2011). The fluorescent pseudomonads also antagonize plant pathogens by producing a range of metabolites like siderophores and other substances such as cyanide (O'Sullivan and O'Gara 1992). They are aggressive colonizers of the rhizosphere of various crop plants, and have a broad spectrum antagonistic activity against different group of plant pathogens. Different mechanisms such as accumulation of phenolic compounds, pathogenesis-related proteins (PR-proteins), lysis of cell wall of the fungal pathogen, and secretion of extracellular lytic enzymes also lead to reduction of plant diseases (O'Sullivan and O'Gara 1992; Saikia et al. 2004). Antagonistic potentiality can be exploited successfully against plant pathogens. So far, several strains of *P. fluorescens* have been exploited for the management of several soil-borne diseases. The different diverse mechanisms of biocontrol include the following:

1. Antibiotic-mediated suppression
2. HCN production
3. Siderophores production
4. Competition for space and nutrients
5. Production of plant growth promoting substances (PGPS)
6. Mineral phosphate solubilization (MPS)
7. Induced systemic resistance

11.3.1 Antibiotic-Mediated Suppression

Antibiotic production has been recognized as an important trait in the biological control of plant diseases by fluorescent pseudomonads (Gurusiddaiah et al. 1986; Homma and Suzui 1989) and has been known for over 150 years. Advances within the past 2 decades have provided new insight to the diversity of antibiotics produced, regulation of synthesis, mode of action, and their functional roles. Compounds important for biological control of plant pathogens, such as phenazines

(Thomashow and Weller 1988), pyoluteorin (Howell and Stipanovic 1980), pyrrolnitrin (Howell and Stipanovic 1979), tropolone (Lindberg 1981), pyocyanin (Dahiya et al. 1988), and 2,4-diacetylphloroglucinol (Shanahan et al. 1992) have been isolated from rhizospheric fluorescent pseudomonads. Naik et al. (2008) successfully screened *P. fluorescens* isolates using pyrrolnitrin gene-specific primers which during successive studies showed antagonistic effects against several fungal plant pathogens including *Pyricularia grisea, Fusarium oxysporum* f. sp. *cubense, Macrophomina phaseolina, Colletotrichum falcatum*, and *C. Capsici*. Homma and Suzui (1989) correlated antibiotic production (purified pyrrolnitrin and pyoluteorin) with disease suppression. Similarly DAPG, the antibiotic compound produced by fluorescent pseudomonad, was found effective to suppress plant pathogens (Shanahan et al. 1992). Hill et al. (1994) also correlated pyrrolnitrin synthesis by *P. fluorescens* BL915 with biological control activity against *Rhizoctonia solani*-incited disease of cotton. Antibiotic production indicated the involvement of DAPG, pyrrolnitrin, and pyoluteorin in the natural antagonism between *P. fluorescens* and pathogens (Rosales et al. 1995; El-Banna and Winkelmann 1998; Haas and Keel 2003; Brodhajen et al. 2005).

11.3.1.1 Screening of *P. fluorescens* Based on Amplification by Antibiotic Biosynthetic Gene-Specific Primers

Different antibiotics have been found to be associated with inhibition of pathogen growth by fluorescent pseudomonads. The biosynthetic genes for phenazine-1-carboxylic acid (PCA), 2,4-DAPG, pyrrolnitrin, pyoluteorin, and the zwittermicin (a self-resistance gene) have been sequenced (Hammer et al. 1997; Nowak-Thompson et al. 1999; Stohl et al. 1999; Mavrodi et al. 1998; Bangera and Thomashow 1996). Sequencing has enabled PCR-based detection of antibiotic-producing strains (Bangera and Thomashow 1999; de Souza and Raaijmakers 2003; Raaijmakers et al. 1997; McSpadden Gardener et al. 2001; Raffel et al. 1996; Picard et al. 2000).

Svercel et al. (2007) selected strains of *Pseudomonas* spp. through PCR-RFLP analysis that produced DAPG and HCN in Phl– HCN+. Zhang et al. (2006) used thirty primers to amplify antibiotic biosynthetic genes encoding phenazine-1-carboxylic acid, 2,4-diacetylphloroglucinol, pyoluteorin, pyrrolnitrin, the zwittermicin in *P. chlororaphis* PA23, *Pseudomonas* spp. strain DF41, and *Bacillus amyloliquefaciens* BS6 (Table 11.1). The presence of antibiotic biosynthetic or self-resistance genes in rhizobacterial strains can be investigated with polymerase chain reaction and by Southern blotting.

11.3.2 HCN Production

HCN produced by rhizobacteria plays a role in biological control of phytopathogens (Voisard et al. 1989; Defago et al. 1990). HCN inhibits the electron transport there by the energy supply to the cell gets disturbed resulting in death of the organism.

Table 11.1 Polymerase chain reaction primers and amplification products from genes encoding enzymes involved in the biosynthesis of several antibiotics (Data from Zhang et al. 2006)

Primer	Sequence	Gene (control strain)	Expected size of PCR product
Phenazine			
PHZ1[a]	GGCGACATGGTCAACGG	*phzCD* (*P. fluorescens* 2-79)	1,400 bp (PA23)
PHZ2[a]	CGGCTGGCGGCGTATAT	*phzCD* (*P. fluorescens* 2-79)	1,400 bp (PA23)
PCA2a[b]	TTGCCAAGCCTCGCTCCAAC	*phzCD* (*P. fluorescens* 2-79)	1,400 bp (not detected)
PCA3b[b]	CCGCGTTGTTCCTCGTTCAT	*phzCD* (*P. fluorescens* 2-79)	1,400 bp (not detected)
2,4-Diacetylphloroglucinol			
Phl2a*c*	GAGGACGTCGAAGACCACCA	*phlD* (*P. fluorescens* CHA0; Pf-5; Q8r11-96; 1 M1-96; Q2-87)	745 bp (not detected)
Phl2b*c*	ACCGCAGCATCGTGTATGAG	*phlD*	745 bp (not detected)
BPF2*d*	ACATCGTGCACCGGTTTCATGATG	*phlD*	~470 bp (not detected)
B2BF*d*	ACCCACCGCAGCATCGTTTATGAGC	*phlD*	~470 bp (not detected)
BPF3*d*	ACTTGATCAATGACCTGGGCCTGC	*phlD*	~470 bp (PA23)
BPR2*d*	GAGCGCAATGTTGATTGAAGGTCTC	*phlD*	~470 bp (PA23)
BPR3*d*	GGTGCGACATCTTTAATGGAGTTC	*phlD*	~470 bp (PA23)
BPR4*d*	CCGCCGGTATGGAAGATGAAAAAGTC	*phlD*	~470 bp (PA23)
Pyrrolnitrin			
PrnAF*e*	GTGTTCTTCGACTTCCTCGG	*prnA* (*P. fluorescens* Pf-5)	1,050 bp (PA23)
PrnAR*e*	TGCCGGTTCGCGAGCCAGA	*prnA* (*P. fluorescens* Pf-5)	1,050 bp (PA23)
PRND1*f*	GGGGCGGGCCGTGGTGATGGA	*prnD P. fluorescens* BL915;	790 bp (PA23)
PRND2*f*	YCCCGCSGCCTGYCTGGTCTG	*prnD P. fluorescens* BL915;	720 bp (PA23)
PrnCf*g*	CCACAAGCCCGGCCAGGAGC	*prnC* (*P. fluorescens* BL915)	720 bp (PA23)
PrnCrg	GAGAAGAGCGGGTCGATGAAGCC	*prnC* (*P. fluorescens BL915)*	720 bp (PA23)

(continued)

Table 11.1 (continued)

Primer	Sequence	Gene (control strain)	Expected size of PCR product
Pyoluteorin			
PltCreg1F*e*	AGGCAATCACTACCATCCGTGCGC	*pltC* (*P. fluorescens* Pf-5)	438 bp (not detected)
PltCreg2R*e*	ATGAGGAGCAGGAGGTGTCGAGCAC	*pltC* (*P. fluorescens* Pf-5)	438 bp (not detected)
PLTC1*f*	AACAGATCGCCCCGGTACAGAACG	*pltC* (*P. fluorescens* Pf-5)	438 bp (not detected)
PLTC2*f*	AGGCCCGGACACTCAAGAAACTCG	*pltC* (*P. fluorescens* Pf-5)	438 bp (not detected)
PltBf*g*	CGGAGCATGGACCCCCAGC	*pltB* (*P. fluorescens* Pf-5)	900 bp (PA23)
PltBr[g]	GTGCCCGATATTGGTCTTGACC	*pltB* (*P. fluorescens* Pf-5)	900 bp (PA23)
plt1[g]	ACTAAACACCCAGTCGAAGG	*pltB* (*P. fluorescens* Pf-5)	440 bp (not detected)
plt2[g]	AGGTAATCCATGCCCAGC	*plt B* (*P. fluorescens* Pf-5)	440 bp (not detected)

Cyanide ions are metabolized mainly to thiocyanate. The cyanide ion is exhaled as HCN and metabolized to lesser degree to other compounds. It inhibits proper functioning of enzymes as well as natural receptors by reversible mechanism of inhibition (Corbett 1974). HCN also inhibits the action of cytochrome oxidase (Gehring et al. 1993). Voisard et al. (1989) reported cyanide production by *P. fluorescens* which helped in suppression of black root rot (*Thielaviopsis basicola*) of tobacco. HCN mutant (obtained by insertional inactivation) of the wild type strain had lost its ability to suppress black root rot of tobacco was also reported. Haas et al. (1991) also reported the role of HCN in biocontrol of take-all (*G. graminis* var. *tritici*) disease of wheat. *P. fluorescens* inhibited pathogens by inducing host defense. Production of HCN by certain fluorescent *Pseudomonas* resulted in suppression of root pathogens (O'Sullivan and O'Gara 1992). Ramette et al. (2003) reported that HCN involved in biological control of root diseases is produced by many plant-associated fluorescent pseudomonads. Ahmad et al. (2006) screened total of 72 rhizobacterial isolates in vitro for different traits like production of IAA, NH_3, HCN, siderophore, phosphate solubility, and antifungal activity. Production of IAA and HCN were found highest in *Pseudomonas* spp., followed by other isolates. Role of HCN and pyocyanin of other fluorescent pseudomonads on the antifungal activity have also been reported by Hassanein et al. (2009) for *F. oxysporum* and *H. oryzae*, while *Aspergillus niger* was not affected. HCN synthase is encoded by three biosynthetic genes viz. *hcn*A, *hcn*B, and *hcn*C (Ramette et al. 2003). However, little is known about the diversity of these genes in fluorescent *Pseudomonas* sp. and other bacteria. But, Bakker and Schippers (1987) suspected that HCN production by some fluorescent pseudomonads might in fact be detrimental to plant growth.

11.3.3 Siderophores

Siderophores (Gr. "iron-bearers") are low molecular weight compound synthesized under iron-deficient concentration by many microorganisms, high affinity ferric-iron chelators that transport iron into the bacterial cell (Mukherjee et al. 2004) via specific outer membrane receptor proteins, thereby providing iron for cellular functions (Leong 1986; Loper and Buyer 1991). Siderophores chelate ferric ions with a high specific activity and serve as vehicles for the transport of ferric iron into microbial cells (Neilands 1981). Transport of iron into the cells is mediated by a membrane receptor that specifically recognizes ferric-siderophore complex (Hemming 1986). Buyer et al. (1986) found that the fluorescent pseudomonads produced yellow green fluorescent siderophores (pyoverdine type) under low iron condition membrane receptor proteins that recognize and take up ferric-pseudobactin complex (Magazin et al. 1986). Siderophore sequesters the trace amount of iron (Fe^{++}) from the rhizosphere and thereby limits the availability of Fe^{++} to the pathogens and ultimately suppresses the pathogen growth (Schroth and Hancock 1981). Siderophore production was postulated to be an important mechanism for the biocontrol activity of PGPR (Neilands 1986; Loper and Buyer 1991; Bakker et al. 1993). Suryakala et al. (2004)

reported that isolates belonging to *P. fluorescens* were reported to produce extracellular siderophore when grown under iron-deficient condition. This siderophore is responsible for the fluorescence of *P. fluorescens*. When iron concentrations are high, pyoverdine is not needed so colonies will not fluoresce under ultraviolet light. Unnamalai and Gnanamanickam (1984) reported that *P. fluorescens* could inhibit the growth of *X. campestris* pv. *citri* and correlated the antagonism to the production of siderophore. Similarly Elad and Baker (1985) reported that siderophore of fluorescent pseudomonads could suppress the chlamydospores germination of *F. oxysporum* in a trace iron condition. Observation of siderophore activity was also made by Becker and Cook (1988) and noted that it inhibits the growth of *Pythium ultimum* and other *Pythium* spp. in the wheat rhizosphere. Ciampi et al. (1997) observed that Pf strain BC 8 produced siderophore like pigmented metabolites, which inhibited the growth of *R. solanacearum* in vitro.

11.3.4 Competition for Space and Nutrients

Rhizosphere competence is a key character of fluorescent pseudomonads, since it determines root colonization potential, competitive ability, and sustenance in the crowded rhizosphere environment. Weller et al. (1985) found out that pseudomonads catabolize diverse nutrients and have a fast generation time in the root zone. Hence, they were projected as logical candidate for biocontrol by competition for nutrients against slow growing plant pathogenic fungi. Similar opinion was also made by Bull et al. (1991) and stated that suppression of take-all of wheat was correlated with colonization of roots by *P. fluorescens* strain 2-79. Genetic work of Anderson et al. (1988) revealed that production of a particular plant glycoprotein called agglutinin was correlated with potential of *P. putida* to colonize the root system. *P. putida* mutants deficient in this ability exhibited reduced capacity to colonize the rhizosphere and a corresponding reduction in Fusarium wilt suppression in cucumber (Tari and Anderson 1988). Mohamed and Caunter (1995) observed *P. fluorescens* to inhibit *Bipolaris maydis* both in vitro and in vivo in infected maize plants but could not detect any inhibitory substances, assayed by a variety of methods, indicating nutrient competition as the operative component of antagonism.

11.3.5 Mineral Phosphate Solubilization

Phosphorus is one of the major nutrients, second only to nitrogen in requirement for plants. Most of phosphorus in soil is present in the form of insoluble phosphates and cannot be utilized by the plants (Pradhan and Sukla 2006). The biological process of conversion of unavailable/fixed form of inorganic phosphorous into primary orthophosphate ($H_2PO_4^-$) and secondary orthophosphate (HPO_4^{-2}) is termed as MPS (Goldstein 1986). Stalstorm (1903) first showed the involvement of microorganisms in the solubilization of insoluble phosphate.

Indian soils are normally deficient in available phosphorus (Johri et al. 2003). To circumvent the phosphorus deficiency, phosphate solubilizing microorganisms play an important role in supplying phosphate to plants in a more environment friendly and sustainable manner. Among phosphate solubilizing bacteria, fluorescent pseudomonads that colonize aggressively at the plant roots have been considered as an important group of bacteria due to their biofertilizing, biocontrol, and its capability of utilizing a wide array of compounds as carbon and energy sources. Fluorescent pseudomonad species such as *P. chlororaphis, P. putida, P. aeruginosa, P. fluorescens, P. trivialis, P. striata*, and *P. poae* have been identified as phosphate solubilizing rhizobacteria (Cattelan et al. 1999; Gaind and Gaur 2002; Bano and Musarrat 2003; Sunish et al. 2005; Gulati et al. 2008). *Pseudomonas fluorescens* were recorded for solubilization of ZnPO4 in the presence of glucose as the carbon source (Di Simine et al. 1998). Phosphate solubilizing *Pseudomonas* sp. showed significant increase in maize plant height after 60 days of growth and an 18 % increase in lettuce shoot fresh matter yield observed in Quebec (Canada) (Chabot et al. 1993). Rhizobacteria can solubilize rock phosphate and calcium phosphate in culture medium (Nahas 1996). Cold-tolerant mutants of *P. fluorescens* were reported as more efficient in tricalcium phosphate solubilization than their respective wild type counterparts at low temperatures (Das et al. 2003). *P. fluorescens* strain Psd isolated from rhizosphere of *Vigna mungo* was also found to solubilize complex phosphates and synthesize phytohormone, IAA. In Iran fluorescent pseudomonads from rice rhizosphere reported for significant phosphate solubilization activity in addition to plant growth promoting properties, IAA and siderophore production (Ramezanpour et al. 2010). Among the different organic acids, gluconic acid seems to be most commonly produced acid by phosphate solubilizing *Pseudomonas fluorescens* (Di Simine et al. 1998)

11.3.6 Production of Plant Growth Promoting Substances

Plant growth promoting rhizobacteria (PGPR) were first defined by Kloepper and Schroth (1978) as the soil bacteria that colonize the roots of plants by following inoculation on to seed and that enhance plant growth. PGPR competitively colonize plant roots, stimulate plant growth, and/or reduce the incidence of plant disease (Kloepper and Schroth 1978). Improvement in plant growth due to the action of PGPR include increase in germination rates, root growth, yield including grain, leaf area, chlorophyll content, magnesium, nitrogen and protein content, hydraulic activity, tolerance to drought and salt stress, shoot and root weights, and delayed leaf senescence. Several mechanisms that involve in the process include phosphate solubilization, production of phytohormones (such as auxin and cytokinin), volatile growth stimulants (such as ethylene and 2,3-butanediol) and nitrogen fixation (Lifshitz et al. 1987; Vessey 2003). PGPR might enhance plant growth by excluding the so-called deleterious rhizobacteria, which are thought to inhibit plant growth without causing root invasion and classical disease (Schroth and Hancock 1982). Pseudomonads make up a dominant population in soil and rhizosphere and exert

Table 11.2 Pseudomonas mediated induced systemic resistance in plant species investigated against fungal pathogens

Plant species	Strain	Challenging pathogens	Disease symptoms	Reference
Bean	Pseudomonas aeruginosa 7NSK2	*Botrytis cinerea*	Grey mold	De Meyer and Hofte (1997)
Cucumber	*Pseudomonas aureofaciens* 25-33	*C. orbiculare*	Anthracnose	Wei et al. (1991)
	Pseudomonas corrugata 13	*Pythium aphanidermatum*	Crown rot	Chen et al. (2000)
Radish	*P. fluorescens* WCS374	*F. oxysporum raphani*	Vascular wilt	Leeman et al. (1995)
	P. fluorescens WCS417	*Alternaria brassicicola*	Necrotic lesions	Ton et al. (2002)
Tomato	*P. fluorescens* WCS417	*F. oxysporum* f. sp. *lycopersici*	Vascular wilt	Duijff et al. (1998)

growth promoting influence on a variety of plant species on account of their strong competitive behavior, colonization potential, and sustainability (Glick 1995). Most of these nonpathogenic strains of *Pseudomonas* found in the rhizospheric soils have multiple traits that make them well suited as PGPR (O'Sullivan and O'Gara 1992).

Fluorescent *Pseudomonas* GRC_2 from potato rhizosphere showed a strong antagonistic effect against *Macrophomina phaseolina*, a charcoal rot pathogen of peanut. Bacterization of peanut seeds with fluorescent *Pseudomonas* GRC_2 resulted in increased seed germination, early seedling growth, fresh nodule weight, grain yield, and reduced charcoal rot disease of peanut in *M. phaseolina*-infested soil (Gupta et al. 2002). The presence of *Pseudomonas fluorescence* inoculant in the combination of microbial fertilizer plays an effective role in stimulating yield and growth traits of chickpea (Rokhzadi et al. 2008). *Pseudomonas fluorescens* B16 isolated from the roots of graminaceous plants has been shown to colonize the roots of various plants, and to increase the height, flower number, fruit number, and total fruit weight of tomato plants (Minorsky 2008). Ten strains of rhizospheric fluorescent pseudomonads isolated from the soils of bajra (*Pennisetum glaucum*), jowar (*Sorghum vulgare*), rice (*Oryza sativa*) and maize (*Zea mays*) showed positive plant growth promoting activity and were found to produce IAA, protease, siderophores, and HCN and were also found to exhibit antagonistic activity against four test fungi viz. *Fusarium oxysporum, Curvularia lunata, Colletotrichum falcatum, Macrophomina phaseolina* in vitro (Suresh et al. 2010).

11.3.7 Induced Systemic Resistance

Some antagonistic PGPR elicit a phenomenon that is known as induced systemic resistance (ISR) in the host plant (Table 11.2). ISR allows plants to withstand pathogen attack to the leaves or roots, without offering total protection (Haas and Defago 2005). ISR triggered by PGPR fortifies plant cell wall strength and alters host physiology, metabolic responses, leading to enhanced synthesis of plant defense chemicals against pathogens and/or abiotic stress factors. ISR was discovered as a mode of action of disease suppression by PGPR *Pseudomonas* spp. independently by two research groups (Van Peer et al. 1991; Wei et al. 1991). In tomato, seed treatment with *P. fluorescens* strain 63–28 resulted in induced resistance against *Fusarium oxysporum* f. sp. *lycopersici* by triggering the host to synthesize more phenolic substances (M'Piga et al. 1997). In tomato and hot pepper, enhanced resistance against invasion of *Pythium* was found due to induction of defense-related enzymes involved in the phenyl propanoid pathway and also due to direct antagonism and plant growth promotion by fluorescent pseudomonads (Ramamoorthy et al. 2002). Fallahzadeh et al. (2009) evaluated fluorescent pseudomonads from cotton rhizosphere for induction of systemic resistance (ISR) against bacterial blight of cotton. After inoculation with the pathogen, PO and PAL activity of all *Pseudomonas*-treated plants drastically increased and level of infected area on leaves lower down. *Pseudomonas* metabolite salicylic acid (SA) was suggested to trigger induced resistance (Leeman et al. 1996; De Meyer and Hofte 1997; Maurhofer et al. 1998) and SA has been mostly observed under iron-limited conditions. ISR was also triggered by *P. fluorescens* EP1 against red rot caused by *Colletotrichum falcatum* on sugarcane (182), *P. fluorescens* 63-28 against *F. oxysporum* f. sp. *radicis-lycopersici* on tomato and *Pythium ultimum* and *F. oxysporum* f. sp. *pisi* on pea roots.

A recent study on bacterial inoculation of wheat roots with *P. fluorescens* Pf29Arp strain reduced the development of *Gaeumannomyces graminis* var. *tritici* (*Ggt*)-induced disease (Daval et al. 2011). The plant host glutathione-*S*-transferase gene was induced by *Ggt* alone and up-regulated by Pf29Arp bacteria in interaction with the pathogen. Finally the study concluded that Pf29Arp antagonism acts through the alteration of fungal pathogenesis and probably through the activation of host defenses (Daval et al. 2011).

11.3.8 Enzyme Activity of the Bacteria

P. fluorescens can produce hydrolytic enzymes, i.e., chitinases and β-1,3-glucanases (Fridlender et al. 1993). *P. fluorescens* strain 114 is known for production of a protease with molecular weight of 47,000 Da which was stable in the pH range of 5–9 and worked optimally between pH 6.5–10 (Hamamoto et al. 1994). Siddiqui et al. (2005) also reported production of extracellular protease by *P. fluorescens* CHA0, a biocontrol factor with activity against the root-knot nematode *Meloidogyne incognita*. *P. fluorescens* strain Pf-5 possesses many extracellular hydrolytic enzymes that degrade polymers found in soil as well as hydrolases used on plant-derived carbohydrates. These enzymes are also capable of degrading and using components of plant tissues like hydrocarbon molecules, fatty acids, and oils (Paulsen et al. 2005). *Pseudomonas fluorescens* KD strain reduces the activity level of the pectinase polygalacturonase (a key pathogenicity factor) from *Pythium ultimum* on cucumber (Rezzonico et al.

2005). In contrast, *P. fluorescens* strains induce laccase activity, enzymes putatively involved in the pathogenicity of *Rhizoctonia solani* (Crowe and Olsson 2001). *P. fluorescens* produces viscosin which is a peptidolipid that enhances antivirality.

Pseudomonas GRC_2, fluorescent *Pseudomonas* strain showed necrotrophic antibiosis in vitro against *Macrophomina phaseolina* and *Sclerotinia sclerotiorum* (Gupta et al. 2001) Scanning electron photomicrographs of zone of interaction showed loss of sclerotial integrity, hyphal shrivelling, mycelial and sclerotial deformities, and hyphal lysis in *M. phaseolina*, whereas hyphal perforations, lysis and fragmentation, were observed in case of *S. sclerotiorum.* It was observed that *Pseudomonas* GRC2 produced enzyme chitinase along with other secondary metabolites, antibiotic substances (unidentified), siderophores volatile compound HCN, and IAA.

11.4 Successful Antagonism by *P. fluorescens*

P. fluorescens is becoming a promising biocontrol agent for its noble antagonistic property against a wide variety of phytopathogenic bacteria and fungi particularly the soil-borne pathogens (Weller 1988). Sarathchandra et al. (1993) reported that antagonists competed with pathogen for nutrients and thereby reduced diseases. Biocontrol strains can also promote plant growth.

The usefulness of fluorescent *Pseudomonas* as biocontrol agent has attracted the attention of researchers for its effectiveness to colonize the rhizosphere of many crop plants with the ability to inhibit the growth of a number of phytopathogens. Some examples are given below (Tables 11.3 and 11.4).

11.5 Genomic Sequence of *Pseudomonas fluorescens*: Insight into Biological Control

Currently, two strains of *P. fluorescens* viz., *P. fluorescens* Pf-5 and *P. fluorescens* PfO-1 have their genomes sequenced completely (Paulsen et al. 2005). *P. fluorescens* Pf-5 is a rhizosphere bacterium that suppresses seedling emergence diseases and produces a spectrum of antibiotics toxic to plant-pathogenic fungi and oomycetes. The genome of Pf-5 is the largest *Pseudomonas* genome sequenced (Loper et al. 2007) to date (including plant pathogens). Complete genome of *P. fluorescens* Pf-5 is 7.1 Mbp, a circular chromosome has GC content of 63.3 % with 6,144 predicted genes. It contains 87 RNAs and 6,137 proteins. The study revealed that 5.7 % of its genome contributes to secondary metabolism which is the largest of the *Pseudomonas* (Paulsen et al. 2005). In addition to six known secondary metabolites produced by Pf-5, three novel secondary metabolite biosynthesis gene clusters identified in the genome could also contribute to biological control.

Table 11.3 Effect of fluorescent pseudomonads against fungal pathogens of crop plants

Sl. No.	*Pseudomonas* strains	Name of the disease	Fungal pathogen	References
1.	*P. fluorescens*	Stem canker potato seeds bacterization	*R. solani*	Burr et al. (1978)
2.	*P. fluorescens* strain Pf-5	Seedling diseases of cotton	*R. solani*, *P. ultimum*	Howell and Stipanovic (1979), Howell and Stipanovic (1980)
3.	*P. fluorescens*	Black root rot of tobacco	*Thielaviopsis basicola*	Voisard et al. (1989)
4.	*Pseudomonas fluorescens* Pf-5	*Pythium* damping off of cucumber	*Pythium* spp.	Kraus and Loper (1992)
5.	*P. fluorescens*	Tan spot of wheat	*Pyrenophora tritici-repentis*	Pfender et al. (1993)
6.	*P. fluorescens*	Fusarium wilt in radish	*F. oxysporum* f. sp. *raphani*	Leeman et al. (1995)
7.	*P. fluorescens* strain Pf-5	Fusarium crown and root rot of tomato	*F. oxysporum* f. sp. *radicis-lycopersici*	Sharifi-Tehrani et al. (1998)
8.	*P. fluorescens*	Downy mildew in pearl millet	*Sclerospora graminicola*	Umesha et al. (1998)
9.	*P. fluorescens* PF-1	Banded leaf and sheath blight of maize	*R. solani* f. sp. *sasakii*	Sivakumar et al. (2000)
10.	*P. fluorescens*	Sheath blight of maize	*R. solani*	Tripathi and Joshi (2002)
11.	*P. fluorescens* IISR-51	Foot rot of black pepper	*P. capsici*	Diby et al. (2004)
12.	*Pseudomonas* culture	Damping off in tomato	*P. aphanidermatum*	Martin and Loper (1999); Srivastava et al. (2004)
13.	*P. fluorescens* (in combination with *T. viride*)	Stem rot of groundnut	*Sclerotium rolfsii*	Manjula et al. (2004)
14.	*P. fluorescens*	Wilt of citrange troyer plant	*F. solani* and *Phoma tracheiphila*	Cirvilleri et al. (2005)
15	Six isolates of fluorescent pseudomonads	Wilt disease of pea complex	*F. oxysporum* f. sp. *pisi*	Siddiqui et al. (2005)
16.	Fluorescent pseudomonads	White mould	*S. sclerotiorum*	Behboudi et al. (2005)
17.	*P. fluorescens*	Root and stem end rot of chick pea	*Macrophomina phaseolina*	Kumar et al. (2007)
18.	*P. fluorescens* Biotype F isolate DF37	Wilt of potato	*Verticillium albo-atrum*	Uppal et al. (2008)

(continued)

Table 11.3 (continued)

Sl. No.	*Pseudomonas* strains	Name of the disease	Fungal pathogen	References
19.	*P. fluorescens*	Sheath blight of rice	*R. solani*	Singh and Sinha (2009)
20.	*Pseudomonas* spp. FC-7B, FC-9B, and FC-24B together with Achromobacter sp. AM1 and *Serratia* sp. DM1	Wilt of tomato	*F. oxysporum* f. sp. *lycopersici*	Srinivasan et al. (2009)
21.	*P. fluorescens*	Rice fungal pathogen	*R. solani*	Battu and Reddy (2009)
22.	*P. fluorescens*	Rice blast and sheath blight	*P. oryzae* and *R. solani*	Reddy and Reddy (2009)
23.	*P. fluorescens*	Sheath blight of rice	*R. solani*	Singh and Sinha (2009)
24.	*Endophytic P. fluorescens* strains EBC5, EBC7, and EBC6	Damping-off of chilli	*P. aphanidermatum*	Muthukumar et al. (2010)
25.	*Pseudomonas* strain PCI2	Damping-off of tomato	*S. rolfsii*	Pastor et al. (2010)
26.	*P. fluorescens* strain Pf29Arp	Take-all disease of wheat	*Gaeumannomyces graminis* var. *tritici*	Daval et al. (2011)

Table 11.4 Effect of fluorescent pseudomonads against bacterial pathogens of crop plants

Sl. No.	*Pseudomonas* strains	Name of the disease	Bacterial pathogen	References
1.	*P. fluorescens*	Banana, brinjal, and tomato	*R. solanacearum*	Chand and Logan (1984)
2.	*P. fluorescens* strain Pf-5	Seed piece decay of potato	*E. carotovora*	Xu and Gross (1986)
3.	*P. fluorescens*	Wilt of tomato	*R. solanacearum*	Kalita (1994)
4.	*P. fluorescens*	Wilt of tomato	*R. solanacearum*	Karuna and Khan (199)
	P. fluorescens	Bacterial wilt disease in tomato	*R. solanacearum*	Kumar and Sood (2001)
5.	*P. fluorescens*	Bacterial blight of rice	*X. oryzae* pv. *oryzae*	Vidhyasekaran et al. (2001)
6.	Fluorescent *Pseudomonas* strains	Tomato bacterial spot	*X. vesicatoria*	Shukla and Gupta (2005)
7.	*P. fluorescens*	Bacterial wilt in chilli	*R. solanacearum*	Umesha et al. (2005)
8.	*P. fluorescens strain* PfG32R	Bacterial wilt of tomato	*R. solanacearum*	Alit-Susanta and Takikawa (2006)
9.	*P. fluorescens*	Eggplant	*R. solanacearum*	Ramesh et al. (2009)
10.	*P. fluorescens Pf 32, Pf93*	Bacterial blight of cotton	*X. campestris* pv. *malvacearum*	Salaheddin et al. (2010)
11.	*P. fluorescens*	Bacterial blight and leaf spots diseases of cotton	*X. campestris* pv. *malvacearum*, *X. vesicatoria*	Bhattiprolu (2010)
12.	*P. fluorescens*	Wilt of tomato	*R. solanacearum*	Bora et al. (2010)
13.	Endophytic *P. fluorescens strain PF-1*	Black rot of cauliflower	*X. campestris pv. campesteris*	Singh et al. (2010)

The genomic sequence provides numerous information regarding catabolic and transport capabilities for utilizing seed and root exudates; an expanded collection of efflux systems for defense against environmental stress and microbial competition and the presence of 45 outer membrane receptors that should allow for the uptake of iron from a wide array of siderophores produced by soil microorganisms. Pf-5 has an extensive collection of regulatory genes, as expected for a large genome bacterium that lives in a rapidly changing environment. Only some of which have been characterized for their roles in regulation of secondary metabolite production or biological control in turn (Loper et al. 2007). Brief overview of the general features of the Pf-5 genomic sequence (Loper et al. 2007) data discussed below:

Attachment: The Pf-5 genome contains several genes that have been implicated in the attachment of *Pseudomonas* spp. to surfaces, including genes encoding for predicted hemagglutinins, hemolysins, and other adhesion-related proteins.

Environmental fitness: Pf-5 has the capacity to colonize seed and root surfaces, for acquisition of diverse compounds for nutrition, attachment to surfaces, as well as defense from environmental stress and rhizospheric microbial competition.

Catabolism: Genomic sequence of Pf-5 specifies a broad metabolic capacity that is shared with other species of *Pseudomonas* and is consistent with a saprophytic lifestyle in soil (dos Santos et al. 2004). The Pf-5 genome contains genes encoding for utilization of a broad spectrum of organic acids, sugars, and amino acids, including those typically found in seed or root exudates. The Pf-5 genome has also genes for the metabolism of a number of plant-derived carbohydrates such as maltose, sucrose, trehalose, and xylose. Presence of genes for utilization of more complex plant-derived molecules such as the aromatic compounds vanillate, benzoate, and hydroxybenzoate, long chain fatty acids, and hydrocarbons also observed.

Rhizosphere colonization: Rhizosphere competence of *P. fluorescens* is governed by genes and presence of which was detected in *P. fluorescens* strain Q8r1-96 but absent in a less rhizosphere competent strain as detected by subtractive hybridization (Mavrodi et al. 2002). The putative rhizosphere competence genes identified in that study were not present in *P. fluorescens* Pf-5 (Paulsen et al. 2005). Moreover, the population size established by Pf-5 in the rhizosphere of pea or wheat was neither as large nor as persistent as populations established by the most rhizosphere competent strains of *P. fluorescens*, as represented by strain Q8r1-96 (Landa et al. 2002, 2003).

Siderophore-mediated iron acquisition: Plant pathogens are subject to iron competition by siderophore-producing biological control agents (Loper and Buyer 1991) such as Pf-5. The Pf-5 genome specifies the biosynthesis of two siderophores—pyoverdine and pyochelin (or related compounds). Genes required for pyoverdine biosynthesis and uptake are organized in three gene clusters in the Pf-5 genome (Ravel and Cornelis 2003). A putative pyochelin biosynthesis gene cluster is present in the Pf-5 genome, although its organization differs from the well-characterized pyochelin gene cluster of *P. aeruginosa* (Michel et al. 2005). Genes encoding outer membrane receptors for 20–30 heterologous siderophores exist in the genomes of all *Pseudomonas* spp. sequenced to date. In Pf-5 genome 45 genes predicted to encode outer membrane proteins that bind the transmembrane protein TonB, a characteristic of ferric siderophore receptors. Outer membrane receptors allow Pf-5 to utilize a wide array of siderophores produced by soil microorganisms.

Self-defense: Genomic sequence of Pf-5 reveals an expanded collection of efflux systems, which typically confer protection against a range of toxic metabolites. Within the genome of Pf-5, 13 regions contain resistance-nodulation-cell division (RND) homologs, with neighboring genes encoding partner proteins of the efflux systems. Other genes are predicted to confer resistance to specific toxins, including tabtoxin, a phytotoxin produced by *P. syringae* pv. *tabaci*, and fusaric acid, a toxin produced by the soil-borne plant pathogen *F. oxysporum* that serves as a signal repressing the production of 2,4 diacetylphloroglucinol by *P. fluorescens* CHA0 (Notz et al. 2002).

Secondary metabolites and other secreted products: Based on the sizes of the nine biosynthetic genes clusters identified to date, it has been revealed that approximately 6 % of the Pf-5 genome is devoted to the production of secondary metabolites. The nine gene clusters are distributed over a large portion of the genome. Six of the gene clusters encode for the biosynthesis of compounds that were known to be produced by Pf-5 before the genomic sequencing project was initiated (Loper et al. 2007). Gene clusters for three other secondary metabolites were identified in the genome of Pf-5 based upon characteristic sequences of polyketide or peptide synthases. These enzymes catalyze the formation of secondary metabolites through a non-ribosomal mechanism of biosynthesis (Wenzel and Muller 2005). Enzymes involved in biosynthesis of these secondary metabolites have characteristic functional domains. These domains are encoded by highly conserved sequences that can be used to identify biosynthetic gene clusters containing polyketide or peptide synthases. This approach was used to identify three gene clusters in the Pf-5 genome that presumably encode for secondary metabolites. The three metabolites discovered through genomic sequencing have not yet been characterized with respect to their toxicities to plant pathogens or their roles in biological control. Their discovery provides new directions for research evaluating mechanisms of biological control. In addition to secondary metabolites, Pf-5 produces other products, including exoenzymes and at least one bacteriocin. Many of these products, such as an extracellular alkaline protease, suppress the root-knot nematode *Meloidogyne incognita* (Siddiqui et al. 2005).

Lack of key pathogenicity factors: Consistent with its commensal lifestyle, Pf-5 lacks a number of virulence factors found in plant pathogens. No evidence for a type III secretion system was found in the genomic sequence of Pf-5, although genes for these export systems have been found in other nonpathogenic strains of *Pseudomonas* spp. (including *P. fluorescens*) associated with plants (Preston et al. 2001; Mazurier et al. 2004; Rezzonico et al. 2005). There is no evidence in the Pf-5 genome for the biosynthesis of the known *P. syringae* phytotoxins, tabtoxin, syringomycin, syringotoxin, syringopeptin, or coronatine. Pf-5 lacks amylase, consistent with an inability to utilize starch, and lacks cellulases and other exoenzymes often associated with degradation of plant cell walls and cell wall components. Therefore, many genes required for pathogenicity or virulence of plant or animal pathogens do not have clear counterparts in the genome of this commensal bacterium.

As expected for a bacterium with a large genome that lives in a rapidly changing environment, Pf-5 has an extensive collection of regulatory genes, only some of which have been characterized for their roles in regulation of secondary metabolite production or biological control. Consistent with its commensal lifestyle, Pf-5 appears to lack a number of virulence and pathogenicity factors found in plant pathogens.

Pseudomonas fluorescens F113, a PGPR strain that has biocontrol activity against fungal plant pathogens, is considered as a model for rhizosphere colonization. Miguel Redondo-Nieto et al. (2012) presented complete genome sequence of *Pseudomonas fluorescens* F113, which showed that besides a core genome very similar to those of other strains sequenced within this species, F113 possesses a wide array of genes encoding specialized functions for thriving in the rhizosphere and interacting with eukaryotic organisms.

The genome of *Pseudomonas fluorescens* PfO-1 has one chromosome with 6.43841 Mbp and 60.5 % GC content. There are 95 RNAs and 5736 proteins. The genome sequencing of *P. fluorescens SBW25* is still in progress (http://www.ncbi.nlm.nih.gov/entrez/query.fcgi). The genomic sequence available, coupled with detailed knowledge of its biology provides a framework for distinguishing genes governing growth, persistence, and activity of plant-associated bacteria.

11.6 Concluding Remarks

Fluorescent pseudomonads as biological control agent of plant pathogens gaining importance in modern and organic agriculture as eco-friendly alternatives of chemicals because it offers disease management with different mechanisms of action unlike chemical pesticides. Although the number of biocontrol formulations is increasing day by day, still it represents only about 1 % of agricultural chemicals. Despite several studies on the complex regulatory gene network controlling the production of effectors by rhizobacteria, very little information is available concerning their effects on fungal pathogenicity and virulence. Very little is known about the frequency and ecology of naturally occurring antibiotic-producing fluorescent *Pseudomonas* spp. The availability of cloned and sequenced antibiotic-biosynthetic genes has facilitated the development of specific primers and probes that can be used to detect naturally occurring antibiotic-producing *Pseudomonas* spp. It can also expedite the search for native antibiotic-producing strains that are better adapted to local soil and environmental conditions and more effective in specific crop-pathogen ecosystems. Sequencing the genome provided further information of its environmental interaction and its metabolic capabilities, which can be used against agricultural disease control. For attaining popularity and better commercial application of biocontrol agents against plant pathogens, it requires increased emphasis on combining various effective compatible strains of different antagonists with each other and with other control methods as well as integration of biocontrol into an overall system.

References

Ahmad F, Ahmad I, Khan NS (2006) Screening of free living rhizospheric bacteria for their multiple plant growth promoting activities. Microbiol Res 163(2):173–181

Alit-Susanta WGN, Takikawa Y (2006) Phenotypic characterization of Pseudomonas fluorescens PfG32R and its spontaneous gacS mutants and biocontrol activity against bacterial wilt disease of tomato. J Genet Plant Pathol 72(3):168–175

Anderson AJ, Tari PH, Tepper CS (1988) Genetic studies on the role of an agglutinin in root colonization by *Pseudomonas putida*. Appl Environ Microbiol 54:375–380

Bakker AW, Schippers B (1987) Microbial cyanide production in the rhizosphere in relation to potato yield reduction and *Pseudomonas* spp. mediated plant growth stimulation. Soil Biochem 19:451–457

Bakker PAHM, Raaijamakers M, Schippers B (1993) Role of iron in the suppression of bacterial plant pathogens by fluorescent pseudomonads. In: Barton LL, Hemming BC (eds) Iron chelating in plant and soil micro-organism. Academic, San Diego, pp 269–281

Bangera MG, Thomashaw LS (1996) Characterization of a genomiclocus required for synthesis of the antibiotic 2,4-diacetylphloroglucinol by the biological control agent *Pseudomonas fluorescens* Q2-87. Mol Plant-Microb Interact 9:83–90

Bangera MG, Thomashow LS (1999) Identification and characterization of a gene cluster for synthesis of the polyketide antibiotic 2,4-diacetylphloroglucinol from *Pseudomonas fluorescens* Q2-87. J Bacteriol 181:3155–3163

Bano N, Musarrat J (2003) Characterization of a new *Pseudomonas aeruginosa* strain NJ-15 as a potential biocontrol agent. Curr Microbiol 46(5):324–328

Battu PR, Reddy MS (2009) Siderophore-mediated antibiosis of rhizobacterial fluorescent Pseudomonads against rice fungal pathogens. Int J PharmTech Res 1(2):227–229

Becker JO, Cook RJ (1988) Role of siderophores in suppression of *Pythium* species and production of increased growth response of wheat by fluorescent pseudomonads. Phytopathology 78:778–782

Behboudi K, Sharifi TA, Hedjaroude GA, Zad J, Mohammadi M, Rahimian H (2005) Effects of fluorescent pseudomonads on *Sclerotinia sclerotiorum*, the causal agent of sunflower root rot. Iranian J Agric Sci 36(4):791–803

Bhattiprolu SL (2010) Efficacy of *Pseudomonas fluorescens* against bacterial blight and leaf spot diseases of cotton (*Gossypium hirsutum*). Indian J Agric Sci 80(3):235–237

Bossis E, Lemanceau P, Latour X, Garden L (2000) The taxonomy of *Pseudomonas fluorescens* and *Pseudomonas putida:* current status and need for revision. Agronomie 20:51–63

Brodhajen M, Paulsen I, Loper JE (2005) Reciprocal regulation of pyoluteorin production with membrane transporter gene expression in *Pseudomonas fluorescens* Pf-5. Appl Environ Microbiol 71:6900–6909

Bull CT, Weller DM, Thomashow LS (1991) Relationship between root colonization and suppression of *Gaeumannomyces graminis* var. *tritici* by *Pseudomonas fluorescens* strain 2-79. Phytopathology 81:954–959

Burr TJ, Schroth MN, Suslow TV (1978) Increased potato yields by treatment of seed pieces with specific strains of *Pseudomonas fluorescens* and *Pseudomonas putida*. Phytopathology 68:1377–1383

Buyer JS, Wright JM, Leong J (1986) Structure of pseudobactin A214, a siderophore from a bean-deleterious Pseudomonas. Biochemistry 25:5492–5499
Cattelan AJ, Hartel PG, Furhmann JJ (1999) Screening for plant growth-promoting rhizobacteria to promote early soybean growth. Soil Sci Soc Am J 63:1670–1680
Chabot R, Antoun H, Cescas M (1993) Stimulation de la croissance du maïset de la laitue romaine par des microorganismesdissolvant le phosphoreinorganique. Can J Microbiol 39:941–947
Chand T, Logan C (1984) Antagonists and parasites of *Rhizoctonia solani* and their efficacy in reducing stem canker of potato under controlled conditions. Trans Br Mycol Soc 38:107–112
Chen C, Belanger R, Benhamou N, Paulitz TC (2000) Defense enzymes induced in cucumber roots by treatment with plant growth-promoting rhizobacteria (PGPR) and *Pythium aphanidermatum*. Physiol Mol Plant Pathol 56:13–23
Ciampi PL, Burzio LO, Burzio LA (1997) Carriers for *Pseudomonas fluorescens*, antagonistic to *Pseudomonas (Ralstonia) solanacearum*, causal agent of bacterial wilt. Fitopatologia 32(1):64–70
Cirvilleri G, Spina S, Scuderi G, Gentile A, Catara A (2005) Characterization of antagonistic root-associated fluorescent Pseudomonads of transgenic and non-transgenic citrange troyer plants. J Plant Pathol 87(3):179–186
Corbett JR (1974) Pesticide design. In: The biochemical mode of action of pesticides. Academic, London, pp. 44–86
Crowe JD, Olsson S (2001) Induction of laccase activity in *Rhizoctonia solani* by antagonistic *Pseudomonas fluorescens* strains and a range of chemical treatments. Appl Environ Microbiol 67:2088–2094
Dahiya JS, Woods DL, Tewari JP (1988) Control of *Rhizoctonia solani*, causal agent of brown girdling root rot of rapeseed, by *Pseudomonas fluorescens*. Bot Bull Acad Sin 29:135–142
Das K, Katiyar V, Goel R (2003) P-solubilization potential of plant growth promoting *Pseudomonas* mutant at low temperature. Microbiol Res 158:559–562
Daval S, Lebreton L, Gazengel K, Boutin M, Guillerm A, Sarniguet A (2011) The biocontrol bacterium *Pseudomonas fluorescens* Pf29Arp strain affects the pathogenesis-related gene expression of the take-all fungus *Gaeumannomyces graminis* var. *tritici* on wheat roots. Mol Plant Pathol 12(9):839–854
De Meyer G, Hofte M (1997) Salicylic acid produced by the rhizobacterium *Pseudomonas aeruginosa* 7NSK2 induces resistance to leaf infection by *Botrytis cinerea* on bean. Phytopathology 87:588–593
de Souza JT, Raaijmakers JM (2003) Polymorphisms within the *prnD* and *pltC* genes from pyrrolnitrin and pyoluteorin producing *Pseudomonas* and *Burkholderia* spp. FEMS Microbiol Ecol 43:21–34
Defago G, Berling CH, Borger U, Keel C, Voisard C (1990) Suppression of black rot of tobacco by a *Pseudomonas* strain: Potential Applications and Mechanisms. In: Cook RJ, Henis Y (eds) Hornby D. Biological Control of Soil Borne Plant Pathogen, CAB International, pp 93–108
Di Simine CD, Sayer JA, Gadd GM (1998) Solubilization of zinc phosphate by a strain of *Pseudomonas fluorescens* isolated from forest soil. Biol Fertil Soils 28:87–94
Diby P, Saju KA, Jisha PJ, Sarma YR, Kumar A, Anandaraj M (2004) Mycolytic enzymes produced by *Pseudomonas fluorescens* and *Trichoderma* spp. against *Phytophthora capsici*, the foot rot pathogen of black pepper (*Piper nigrum* L.). Ann Microbiol 55:129–133
dos Santos VAPM, Heim S, Moore ERB, Strätz M, Timmis KN (2004) Insights into the genomic basis of niche specificity of *Pseudomonas putida* KT2440. Environ Microbiol 6:1264–1286
Duijff BJ, Pouhair D, Alivian C, Alabouvette C, Lemanceau P (1998) Implication of systemic induced resistance in the suppression of Fusarium wilt of tomato by *P. fluorescente* WCS417r and by non pathogenic *Fusarium oxysporum* by Fo47. Eur J Plant Pathol 104:903–910
Dwivedi D, Johri BN (2003) Antifungals from fluorescent pseudomonads: biosynthesis and regulation. Curr Sci 85(12):1693–1702
Elad Y, Baker R (1985) The role of competition for iron and carbon in suppression of chlamydospore germination of Fusarium sp. By Pseudomonas spp. Ecol Epidemiol 75:1053–1059
El-Banna N, Winkelmann G (1998) Pyrrolnitrin from *Burkholderia cepacia*: antibiotic activity against fungi and novel activities against streptomyces. J Appl Microbiol 85:69–78

Fallahzadeh V, Ahmadzadeh M, Marefat A, Ghazanfary K (2009) Application of rhizobacteria for induction of systemic resistance to bacterial blight of cotton caused by *Xanthomonas campestris* pv. *malvacearum* using fluorescent pseudomonads of rhizosphere. J Plant Prot Res 49(4):416–42

Fridlender M, Inbar J, Chet I (1993) Biological control of soil borne plant pathogens by a B 1,3-glucanase-producing *Pseudomonas cepacia*. Soil Biol Biochem 25:1211–1221

Gaind S, Gaur AC (2002) Impact of fly ash and phosphate solubilising bacteria on soybean productivity. Bioresour Technol 58:313–315

Garrett SD, Jackson RM, Katznelson H, Rovira AD (1965) Biocontrol mechanisms that operate in the rhizosphere. In: Baker KF, Snyder WC (eds) Ecology of soil-borne plant pathogens: prelude to biological control. University of California Press, Berkeley

Gehring PJ, Nolan RJ, Watanabe PG (1993) Solvents, fumigants and related compounds. In: Hayes WJ, Laws ER (eds) Handbook of pesticide toxicology, vol 2. Academic, San Diego, pp 646–649

Glick BR (1995) The enhancement of plant growth by free living bacteria. Can J Microbiol 41:109–117

Glick BR, Karaturovic DM, Newell PC (1995) A novel procedure for rapid isolation of plant growth promoting pseudomonads. Can J Microbiol 41:533–536

Goldstein AH (1986) Bacterial solubilization of mineral phosphates: historical perspective and future prospects. Am J Altern Agric 1:51–57

Gulati A, Rahli P, Pratibha V (2008) Characterization of phosphate solubilizing fluorescent pseudomonads from the rhizosphere of sea buckthorn growing in the cold deserts of Himalayas. Curr Microbiol 56:73–79

Gupta CP, Dubey RC, Kamng SC, Maheshwari DK (2001) Antibiosis-mediated necrotrophic effect of *Pseudomonas* GRC2 against two fungal plant pathogens. Curr Sci 81(1):91–94

Gupta A, Meyer JM, Goel R (2002) Development of heavy metal resistant mutants of phosphate solubilizing *Pseudomonas* sp. NBRI 4014 and their characterizaiton. Curr Microbiol 45: 323–327

Gurusiddaiah S, Weller MD, Sarkar A, Cook JR (1986) Characterization of an antibiotic produced by a strain of *Pseudomonas fluorescens* inhibitory to *Gaeumannomyces graminis* var. *tritici* and spp. Antimicrob Agents Chemother 29:488–495

Haas D, Defago G (2005) Biological control of soil borne pathogens by fluorescent Pseudomonads. Nat Rev Microbiol 3:307–319

Haas D, Keel C (2003) Regulation of antibiotic production in root-colonizing *Pseudomonas* spp. and relevance for biological control of plant disease. Annu Rev Phytopathol 41:117–153

Haas D, Keel C, Laville J, Maurhofer M, Oberhansli T, Schnider U, Voisard C, Wuthrich B, Defago G (1991) Secondary metabolites of *Pseudomonas fluorescens* strain CHAO involved in the suppression of root diseases. In: Hennecke H, Verma DPS (eds) Advances in molecular genetics of plant-microbe interactions. Kluwer Academic, Dordrect

Hamamoto T, Kaned M, Horikoshi K, Kudo T (1994) Characterization of a protease from a psychrotroph, *Pseudomonas fluorescens* 114. Appl Environ Microbiol 60:3878–3880

Hammer PE, Hill DS, Lam ST, Pee KH, Ligon JM (1997) Four genes from *Pseudomonas fluorescens* that present the biosynthesis of pyrrolnitrin. Appl Environ Microbiol 63:2147–2154

Hassanein WA, Awny NM, El-Mougith AA, Salah El-Dien SH (2009) The antagonistic activities of some metabolites produced by *Pseudomonas aeruginosa* Sha8. J Appl Sci Res 5(4):404–414

Hemming BC (1986) Siderophore receptors of root colonizing *Pseudomonas* sp. J Plant Nutr 9:505–507

Hill DS, Stein JI, Torkewitz NR, Morse AM, Howell CR, Pachlatka JP, Becker JO, Ligon JM (1994) Cloning of genes involved in the synthesis of pyrrolnitrin from *Pseudomonas fluorescens* and role of pyrrolnitrin synthesis in biological control of plant disease. Appl Environ Microbiol 60:78–85

Holt JG, Krieg NR, Sneath PHA, Staley JT, Williams ST (1994) Bergey's manual of determinative bacteriology, 9th edn. Lippincott Williams & Wilkins, Baltimore

Homma Y, Suzui T (1989) Role of antibiotic production in suppression of radish dampingoff by seed bacterization with *Pseudomonas cepacia*. Ann Phytopathol Soc Jpn 55:643–652

Howell CR, Stipanovic RD (1979) Control of *Rhizoctonia solani* on cotton seedlings with *Pseudomonas fluorescens* and with an antibiotic produced by the bacterium. Phytopathology 69:480–482

Howell CR, Stipanovic RD (1980) Suppression of *Pythium ultimum*-induced damping off of cotton seedling by *Pseudomonas fluorescens* and its antibiotic pyoluteorin. Phytopathology 70:712–715

Johri BN, Sharma A, Virdi JS (2003) Rhizobacterial diversity in India and its influence on soil and plant health. Adv Biochem Eng Biotechnol 84:49–89

Kalita BC (1994) Epidemiology and management of bacterial wilt of tomato caused by *P. solanacearum* (E.F. Smith). M.Sc. (Agri.). Thesis, Assam Agri. Univ, Jorhat, pp 49–50

Karuna K, Khan ANA (1994) Biological control of wilt of tomato caused by *P. solanacearum* using antagonistic bacteria. Indian Phytopathol 47:326

King EO, Ward MK, Raney DE (1954) Two simple media for the demonstration of pyocyanin and fluorescein. J Lab Clin Med 44:301–307

Kloepper JW, Schroth MN (1978) Plant growth promoting rhizobacteria on radish. In: Proceedings of the fourth international conference on plant pathogenic bacteria, vol. 2. Station de Pathologie Vegetale et Phytobacteriologie, INRA, Angers. pp 879–882

Kraus J, Loper JE (1992) Lack of evidence for a role of antifungal metabolite production by *Pseudomonas fluorescens* Pf-5 in biological control of *Pythium* damping-off of cucumber. Phytopathology 82:264–271

Kumar P, Sood AK (2001) Integration of antagonistic rhizobacteria and soil solarization for the management of bacterial wilt of tomato caused by *Ralstonia solanacearum*. Indian Phytopathol 54(1):12–15

Kumar V, Kumar A, Verma VC, Gond SK, Kharwar RN (2007) Induction of defense enzymes in *Pseudomonas fluorescens* treated chickpea roots against *Macrophomina phaseolina*. Indian Phytopathol 60(3):289–295

Landa BB, Mavrodi OV, Raaijmakers JM, Mcspadden-Gardener BB, Thomashow LS, Weller DM (2002) Differential ability of genotypes of 2,4- diacetylphloroglucinol-producing *Pseudomonas fluorescens* strains to colonize the roots of pea plants. Appl Environ Microbiol 68:3226–3237

Landa BB, Mavrodi DM, Thomashow LS, Weller DM (2003) Interactions between strains of 2,4-diacetylphloroglucinol-producing *Pseudomonas fluorescens* in the rhizosphere of wheat. Phytopathology 93:982–994

Leeman M, Van Pelt JA, Hendrickx MJ, Scheffer RJ, Bakker PAHM, Schippers B (1995) Biocontrol of Fusarium wilt of radish in commercial greenhouse trials by seed treatment with *Pseudomonas fluorescens* WCS374. Phytopathology 85:1301–1305

Leeman M, Den Ouden FM, Van Pelt JA, Dirkx FPM, Steijl H, Bakker PAHM, Schippers B (1996) Iron availability affects induction of systemic resistance against Fusarium wilt of radish by *Pseudomonas fluorescens*. Phytopathology 86:149–155

Leong J (1986) Siderophores: their biochemistry and possible role in the biocontrol of plant pathogens. Annu Rev Phytopathol 24:187–209

Lifshitz R, Kloepper JW, Kozlowski M, Simonson C, Carlson J, Tipping EM, Zaleska I (1987) Growth promotion of canola (rapeseed) seedlings by a strain of Pseudomonas putida under gnotobiotic conditions. Can J Microbiol 33:390–395

Lindberg GD (1981) An antibiotic lethal to fungi. Plant Dis 65:680–683

Loper JE, Buyer JS (1991) Siderophores in microbial interactions on plant surfaces. Mol Plant Microbe Interact 4:5–13

Loper JE, Kobayashi DY, Paulsen IT (2007) The genomic sequence of *Pseudomonas fluorescens* Pf-5, insights into biological control. Phytopathology 97:233–238

M'Piga P, Belanger RR, Paulitz TC, Benhamou N (1997) Increased resistance to *Fusarium oxysporum* f. *sp. radicis-lycopersici* in tomato plant treated with the endophytotic bacterium Pseudomonas fluorescens strain 63-28. Mol Plant Pathol 50:301–320

Magazin MD, Moores JC, Leong J (1986) Cloning of gene coding for the outer membrane receptor protein for ferric Pseudobactin, a siderophore from plant growth promoting *Pseudomonas* strain. J Biol Chem 261:795–799

Manjula K, Krishna G, Kishore GAG, Singh SD (2004) Combined application of *P. fluorescens* and *Trichoderma viride* has an improved biocontrol activity against stem rot in groundnut. Plant Pathol J 20(1):75–80

Martin FN, Loper JE (1999) Soilborne plant diseases caused by *Pythium* spp.: Ecology, epidemiology, and prospects for biological control. Crit. Rev. Plant Sci 18:111–181

Maurhofer M, Reimmann C, Schmidli-Sacherer P, Heeb S, Haas D, Défago G (1998) Salicylic acid biosynthetic genes expressed in *Pseudomonas fluorescens* strain P3 improve the induction of systemic resistance in tobacco against tobacco necrosis virus. Phytopathology 88:678–684

Mavrodi DV, Ksenzenko VN, Bonsall RF, Cook RJ, Boronin AM, Thomashow LS (1998) A seven-gene locus for synthesis of phenazine-1-carboxylic acid by *Pseudomonas fluorescens* 2-79. J Bacteriol 180:2541–2548

Mavrodi DV, Mavrodi OV, McSpadden-Gardener BB, Landa BB, Weller DM, Thomashow LS (2002) Identification of differences in genome content among *phlD*-positive *Pseudomonas fluorescens* strains by using PCR-based substractive hybridization. Appl Environ Microbiol 68:5170–5176

Mazurier S, Lemunier M, Siblot S, Mougel C, Lemanceau P (2004) Distribution and diversity of type III secretion system-like genes in saprophytic and phytopathogenic fluorescent pseudomonads. FEMS Microbiol Ecol 49:455–467

McSpadden Gardener BB, Mavrodi DV, Thomashow LS, Weller DM (2001) A rapid polymerase chain reaction-based assay characterizing rhizosphere populations of 2,4-diacetylphloroglucinol-producing bacteria. Phytopathology 91:44–54

Michel L, Gonzalez N, Jagdeep S, Nguyen-Ngoc T, Reimmann C (2005) PchR-box recognition by the AraC-type regulator PchR of *Pseudomonas aeruginosa* requires the siderophore pyochelin as an effector. Mol Microbiol 58:495–509

Minorsky PV (2008) On the inside. Plant Physiol 146:323–324

Mohamed S, Caunter IG (1995) Isolation and characterization of a *Pseudomonas fluorescens* strain suppressive to *Bipolaris maydis*. J Phytopathol 143:111–114

Mukherjee PK, Latha J, Hadar R, Horwtiz BA (2004) Role of two G-protein alpha subunits, TgaA and TgaB, in the antagonism of plant pathogens by *Trichderma virens*. Appl Environ Microbiol 70:542–549

Muthukumar A, Bhaskaran R, Sanjeevkumar K (2010) Efficacy of endophytic *P. fluorescens* (Trevisan) migula against chilli damping-off. J Biopest 3(1):105–109

Nahas E (1996) Factors determining rock phosphate solubilization by micro organisms isolated from soil. World J Microbiol Biotechnol 12:567–572

Naik PR, Raman G, Narayanan KB, Sakthivel N (2008) Assessment of genetic and functional diversity of phosphate solubilizing fluorescent pseudomonads isolated from Rhizospheric soil. BMC Microbiol 8(230):1–14

Neilands JB (1981) Microbial iron compounds. Annu Rev Biochem 50:715–731

Neilands JB (1986) Siderophores in relation to plant growth and disease. Annu Rev Plant Physiol 37:187–208

Nielsen TH, Thrane C, Christophersen C, Anthoni U, Sorensen J (2000) Structure, production, characteristics and fungal antagonism of tensin- a new antifungal cyclic lipopeptide from *Pseudomonas fluorescens* strain 95.578. J Appl Microbiol 89:992–1001

Notz R, Maurhofer M, Dubach H, Haas D, Defago G (2002) Fusaric acid-producing strains of *Fusarium oxysporum* alter 2,4-diacetylphloroglucinol biosynthetic gene expression in *Pseudomonas fluorescens* CHA0 in vitro and in the rhizosphere of wheat. Appl Environ Microbiol 68:2229–2235

Nowak-Thompson B, Chaney N, Wing JS, Gould SJ, Loper JE (1999) Characterization of the pyoluteorin biosynthetic gene cluster of *Pseudomonas fluorescens* Pf-5. J Bacteriol 181: 2166–2174

O'Sullivan DJ, O'Gara F (1992) Traits of fluorescent *Pseudomonas* spp. involved in suppression of plant root pathogens. Microbiol Rev 56:662–676

Palleroni NJ (1975) General properties and taxonomy of the genus *Pseudomonas*. pi-36. In: Clarke PH, Richmond MH (eds) Genetics and biochemistry of *Pseudomonas*. Wiley, Baltimore

Palleroni NJ (1984) Family I: Pseudomonadaceae. In: Tansil B (ed) Bergeys Manual of Systematic Bacteriology. Williams and Wilkins, Baltimore, USA, pp 141–161

Palleroni S, Kunisawa R, Contopoolou R, Doudoroff IM (1973) Nucleic acid homologies in the genus *Pseudomonas*. Int J Syst Evol Microbiol 23:333–339

Pastor NA, Reynoso MM, Tonelli ML, Masciarelli O, Rosas SB, Rovera M (2010) Potential biological control *Pseudomonas* sp. pci2 against damping-off of tomato caused by *Sclerotium rolfsii*. J Plant Pathol 92(3):737–745

Paulsen IT, Press CM, Ravel J (2005) Complete genome sequence of the plant commensal *Pseudomonas fluorescens* Pf-5. Nat Biotechnol 23:873–878

Pfender WF, Kraus J, Loper JE (1993) A genomic region from *Pseudomonas fluorescens* Pf-5 required for pyrrolnitrin production and inhibition of *Pyrenophora tritici-repentis* in wheat straw. Phytopathology 83:1223–1228

Picard C, Cello FD, Ventura M, Fan R, Gucket A (2000) Frequency and diversity of 2, 4-diacetyl phloroglucinol producing bacteria isolated from the maize rhizosphere at different stages of plant growth. Appl Environ Microbiol 66:948–955

Pradhan N, Sukla LB (2006) Solubilization of inorganic phosphates by fungi isolated from agriculture soil. Afr J Biotechnol 5:850–854

Preston GM, Bertrand N, Rainey P (2001) Type III secretion inplant growth-promoting *Pseudomonas fluorescens* SBW25. Mol Microbiol 41:999–1014

Raaijmakers JM, Weller DM, Thomashow LS (1997) Frequency of antibiotic-producing *Pseudomonas* spp. in natural environments. Appl Environ Microbiol 63:881–887

Raffel SJ, Stabb EV, Milner JL, Handelsman J (1996) Genotypic and phenotypic analysis of zwittermicin A-producing strains of *Bacillus cereus*. Microbiology 142:3425–3436

Ramamoorthy V, Raguchander T, Samiyappan R (2001) Induction of systemic resistance by plant growth promoting rhizobacteria in crop plants against pests and diseases. Crop Prot 20:1–11

Ramamoorthy V, Raguchander T, Samiyappan R (2002) Enhancing resistance of tomato and hot pepper to Pythium diseases by seed treatment with fluorescent pseudomonads. Eur J Plant Pathol 108:429–441

Ramesh R, Joshi AA, Ghanekar MP (2009) Pseudomonads: Major antagonistic endophytic bacteria to suppress bacterial wilt pathogen, *Ralstonia solanacearum* in the egg-plant (*Solanum melongena* L.). World J Microbial Biotechnol 25(1):47–55

Ramette A, Frapolli M, Défago G, Moënne-Loccoz Y (2003) Phylogeny of HCN synthase-encoding hcnBC genes in biocontrol fluorescent pseudomonads and its relationship with host plant species and HCN synthesis ability. Mol Plant Microbe Interact 16:525–535

Ramezanpour MR, Popov Y, Khavazi K, Rahmani HA (2010) Molecular geno systematic and physiological characteristics of fluorescent pseudomonads isolated from the rice rhizosphere of Iranian paddy fields. Afr J Agric Res 6(1):145–151

Ravel J, Cornelis P (2003) Genomics of pyoverdine-mediated iron uptake in pseudomonads. Trends Microbiol 11:195–200

Reddy PK, Reddy MS (2009) Biochemical and PCR-RAPD characterization of *Pseudomonas fluorescens* produced antifungal compounds inhibit the rice fungal pathogens *in vitro*. J Pure Appl Microbiol 3:347–350

Redondo-Nieto M, Matthieu B, John PM, Kieran G, Martínez-Granero F, Emma B, Ana N, Sánchez-Contreras M, Jennifer AM, Stephen RG, Eric RC, Candela M, Willem JS, Paul BR, David D, O'Gara F, Marta M, Rafael R (2012) Genome Sequence of the Biocontrol Strain *Pseudomonas fluorescens* F113. J Bacteriol 194(5):1273–1274

Rezzonico F, Binder C, Défago G, Moënne-Loccoz Y (2005) The type III secretion system of biocontrol *Pseudomonas fluorescens* KD targets the phytopathogenic chromista *Pythium ultimum* and promotes cucumber protection. Mol Plant Microbe Interact 18:991–1001

Rokhzadi A, Asgharzadeh A, Darvish F, Nourmohammadi G, Majidi E (2008) Influence of plant growth-promoting rhizobacteria on dry matter accumulation and yield of chickpea (*Cicer arietinum* L.) under field condition. Am-Euras J Agric Environ Sci 3(2):253–257

Rosales AM, Thomashow L, Cook RJ, Mew TW (1995) Isolation and identification of antifungal metabolites produced by rice-associated antagonistic *Pseudomonas* spp. Phytopathology 85:1028–1032

Saikia R, Kumar R, Singh T, Srivastava AK, Arora DK, Lee MW (2004) Induction of defense enzymes and pathogenesis related proteins in *Pseudomonas fluorescens* treated chickpea in response to infection by *Fusarium oxysporum* f. sp. *ciceri*. Microbiology 32:47–53

Salaheddin K, Valluvaparidasan V, Ladhalakshmi D, Velezhahan R (2010) Management of bacterial blight of cotton using a mixture of *Pseudomonas fluorescens* and *Bacillus subtilis*. Plant Protect Sci 46(2):41–50

Sarathchandra U, Duganzich D, Burch G (1993) Occurrence of antifungal fluorescent *Pseudomonas* spp. on some horticultural and pastoral plants. N Z J Crop Horticult Sci 21(3):267–272

Schroth MN, Hancock JG (1981) Selected topics in biological control. Annu Rev Microbial 35:453–476

Schroth MN, Hancock JG (1982) Disease suppressive soil and root-colonizing bacteria. Science 216(25):1376–1381

Shanahan P, O'Sullivan DJ, Glennon JD, O'Gara F (1992) Isolation and characterization of an antibiotic-like compound from a fluorescent pseudomonad and investigation of physiological parameters influencing its production. Appl Environ Microbiol 58:353–358

Sharifi-Tehrani A, Zala M, Natsch A, Moenne-Loccoz Y, Defago G (1998) Biocontrol of soil-borne fungal plant diseases by 2,4- diacetylphloroglucinol-producing fluorescent pseudomonads with different restriction profiles of amplified 16S rDNA. Eur J Plant Pathol 104:631–643

Shukla A, Gupta SK (2005) Integrated approach to management of tomato bacterial spot (*Xanthomonas vesicatoria*). J Mycol Plant Pathol 35(2):317–319

Siddiqui S, Siddiqui ZA, Ahmad I (2005) Evaluation of fluorescent Pseudomonads and *Bacillus* isolates for the biocontrol of a wilt disease complex of pigeon pea. World J Microbiol Biotechnol 21(5):729–732

Singh R, Sinha AP (2009) Biological control of sheath blight of rice with *Pseudomonas fluorescens*. Indian Phytopathol 62(3):381–383

Singh D, Dhar S, Yadav DK (2010) Effect of endophytic bacterial antagonist against black rot disease of cauliflower caused by *Xanthomonas campestris* pv. *campestris*. Indian Phytopathol 63(2):122–126

Sivakumar G, Sharma RC, Rai SN (2000) Biocontrol of banded leaf and sheath blight of maize by peat based *Psuedomonas fluorescens* formulation. Indian Phytopathol 53(2):190–192

Spiers AJ, Bukling A, Rainey PB (2005) The causes of Pseudomonas diversity. Microbiology 146(10):2–9

Srinivasan K, Gilardi G, Garibaldi A, Gullino ML (2009) Bacterial antagonists from used rockwool soilless substrates suppress *Fusarium* wilt of tomato. J Plant Pathol 91(1):147–154

Srivastava S, Yadav KS, Kundu BS (2004) P0hosphate solubilizing pseudomonads suppress damping-off disease in tomato. J Mycol Plant Pathol 34:662–664

Stalstorm YA (1903) Beitrag zur kennturs der ein-wisking sterilia and in ha hung botindlichen strolte amt dil torlichkeit der phosphorsen der tricalcium phosphate. Zentralbl Baketeriol 11:724–732

Stohl EA, Milner JL, Handelsman J (1999) Zwittermicin A biosynthetic cluster. Gene 237: 403–411

Sunish KR, Ayyadurai N, Pandiaraja P, Reddy AV, Venkateswarku Y, Prakash O, Sakthival N (2005) Characterization of antifungal metabolite produced by a new strain *Pseudomonas aeruginosa* PUPa3 that exhibits broad spectrum antifungal activity and biofertilizing traits. J Appl Microbiol 98(1):145–154

Suresh A, Pallavi P, Srinivas P, Praveen Kumar V, Jeevan Chandra S, Ram Reddy S (2010) Plant growth promoting activities of fluorescent pseudomonads associated with some crop plants. Afr J Microbiol Res 4(14):1491–1494

Suryakala D, Maheshwaridevi PV, Lakshmi KV (2004) Chemical characterization and *in vitro* antibiosis of siderophores of rhizosphere fluorescent pseudomonads. Indian J Microbiol 44(2):105–108

Svercel M, Duffy B, Defago G (2007) PCR amplification of hydrogen cyanide biosynthetic locus *hcn*AB in *Pseudomonas* spp. J Microbiol Methods 70:209–213

Tari PH, Anderson AJ (1988) Fusarium wilt suppression and agglutinability of *Pseudomonas putida*. Appl Environ Microbiol 54:2037–2041

Thomashow LS, Weller DM (1988) Role of a phenazine antibiotic from *Pseudomonas fluorescens* in biological control of *Gaeumannomyces graminis* var. *tritici*. J Bacteriol 170:3499–3508

Ton J, Van Pelt JA, Van Loon LC, Pieterse CMJ (2002) Differential effectiveness of salicylate-dependent and jasmonate/ethylene-dependent induced resistance in Arabidopsis. Mol Plant Microbe Interact 15:27–34

Tripathi M, Joshi BN (2002) *In vitro* antagonistic potential of pseudomonads and control of sheath blight of maize caused by Rhizoctonia solani. Indian J Microbiol 42:207–214

Umesha S, Dharmesh SM, Shetty SA, Krishnappa M, Shetty HS (1998) Biocontrol of downy mildew disease of pearl millet using *Pseudomonas fluorescens*. Crop Prot 17(5):387–392

Umesha S, Kavitha R, Shetty HS (2005) Transmission of seed-borne infection of chilli by *Burkholderia solanacearum* and effect of biological seed treatment on disease incidence. Arch Phytopathol Plant Prot 38(4):281–293

Unnamalai N, Gnanamanickam SS (1984) *Pseudomonas fluorescens* is an antagonistic to *Xanthomonas citri* (Hasse) Dye. the incitant of citrus canker. Curr Sci 53:703–704

Uppal AK, El-Hadrami A, Adam LR, Tenuta M, Daaye F (2008) Biological control of potato Verticillium wilt under controlled and field conditions using selected bacterial antagonists and plant extracts. Biol Ctrl 44(1):90–100

Van Peer R, Niemann GJ, Schippers B (1991) Induced resistance andphytoalexin accumulation in biological control of fusariumwilt of carnation by *Pseudomonas* sp. strain WCS417r. Phytopathology 81:728–734

Vessey KJ (2003) Plant growth promoting rhizobacteria as biofertilizers. Plant Soil 255:571–586

Vidhyasekaran P, Kamala N, Ramanathan A, Rajappan K, Paranidharan V, Velazhahan R (2001) Induction of systemic resistance by *Pseudomonas fluorescens* Pf1 against *Xanthomonas oryzae* pv. *oryzae* in rice leaves. Phytoparasitica 29:155–166

Voisard C, Keel C, Haas D, Defago G (1989) Cyanide production by *Pseudomonas fluorescens* helps to suppress black root rot of tobacco under gnotobiotic conditions. EMBO J 8:351–358

Wei G, Kloepper JW, Tuzum S (1991) Induction of systemic resistance of cucumber to *Colletotrichum orbiculare* by selected strains of plant growth promoting rhizobacteria. Phytopathology 81:1508–1512

Weller DM (1988) Biological control of soil borne plant pathogens in the rhizosphere with bacteria. Annu Rev Phytopathol 26:379–407

Weller DM, Zhang BX, Cook RJ (1985) Application of a rapid screening test for selection of bacteria suppressive to take-all of wheat. Plant Dis 69:710–713

Wenzel SC, Muller R (2005) Formation of novel secondary metabolites by bacterial multimodular assembly lines: deviations from textbook biosynthetic logic. Curr Opin Chem Biol 9:447–458

Xu GW, Gross DC (1986) Field evaluations of the interactions among fluorescent pseudomonads, *Erwinia caratovora* and potato yields. Phytopathology 76:432–430

Zhang Y, Fernando WGD, de Kievit TR, Berry C, Daayf F, Paulitz TC (2006) Detection of antibiotic-related genes from bacterial biocontrol agents with polymerase chain Reaction. Can J Microbiol 52:476–481

Chapter 12
Oat Fungal Diseases and the Application of Molecular Marker Technology for Their Control

Adrian Lester Cabral, Belaghihalli N. Gnanesh, Jennifer Mitchell Fetch, Curt McCartney, Thomas Fetch, Robert F. Park, James G. Menzies, Brent McCallum, Ganapathy Kuyyamudi Nanaiah, and Aakash Goyal

12.1 Introduction

Oat (*Avena* spp.) is an important cereal crop grown worldwide for grain and fodder. Fungal diseases are an important constraint to oat production. Significant yield losses have been attributed to crown rust (caused by *Puccinia coronata* Corda f. sp. *avenae* Eriks. [*Pca*]) and stem rust (caused by *Puccinia graminis* Pers. f. sp. *avenae* Eriks. and Henn. [*Pga*]), which persist in all major oat growing regions of the world. Over the past century, severe rust epidemics have occurred in many of the major oat growing regions of the world. Numerous stem rust epidemics occurred between the early 1900s and the late 1970s in the Midwestern region of the USA, with yield losses as high as 25 % (Roelfs 1978). Similar epidemics occurred in Saskatchewan

A.L. Cabral, Ph.D. • B.N. Gnanesh, Ph.D. (✉) • C. McCartney, Ph.D.
T. Fetch, Ph.D. • J.G. Menzies, B.Sc., M.Sc., Ph.D. • B. McCallum, Ph.D.
Agriculture and Agri-Food Canada, 101 Route 100, Morden, MB, R6M 1Y5, Canada
e-mail: gnaneshbn@gmail.com; Tom.Fetch@agr.gc.ca

J. Mitchell Fetch, Ph.D.
Agriculture and Agri-Food Canada, 2701 Grand Valley Road, Brandon, MB, R7A 5Y3, Canada

R.F. Park, B.Sc. (Hons.), Ph.D.
Plant Breeding Institute, The University of Sydney, Narellan, NSW, Australia

G.K. Nanaiah, Ph.D.
Directorate of Sorghum Research, Hyderabad, India

A. Goyal, Ph.D., F.I.C.N., F.S.A.B.
Biodiversity and Integrated Gene Management Program,
International Center for Agriculture Research in the Dry Area,
Rabat, Morocco

A. Goyal and C. Manoharachary (eds.), *Future Challenges in Crop Protection Against Fungal Pathogens*, Fungal Biology, DOI 10.1007/978-1-4939-1188-2_12,

and Manitoba in Canada (Martens 1985), with a severe epidemic causing yield losses of 35 % in 1977 (Martens 1978). In 2002, losses of nearly 7 % were reported in the prairie provinces of Canada, likely the result of a new virulent race NA67 (Fetch 2005). During the first half of the twentieth century, rusts wreaked havoc on wild oats in coastal areas of Australia during years favorable for the proliferation of the fungus (Waterhouse 1952).

Crown rust is the most serious threat to oat production in North America. An epidemic that occurred in 1904 in North Dakota, South Dakota, and Minnesota, USA and in the neighbouring prairie provinces of Canada was favored by unusually high quantities of moisture that prevailed at the grain-filling stages (Carleton 1905). Between 1918 and 1930, annual estimated losses in the USA averaged greater than 13.7 million bushels (Murphy 1935). Simons and Murphy (1968) observed annual losses of up to 30 % during years of severe infestation of crown rust.

Powdery mildew (caused by *Blumeria graminis* (DC.) Speer f. sp. *avenae* Em. Marchal) and *Fusarium* head blight (FHB; *Fusarium* spp.) can also cause significant losses depending on the climate and geographical region. Powdery mildew is the most important foliar disease of oat in the cooler humid regions of Europe (Roderick et al. 2000), including Poland (Sebesta et al. 1991). Crop losses caused by powdery mildew are significant, with estimates up to 39 % from comparisons between susceptible and resistant near-isogenic lines (Lawes and Hayes 1965). Similarly, Jones (1977) found a yield reduction of 20 % in the susceptible cv. Sun II, but only 9 % in the moderately resistant cv. Maldwyn. Clifford (1995) reported annual crop losses from 5 to 10 % in Great Britain, with much higher reductions in small plot trials. Hsam et al. (1997) investigated powdery-mildew resistance in 259 common oat cultivars and breeding lines. They found 67 % of the investigated plants were susceptible to this disease. In contrast to wheat and barley, FHB has not caused large epidemics or yield losses in oat. This may be attributable to a lack of visual symptoms associated with infection (Tekauz et al. 2004). Nevertheless, FHB can impact significantly on oat quality due to the production of mycotoxins.

One highly effective method to control fungal diseases is through the use of fungicides. While fungicides have been shown to limit yield losses in oat during rust outbreaks (Brink and Belay 2006), they do not necessarily provide complete protection of the crop and are less effective than host resistance in reducing the overall size of rust populations (Park 2008). Fungicides are an added cost and their use may not be economic in oat production. Additionally, they can have environment impacts from residues in soil or runoff into groundwater. Harder and Haber (1992) cited ecological concerns linked to their use, in addition to difficulties in obtaining fungicide registration for smaller acreage crops such as oat. Other methods of control include cropping strategies such as early planting (Fleischmann and McKenzie 1965), or the use of early maturing cultivars to escape disease (Simons and Michel 1968; Suttie and Reynolds 2004), maintaining an optimum plant density (Burdon and Chilvers 1982), and crop rotations to avoid inoculum of stubble-borne diseases such as FHB. However, the deployment of resistant cultivars is the most economic and environmentally safe means to control diseases in oat.

12.2 Genetic Resistance

The incorporation of plant resistance genes to provide protection from various diseases is used by oat breeders worldwide, primarily for the rust and smut diseases (Ohm and Shaner 1992). Resistance genes conditioning hypersensitive responses at the seedling stage (i.e. "major" genes) have been and continue to be the chief method of control for crown rust and stem rust (Gnanesh et al. 2013). Although major genes provide complete resistance against specific races of rust via a gene-for-gene interaction (Flor 1955), they can be short lived, remaining effective for between 3 and 7 years when deployed singly (Frey and Browning 1971). Adult plant resistance (APR) genes (i.e. "minor" or slow-rusting resistances) can also be effective in reducing the severity of rust outbreaks. APR can be durable (Johnson and Law 1975; Johnson 1984) or non-durable (Dyck et al. 1966; Knott 1968), and the resistance conferred is only expressed at the mature stages of plant growth (Hare 1997). Although previously reported only in hexaploid oat, APR was identified in diploid and tetraploid oat accessions by Cabral et al. (2011). Cultivars with durable APR remain resistant to a pathogen for longer periods of time compared to cultivars with only major gene resistance. Carson (2009a) reported oat cultivars with major gene resistance to crown rust in the USA generally succumbed to the pathogen 5 years or less after release. Besides using major and APR genes, several other strategies can be applied in breeding for resistance to crown rust. These include gene pyramiding, use of multiline cultivars, line or varietal mixtures, and selection for partial resistance (Cabral 2009).

Multilines are a mixture of resistant genotypes that are known to reduce the severity of rust infections. Jensen (1965) defined multiline varieties as composites, which are phenotypically similar but genetically different. A line or varietal mixture is broadly defined as a heterogeneous crop of a single species (Wolfe 1985). Browning and Frey (1969) outlined the main relative advantages of multiline cultivars over pure line cultivars in disease control. These include an extended life span for a useful resistance gene(s), and a decreased severity of rust infection on host plants resulting in optimum crop yields. Carson (2009b) suggested that although multilines or varietal mixtures might provide some initial protection against *Pca*, they permit selection of complex races of the pathogen that reduce the long-term durability of the *Pc* genes in the multiline or varietal mixture.

Partial resistance is a form of incomplete resistance, considered to be durable (Long et al. 2006) and polygenic in inheritance. The use of recurrent selection for partial resistance in oat as a means to provide protection against *Pca* has been reported in several studies (Díaz-Lago et al. 2002; Long et al. 2006). Among the various resistance breeding strategies, Holland and Munkvold (2001) considered gene pyramiding and selection for partial resistance to be more advantageous than single gene resistance and complete resistance (race-specific). Nevertheless, the use of major genes in developing resistant oat cultivars is the most common because it is easiest, and when effective, results in crops that are nearly entirely free from disease, while the resistance gene endures.

12.3 Crown Rust (*Pca*) and Stem Rust (*Pga*) Resistance Genes

Previous research has identified numerous major resistance genes that protect cultivated oats against *Pca* and *Pga*. There are currently more than 100 described *Pc* genes (Sanz et al. 2013; Gnanesh et al. 2014), but because their chromosomal locations have not been determined, many could be the same or allelic. Most *Pc* genes are inherited dominantly, but some are either partially dominant or recessive (Simons et al. 1978). Many genes were derived from accessions of hexaploid *A. sterilis* collected in Israel and other Mediterranean countries during the 1960s and the early 1970s (Leonard et al. 2004) and include *Pc34*, *Pc35*, *Pc36*, *Pc38* *Pc39*, *Pc40*, *Pc41*, *Pc42*, *Pc43*, and *Pc45–Pc77*. Another source of many resistance genes is the diploid species *A. strigosa* and include *Pc15*, *Pc16*, *Pc17*, *Pc19*, *Pc23*, *Pc30*, *Pc37*, *Pc81*, *Pc82*, *Pc83*, *Pc84*, *Pc85*, *Pc86*, *Pc87*, *Pc88*, *Pc89*, *Pc90*, and *Pc94* (CDL 2013). Recently, Carson (2009a) identified 48 accessions of *A. barbata* in a buckthorn nursery at St. Paul, Minnesota that were resistant to a diverse bulk mixture of *Pca* races, which may be useful as new sources of resistance.

Several *A. sterilis*-derived resistance genes, such as *Pc38*, *Pc39*, *Pc48*, *Pc58*, *Pc59*, *Pc60*, *Pc61*, and *Pc68*, have been used in breeding programs in North America. Genes *Pc38* and *Pc39* were the first to be deployed widely in commercial cultivars in Canada (Chong and Kolmer 1993; McCallum et al. 2007). The widespread deployment of *Pc38* and *Pc39* in oat cultivars resulted in quickly increased frequencies of virulence for these genes in the *Pca* population, thereby rendering them ineffective (Chong and Seaman 1997). Similarly, virulence to *Pc48* arose quickly upon the release of the cultivar "Triple Crown" (Chong and Zegeye 2004) because its resistance relied on only one effective gene. Virulence to gene *Pc68* developed less rapidly in the pathogen population (10 years), perhaps due to its presence in combination with *Pc38*+*Pc39*, but this gene combination was eventually defeated in 2005 (Chong et al. 2008). While the resistance in "Leggett" (*Pc68*+*Pc94*) and "HiFi" (*Pc91*) is effective (Chong et al. 2011), efforts to develop new cultivars with different combinations of *Pc* genes is needed because both cultivars also rely on only one currently effective gene. In Australia, "Drover" (*Pc91*) was for several years the only oat cultivar with effective resistance to *Pca* (Park and Kavanagh 2009). However, virulence for *Pc91* was detected in late 2012 in Queensland. The cultivar "Aladdin" is hypothesised to carry *Pc91*+*Pc50* and remains resistant to this new pathotype, possibly being protected by *Pc50* (Park 2013).

While most *Pc* resistant genes have been identified from *A. sterilis* and *A. strigosa*, most numbered *Pg* resistant genes are derived from the cultivated hexaploid *A. sativa*. Genes *Pg-1*, *Pg-2*, *Pg-3*, *Pg-4*, *Pg-5*, *Pg-8*, *Pg-9*, *Pg-10*, *Pg-11*, *Pg-12*, and *Pg-14* were all derived from *A. sativa*, *Pg-6* and *Pg-7* from *A. strigosa*, *Pg-13*, *Pg-15*, and *Pg-17* from *A. sterilis*, and *Pg-16* from *A. barbata* (Fetch and Jin 2007; Gnanesh et al. 2014). Five *Pg* genes (*Pg-1*, *Pg-2*, *Pg-4*, *Pg-9*, and *Pg-13*) have been deployed in Canadian oat cultivars. Genes *Pg-2* and *Pg-13* are thought to be present in most currently resistant cultivars (Fetch and Dunsmore 2004), and were also confirmed to be in "AC Ronald" and "AC Gwen" (Mitchell Fetch and Fetch 2011). However, virulence for these two genes was detected in 1998 with the emergence of races NA67 (TJJ) and NA76 (TJG), thereby rendering all current cultivars susceptible to stem rust (McCallum et al. 2007). Nine *Pg* genes (*Pg-1*, *Pg-2*, *Pg-3*, *Pg-4*, *Pg-8*, *Pg-9*, *Pg-13*, *Pg-a*, and *Pg-Sa*) have been deployed in Australian commercial varieties (Adhikari et al. 2000). Of these *Pg-8*, *Pg-13*, and *Pg-a* were reported to be effective against *Pga* in Australia up until the late 1980s (Oates 1992). No current Australian oat cultivar is resistant to stem rust (Park 2008).

12.4 Powdery Mildew Resistance Genes

Based on the reaction of differential oat cultivars and lines to various pathotypes of *Blumeria graminis avenae*, resistance to powdery mildew in oat is governed by major genes that have been characterized as oat mildew resistance (OMR) groups (Jones and Jones 1979). Several sources with major gene resistance to powdery mildew, including common oats (Jones 1983; Hsam et al. 1997), wild oat species such as *A. barbata* (Aung et al. 1977; Thomas et al. 1980), *A. strigosa*, *A. occidentalis* (Herrmann and Roderick 1996), *A. pilosa* (Hoppe and Kummer 1991), and *A. sterilis* (Hayes and Jones 1966), have been reported. Two main sources of powdery mildew resistance are currently deployed in European oat breeding programs (Roderick et al. 2000). The first is derived from the *A. sterilis* line CAV1832 with *Pc54* (Sebesta et al. 1993). The second is from the lines APR 122 and APR 166, derived from *A. eriantha*, controlled by a single gene. To date, six OMR groups have been characterized in oat, but only three (OMR1, OMR2, and OMR3) have been used commonly in breeding programmes (Kowalczyk et al. 2004; Okoń 2012).

12.5 *Fusarium* Head Blight Resistance Genes

Fusarium head blight (FHB) is a devastating fungal disease of cereal crops that is prevalent in North America and northern Europe. FHB is caused by any 1 of the 17 species of the *Fusarium* fungus (Parry et al. 1995) that infect oat, wheat, barley, and other grasses. In North America, *Fusarium graminearum* is the main causal organism of FHB in wheat (Schroeder and Christensen 1963), and results in shrivelled and chalky kernels that have a low germination percentage (Gilbert and Tekauz 1995). In oat, symptoms are not as obvious as they are in wheat and barley, but occasionally bleached spikelets are found (Tekauz et al. 2004). In areas with FHB, oat was found to be infected with *Fusarium* species, but at a lower frequency than either wheat or barley (Tekauz et al. 2004). The main causal species in Manitoba, Canada are *Fusarium graminearum*, *Fusarium poae*, *Fusarium sporotrichiodes*, and *Fusarium avenaceum*, in the order of their prevalence when isolated from oat groats collected from commercially produced crops (Tekauz et al. 2008).

However, *F. avenaceum* predominated in the neighbouring province of Saskatchewan and in Manitoba in years prior to major FHB epidemics during the early 2000s (Tekauz et al. 2008). A majority of the commonly grown oat cultivars do not possess good resistance to FHB (Mielniczuk et al. 2004; Tekauz et al. 2008). Additionally, *Gibberella zeae*, the teleomorph (sexual stage) of *F. graminearum*, produces large amounts of the mycotoxin deoxynivalenol (DON) on oat seed (Tekauz et al. 2008), besides other toxins like acetyl-deoxynivalenols (3-ADON or 15-ADON) and nivalenol (NIV). At present, information on the resistance of oats to *Fusarium* is limited but genetic differences in the resistance levels of cultivars do exist and progress can be made by breeding for resistance (Tekauz et al. 2008). Since visual symptoms of FHB are so difficult to observe in oat, other methods of assessment, such as the evaluation of groat colonization by the fungus and mycotoxin measurement, need to be employed to effectively develop resistant oat cultivars.

12.6 The Role of Molecular Markers

Genetic markers are pivotal to modern genomics research. Molecular markers are useful tools in plant disease resistance breeding programs for the evaluation of germplasm for disease resistance, mapping major and minor resistance genes, postulation of resistance genes, marker-assisted selection, and map-based cloning of resistance genes. The relatively small and scattered international oat research community is continually challenged by the large and complex genome of hexaploid oat and by a scarcity of genetic sequence data.

Molecular markers, mainly restriction fragment length polymorphisms (RFLP), amplified fragment length polymorphisms (AFLP), simple sequence repeats (SSR), and diversity array (DArT) markers, have played a major role in the genetic mapping of economically important disease resistance genes to linkage groups. The construction of a genetic map of the Kanota × Ogle population (*A. sativa*) using restriction fragment length polymorphisms (RFLP) (O'Donoughue et al. 1995) provided a reference genetic map that has aided comparative mapping studies and will enable future development of molecular markers for other traits (Wight et al. 2003). More recently, DArT markers were placed on the Kanota × Ogle map (Tinker et al. 2009). Microsatellite (SSR) markers are broadly used for marker-assisted selection in crop breeding because of ease of implementation and codominance. EST-derived SSRs (Becher 2007) and genomic SSRs (Li et al. 2000; Pal et al. 2002) have been developed and screened in oat cultivars. A total of 216 SSR primer sequences were developed from ESTs (Becher 2007). These *Avena* EST-derived microsatellite loci (AME) were developed by screening a set of 7,021 EST sequences that were published earlier by Rines et al. (2004). Forty-five AME loci have been placed onto 24 of the 29 linkage groups of the Kanota × Ogle map developed by Wight et al. (2003). Recently, Gutierrez-Gonzalez et al. (2013) developed gene-derived SSR markers with the potential to be used in oat breeding programs. In total, 4,639 SSRs were found within 4,128 different transcripts. The most abundant SSRs were trinucleotide repeats (2,841; 61.2 %).

Oliver and coworkers (2011) provided the first set of oat-based single nucleotide polymorphism (SNP) markers, and a pipeline for the large scale development of a much-needed genomic resource. The SNP discovery and validation pipeline is an effective method to identify SNP markers in oat. These markers have a high assay validation rate and proven utility in a variety of applications. Collaborative initiatives, such as the Collaborative Oat Research Enterprise (CORE) (http://wheat.pw.usda.gov/CORE600/home), are now delivering genomics platforms for oat that are revolutionizing the possibilities for research and innovation. The CORE is a global research partnership represented by 28 oat research sites that was funded by the United States Department of Agriculture, the North American Millers Association, and the Prairie Oat Growers Association. Ever since the first publication of a diploid oat map (O'Donoughue et al. 1992), the oat research community has made efforts in developing oat linkage maps, which include Kanota × Ogle (O'Donoughue et al. 1995; Tinker et al. 2009; Wight et al. 2003), Ogle × TAM O-301 (Portyanko et al. 2001), Ogle × MAM17-5 (Zhu and Kaeppler 2003), Terra × Marion (DeKoeyer et al. 2004), and Aslak × Matilda (Tanhuanpää et al. 2008). However, all of these maps are incomplete and fragmented, and a map with 21 linkage groups (LG), well defined by chromosomal assignments was achieved only recently when Oliver et al. (2013) developed a physically anchored consensus map with 21 LGs through SNP mapping in six hexaploid oat populations and SNP deletion analysis in a set of monosomic stocks. This map provides substantial improvements over previous maps because of low error rates in the scoring of SNP markers, and the joint mapping of markers assayed in parallel across multiple populations. The current availability of a large number of SNP markers and relatively inexpensive high and low-plexity SNP assays (e.g. Infinium SNP assay from Illumina, Inc. and KASP™ar assays from LGC Genomics Ltd.) have made it possible to assign resistance genes to oat chromosomes rapidly (Gnanesh et al. 2013). Impacts of CORE work will be seen in areas of QTL and association mapping, and studies of genome structure and evolution, leading to the accelerated improvement of oat through marker-assisted breeding.

12.7 Mapping of Resistance Genes/QTLs

12.7.1 Rust

A number of race-specific resistance genes, such as *Pc68* (Chen et al. 2006; Satheeskumar et al. 2011), *Pc71* (Bush and Wise 1998), *Pc92* (Rooney et al. 1994), *Pc94* (Chong et al. 2004) and *Pc91* (McCartney et al. 2011), have been mapped and markers that are closely linked to crown and stem rust genes have been identified (reviewed in detail by Gnanesh et al. 2014). Currently, the most commonly utilized *Pc* genes in North America oat breeding programs are *Pc: 38*, *39*, *48*, *58*, *59*, *60*, *61*, *62*, *68*, *91*, *94*, *96*, *97*, *101*, and *103*. Similarly, for stem rust *Pg: a*, *2*, *9*, *10*, *11*, *12*, *13*, and *16* are the most commonly utilized resistance genes. A few of these genes have been defeated, but they can be very effective when they are pyramided in a single variety. Markers have been developed for many of the commonly utilized *Pc* genes like *Pc68*, *Pc91*, and *Pc94*. However, the use of these markers in oat breeding has been limited, because most of the reported markers (RAPDs, AFLPs, and DArTs) are dominant in nature and are not suitable for high-throughput genotyping. Also, their use in early generation MAS is limited as they cannot reliably detect heterozygous genotypes. Hence, KASPar markers linked to resistance genes will be very effective and they have been used successfully in high-throughput marker-assisted selection of oat crown rust gene *Pc91* (Gnanesh et al. 2013). Allele-specific markers for the crown rust resistant genes, *Pc68*, *Pc94* and *PcKM* have also been developed (Gnanesh et al. unpublished). Further, it would be useful to find markers for the newly identified genes *Pc101* and *Pc103*.

Host resistance is usually based on single major genes conferring complete resistance, but with the emergence of new pathogen races this resistance may be easily overcome. Partial resistance is believed to be more effective in controlling plant diseases because it allows sporulation of the pathogen, reduces selection pressure for virulence, and thus slows the evolution of pathogen virulence. A number of breeding techniques are recommended to acquire durable resistance. Among these are recurrent selection, which, by definition, is increasing the frequency of desirable alleles and consequently gains in the population due to repeated cycles of selection and recombination of selected lines (Díaz-Lago et al. 2002). The authors demonstrated the usefulness of rapid cycle recurrent selection as a population improvement procedure capable of effectively increasing the level of partial resistance to crown rust in an adapted and high-yielding oat population. Their results indicated that selection for partial crown rust resistance in early generations can produce adequate gains per recurrent selection cycle with a minimum cycle length. Another breeding technique is advanced backcross—QTL (AB-QTL), which incorporates genes located in different linkage groups into a single variety by successive backcrosses with selection based on the use of molecular markers (Lambalk et al. 2004). Several studies have identified QTLs conferring durable crown rust resistance in oat. For example Portyanko et al. (2005) utilized 230 markers, a majority of which were RFLP and AFLP, and identified four major and three minor QTLs for partial resistance to *Pca* in the oat line MN841801-1. Hoffman et al. (2006) used six crown rust isolates, avirulent on TAM O-301 and virulent on Ogle, to test the parents and TAM O-301 × Ogle RILs. Genetic analyses of the segregation data for each of the six isolates indicated that the resistance was conditioned by a complex of three loci (*Pc58a*, *58b*, and *58c*), two of which (*Pc58a* and *Pc58c*) were tightly linked. Zhu and Kaeppler (2003) found two consistent QTLs (*Pcq1* and *Pcq2*) for crown rust resistance in Ogle × MAM 17-5 that explained 48.5–70.1 % and 9.6–14.0 % of resistance, respectively. Similarly, 4–8 QTLs using AFLP, RFLP, SCAR markers for crown rust APR were detected by Jackson et al. (2007, 2008, 2010) and Acevedo et al. (2010). Most of the QTLs for APR in oat reported so far have not been successfully validated and used in MAS. The crown rust resistance in the oat line MN841801 has been effective for more than 30 years, but it is not known whether or not this resistance has been widely deployed in oat varieties. Lin et al. (unpublished) identified a major APR QTL explaining up to 74 % of the phenotypic variance in a RIL population of the cross AC Assiniboia × MN841801. A single QTL segregating in this population makes it a suitable candidate for use in marker-assisted breeding and also an ideal target for map-based cloning of the gene underlying the QTL.

Lange (2012) evaluated winter and spring oat resistance to crown rust in four field environments in Texas, Louisiana, Minnesota, and North Dakota during a 2-year study in 2010 and 2011. Plants representing 702 elite lines of oat phenotyped for crown rust resistance were found to have highly diverse responses. The winter oat lines demonstrated the best crown rust resistance and are expected to yield the most genes/QTL to be used in developing durable crown rust resistance. Oat lines developed in the US states along the *Puccinia* pathway in Texas, Louisiana, Minnesota, North Dakota, and Wisconsin on average exhibited the best crown rust resistance as compared to other areas in the country. GGE biplot analysis indicated that Castroville, TX was the most representative and most ideal testing location (Lange 2012). The above results are expected to increase knowledge of the genetic diversity of the oat germplasm, and yield comprehensive genotyping and phenotyping information for North American oat breeding programs. Another approach is the cloning and study of resistance gene analogs (RGAs) (Irigoyen et al. 2004; Kremer et al. 2001; Portyanko et al. 2005; Satheeskumar et al. 2011). Cloning and study of members of defined resistant gene families, such as wheat *Lrk10*-type genes in oats (Cheng et al. 2003). Sanz et al. (2013) found that nucleotide binding site (NBS) and protein kinase (PK) based markers cover partly complementary regions of oat genomes. Markers of the different classes obtained were found to be associated with the two resistance loci, *PcA* and R-284B-2, mapped on *A. strigosa* × *A. wiestii* (Asw map), and with five out of eight QTLs for partial resistance in the MN841801-1 × Noble-2 (MN) map. Fifty-three RGA-RFLPs and 187 NBS/PK profiling markers were also mapped on the hexaploid map *A. byzantina* cv. Kanota × *A. sativa* cv. Ogle.

12.7.2 Powdery Mildew and FHB

Very few linked molecular markers have been developed for oat powdery mildew and FHB resistances. Therefore, developing molecular markers tightly linked to the resistance genes would be beneficial both for oat breeding purposes and to investigate oat genomic regions containing interesting resistance genes.

Yu and Herrmann (2006) were the first to map powdery-mildew resistance in hexaploid oat. A resistance source from *A. macrostachya* was successfully introgressed into hexaploid oat and their work revealed the resistance is controlled by a dominant gene, designated *Eg-5*. One SSR marker AM102 and four AFLP-derived PCR-based markers were identified linked to *Eg-5*. The powdery mildew resistance gene *Eg-3* was mapped with RFLP markers from *Triticeae* group-1 chromosomes using an F_3 population from a cross between *A. byzantina* cv. Kanota and *A. sativa* cv. Rollo (Mohler et al. 2012). This comparative mapping approach positioned *Eg-3* between cDNA-RFLP marker loci cmwg706 and cmwg733. Okoń and Kowalczyk (2012) identified a SCAR-BG8 marker linked to oat mildew resistance group 2 (OMR2) and this marker might be used to select genotypes that are resistant to OMR2. Hagmann et al. (2012) developed a linkage map with 366 DArT markers and detected a major QTL for adult plant/field powdery mildew resistance, located close to the DArT marker oPt-6125 in a region corresponding to linkage group 5_30 on the K×O map. This QTL accounted for 31 % of the phenotypic variation in the mapping population. A BLAST search of the oPt-6125 clone sequence revealed a high degree of similarity to a wheat mRNA expressed under mildew infection pressure. By contrast, no major QTL were found for seedling-stage resistance under artificial infection of powdery mildew.

Breeding FHB resistant cultivars is an economically and environmentally friendly way to reduce mycotoxins on grain (DON), either by the identification of resistance QTL or phenotypic evaluation. A DON resistance QTL was identified in the recombinant inbred line population Hurdal×Z595-7 (He et al. 2013). The QTL *Qdon.umb-17A/7C*, located on chromosome 17A/7C, was detected in all experiments using composite interval mapping, with phenotypic effects of 12.2–26.6 %. In addition, QTL for DON were also found on chromosomes 5C, 9D, 13A, 14D and unknown_3, while a QTL for FHB was found on 11A. A half-sib population of HZ595, Hurdal×Z615-4, was phenotyped in 2011 for validation of QTL found in HZ595, and *Qdon.umb-17A/7C* was again localized with a phenotypic effect of 12.4 %. Three SNPs closely linked to *Qdon.umb-17A/7C* were identified in both populations, and one each for QTL on 5C, 11A, and 13A were identified in HZ595. These SNPs, together with those yet to be identified, could be useful in marker-assisted selection of resistance QTL.

12.8 The Future of Molecular Approaches in Disease Resistance Breeding in Oats

Advances in sequencing technologies, collectively known as next generation sequencing (NGS), provides new opportunities to explore transcriptomes. However, the genome and transcriptome of oats are among the least explored of cereal grain crops. While the complexity associated with its large and repetitive genome (allo-hexaploid, $2n=6\times=42$) is an impediment, it is clear that less effort has been devoted to oat genome research (Gutierrez-Gonzalez et al. 2013). NGS has made possible high-throughput transcriptome sequencing (RNA-Seq), giving rise to a multitude of transcriptomes and transcript profiling studies in many organisms, including numerous plant species. In oat, Gutierrez-Gonzalez et al. (2013) employed RNA-Seq to generate and characterize the first gene expression atlas for hexaploid oat. Huang et al. (2013) used genotyping-by-sequencing (GBS) to genotype 746 diverse oat varieties and 622 progenies from 8 biparental populations. The number of segregating SNPs observed inside the subpopulations reflects the intrapopulation diversity, both for biparental populations and diversity panels used in different breeding programs.

The availability of these data will improve the oat consensus map, association mapping studies, and generate predictions regarding breeding value and QTL present in North American oat germplasm. Oat genetic and genomic research has advanced rapidly in recent years and further progress is needed to keep oat competitive with other cereal crops. The development of new selection tools for developing disease resistant oat varieties will play an important role.

Acknowledgement The authors thank Prairie Oat Growers Association (POGA), Regina, Saskatchewan, Canada for supporting oat research.

References

Acevedo M, Jackson EW, Chong J, Rines HW, Harrison S, Bonman JM (2010) Identification and validation of quantitative trait loci for partial resistance to crown rust in Oat. Phytopathology 100(5):511–521

Adhikari KN, McIntosh RA, Oates JD (2000) Distribution and temperature sensitivities of genes for stem rust resistance in Australian oat cultivars and selected germplasm. Aust J Agr Res 51:75–83

Aung T, Thomas H, Jones IT (1977) The transfer of the gene for mildew resistance from *Avena barbata* (4×) into the cultivated oat *A. sativa* by an induced translocation. Euphytica 26(3):623–632

Becher R (2007) EST-derived microsatellites as a rich source of molecular markers for oats. Plant Breed 126:274–278

Brink M, Belay G (2006) Plant resources of tropical Africa, vol I, Cereals and pulses. PROTA Foundation, Wageningen

Browning JA, Frey KJ (1969) Multiline cultivars as a means of disease control. Annu Rev Phytopathol 7:355–382

Burdon JJ, Chilvers GA (1982) Host-density as a factor in plant disease ecology. Annu Rev Phytopathol 20:143–166

Bush AL, Wise RP (1998) High-resolution mapping adjacent to the *Pc71* crown-rust resistance locus in hexaploid oat. Mol Breed 4(1):13–21

Cabral AL (2009) The genetics of host: pathogen interactions in wild and cultivated *Avena*: oat rust pathogens *Puccinia coronata* f. sp. *avenae* and *Puccinia graminis* f. sp. *avenae* [Ph.D. dissertation]. University of Sydney, Plant Breeding Institute, Cobbitty

Cabral AL, Singh D, Park RF (2011) Identification and genetic characterization of adult plant resistance to crown rust in diploid and tetraploid accessions of *Avena*. Ann Appl Biol 159(2):220–228

Carleton MA (1905) Lessons from the grain-rust epidemic of 1904. USDA Farmers Bull 219

Carson ML (2009a) Broad-spectrum resistance to crown rust, *Puccinia coronata* f. sp. *avenae*, in accessions of the tetraploid slender oat, *Avena barbata*. Plant Dis 93:63–366

Carson ML (2009b) Crown rust development and selection for virulence in *Puccinia coronata* f. sp. *avenae* in an oat multiline cultivar. Plant Dis 93:347–353

CDL (2013) [Internet]. Cereal Disease Laboratory, St. Paul [cited 2013 Sept 6]. Available from http://www.ars.usda.gov/Main/docs.htm?docid=10342

Chen G, Chong J, Gray M, Prashar S, Procunier DJ (2006) Identification of single-nucleotide polymorphisms linked to resistance gene Pc68 to crown rust in cultivated oat. Can J Plant Pathol 28(2):214–222

Cheng DW, Armstrong KC, Drouin G, McElroy A, Fedak G, Molnar SD (2003) Isolation and identification of *Triticeae* chromosome 1 receptor-like kinase genes (*Lrk10*) from diploid, tetraploid, and hexaploid species of the genus *Avena*. Genome 46(1):119–127

Chong J, Kolmer JA (1993) Virulence dynamics and phenotypic diversity of *Puccinia coronata* f. sp. *avenae* in Canada from 1974 to 1990. Can J Bot 71:248–255

Chong J, Seaman WL (1997) Incidence and virulence of *Puccinia coronata* f. sp. *avenae* in Canada in 1995. Can J Plant Pathol 19:176–180

Chong J, Zegeye T (2004) Physiologic specialization of *Puccinia coronata* f. sp. *avenae*, the cause of oat crown rust, in Canada from 1999 to 2001. Can J Plant Pathol 26(1):97–108

Chong J, Reimer E, Somers D, Aung T, Penner GA (2004) Development of sequence-characterized amplified region (SCAR) markers for resistance gene *Pc94* to crown rust in oat. Can J Plant Pathol 26(1):89–96

Chong J, Gruenke J, Dueck R, Mayert W, Woods S (2008) Virulence of oat crown rust [*Puccinia coronata* f. sp. *avenae*] in Canada during 2002–2006. Can J Plant Pathol 30:115–123

Chong J, Gruenke J, Dueck R, Mayert W, Mitchell Fetch J, McCartney CA (2011) Virulence of *Puccinia coronata* f. sp. *avenae* in the eastern prairie region of Canada during 2007–2009. Can J Plant Pathol 33(1):77–87

Clifford BC (1995) Diseases, pest and disorders of oat. In: Welch RW (ed) The oat crop. Chapman & Hall, London, pp 252–278

DeKoeyer DL, Tinker NA, Wight CP, Deyl J, Burrows VD, O'Donoughue LS, Lybaert A, Molnar SJ, Armstrong KC, Fedak G, Wesenberg DM, Rossnagel BG, McElroy AR (2004) A molecular linkage map with associated QTLs from a hulless × covered spring oat population. Theor Appl Genet 108(7):1285–1298

Díaz-Lago JE, Stuthman DD, Abadie TE (2002) Recurrent selection for partial resistance to crown rust in oat. Crop Sci 42:1475–1482

Dyck PL, Samborski DJ, Anderson RG (1966) Inheritance of adult-plant leaf rust resistance derived from the common wheat varieties Exchange and Frontana. Can J Genet Cytol 8:665–671

Fetch TG (2005) Races of *Puccinia graminis* on wheat, barley, and oat in Canada, in 2002 and 2003. Can J Plant Pathol 27(4):572–580

Fetch TG Jr, Dunsmore KM (2004) Physiologic specialization of *Puccinia graminis* on wheat, barley and oat in Canada in 2001. Can J Plant Pathol 26:148–155

Fetch TG Jr, Jin Y (2007) Letter code system of nomenclature for *Puccinia graminis* f. sp. *avenae*. Plant Dis 91:763–766

Fleischmann G, Mckenzie RIH (1965) Yield losses in Gary oats infected with crown rust. Phytopathology 55:767–770

Flor HH (1955) Host parasite interactions in flax rust: its genetics and other implications. Phytopathology 45:680–685

Frey K, Browning JA (1971) Breeding crop plants for disease resistance. In: Mutation breeding for disease resistance. Proc. Int. Atomic Energy Assoc., 12–16 Oct. IAEA Vienna, pp 45–54

Gilbert J, Tekauz A (1995) Effects of Fusarium head blight and seed treatment on germination, emergence, and seedling vigour of spring wheat. Can J Plant Pathol 17:252–259

Gnanesh BN, Mitchell Fetch J, Menzies JG, Beattie AD, Eckstein PE, McCartney CA (2013) Chromosome location and allele-specific PCR markers for marker-assisted selection of the oat crown rust resistance gene *Pc91*. Mol Breed 32(3):679–686. doi:10.1007/s11032-013-9900-6

Gnanesh BN, Mitchell Fetch J, Zegeye T, McCartney C, Fetch T (2014) Oat. In: Pratap A, Kumar J (eds) Alien Gene Transfer in Crop Plants, vol. 2. Springer New York, pp 51–73

Gutierrez-Gonzalez J, Tu Z, Garvin D (2013) Analysis and annotation of the hexaploid oat seed transcriptome. BMC Genomics 14(1):471

Hagmann E, von Post L, von Post R, Eklund M, Larsson C-T, Tuvesson S, Vallenback P, Ceplitis A (2012) QTL mapping of powdery mildew resistance in oats using DArT markers. In: Proceedings of 15th International EWAC Conference, 7–11 Nov 2011, Novi Sad, Serbia, European Cereals Genetics Co-operative Newsletter, pp 91–94

Harder DE, Haber S (1992) Oat diseases and pathologic techniques. In: Marshall HG, Sorrells ME (eds) Oat science and technology, vol 33, ASA monograph. ASA-CSSA-SSSA, Madison, pp 307–425

Hare RA (1997) Characterization and inheritance of adult plant stem rust resistance in durum wheat. Crop Sci 37:1094

Hayes JD, Jones IT (1966) Variation in the pathogenicity of *Erysiphe graminis* D.C. f. sp. *avenae*, and its relation to the development of mildew resistant oat cultivars. Euphytica 15:80–86

He X, Skinnes H, Oliver R, Jackson E, Bjørnstad Å (2013) Linkage mapping and identification of QTL affecting deoxynivalenol (DON) content (Fusarium resistance) in oats (*Avena sativa* L.). Theor Appl Genet 126(10):2655–70

Herrmann M, Roderick HW (1996) Characterisation of new oat germplasm for resistance to powdery mildew. Euphytica 89:405–410

Hoffman DL, Chong J, Jackson EW, Obert DE (2006) Characterization and mapping of a crown rust resistance gene complex (*Pc58*) in TAM O-301. Crop Sci 46(6):2630–2635

Holland JB, Munkvold GP (2001) Genetic Relationships of Crown Rust Resistance, Grain Yield, Test Weight, and Seed Weight in Oat. Crop Sci 41:1041

Hoppe HD, Kummer M (1991) New productive hexaploid derivatives after introgression from *A. pilosa* features. Vortr Pflanzenzüchtg 20:56–61

Hsam SLK, Peters N, Paderina EV, Felsenstein F, Oppitz K, Zeller FJ (1997) Genetic studies of powdery mildew resistance in common oat (*Avena sativa* L.). I. Cultivars and breeding lines grown in Western Europe and North America. Euphytica 96:421–427

Huang Y, Poland J, Jackson EW, Tinker N (2013) Genotyping-by-sequencing (GBS) in the real oat world: making the data talk. Proceedings of the XXI plant and animal genome conference, Jan 12–16, San Diego. Available from https://pag.confex.com/pag/xxi/webprogram/Paper6751.html

Irigoyen ML, Loarce Y, Fominaya A, Ferrer E (2004) Isolation and mapping of resistance gene analogs from the *Avena strigosa* genome. Theor Appl Genet 109(4):713–724

Jackson EW, Obert DE, Menz M, Hu G, Avant JB, Chong J, Bonman JM (2007) Characterization and mapping of oat crown rust resistance genes using three assessment methods. Phytopathology 97(9):1063–1070. doi:10.1094/phyto-97-9-1063

Jackson EW, Obert DE, Menz M, Hu G, Bonman JM (2008) Qualitative and quantitative trait loci conditioning resistance to *Puccinia coronata* pathotypes NQMG and LGCG in the oat (*Avena sativa* L.) cultivars Ogle and TAM O-301. Theor Appl Genet 116(4):517–527

Jackson EW, Obert DE, Avant JB, Harrison SA, Chong J, Carson ML, Bonman JM (2010) Quantitative trait loci in the Ogle/TAM O-301 oat mapping population controlling resistance to *Puccinia coronata* in the field. Phytopathology 100(5):484–492. doi:10.1094/phyto-100-5-0484

Jensen NF (1965) Multiline superiority in cereals. Crop Sci 5(6):566–568

Johnson R (1984) A critical analysis of durable resistance. Annu Rev Phytopathol 22(1):309–330

Johnson R, Law CN (1975) Genetic control of durable resistance to yellow rust (*Puccinia striiformis*) in the wheat cultivar Hybride de Bersee. Ann Appl Biol 81:385–91

Jones IT (1977) The effect on grain yield of adult plant resistance to mildew in oats. Ann Appl Biol 86:267–277

Jones IT (1983) Transgressive segregation for enhanced level of adult plant resistance to mildew in the oat cross Mostyn × Maldwyn. Euphytica 32:499–503

Jones IT, Jones ERL (1979) Mildew of oats. UK Cereal Pathogen Virulence Survey 1978. Annual Report, pp 59–63

Knott DR (1968) The inheritance of resistance to stem rust races 56 and 15B-1L (Can.) in the wheat varieties Hope and H-44. Can J Genet Cytol 10:311–320

Kowalczyk K, Hsam SLK, Zeller FJ (2004) Identification of oat powdery mildew resistance group 2 (OMR2) and Polish common oat (*Avena sativa* L.) cultivars. Workshop "Resistance of cereals to biotic stresses," Nov 28–Dec 1, Radzików, pp 122–125

Kremer CA, Lee M, Holland JB (2001) A restriction fragment length polymorphism based linkage map of a diploid *Avena* recombinant inbred line population. Genome 44(2):192–204

Lambalk JJM, Faber NM, Bruijnis AB, Conijn PCJ, Witte IAD, Nieuwenhuis J, De Jong CJ (2004) A method for obtaining a plant with a lasting resistance to a pathogen. United States Patent Application Publication No. US 2004/0226060 A1

Lange C (2012) Reaction of North American Oats (*Avena sativa* L.) to crown rust [dissertation]. Texas A&M University, College Station

Lawes DA, Hayes JD (1965) The effect of mildew (*Erysiphe graminis* f. sp. *avenae*) on spring oats. Plant Pathol 14(3):125–128

Leonard KJ, Anikster Y, Manisterski J (2004) Patterns of virulence in natural populations of *Puccinia coronata* on wild oat in Israel and in agricultural populations on cultivated oat in the United States. Phytopathology 94:505–514

Li CD, Rossnagel BG, Scoles GJ (2000) The development of oat microsatellite markers and their use in identifying relationships among *Avena* species and oat cultivars. Theor Appl Genet 101(8):1259–1268

Long J, Holland JB, Munkvold GP, Jannink JL (2006) Responses to selection for partial resistance to crown rust in oat. Crop Sci 46:1260–1265

Martens JW (1978) Stem rust of oats in Canada in 1977. Can Plant Dis Surv 58:51–52

Martens JW (1985) Stem rust of oats. In: Roelfs AP, Bushnell WR (eds) The cereal rusts, vol II, Diseases, distribution, epidemiology, and control.. Academic Press, New York

McCallum BD, Fetch T, Chong J (2007) Cereal rust control in Canada. Aust J Agr Res 58:639–647

McCartney CA, Stonehouse RG, Rossnagel BG, Eckstein PE, Scoles GJ, Zatorski T, Beattie AD (2011) Chong J Mapping of the oat crown rust resistance gene *Pc91*. Theor Appl Genet 122(2):317–325

Mielniczuk E, Kiecana I, Perkowski J (2004) Susceptibility of oat genotypes to *Fusarium crookwellense* Burgess, Nelson and Toussoun infection and mycotoxin accumulation in kernels. Biologia, Bratislava 59(6):809–816

Mitchell Fetch J, Fetch T Jr (2011) Inheritance of resistance to oat stem rust in the cultivars Ronald and AC Gwen. Can J Plant Sci 91:419–423

Mohler V, Zeller F, Hsam SK (2012) Molecular mapping of powdery mildew resistance gene *Eg-3* in cultivated oat (*Avena sativa* L. cv. Rollo). J Appl Genet 53(2):145–148

Murphy HC (1935) Effects of crown rust infection on yield and water requirement of oats. J Agr Res 50(5):387–411

O'Donoughue LS, Wang Z, Röder M, Kneen B, Leggett M, Sorrells ME, Tanksley SD (1992) An RFLP-based linkage map of oats based on a cross between two diploid taxa (*Avena atlantica*×*A. hirtula*). Genome 35(5):765–771

O'Donoughue LS, Kianian SF, Rayapati PJ, Penner GA, Sorrells ME, Tanksley SD, Phillips RL, Rines HW, Lee M, Fedak G, Molnar SJ, Hoffman D, Salas CA, Wu B, Autrique E, Deynze V (1995) A molecular linkage map of cultivated oat. Genome 38:68–380

Oates JD (1992) Australian Oat Rust Survey, 1991–1992. University of Sydney, Plant Breeding Institute, Cobbitty

Ohm HW, Shaner G (1992) Breeding oat for resistance to diseases. In: Marshall HG, Sorrells ME (eds) Oat science and technology, vol 33, ASA monograph. ASA-CSSA-SSSA, Madison, pp 657–698

Okoń S (2012) Identification of powdery mildew resistance genes in polish common oat (*Avena sativa* L.) cultivars using host-pathogen tests. Acta Agrobot 65(3):63–68

Okoń S, Kowalczyk K (2012) Identification of SCAR markers linked to resistance to powdery mildew in common oat (*Avena sativa*). J Plant Dis Protect 119:179–181

Oliver RE, Lazo GR, Lutz JD, Rubenfield MJ, Tinker NA, Anderson JM, Wisniewski Morehead NH, Adhikary D, Jellen EN, Maughan PJ, Brown Guedira GL, Chao S, Beattie AD, Carson ML, Rines HW, Obert DE, Bonman JM, Jackson EW (2011) Model SNP development for complex genomes based on hexaploid oat using high-throughput 454 sequencing technology. BMC Genomics 12(1):77. doi:10.1186/1471-2164-12-77

Oliver RE, Tinker NA, Lazo GR, Chao S, Jellen EN, Carson ML, Rines HW, Obert DE, Lutz JD, Shackelford I, Korol AB, Wight CP, Gardner KM, Hattori J, Beattie AD, Bjørnstad Å, Bonman JM, Jannink J-L, Sorrells ME, Brown-Guedira GL, Mitchell Fetch JW, Harrison SA, Howarth CJ, Ibrahim A, Kolb FL, McMullen MS, Murphy JP, Ohm HW, Rossnagel BG, Yan W, Miclaus KJ, Hiller J, Maughan PJ, Redman Hulse RR, Anderson JM, Islamovic E, Jackson EW (2013) SNP discovery and chromosome anchoring provide the first physically-anchored hexaploid oat map and reveal synteny with model species. PLoS One 8(3):e58068. doi:10.1371/journal.pone.0058068

Pal N, Sandhu JS, Domier LL, Kolb FL (2002) Development and characterization of microsatellite and RFLP derived PCR markers in oat. Crop Sci 42:912–918

Park RF (2008) Breeding cereals for rust resistance in Australia. Plant Path 57:591–602

Park RF (2013) New oat crown rust pathotype with virulence for *Pc91*. Cereal rust report—season 2013, vol 11(1), Plant Breeding Institute, University of Sydney, Cobbitty

Park RF, Kavanagh P (2009) Cereal rust survey–annual report 2008–2009. Plant Breeding Institute, University of Sydney, Cobbitty

Parry DW, Jenkinson P, McLeod L (1995) Fusarium ear blight (scab) in small grain cereals- a review. Plant Pathol 44:207–238

Portyanko VA, Hoffman DL, Lee M, Holland JB (2001) A linkage map of hexaploid oat based on grass anchor DNA clones and its relationship to other oat maps. Genome 44(2):249–265. doi:10.1139/g01-003

Portyanko VA, Chen G, Rines HW, Phillips RL, Leonard KJ, Ochocki GE, Stuthman DD (2005) Quantitative trait loci for partial resistance to crown rust, *Puccinia coronata*, in cultivated oat, *Avena sativa* L. Theor Appl Genet 111(2):313–324. doi:10.1007/s00122-005-2024-6

Rines HW, Phillips RL, Anderson OD, Vance CP, Crossman CC, Lazo GR, Miller SS, Taller J (2004) ESTs, cytogenetic stocks, and other tools for oat genomics (abstract). Proceedings of the 7th international oat conference, July 17–22, Helsinki

Roderick HW, Jones ERL, Šebesta J (2000) Resistance to oat powdery mildew in Britain and Europe: a review. Ann Appl Biol 136(1):85–91

Roelfs AP (1978) Estimated losses caused by rust in small grains cereals in the United States, 1918-1976. US Dept Agric Misc Publ 1363:85

Rooney WL, Rines HW, Phillips RL (1994) Identification of RFLP markers linked to crown rust resistance gene *Pc91* and *Pc92* in oat. Crop Sci 34:940–944

Sanz M, Loarce Y, Fominaya A, Vossen J, Ferrer E (2013) Identification of RFLP and NBS/PK profiling markers for disease resistance loci in genetic maps of oats. Theor Appl Genet 126(1):203–218

Satheeskumar S, Sharp PJ, Lagudah ES, McIntosh RA, Molnar SJ (2011) Genetic association of crown rust resistance gene *Pc68*, storage protein loci, and resistance gene analogues in oats. Genome 54:484–497

Schroeder HW, Christensen JJ (1963) Factors affecting resistance of wheat to scab caused by *Gibberella zeae*. Phytopathology 53(7):831–838

Sebesta J, Kummer M, Roderick HW, Hoppe HD, Cervenka J, Swierczewski A, Muller K (1991) Breeding oats for resistance to rusts and powdery mildew in central Europe. Ochrana rostlin 27:229–238

Sebesta J, Roderick HW, Chong J (1993) The oat line *Pc54* as a source of resistance to crown rust, stem rust and powdery mildew in Europe. Euphytica 71:91–97

Simons MD, Michel LJ (1968) Oat maturity and crown rust response. Crop Sci 8:254–256

Simons MD, Murphy HC (1968) Oat diseases and their control. ARS-USDA Handbook No. 343. USDA, Washington, DC

Simons MD, Martens JW, McKenzie RIH, Nishiyama I, Sadanaga K, Sebesta J, Thomas H (1978) Oats: a standardized system of nomenclature for genes and chromosomes and catalogue of genes governing characters. USDA Agriculture handbook No. 509. USDA, Washington, DC

Suttie JM, Reynolds SG (2004) Fodder oats: a world overview, vol 33, Plant production and protection series. Food and Agriculture Organization of the United Nations, Rome

Tanhuanpää P, Kalendar R, Schulman AH, Kiviharju E (2008) The first doubled haploid linkage map for cultivated oat. Genome 51(8):560–569

Tekauz A, McCallum B, Ames N, Mitchell FJ (2004) Fusarium head blight of oat—current status in western Canada. Can J Plant Pathol 26(4):473–479. doi:10.1080/07060660409507167

Tekauz A, Mitchell Fetch J, Rossnagel BG, Savard ME (2008) Progress in assessing the impact of Fusarium head blight on oat in western Canada and screening of *Avena* germplasm for resistance. Cereal Research Communications 36:49–56

Thomas H, Powell W, Aung T (1980) Interfering with regular meiotic behaviour in Avena sativa as a method of incorporating the gene for mildew resistance from *A. barbata*. Euphytica 29: 635–640

Tinker NA, Kilian A, Wight CP, Uszynska KH, Wenzl P, Rines HW, Bjørnstad A, Howarth CJ, Jannink JL, Anderson JM, Rossnagel BG, Stuthman DD, Sorrells ME, Jackson EW, Tuvesson S, Kolb FL, Olsson O, Federizzi LC, Carson ML, Ohm HW, Molnar SJ, Scoles GJ, Eckstein PE, Bonman JM, Ceplitis A, Langdon T (2009) New DArT markers for oat provide enhanced map coverage and global germplasm characterization. BMC Genomics 10:39

Waterhouse WL (1952) Australian rust studies. IX. Physiological race determinations and surveys of cereal rusts. P Linn Soc NSW 77:209–58

Wight CP, Tinker NA, Kianian SF, Sorrells ME, O'Donoughue LS, Hoffman D, Groh S, Scoles GJ, Li CD, Webster FH, Phillips RL, Rines HW, Livingston SM, Armstrong KC, Fedak G, Molnar SJ (2003) A molecular marker map in 'Kanota' × 'Ogle' hexaploid oat (*Avena* spp.) enhanced by additional markers and a robust framework. Genome 46(1):28–47

Wolfe MS (1985) The current status and prospects of multiline cultivars and variety mixtures for disease resistance. Annu Rev Phytopathol 23:251–273

Yu J, Herrmann M (2006) Inheritance and mapping of a powdery mildew resistance gene introgressed from *Avena macrostachya* in cultivated oat. Theor Appl Genet 113(3):429–437

Zhu S, Kaeppler HF (2003) Identification of quantitative trait loci for resistance to crown rust in oat line MAM17-5. Crop Sci 43(1):358–366

Index

A. Goyal and C. Manoharachary (eds.), *Future Challenges in Crop Protection Against Fungal Pathogens*, Fungal Biology, DOI 10.1007/978-1-4939-1188-2,

Printed by Printforce, the Netherlands